Electronic Materials

Science and Technology

Electronic Materials

Science and Technology

Shyam P. Murarka

Rensselaer Polytechnic Institute
Center for Integrated Electronics
Troy, New York

Martin C. Peckerar

Microelectronics Processing Facility
Naval Research Laboratory
Washington, DC

and

Department of Electrical Engineering
University of Maryland
College Park, Maryland

ACADEMIC PRESS, INC.

Harcourt Brace Jovanovich, Publishers

Boston San Diego New York
London Sydney Tokyo Toronto

This book is printed on acid-free paper. ♾

ACADEMIC PRESS, INC.
1250 Sixth Avenue, San Diego, CA 92101

United Kingdom Edition published by
ACADEMIC PRESS LIMITED
24-28 Oval Road, London NW1 7DX

Cover design by Nono Kusuma

Library of Congress Cataloging-in-Publication Data

Murarka, S.P.
Electronic materials: science and technology / Shyam P. Murarka
and Martin C. Peckerar.
p. cm.
Bibliography: p.
Includes index.
ISBN 0-12-511120-7
1. Electronics—Materials. 2. Semiconductors. I. Peckerar, Martin
Charles, Date– . II. Title.
TK7871.M87 1989
621.3815'2—dc19 88-34033
CIP

CORRECTED SECOND PRINTING

Printed in the United States of America

91 92 9 8 7 6 5 4 3 2

Table of Contents

Preface

At the time of this writing, the 20th century draws to a close. The intellectual foment that marked its beginnings has changed the way we view the physical world forever. Our new insights into physical processes have given rise to an astounding array of tools and technologies. What these developments will ultimately bring into our daily lives, their originators can only guess. This transformation of concept into implementation is nowhere more evident than in the area of electronics. The "new" electronics, brought about by the semiconductor revolution, have directly embodied the concepts of quantum physics and quantum chemistry developed in the first decades of the century.

Semiconductor electronics must still be regarded as an emerging technology. A person whose life spans its whole significant history would still be in early middle age. Most of the key figures in the field are still alive today. While this developmental status makes for an atmosphere of considerable excitement, it also creates a significant number of academic challenges. How do we prepare engineers for participation in a field largely characterized by "Edisonian" flashes of brilliance? Standard procedures are only now being finalized. In addition, as with most of our current technology, the field is "cross disciplinary." That is, it requires familiarity with many areas of basic science, rather than any single field. In the paragraphs below, we elaborate on our response to this challenge.

First, we note that there are generally two classes of engineers taking academic courses in semiconductor electronics technology: electrical engineers and specialists in materials science. Electrical engineers usually have little specialized knowledge of chemical reactions or solid state mechanics. And yet, our ability to fabricate semiconductor devices depends critically on our understanding of these items. Materials science majors rarely have in-depth coursework in the areas of electronic circuits or device physics, yet such information is essential to understanding the goals of the materials modification these engineers are expected to provide. This necessitates the presentation of a considerable amount of review material in any textbook hoping to train practitioners in the area of semiconductor process technology. In this text, this is accomplished through direct incorporation of review material in individual chapters and through appendices.

This is not to say that we expect our readers to come to class with no prior background in any of the areas mentioned above. The text is aimed at a senior-level/first-year graduate audience. Undergraduate basic chemistry is a prerequisite, as is an introductory semester in device technology. Students should know the basics of transistor action and be able to distinguish between bipolar and MOS device structures. On the other hand, we do not expect the student to be expert in any significant number of these areas. It is the goal of this text to provide some functional technical fluency in this diverse arena.

It is not the intent of this book to provide the "how-to" information needed to establish unit processes in a semiconductor process line. This information is of limited value, as it depends strongly on equipment vendor/production-line personnel interaction. Furthermore, approaches to the associated problems change rapidly on today's fab-lines. What we attempt to provide is a conceptual understanding of the basic physical and chemical processes encountered in fab-line maintenance. Admittedly, in a field this young, some of our views of these processes will change as our understanding matures. We have attempted to present models that illustrate the underlying physical principles of the processes performed. Frequently, these models may not be of sufficient accuracy to truly predict process outcome, but they were selected to provide some understanding of observed trends as key process parameters are varied. It is our hope that the limitations of these models will be clearly understood by the reader.

The book is divided in chapters that follow a generally accepted sequence in processing of the semiconductor devices. Chapter 1 introduces the essentials of semiconductor device processing in a top-down approach. We have chosen a complementary metal-oxide semiconductor (CMOS) device as our vehicle for the introduction and ensuing discussion. This chapter exposes the reader to the elements of device physics and circuits, and prepares the reader to understand the language of device/circuit designers and electrical and process engineers. A pictorial and graphical understanding of the devices and process-induced changes in the materials, which leads to the final device, is provided.

In the second chapter, the preparation, crystal growth, cleaning, and characterization of the semiconductor are described. The defects that influence the properties of crystals are discussed from a thermodynamic viewpoint. The student is acquainted with quantitative methods for calculating defect densities. Basic crystal growth science and its impact on device/circuit technology is presented.

In Chapters 3 through 6, topics discussed include oxidation of semiconductors, diffusion and ion implantation in semiconductors, and metallization of the semiconductor devices. These are essential processes a semiconductor crystal goes through leading to devices and circuits. In each chapter, the properties and characterization of the processed material, the basic understanding of the mechanisms of the process, and the process applications are emphasized. Figures, tables, and calculations are provided to enable the reader to determine the usefulness of the process or the material for his or her specific application.

Chapter 7 covers the fundamentals of the chemical vapor deposition. The presentation in this chapter differs from other textbook accounts of CVD. We attempt to provide a fundamental understanding of the process by not only considering the gas phase boundary layer and chemical equilibria issues, but also the nucleation and growth phenomena and modeling of the CVD process.

Chapters 8, 9, and 10 address the pattern generation and transfer using lithographic tools, resists, and etching methods. Various lithographic tools, physics fundamentals, and tool applications and limitations are discussed in Chapter 8. In Chapter 8, the Fourier-optic approach is emphasized, even though ray-tracing is still the primary tool of the optical systems designer. Fourier optics provides a quick way to estimate intensity distributions in a variety of "real-world" situations. Various resist systems, the basics of resist chemistry, handling, and exposure to radiations—including resist exposure modeling —are presented. Finally in Chapter 10, fundamentals and applicability of various wet and dry etching systems are presented. Dry etching has been the least understood process mainly because of the large variabilities in carrying out the etching. Here we restrict ourselves to models that describe proven effects. These models are somewhat more qualitative than most, but they provide trend indicators for key variables.

Finally in Chapter 11 we examine the processes that are under development and may be adopted in the future for device and circuit fabrication. The concept and applicability of in situ processing (also called integrated processing) is discussed. An attempt is made to introduce the fundamental concepts, such as laser processing, and discuss their limitations.

It should be noted that the text represents considerably more than a semester's work. There are a number of approaches the authors have taken to convey the material presented. One approach is to provide selected readings in most of the chapters and to focus in on a single topic for emphasis. Selected readings would provide an overview of a range of topics. Suggestions for in-depth focus groups are as follows: Chapters 2, 6, 7, and 11 (materials structure and deposition); Chapters 8, 9, and 10 (lithography and pattern transfer); Chapters 3, 4, and 5 (solid state transport). Chapter 1 serves as a general overview for all topics. Problems cover a wide range of difficulties. "Difficult" problems should be assigned only when preceded by in-depth classroom coverage of the material.

Wherever our technology takes us, we feel that the following observations will be valid. We present these observations to you as a conclusion to this preface. Materials science and electronics technology are now in possession of techniques for manipulating matter on an atomic level. We are confident that this will lead to new classes of devices whose basic functions will once again lead us to challenge our fundamental understanding of the physical universe. Our ability to respond to these challenges rests on our ability to break out of the traditional models of narrow specialization. We offer this text as a first step in this process.

Acknowledgments

This book would not have been possible without the encouragement, support, and critical reviews of many of our colleagues. We would especially like to thank Jack Ayers, Timothy Braggins, J. Calvert, John Corelli, Nick Croitorou, Andrew Culhane, Robert Doremus, Pradip Dutta, Richard Fair, David Fraser, Jeffrey Frey, Daniel Friedman, Ping-Tong Ho, John Hudson, Agis Illiadis, Dmitri Iouanou, Justin Kreutzer, Hyman Levinstein, H.L. Lin, Robert Marcus, David Markel, James Neeley, Edward Palik, R.F.W. Pease, R.A. Powell, Rajat Rakhit, Al Rienberg, C.J. Russo, Thomas Seidel, Loretta Shirey, Ashok Sinha, Hank Smith, C. Steinbruchel, Noel Thomas, C.J. Taylor, Richard Wagner, and David Williams. We are very grateful to them for their contributions that made this book possible. Our students and other staff at Naval Research Laboratory, University of Maryland, and Rensselaer Polytechnic Institute are acknowledged for their direct and indirect support. We are grateful to Patti Doyle, Patsy Keehn, Peggy Kroemer, Donna Maple, and Evelyn Smith for their patience and impatience in typing the original and many subsequent versions of the manuscript. Figures and cover are courtesy of N. Kusuma.

We are grateful to the many authors whose papers we have followed closely in various parts of the book and who allowed us to use their work in this book. We are also thankful to the American Institute of Physics, American Physical Society, Institute of Electrical and Electronics Engineers, The Electrochemical Society, Inc., Japanese Journal of Applied Physics, Pergamon Press, Inc., John Wiley and Sons, Inc., Elsevier–North Holland, McGraw-Hill, Inc., Solid State Technology, and Varian Associates for the copyrighted materials.

Most of all we want to thank our wives Saroj and Nancy and our sons Sumeet and Amal, and Andrew and Robby for the love, understanding, patience, and impatience that made the preparation of this book possible.

1 Essentials of Processing: A Top Down Approach

1.1 Introduction

Assume for a moment that you have been hired as a senior process engineer for a semiconductor corporation whose specialty is making high-speed complementary-metal-oxide-semiconductor (CMOS) random access memories (RAMs) [1,2]. As a process engineer, your work will largely be concerned with materials modification and patterning. That is, you will modify the electrical properties of silicon through diffusion or implantation; you will hook differently doped regions of the semiconductor together using metal or polycrystalline silicon thin films; you may electrically isolate conductors with grown or deposited insulating films. Dopants are spatially localized in the semiconductor substrate, and the metal interconnects appear as neatly cut wires too fine for the unaided eye to see. You will have to make use of techniques designed to form patterns of the desired materials on the substrate. This is done using a process called *lithography*, very similar to conventional photography. Most of this book is concerned with materials modification and with lithography. The semiconductor base material is *usually* silicon, although occasional reference is made to other semiconductors (such as gallium arsenide). However, as a process engineer, you should go beyond your background in materials to be fully successful. You must have some understanding of the goals of the process that you are asked to perform. In this chapter, these goals are outlined. CMOS device structures are taken as a processing example that is carried through the book. CMOS is chosen as the vehicle as it is a versatile, low-power technology that is finding use in both the analog and digital areas; in addition, the simpler nMOS process is a subset of the full CMOS process. We begin with a discussion of the end-of-the-line product and follow this with some details of process development. We call this approach *top down.*

There are two types of goals we must distinguish. Any successful semiconductor manufacturer wants to make products that are both good and cheap. By a good product, we mean one that reliably meets rigorous technological standards of performance. These standards might include operating speed, power dissipation, or, for a RAM, how much information it can store. To make a product cheap requires an understanding of product manufacturability. We must minimize the number of steps in the process and be assured that each step will yield a large number of successful outcomes. In the text below, the subjects of performance and manufacturability are covered.

1.2 An Overview of a CMOS Chip

In keeping with the "top down"approach, let us discuss the basic function of a CMOS RAM [3]. A RAM is an information storage device. Information is stored as a 1 or a 0 at storage sites called *cells*. Each cell is designated by a binary address. For example, a small RAM may be able to store 256 individual pieces of information, or *bits*. Loading the binary number 11111111 onto the RAM address port selects memory cell number 255. It does this by an electronic process called decoding. Cells are labelled 0 to 255 (256 bits in all). Once the cell has been selected, it can be read from or written to. This may be accomplished by supplying a signal to a control line called the write-enable line. A high voltage on this line (or a low voltage, depending on the design) selects the cell for writing. The opposite voltage condition means "read the cell."Another control line, called chip select, is usually provided. This activates a single RAM device, or *chip*, from many RAMs possibly present in the system. To write into the cell, an electrical impulse must be sent to the cell to set its state. To read, the cell is connected to a sense amplifier, frequently called a *detection latch*, which amplifies the small signal from the selected cell. Usually, the output capacitance of the sense amplifier is very low. To electrically influence the relatively large capacitances encountered by the signal going "off-chip,"a buffer circuit is usually provided.

To summarize, first the chip is selected. Next, an address is loaded onto the RAM address lines. The RAM decodes the address and a single cell is selected. The information contents of the cell are either set (write operation) or sensed (read operation) through the detection latch and output buffer. The output of the memory can be organized in a number of ways. Either a single bit can be read out (termed *by one* organization), or a group of n bits (termed *by n* organization) can be fed to output ports. A floorplan of a RAM chip is shown in fig. 1.1. In this figure, you can see the main memory block, composed of the individual memory cells, the decoder circuit, which is broken into row and a column decoder, the sense amplifier, and the latch and buffer circuits.

1.3 Logic Devices for CMOS RAMs

In this section, we concentrate on two of the key circuits used in the CMOS RAM: the cell and the latch. Our goal in this chapter is to focus on how materials selection and modification impact on device performance. Since this is the case, some of the fine points of circuit design are neglected, and we ignore complicated circuit structures like the decoder. Complex logic

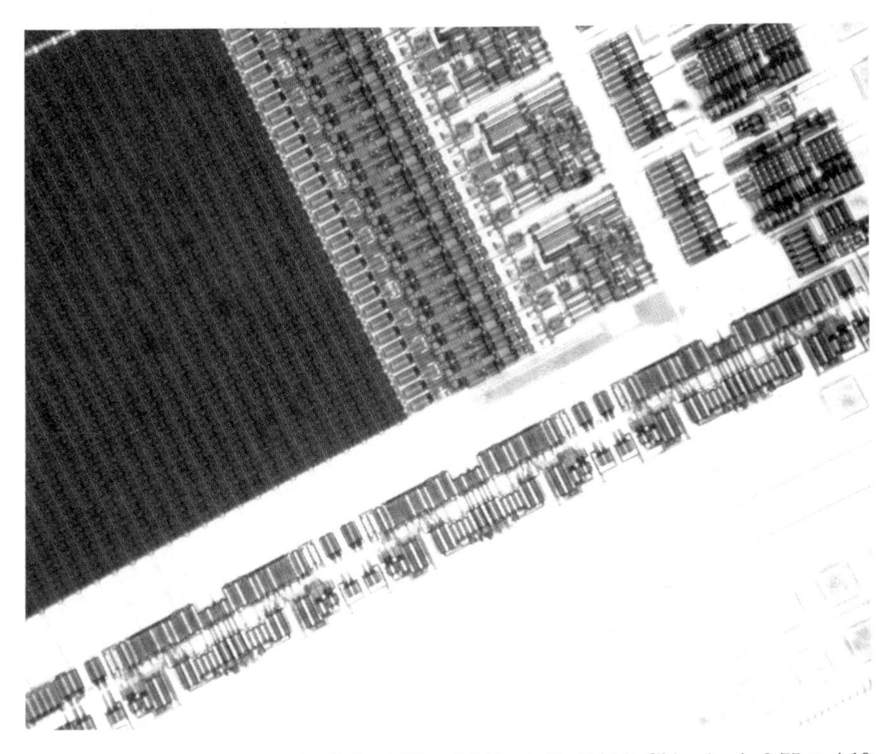

Fig. 1.1. Photomicrograph of the 16K × 16 bit static RAM. Chip size is 3.75 × 4.19 mm.

circuits are, however, built from the simpler circuits. What we learn here carries over into the area of more complex circuits.

Both the latch and the decoder are bistable circuits called *flip-flops*. By bistable, we mean that there are two possible stable states of the circuit. You may have guessed that the two states are the 1 and 0 information states discussed above. By flip-flop, we mean that the circuit can be made to flip-flop back and forth between these states. Both the cell and the latch are flip-flops of the simplest type: R/S (for reset/set) flip-flops. A circuit symbolic description of this device is shown in fig. 1.2. A logic, or "truth,"table for this device is also shown in the figure. This table describes what the output of the device does after various stimuli are applied to the inputs.

Applying high voltage to both R and S inputs leaves the outputs (Q and $\bar{Q}$) in an indeterminate state. That is, we cannot be sure what the circuit will do. Applying one type of stimulus (either high or low) to R and the opposite type of stimulus to S leads to a well-defined output set, Q and $\overline{Q}$. This output combination is stable, even after the initial input input stimulus is removed. Thus, we have information storage. Reversal of the R and S stimuli causes a reversal, or "flipping,"of the Q and $\bar{Q}$ output states. Once

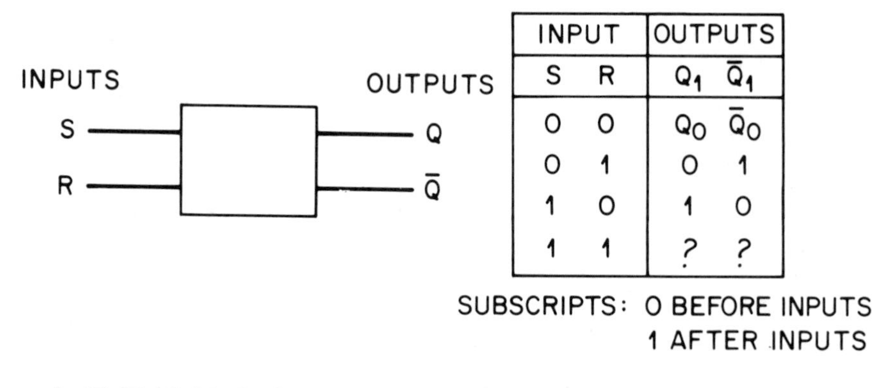

INPUT		OUTPUTS	
S	R	Q_1	$\bar{Q}_1$
0	0	Q_0	$\bar{Q}_0$
0	1	0	1
1	0	1	0
1	1	?	?

Fig. 1.2. A block diagram and truth table for an R/S flip-flop.

INPUT	OUTPUT
A	C
0	1
1	0

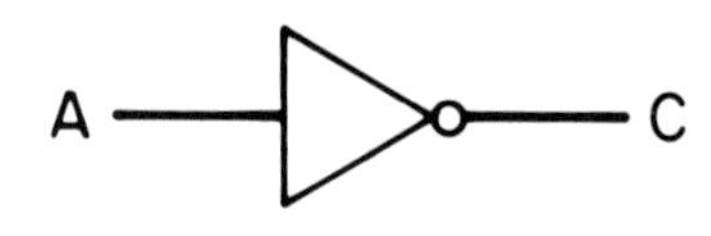

Fig. 1.3. Inverter and inverter truth table.

again, this new state is stable after the input stimuli are removed. Repeated application of the same stimuli to R and S does not change the output state. The write operation is simply the application of the appropriate R and S stimuli to get the output. The output is fixed until we reset the inputs by reversing the input stimuli.

The R/S flip-flop is made up of even simpler logic elements called *inverters*. Individual inverters are represented by the triangular structures shown in fig. 1.3. The truth table for the inverter is also shown in this figure. A high voltage at the input yields a low voltage output (and vice versa). The R/S flip-flop is realized by cross coupling two inverters, as shown in fig. 1.4. As an exercise, you can verify that the truth table of the R/S flip-flop is the only one consistent with inverter operation as summarized in fig. 1.3.

The memory cell is just an R/S flip-flop with "pass transistors" attached, as shown in fig. 1.4. These transistors open the cell to other circuits for reading and for writing. For reading, the cell output is fed into another R/S flip-flop, which is the latch. The latch stores the state of the cell for

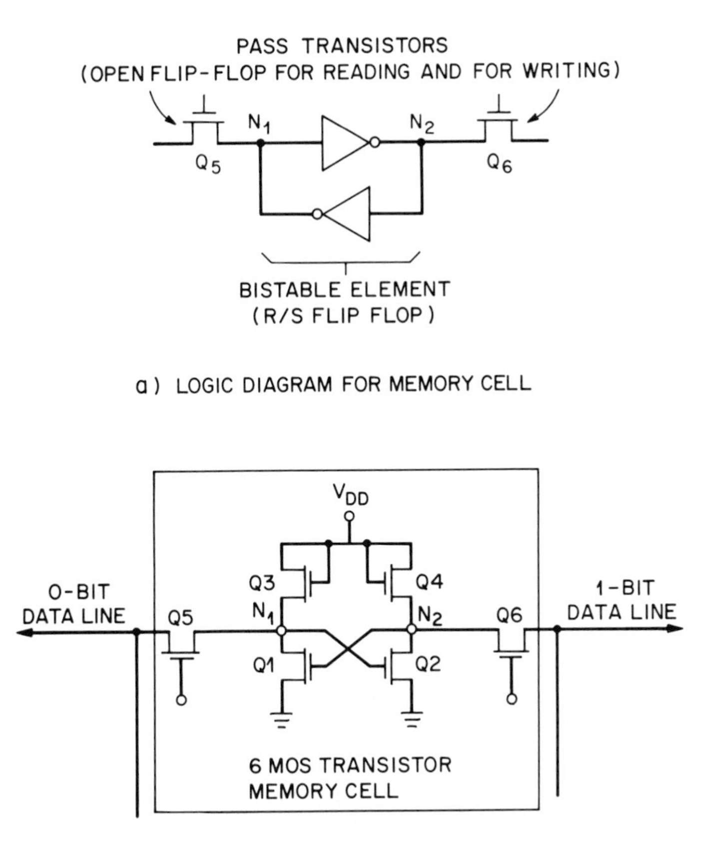

Fig. 1.4. A 6-transistor MOS memory cell. This is, in actuality, two cross coupled invertors, as shown in fig. 1.4a. The pass transistors (in fig. 1.4a) open the cell for reading or for writing.

read-out by external circuitry. The latch is also designed to provide some voltage or current gain to drive buffers or other on-chip circuits. Transistors (in fig. 1.4a) open the cell for reading or for writing.

1.4 Circuit Elements for the RAM

An integrated circuit (IC) is an assemblage of what were formerly called *discrete devices* on a single slice of semiconductor. These discrete devices are resistors, transistors, capacitors, etc. Small inductors can also be fabricated on ICs, but this is rarely done as it is generally wasteful of space to integrate high-value inductors. Other circuit elements using active devices are used

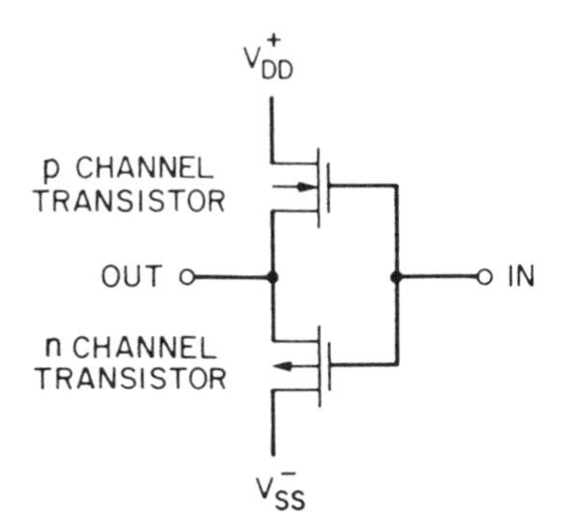

Fig. 1.5. p-channel transistor/n-channel transistor series combination used in a CMOS inverter.

to assume the inductor function. There are a number of ways to build an integrated circuit logic element. We can use bipolar devices or metal-oxide-semiconductor (MOS) structures. There are further subdivisions possible within these two groupings. For example, in MOS technology, we can build both p and n channel transistors. Our design example calls for CMOS. This is an MOS technology, which makes use of both of these types of transistors. Let us now look at the circuit for the CMOS inverter.

This circuit is shown in fig. 1.5. When a positive voltage (V_{dd}) is applied to the input terminal, the n-channel transistor turns on. V_{dd} is called the "high" voltage level and is defined as the logical "1" level. The n-channel transistor then becomes the low-resistance element in the voltage divider comprised of the two transistors in series. As a result, the voltage of the output node approaches negative V_{ss} (the negative supply voltage). V_{ss} is the "low" level, or logical O. When V_{ss} is applied to input, the p-channel device turns on. By the same reasoning used above, the output voltage rises to the V_{dd} level. This behavior is summarized in the truth table of the inverter, as shown in fig. 1.3.

Note that one transistor is always off at the end of the switching cycle. There is never a clear current path between the V_{dd} and V_{ss} supplies for the device in the switched state. This makes for a very low power circuit, since power dissipation in any element of a circuit path goes as I^2R, as is the case in the on-cycle phase of a n-MOS device. Here, I is the current between the supplies and R is the resistance of this path. Since current is near zero, power dissipation is small *at the end of the switching cycle.* During the switching cycle, there is current flow as charge distributions readjust and load capacitances are charged and discharged. It can be shown that the average power dissipation is

$$P \approx fC_LV^2, \tag{1.1}$$

where f = frequency of operation, V = maximum voltage excursion ($V_{dd} - V_{ss}$), and C_L = gate capacitance. At high speeds, when f is large, CMOS

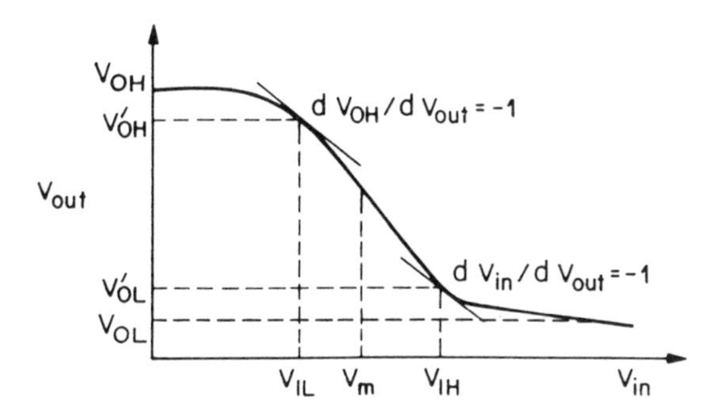

Fig. 1.6. The inverter voltage transfer curve.

may dissipate a large amount of power. In the low frequency, or quiescent state, power dissipation is minimal.

Let us now proceed to analyze the inverter to get detailed information on how it changes states [4]. A useful summary of inverter operation is contained in the voltage transfer characteristic curve. This is a plot of voltage at the output node as a function of voltage at the input node, as shown in fig. 1.6. Important parameters identified on this figure are:

- V_{OH} : highest output voltage
- V_{OL} : lowest ouput voltage
- V_{IL} : first input voltage encountered when moving left to right on the input voltage axis at which $(dV_{\mathrm{OUT}}/dV_{\mathrm{IN}})$ = -1
- V_{IH} : second input voltage at which this derivative is -1
- V_{m} : point at which $V_{\mathrm{in}} = V_{\mathrm{out}}$

For a given logic gate, there are allowable tolerances for both the logic 0 and the logic 1 voltage levels. The logic state will be unambiguiously defined within these tolerances. Let us refer to fig. 1.6. On this curve, there are two input bias points at which $dV_{\mathrm{out}}/dV_{\mathrm{in}}$ = -1: at V_{IL} and at V_{IH}. The corresponding output voltages associated with V_{IL} and V_{IH} are V'_{OH} and V'_{OL}, respectively.

If the inverter gate described by this transfer curve is used to drive an identical gate, the low noise margin, NM_{L}, is defined as

$$NM_{\mathrm{L}} = (V_{\mathrm{IL}} - V'_{\mathrm{OL}}). \tag{1.2a}$$

When the first gate (the drive gate) is in its output low level, the second gate (the driven gate) should be in its output high level. NM_{L} is the largest amount of noise that can be imposed on the driven gate input without causing an erroneous logic-zero to appear at its output. The validity of eq. (1.2a) can be verified as follows. Any output voltage between V'_{OL} and V_{OL} emanating from the drive gate will unambiguously be considered a logic-zero. Outputs greater than V'_{OL} imply that the inverter is operating either in the unstable, steeply varying portion of the transfer curve (between V_{IL} and V_{IH}) or it is in the output-high state. In either case, the state of the driven gate would be ambiguous or clearly erroneous. Assume that the output of the drive gate is *just* at V'_{OL}. This is now the input voltage to the driven gate. An amount of noise equal to NM_{L} added to the drive-gate output puts the driven gate into its unstable condition. Similarly, the high noise margin, NM_{H}, is defined as

$$NM_{\mathrm{H}} = (V'_{\mathrm{OH}} - V_{\mathrm{IH}}). \tag{1.2b}$$

In other words, NM_{H} is the maximum noise that can be imposed on the input of the next gate without causing an erroneous output 1.

If we know the p- and n-channel current-voltage (*I-V*) characteristics, we can derive the voltage transfer curve. By the *I-V* plot, we mean the drain current plotted as a function of source-to-drain voltage for a variety of possible gate voltages. From fundamental device physics considerations, the following relationships are derived [4,5]:

$$I_{\mathrm{d}} = \left(\frac{K_1}{2}\right)\left[2(V_{\mathrm{gs}} - V_{\mathrm{T}})V_{\mathrm{ds}} - V_{\mathrm{ds}}^2\right] \tag{1.3}$$

when

$$V_{\mathrm{gs}} \geq V_{\mathrm{T}};\ V_{\mathrm{ds}} \leq (V_{\mathrm{gs}} - V_{\mathrm{T}}),$$

and

$$I_{\mathrm{d}} = \left(\frac{K_1}{2}\right)(V_{\mathrm{gs}} - V_{\mathrm{T}})^2 \tag{1.4}$$

when

$$V_{\mathrm{gs}} \geq V_{\mathrm{T}};\ V_{\mathrm{ds}} \geq (V_{\mathrm{gs}} - V_{\mathrm{T}}),$$

where I_{d} = drain current, V_{ds} = drain-to-source voltage, V_{gs} = gate-to-source voltage, V_{T} = threshold voltage, $K_1 = (W/L)\mu C_{\mathrm{ox}}$, W = width of the transistor gate, L = length of the transistor gate, μ = mobility of the channel inversion charge, and C_{ox} = oxide capacitance per unit area. The relevant geometric parameters of the device and a cross section of the device itself are shown in fig. 1.7. Equation (1.3) refers to the devices in the "linear region" of operation. Here, every portion of the channel exists in strong inversion. The *I-V* plots are approximately straight lines whose slope increases as the gate-to-source voltage increases. When the drain voltage is within a threshold voltage of the gate voltage, the strong inversion condition no longer exists at the drain. At this point, the drain current ceases to be a function of drain voltage. We say the transistor is in *saturation*. In this region, eq. (1.4) holds. Care must be taken to assure that sign convention difficulties do not arise in applying (1.3) and (1.4). In both the p- and n-channel devices, inversion charge flows from source to drain. Thus, for the n channel, the

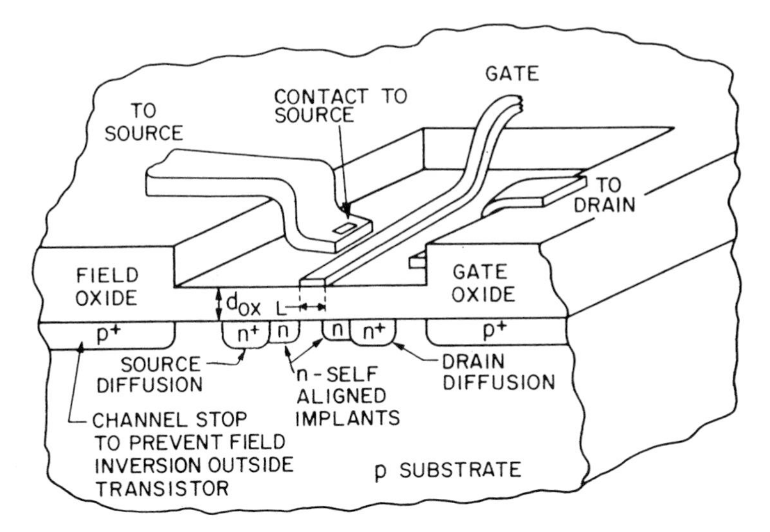

Fig. 1.7. A cross section of the MOS n-channel transistor. Please note the following: the MOS transistor active region is in the device window (thin oxide region). Thick oxide and first field channel prevent formation of unwanted inversion layer conducting paths. The edge of the space charges associated with the n self-aligned implant boundaries define the transistor channel length. The width of the transistor is determined by the field oxide boundary.

current flow is, by convention, from drain to source. In the p channel, current flow is from source to drain. V_T is negative for a p-channel device and positive for an n-channel device. Device *I-V* characteristics based on the above analysis are shown in fig. 1.8.

We derive the inverter transfer curve as follows. Note that the gate voltage on both the p channel and n-channel transistors is V_{in}. The output voltage determines V_{ds} for both transistors:

$$V_{ds}^{p} = V_{dd} - V_{out}, \tag{1.5a}$$

and

$$V_{ds}^{n} = V_{ss} - V_{out}. \tag{1.5b}$$

If we choose the n-channel drain-to-source voltage drop, V_{ds}^{n}, as the voltage parameter, we can plot *I-V* characteristics for both transistors on the same graph for a given V_{in}. This is shown in fig. 1.9. The point of intersection of these curves gives V_{ds}^{n} for a given V_{in}. From this, we can calculate V_{out}.

It should be noted that the above analysis is highly simplified in a number of ways. First, eqs. (1.3) and (1.4) must be modified to account for second-order effects. These effects appear as the device dimensions are scaled down for very-large-scale-integration (VLSI). In addition, we have come to the realization that mobility is a function of local electric field in the conducting channel. The electrical channel length itself is a function of drain voltage.

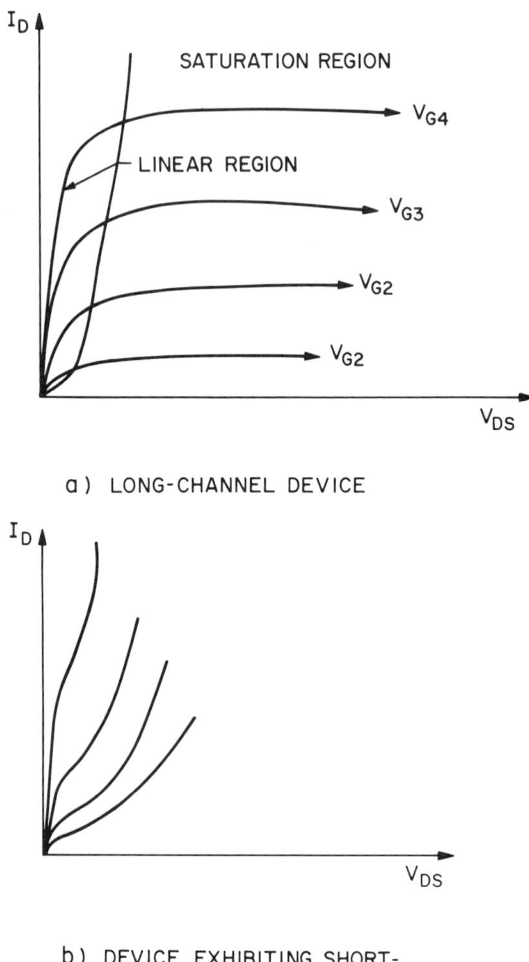

Fig. 1.8. Some typical transistor *I-V* plots. Part (a) indicates an "ideal" *I-V* characteristic. Part (b) exhibits excessive slope in the saturation region and punch-through due to the spreading of the source-drain depletions.

Furthermore, the analysis that gives rise to these equations is a static or DC analysis. Logic devices are frequently changing states. The highest speed at which this changing of states can be done is limited by the rate at which capactive elements can be charged and discharged through resistive interconnects. This complicates the appearance of the *I-V* response in the voltage or current transient state.

The second-order effects are of great interest to the process engineer. First of all, as device dimensions are scaled, these second-order effects rapidly become first-order effects. In order to optimize device performance in light

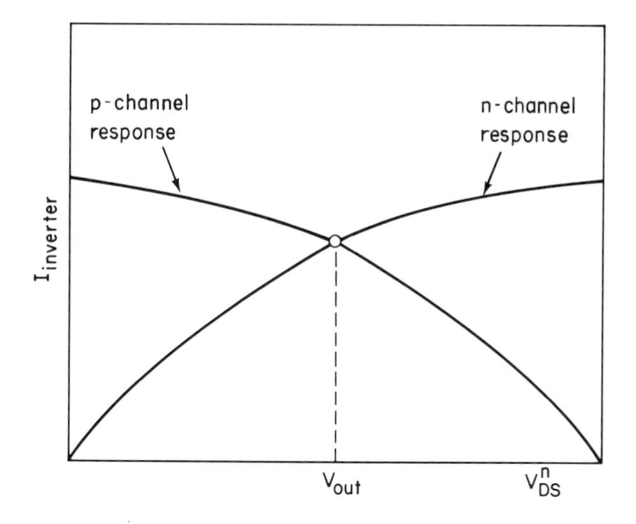

Fig. 1.9. Graphical solution to inverter transistor curve.

of this behavior, a variety of new processes have been developed. The process engineer is called on to implement these processes. In addition, materials engineers must provide low-resistance interconnects and contacts. They must also be able to predict the sheet resistivity and junction capacitances of diffused structures. Even such effects as high-field mobility degradation are, to some degree, process dependent. This close coupling of materials and device performance is the major theme of this book.

1.5 Discrete Devices

In keeping with the top-down approach, first consider the MOS transistor—a four-terminal device [5,6]. The device cross section is shown in fig. 1.7. The n-type diffusions into p substrates give an n channel device; p diffusions into an n substrate give a p-channel device. Source, drain, and gate represent three of the four terminal contacts. The fourth is the substrate.

Equations (1.3) and (1.4) indicate that the important process-related parameters defining device performance are K_1 and V_T. The K_1 constant is a factor that is both geometry and process dependent. C_{ox} depends on the thickness and dielectric constant of the oxide beneath the transistor gate. The formula for this quantity is

$$C_{ox} = \frac{\epsilon_{ox}}{t_{ox}}, \tag{1.6}$$

where ϵ_{ox} = oxide dielectric permittivity, a materials property and t_{ox} =

oxide thickness. W and L, the width and length of the capacitor, depend on the gate layout geometry. Mobility, μ is also a materials-related property [7]. It gives us the carrier speed for a given electric field:

$$v = \mu E, \tag{1.7}$$

where v = carrier speed and E = applied field. In free-space acceleration, the mobile carrier speed increases linearly with time if the force on the carrier is constant. In the solid, mobile carriers are frequently scattered. This scattering prevents the carrier from moving in the direction of the applied field. Some of the energy of motion can be dissipated through inelastic scattering collisions. It is left as an exercise to determine how scattering can lead to an expression like (1.7).

Let us now consider what mechanisms create mobile carrier scattering. One of the most important scattering mechanisms involves phonons. A phonon is a thermally induced lattice vibration in the solid. As this wave of atomic displacements from equilibrium positions passes through the solid, a fluctuation in the local electrostatic potential travels with it. This fluctuation can scatter the carrier. Cooling obviously can reduce, if not eliminate, this type of scattering. Bulk defects (vacancies, interstitials, dislocations, stacking faults, etc.) also give rise to scattering. Surface roughness at the oxide-semiconductor interface has been shown to degrade mobility. Extrinsic defects are also a source of scattering, as shown in fig. 1.10. As the background doping of the semiconductor substrate increases, mobility is degraded. The dopant ions themselves cause potential fluctuations that interact with carriers. More importantly, doping by diffusion or by implantation processes creates interstitials and vacancies that impede transport. In addition, fixed charges present at or near the oxide-semiconductor interface (the interface fixed charge or interface states) cause Coulombic scattering to occur. As the gate voltage increases, the mobile channel charges are brought closer to these interfacial scatterers. Thus, a gate-voltage dependent mobility is observed [7]:

$$\mu = \frac{\mu_0}{(1 + V_{gs}\theta)}, \tag{1.8}$$

where μ_0 = channel mobility with no gate bias applied, V_{gs} = gate to source voltage, and θ = a constant. This effect is termed *transverse field mobility degradation*, since the field responsible for it is perpendicular to the transport direction.

In addition to the transverse field mobility degradation, mobility also falls off as fields increase in the transport direction. This is shown in fig. 1.11. In silicon, this turnover in the velocity-field curve occurs for a number of reasons. Above a certain energy, new pathways for interaction of the carriers with the lattice form occur. Phonons may be created through direct interaction of the mobile carrier with the lattice. In III-V compounds, like

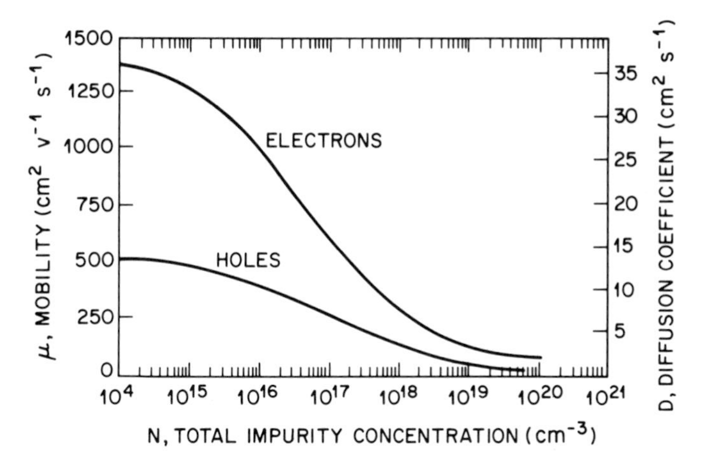

Fig. 1.10. Carrier mobility and diffusion coefficient as a function of total doping concentration in silicon at 300°K.

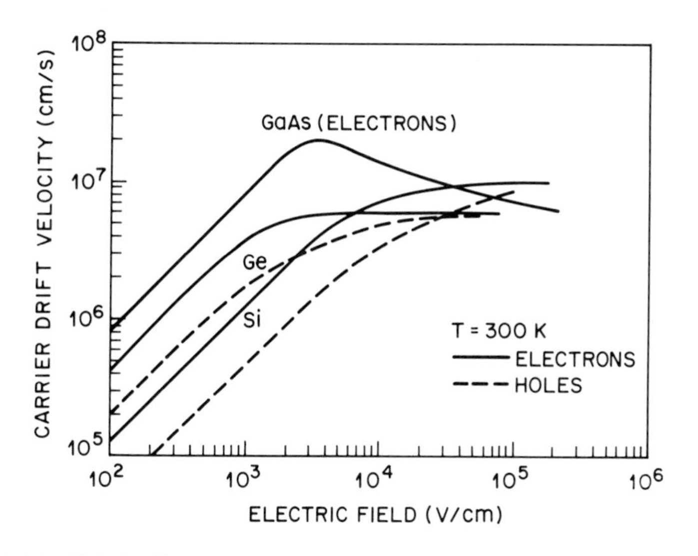

Fig. 1.11. Velocity/field curves for Si and GaAs. Velocity saturation effects result from phonon mediated carrier scattering at high electric fields. In silicon, there is a velocity versus field curve level as the carriers produce phonons. In GaAs, there is a velocity drop-back as carriers scatter into low-mobility valleys.

GaAs, the electron can be scattered into higher valleys in the semiconductor band structure. These higher valleys usually have higher effective carrier mass. Even for the case of transport in a single valley, as the carrier energy increases, the parabolic band approximation breaks down and the effective mass of the carrier increases. All of these processes degrade mobility [8].

To summarize, there are four factors that determine mobility. They are

1. Scattering of carriers by lattice vibrations (phonons) in the solid
2. Scattering of carriers by rough spots, impurities, or defects in the channel
3. Scattering of carriers by fixed electronic charge in the oxide near the channel
4. Electric field

Factor 1 is not process dependent. All solids at temperatures above absolute zero have some phonon spectrum associated with them. Factors 2 and 3 are exceptionally process dependent. Surface roughness is a function of the types of cleans and etches used on the surface. Surface defects are a function of device fabrication temperature cycles (as shown in Chapter 2). Electronic fixed charge is a function of the oxidation process employed. This is discussed in depth in Chapter 3. Processing has some influence on the electric field dependent mobility. Reduction of intrinsic and extrinsic defects and surface roughness will certainly reduce the transverse field mobility degradation. Intervalley scattering is largely a property inherent in the crystal.

It is interesting to note that one of the driving forces for feature size reduction in VLSI is the reduction of MOS gate-length transit time. This can lower the signal propagation time across the circuit. Increasing the mobility, however, has the same effect. Mobilities in MOS surface channels can range from 400 to close to 900 cm^2/V-sec. It may be easier to develop high-mobility processes than to shrink gate feature sizes into the submicron domain!

Now let us turn our attention to the factors that affect the threshold voltage, V_T. For the moment, we ignore second-order effects and consider the case of a grounded source. We list the following as determining V_T:

1. Doping of the substrate beneath the gate
2. Gate-to-semiconductor work-function difference
3. Oxide thickness
4. Insulator and interface charge

All of these terms, except work-function difference, are determined by the fabrication process. The general expression for threshold voltage is [5]

$$V_T = \phi_{ms} - 2\phi_b \pm \left(\frac{1}{C_{ox}}\right)\sqrt{4qN_s\epsilon_s \mid \phi_b \mid} - \frac{Q_{ox}}{C_{ox}} - \frac{Q_{it}^{occ}}{C_{ox}}, \tag{1.9}$$

where

ϕ_{ms} = metal-semiconductor work-function difference,
$\phi_b = (kT/q)\ ln(N_s/n_i)$,
k = Boltzmann constant,
N_s = substrate doping,
T = temperature,
q = electron charge,
n_i = semiconductor intrinsic carrier concentration,
Q_{ox} = effective oxide charge (coul/cm^2), and
Q_{it}^{occ} = *occupied* interface stage charge (coul/cm^2).

The detailed nature of the oxide fixed charge and methods for evaluating Q_{it}^{occ} are given in section 3.2 of this text. It should also be noted that Q_{ox}, as defined here, represents the *effective* oxide charge at the oxide/semiconductor interface. By effective, we mean those changes capable of producing band-bending in the semiconductor. Q_{ox} is actually a sum of four other terms: oxide fixed charge, bulk trap charge, oxide mobile charge, and interface-state charge. Methods for computing effective oxide charge are given in Section 3.2. This equation can be used for n- and p-channel MOSFETs if the following conventions are used: For both n and pMOS, ϕ_{ms} is positive for p-type polysilicon gates; ϕ_{ms} is negative for n-type polysilicon gates and for most metal gates. The ϕ_b term is negative for nMOS and positive for pMOS. The third term in eq. (1.9) takes a negative sign for nMOS and a positive sign for pMOS.

The fixed and mobile charge densities can be positive or negative. The threshold shift caused by a positive charge is negative. The threshold shift caused by a negative charge is positive. The same is true for surface states. Remember, even if the density of interface states (D_{it}) is constant, the number of occupied surface states per-square-centimeter, Q_{it}^{occ}, changes with gate bias. This is because the surface occupancy depends on the position of the Fermi level in the bandgap. Acceptor states are neutral when they are above the Fermi level, negative when they are below. Donors are neutral when they are below the Fermi level, positive when they are above. Thus, to compute threshold in the presence of surface states, the nature and distribution of these states in the semiconductor energy gap must be specified.

Variation of the substrate doping (N_s) through ion-implantation (Chapter 5) or diffusion (Chapter 4) is the primary method of adjusting V_T. Oxide thickness is also important since it appears in three of the terms of (1.9) through C_{ox}. The oxide fixed charge and interface charge are process related. However, as indicated in Chapter 3, these terms are less easily controlled than N_s and C_{ox}.

Equation (1.9) holds for the case of a grounded source. Manipulation of the source-to-substrate bias voltage also influences the threshold voltage. This is called *body effect* [9]. It is explained as follows. The total charge in the semiconductor beneath the MOSFET gate is the sum of two terms: the

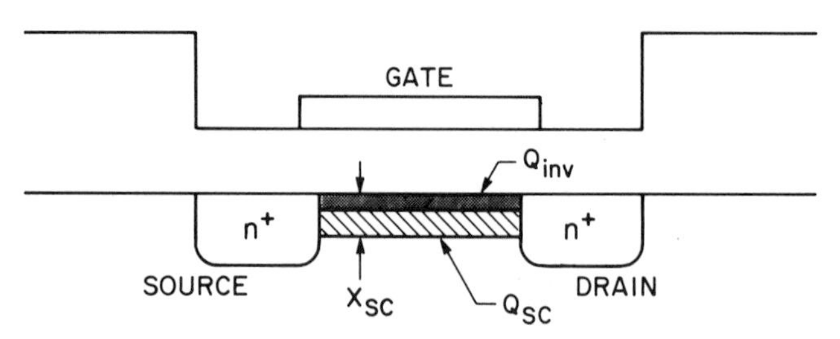

Fig. 1.12. Origin of substrate sensitivity (body effect). Substrate bias increases Q_{sc} at the expense of Q_{inv}.

bulk semiconductor space charge, Q_{sc}; and the mobile inversion layer charge, Q_{inv}. We usually define the threshold voltage as the point at which the strong inversion condition (semiconductor surface potential equal to $2\phi_b$) is reached. But this condition can also be expressed in terms of the sum $Q_{sc} + Q_{inv}$. We say that when inversion is reached, this sum equals some critical charge, Q_{crit}. Consider the effect of keeping the source grounded and applying a reverse bias to the substrate. The mobile inversion layer, which can be viewed as a shorting bar across the channel between the source and drain, extends the effect of the reverse bias over the whole channel. The reverse bias causes the semiconductor space charge to grow around the source-substrate junction *and* under the gate in the active channel. Thus, the Q_{sc} term grows but Q_{crit} remains fixed. As a result, Q_{inv} is lowered for a given gate bias when the source-to-substrate diode is reverse biased. Q_{sc} grows at the expense of Q_{inv}. This process is illustrated in fig. 1.12. To bring the device to the strong inversion condition, a larger gate bias must be applied. The threshold voltage appears to increase. The amount of threshold shift is obtained by computing the change in gate voltage that would give rise to the increase in Q_{sc}:

$$\Delta V_T = \frac{1}{C_{ox}} \left[\sqrt{[2qN_s\epsilon_s(V_R + 2 \mid \phi_b \mid)]} - \sqrt{N_s\epsilon_s 2 \mid \phi_b \mid} \right] \tag{1.10}$$

where V_R = substrate reverse bias. Once again, we note that the substrate sensitivity is process dependent through N_s, the substrate background concentration.

Current emphasis in very-large-scale-integrated circuit research is on *scaling* [10,11]. Scaling refers to the shrinkage of device dimensions to achieve higher speed and circuit density. Scaling is done in such a way as to maintain either constant electric field (CEF scaling), constant output and supply voltages (CV scaling), or some degree of constancy in both areas (usually called *quasiconstant voltage*, or QCV scaling). In CEF scaling, the same basic device operating characteristics are obtained at lower voltage, and the reliability of the device (which is dependent on operating fields) is unaffected. Device or circuit speeds are faster, due to reduced gate dimensions

TABLE 1.1: Scaling relationships of the MOSFET parameters as a function of the fundamental dimensional scaling constant, k

Device parameters	CEF	CV	QCV
Voltage, V	1/k	1	$1/k^{-2}$
Lateral dimension, L, W	1/k	1/k	1/k
Vertical dimension, t_{ox}	$1/k$	$1/k^{1/2}$	$1/k$
Doping concentration, N_a, N_d	k	k	k

in CV scaling. Device operating characteristics are altered, but the input and output voltages are compatible with other unscaled devices. In QCV, output voltages are held fixed, but transistor device capability is enhanced. For each of these approaches, materials scaling rules are given in table 1.1 in terms of a fundamental scaling constant, k.

Scaling down the device dimensions brings a host of process-dependent factors into play. The most familiar of these are the short- and narrow-channel phenomena. As channel length, L, is reduced, V_T appears to decrease (if all other process and geometry parameters are fixed). In addition, the source-drain depletions will touch if L is too small. This does *not* lead to an immediate shorting of the source to the drain. There is still an electrostatic potential barrier between the source and the drain that prevents current flow between these nodes. However, this barrier decreases as the source-drain depletions overlap. The barrier is further lowered as the drain bias is increased, as illustrated in fig. 1.13. This leads to an increase in drain current for gate biases lower than V_T. We say the *subthreshold leakage current* increases as a result of *drain-induced barrier lowering* (DIBL) [12]. The subthreshold leakage leads to increased power dissipation—a potential performance limiter. In the worst case, these devices may appear "on" to other units in the logic chain.

In the following paragraphs, we consider the effects of scaling on device performance and how choice of process influences these effects. First, consider the short-channel problem, as illustrated in fig. 1.14. The distance, x_{max}, is the extent of the semiconductor space charge beneath the gate when the strong inversion condition is met (semiconductor surface potential is $2\phi_b$). Here, x_{max} appears constant, since we are assuming the drain bias is small. Increasing the drain bias *will* increase the extent of the drain depletion under the gate. The junction space charge associated with the source-drain diodes is characterized by a length x_j. The x_j appears equal for the source and drain, also due to the small drain bias. Note that there is a region of overlap over which both space charges are common. This is referred to as the *charge-sharing region* [13]. In order to achieve transistor turn-on, we

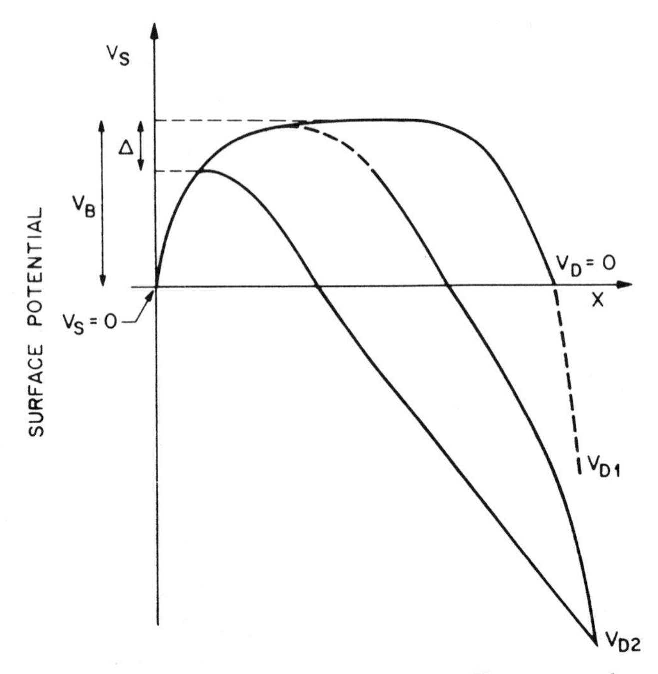

Fig. 1.13. Origin of drain-induced barrier lowering. Here, we see the semiconductor surface potential plotted as a function of position and drain bias. When V_D is low, a well formed potential barrier exists between source and drain. This prevents the drain bias from pulling mobile charge directly from the source. When V_D gets large, this barrier reduces by some amount, Δ. When V_D gets very large, it disappears.

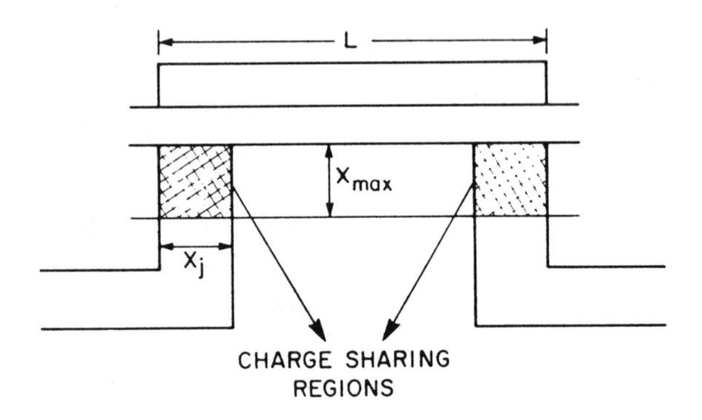

Fig. 1.14. Short channel effect. Charge-sharing boxes are depleted *without* application of gate-bias. This lowers threshold.

have the following condition on x_{max}:

$$x_{max} = \sqrt{\frac{4\epsilon_s \phi_b}{qN_s}}. \qquad (1.11)$$

Similarly,

$$x_{\mathrm{j}} = \sqrt{\frac{2\epsilon_{\mathrm{s}}\phi_{\mathrm{bi}}}{qN_{\mathrm{s}}}}, \tag{1.12}$$

where ϕ_{bi} is the built-in voltage of the source/drain to substrate diodes. As seen in fig. 1.14, some portion of the channel is depleted *even without gate bias applied.* This portion is just the sum of the source and drain charge-sharing regions. If we assume a uniform bulk doping density and a grounded source, drain, and substrate, the total charge in these regions is

$$Q = 2\mathrm{q}N_{\mathrm{s}}x_{\mathrm{max}}x_{\mathrm{j}}W, \tag{1.13}$$

where W = channel width. As a result of this, the Q_{crit} for turn-on is lowered by an amount Q. Thus, V_{T} is lowered. The amount of V_{T} lowering is found by converting Q into a charge per unit area and by dividing the result by C_{ox}

$$\Delta V_{\mathrm{T}}^{\mathrm{sc}} = \frac{Q}{(WLC_{\mathrm{ox}})} = \left(\frac{1}{C_{\mathrm{ox}}}\right)\left(\frac{\epsilon_{\mathrm{s}}}{L}\right)\sqrt{2\phi_{\mathrm{bi}}\phi_{\mathrm{b}}}. \tag{1.14}$$

The superscript, sc, refers to the fact that this is a short channel effect. In this approximation, the short channel threshold shift varies inversely with L. In practice, we find that (1.14) overestimates $\Delta V_{\mathrm{t}}^{\mathrm{sc}}$. This is because not all of the source-drain depletion charge is effective in influencing the surface potential. That is, only a fraction of Q bends the bands near the surface toward the strong inversion condition. In some approximations, we say only half of Q is effective. A better approximation is obtained by saying

$$V_{\mathrm{t}}^{\mathrm{observed}} = V_{\mathrm{T0}} - f\Delta V_{\mathrm{t}}^{\mathrm{sc}}, \tag{1.15}$$

where f = a fraction (usually approximately 1/2) and V_{T0} = device threshold in the absence of short-channel effects. The f constant can be arrived at by measurement. It should be kept in mind that increasing the drain bias widens the junction space charge, further lowering the apparent threshold. In addition, all of these approximations are essentially one dimensional. They do not account for the detailed three dimensional nature of the charge-sharing regions. The model does illustrate the basic physics of the effect.

Scaling in the width direction also influences threshold [14] (the narrow-channel effect). To see this, consider fig. 1.15. The semiconductor bulk space-charge does not terminate abruptly at the width extremes (points $p1$ and $p2$). Rather, $p1$ and $p2$ can be viewed as bias points for the surrounding bulk. At turn-on the surface potential at $p1$ and $p2$ is $2\phi_{\mathrm{b}}$. This bias at $p1$ and $p2$ causes the bulk depletions to swing out from under the gate in the quarter circles shown in this figure. The radius, r, of these quarter circles is

$$r = \sqrt{\frac{4\epsilon_{\mathrm{s}}\phi_{\mathrm{b}}}{qN_{\mathrm{s}}}}. \tag{1.16}$$

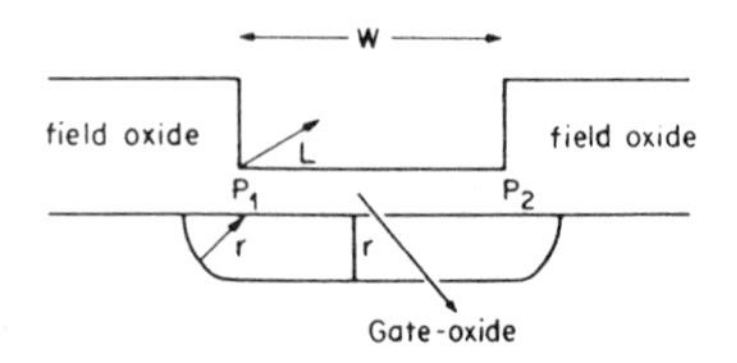

Fig. 1.15. Narrow channel effect. Hemi-cylinders of depletion under field oxide must be created by gate-bias. This *raises* threshold.

This space charge extends back in the length extension to form a quarter cylinder. An extra charge, amounting to twice the charge in each quarter cylinder, must be supplied to the gate to achieve the turn-on condition. In this case, the threshold appears to have increased. Using reasoning similar to that applied to the short channel effect, we would expect that this increase would be

$$\Delta V_{\mathrm{T}}^{\mathrm{nc}} = \frac{\pi r^2 q N_{\mathrm{s}}}{2C_{\mathrm{ox}}W} \tag{1.17}$$

and

$$V_{\mathrm{T}}^{\mathrm{observed}} = V_{\mathrm{T0}} + \Delta V_{\mathrm{T}}. \tag{1.18}$$

Here, the increase in threshold goes inversely with W. In certain scaling approaches, the narrow-channel effect can be used to partially offset the short-channel effect. However, in recent years, ion implantation has been used more effectively to control these problems.

All of the above effects (short- and narrow-channel effects, as well as DIBL) are strongly influenced by substrate doping. The encroachment of the source and drain depletions under the gate is reduced by increasing N_{s}. Raising N_{s} also increases V_{T}, which offsets the short channel effect. Relatively recent developments in the area of implantation technology minimize these effects and have improved the situation. Specifically, the development of the *self-aligned* gate technique is of major importance in this area. This technique is illustrated in fig. 1.16. The gate itself serves as a mask, protecting the active channel region from implantation of dopant shot at it by an ion-beam particle accelerator. This has two advantages. First, the source-drain junction edges line up almost exactly with the gate edge. Any heavy doping that reaches under the gate has a conductivity that is not modulated by the gate. This heavily doped region is not part of the active channel. However, it does form a gate-oxide-conductor capacitor. This *overlap capacitance* is an unwanted parasitic that slows device response down. Note that there is some encroachment of the implant under the gate as a result of ion scattering. This is discussed in greater depth in Chapter 5. In addition, as shown in the figure, the junction is *shallow*. The depletion associated with it terminates below x_{max}. This minimizes charge sharing.

The four-terminal device is the most complicated device in an IC, but it is not the only device. Diodes, capacitors, and resistors are also used in the

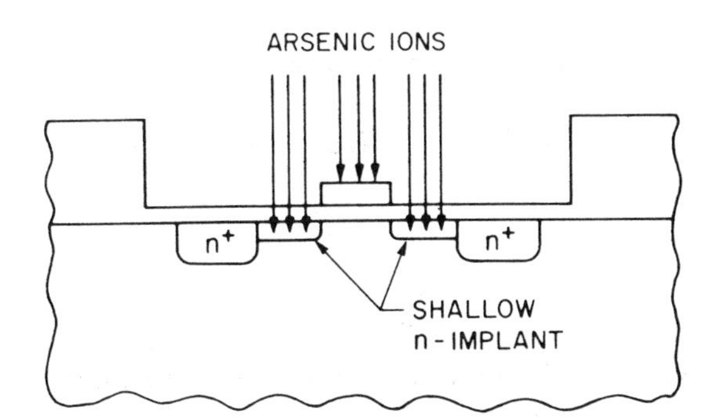

Fig. 1.16. Self-aligned source-drain process.

completed circuit. We now turn our attention to these two terminal devices, beginning with the diode. From Shockley's "law of the junction," elementary device physics tells us that the current flow through a p-type/n-type silicon abrupt junction is given by [5]

$$J = J_o \left[\exp\left(\frac{qV_b}{kT} \right) - 1 \right], \tag{1.19}$$

where J_o = the saturation current and V_b = applied bias. The saturation current is given by

$$J_o = qn_i^2 \left[\frac{D_p^{1/2}}{N_d \tau_n^{1/2}} + \frac{D_n^{1/2}}{N_a \tau_n^{1/2}} \right], \tag{1.19a}$$

where n_i = intrinsic carrier concentration; $D_{p,n}$ = diffusion coefficient for holes, electrons; $N_{d,a}$ = donor, acceptor carrier concentrations; and $\tau_{p,n}$ = hole and electron minority carrier lifetimes.

The physical picture of the diode transport process is this. Forward or reverse bias applied to the diode terminals changes the mobile charge stores at the junction space-charge edges. Specifically, the excess minority charge stores are either enhanced (forward bias) or suppressed (reverse bias). These excess charge stores decay exponentially as we move from the space-charge edge away from this metallurgical junction. The decay is created by GR processes. The electric field at the space-charge edge is small and transport is driven by *diffusion*. Thus, the current density given in eq. 1.19 is obtained by taking the position derivative of the excess minority carrier distribution at the point where this derivative is largest (the space-charge edge). It is assumed that at this point the sole transport mechanism giving rise to diode current is diffusion. As we move farther from the space-charge edge, GR currents start becoming important. Far from the junction, deep in the semiconductor bulk, transport is ohmic.

Equation (1.19) is derived without accounting for the generation and recombination (GR) processes that occur within the junction depletion layer. These effects are dealt with by modifying the equation as follows:

$$J = J_o \left[\exp\left(\frac{qV_b}{nkT} \right) - 1 \right], \tag{1.20}$$

where n is called the *diode ideality factor.* It is a constant that ranges from 1 to 2. As n approaches 2, the GR processes becomes more important in determining the diode's current-voltage characteristics. GR centers are the result of defects in the crystal structure. As the forward-biased junction depletion floods with mobile charge, these centers act as intermediate states, assisting the mobile charge store to equilibrium. In the case of forward bias, this means carrier *recombination*, and the diode current's voltage-sensitivity lowers. In reverse bias, the mobile charge stores are suppressed, and mobile carrier generation dominates. Defects *increase* the diode reverse leakage current. These defects can be mechanical (dislocations, stacking faults, etc.) or extrinsic heavy metal impurities. These same defects are responsible for reducing the minority carrier lifetimes. Heavy metals can be rendered electrically inactive by anneals described in the next chapter. The presence of mechanical defects is a result of stress on the wafer applied during temperature cycles and through doping. Minimization of these stresses minimizes mechanical defects.

Capacitors are of interest for two reasons. First, they are necessary circuit elements in most ICs. In addition, they are among the most useful electrical diagnostic tools available to the process engineer. To see this, consider the typical MOS IC capacitor (fig. 1.17). This device provides a voltage-variable capacitance, that is, when the voltage applied to the metal (or polysilicon) gate changes, the small-signal capacitance changes [15]. To see how this occurs consider the figure. The total capacitance is a series sum of two capacitances: the oxide capacitance and the depletion-layer capacitance. The oxide capacitance is fixed, and is given by eq. (1.6). The depletion capacitance is

$$C_{\mathrm{dl}} = \frac{\epsilon_{\mathrm{s}}}{d_{\mathrm{dl}}}, \tag{1.21}$$

where the d_{dl} is the depletion layer thickness. Note that the capacitances expressed in eqs. (1.6) and (1.21) are capacitances *per unit area.*

Since d_{dl} changes as a function of gate bias, the total measured capacitance changes. This is summarized in fig. 1.18 for MOS capacitors on p-type or n-type substrates. The effect of the bias on the conduction and valence band edge as a function of position are shown as fig. 1.19. Referring to figure 1.18 (the p-type substrate), when the voltage is strongly negative, the majority carrier concentration is enriched. The surface is *accumulated.* The accumulation layer responds to the small high- (or low-) frequency signal

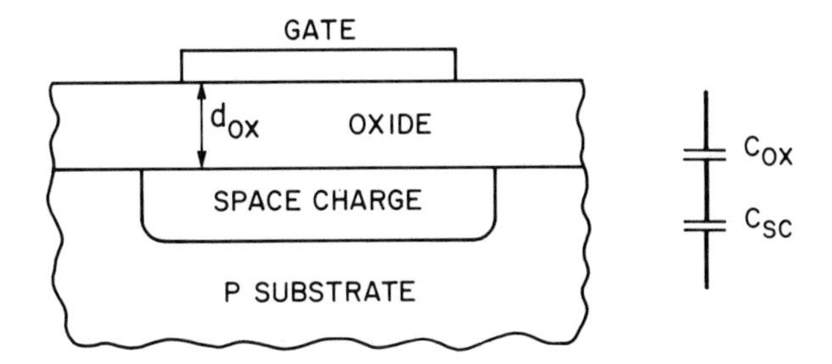

Fig. 1.17. MOS capacitor and equivalent circuit.

applied to make the capacitance measurement. The layer acts as a metal field plate. Thus, the measured capacitance is due to the gate, the oxide dielectric, and the thin metal-like accumulation layer. The measured capacitance is C_{ox}. As the gate bias becomes increasingly more positive, the majority carriers are repelled and a depletion layer appears in the silicon. The total capacitance is

$$C_{tot} = \frac{C_{ox}C_{dl}}{(C_{ox} + C_{dl})}, \tag{1.22}$$

which is smaller than C_{ox} alone.

Larger levels of positive bias result in the field generation of large amounts of mobile minority charge. An *inversion layer* is formed [16]. Two types of behavior are now possible, depending on the frequency of the small signal. At low frequency (e.g., $\lesssim$ 10 Hz), the inversion layer changes charge density as dictated by the small signal. The layer acts as a metal field plate and the measured capacitance rises to C_{ox}. Now, consider the high-frequency ($\gtrsim$ 10 Hz) case. The ability of the inversion layer to respond to the small signal depends on the rate at which the surface can generate and annihilate mobile minority charge. In well-prepared surfaces, this process can be slow. While the inversion layer responds to changes in the DC bias, it cannot respond to changes in the small amplitude capacitance-measuring signal. The formation of the inversion layer inhibits growth of the depletion layer. Field lines originating on the metal field plate (as a result of DC bias) terminate in the inversion layer. They cannot reach down into the silicon bulk to uncover more space charge. The value of d_{dl} becomes fixed at x_{max}, as given in eq. (1.11). This fixes the value of the depletion capacitance. The edge of the depletion can respond to the small signal. Thus, the capacitance reaches some minimum value and stays fixed as the DC bias is increased. This behavior is summarized in fig. 1.18. The value of this C_{min} is given by eq. (1.22), with C_{dl} determined by x_{max}.

There is another parameter that is useful in characterizing CV plots: the flat-band capacitance (C_{fb}). The voltage at which this capacitance occurs is called the flat-band voltage, V_{fb}. When the flat-band condition is reached, there are no external-bias induced electric fields in the semiconductor. The

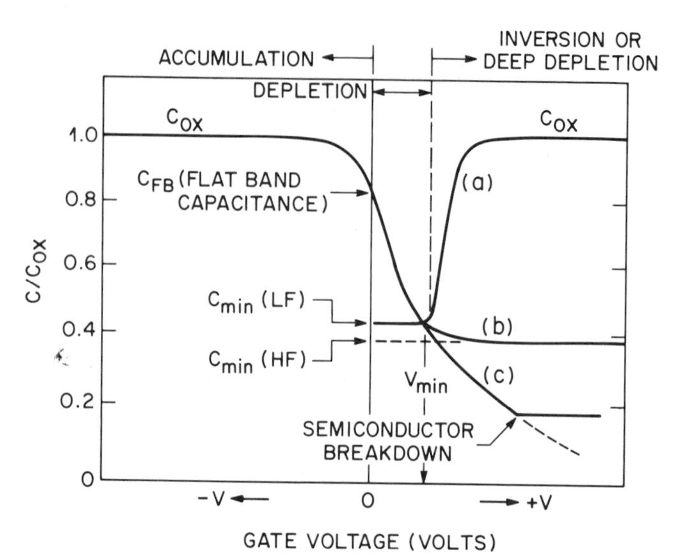

Fig. 1.18. Three regions of operation for capacitor shown in fig. 1.17. High- and low-frequency capacitive response as a function of frequency is shown. The silicon substrate is p-type.

flatband capacitance of the bulk silicon is given by

$$C_{\text{fb}}\ (\text{bulk}) = \frac{\epsilon_{\text{Si}}}{L_{\text{D}}}, \tag{1.23}$$

where L_{D} is the extrinsic Debye screening length and ϵ_{Si} is the permittivity of silicon. Again, C_{fb} (bulk) is a capacitance per unit area. For a p-type semiconductor, L_{D} is given by

$$L_{\text{D}} = \left(\frac{kT\epsilon_{\text{Si}}}{pq^2}\right)^{1/2}, \tag{1.24}$$

where p is the hole concentration and q is electronic charge. Debye length is a measure of the depth in the semiconductor below which bulk free carriers are shielded from voltage variations applied at the surface. Debye length increases as the free carrier concentration decreases and is, therefore, largest for the intrinsic semiconductor and is given by

$$L_{\text{Di}} = \left(\frac{kT\epsilon_{\text{Si}}}{2n_{\text{i}}q^2}\right)^{1/2}, \tag{1.25}$$

where n_{i} is the intrinsic electron or hole concentration. Once again, the total capacitance is given by the series combination of the oxide and bulk terms

$$C_{\text{fb}} = \frac{C_{\text{fb}}(\text{bulk})\ C_{\text{ox}}}{C_{\text{fb}}(\text{bulk}) + C_{\text{ox}}}. \tag{1.26}$$

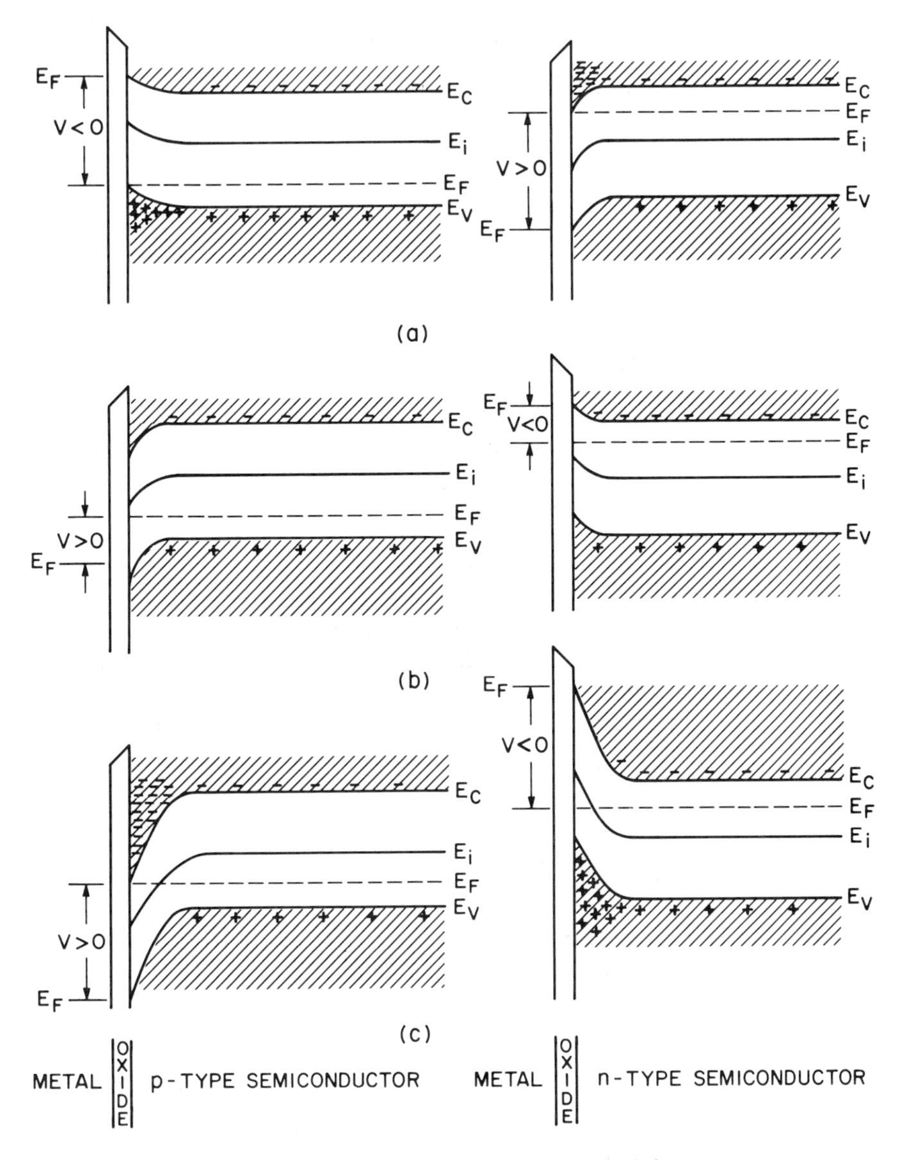

Fig. 1.19. Departure from ideal *C-V* behavior due to oxide defects.

As mentioned above, capacitance-voltage (CV) measurements represent a valuable diagnostic tool. Unwanted charges in the insulator shift the position of the CV curve along the voltage axis. Interface states distort the curve, apparently stretching it out. These effects are discussed in depth in Chapter 3.

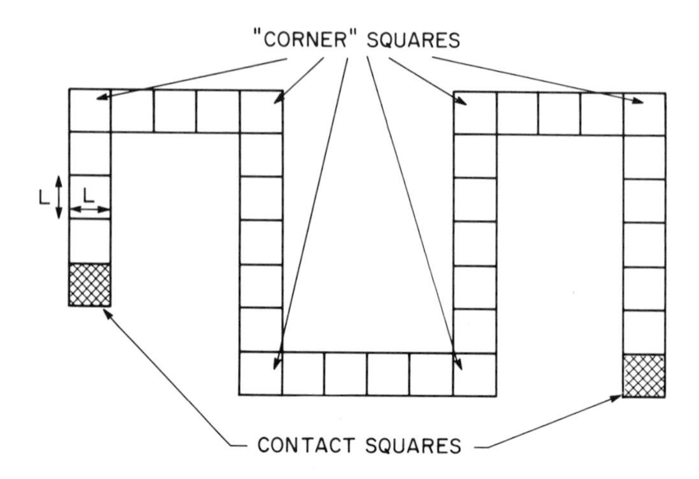

Fig. 1.20. A serpentine resistor pattern.

The final device for consideration here is the IC resistor [17,18]. There are a number of designs and configurations for such resistors. Thin, resistive metals like nichrome can serve as a resistor element. More commonly, diffusion tubs contacted at either end are employed. Such a resistor is shown in fig. 1.20. The folded configuration, known as a serpentine design, gives maximum resistance for minimum area. To compute the resistance of such a structure, we must know the *resistivity* of the diffusion. Resistivity of a doped layer is given by

$$\rho = \frac{1}{qc\mu}(\text{ohm} - \text{cm}), \tag{1.27}$$

where c is the number of mobile carriers per unit volume in the doped layer. The resistance of this layer is (assuming uniform doping)

$$R = \rho\left(\frac{l}{wx}\right), \tag{1.28}$$

which is the resistivity multiplied by the length of the resistor, divided by its cross-sectional area. The question of how c is determined must be addressed. To first approximation, c is the magnitude of the difference between electrically active acceptor and donor impurity concentrations: $N_\text{A} - N_\text{D}$. Frequently, in processing, we specify a quantity related to the resistivity: the *sheet resistivity*. This is the resistivity divided by the thickness of the layer. The units are given as *ohms per square*, even though it appears that the units should simply be ohms. The reason for this is as follows. First, break the diffusion region into squares, as shown in fig. 1.20. This is as if we are measuring the resistance of a square solid with $\ell = w$, giving $R_\square = \rho/x$. Of

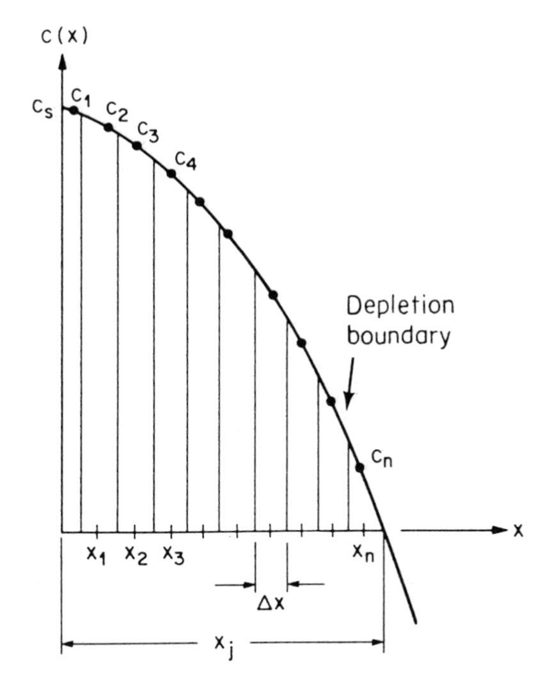

Fig. 1.21. Differential slab construction for evaluating resistivity in the case of a position-dependent net mobile carrier concentration, c_n.

course, in computing sheet resistivity, the extent of the depletion region must be accounted for, since the mobile carrier concentration is zero in the depletion. To compute the total resistance of some arbitrarily shaped diffusion, we break the conducting path into squares and multiply $R_\square$ by the number of squares obtained. The net carrier is not constant in the case of diffused resistors. Concentration varies as a function of depth from the surface. The problem can then be addressed by breaking the resistor into slices of constant doping concentration, shown in fig. 1.21, and by forming the parallel- resistor sum. In this case, we can write:

$$R_\square \approx \frac{1}{\sum\limits_{n} q\mu_n c_n \Delta x} \approx \left[\int_0^{x_j'} q\mu(x)c(x)dx \right]^{-1}, \tag{1.29}$$

where the sum extends over slabs from the surface to the depletion boundary, and x_j' is the depth of undepleted material. Note that the mobility is position dependent as it is a function of doping concentration.

Not all the squares in fig. 1.20 count equally. For example, the first and last squares of the serpentine are where metal contacts are made to the diffusion. These squares can have higher sheet resistivities than most of the squares as a result of interface and material interactions. This is because metal-to-silicon contacts are not perfect. They may not even be ohmic (i.e., exhibit linear current/voltage characteristics). This is discussed in Chapter 6. The contact resistance can be measured in a number of ways. For example, a number of lines with different numbers of squares can be made. Resistance can be plotted as a function of line length (in "squares"). The slope of the line is the sheet resistivity. When extrapolated to zero "squares", we find the resistance is not zero. But rather, the nonzero intercept on the resistance axis is twice the excess resistance associated with the contact. Contact resistance is a function of the geometry of the metal-semiconductor contact window.

The corner squares (where the line bends) also do not count as a simple square. This is because the field lines through these squares are not straight lines. The difference in the amount of the contribution can be calculated. We do this by finding the average length of the field line from the input face of the square to the output face of the square. The square will contribute more or less to the line resistance depending on the ratio of this line length to the square dimension, l. The mathematical technique of conformal maps is frequently useful here [19]. In fig. 1.22, we work out an approximation to the corner square problem by rounding the edge of the square and by assuming the field lines are quarter circles centered about the bend point. The approximation is slightly higher than the actual value, due to bunching of field lines near a sharp corner. This is accounted for in a true conformal map. A problem relating to this approach in which there are no corners is given below.

This concludes our discussion of the major circuit elements used in ICs. In this discussion, we have focused on the way processing impacts device performance. In subsequent chapters, we give detailed methods for modeling and controlling materials parameters to optimize device performance. We conclude this chapter with a detailed process flow for producing the CMOS structure discussed above.

1.6 A CMOS Process Flow

In fig. 1.23 and in frames I through IX, cross sections and plane views of the CMOS device are shown at various stages of the process [20]. Note that this is a basic exposition of a final process. It will be elaborated in the text as we study the basics of each process step. The reader should, at this point, review the accompanying frames to get a feeling for process flow. The following paragraphs emphasize key points and do not represent a running commentary on this flow.

The initial phases of the process involve installation of the diffusion tubs, source-drain diffusions, and isolation implants. It should be noted that oxides are grown prior to implantation (step 4). These oxides randomize

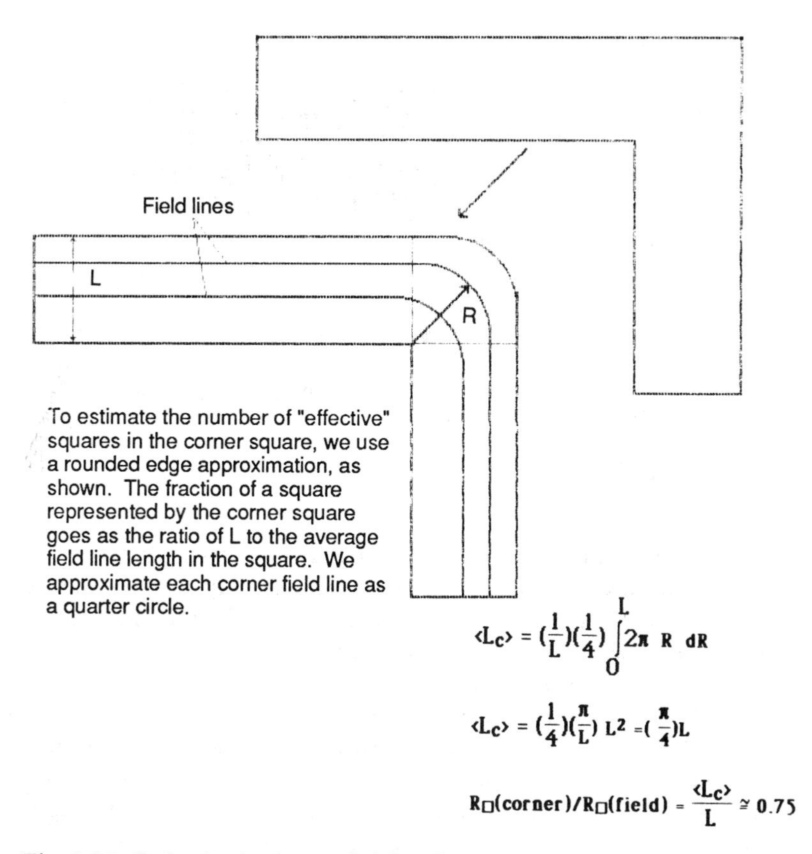

Fig. 1.22. Reduction in sheet resistivity of corner squares.

the implant ion current and prevent "channeling" (Section 5.2.3). The oxide over the implant is thickened during the n-tub "drive" anneal (step 6) to prevent dopant loss. This is followed by field and gate oxide growth, contact window, and via window definition. Field oxidations are "self-aligned" using a process described in Section 7.7. A contact window is formed between an interconnect and the silicon substrate. A via is formed between interconnect layers. Next, the interconnect layers are formed. These are metal or polysilicon lines that hook different parts of the circuit together. In this process, polysilicon is the material from which the active transistor gates are formed. The source and drain regions are self-aligned to the gate by implanting ions on either side of the polysilicon gate after the gate is defined (steps 29 and 32 of the process). Actually, this is a two-step process in CMOS. First, the region above the n channel is exposed for implant (as shown in figure associated with frame 8); that is, a hole is cut in resist to let the n-dopant ions in. Next, the n channels are protected by resist and the p-channel source/drain self-aligned extensions are implanted. In the process

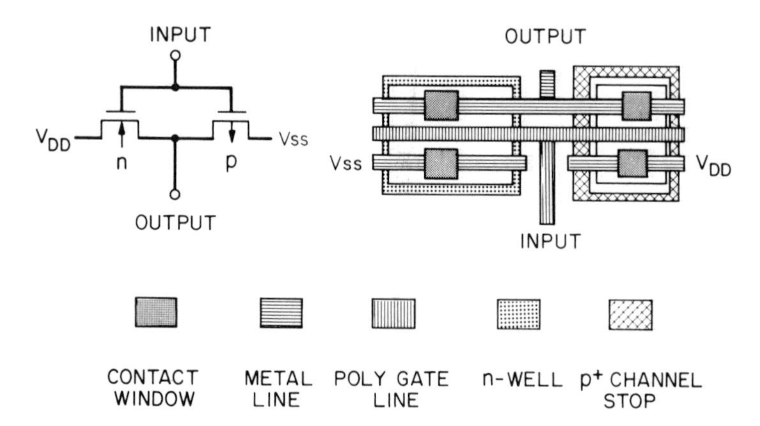

Fig. 1.23. CMOS inverter: circuit and physical layout.

shown, the self-aligned implants join deeper diffusions some distance from the gate edge. The deep diffusions are easily contacted by polysilicon or by metal. All-implanted source/drains are possible. But the resulting shallow junctions are easily destroyed in the contacting process. This is discussed in greater depth in Chapters 5 and 6.

As a final step, a deposited glass is frequently laid over the completed device, and vias are etched to the contact pads. This thick deposited glass, called a *passivating layer*, protects the metal lines from getting scratched. It also protects the device from environmental contaminants. This last step is not shown. Processing up to and including the polysilicon gate definition is called the *front end of the line.* The remaining processes are the *back end of the line.*

The process steps outlined above are generic to all IC processes (nMOS and bipolar, as well as CMOS). There are a number of nuances peculiar to the CMOS process alone. The deep well diffusion (n well, in this case) is certainly unique to CMOS, as is the fabrication of devices within this well. What problems must be addressed as a result of the CMOS well? First, the well dimensions set a limit on how close the p- and n-channel devices can be placed. This can create a limit on the number of devices per square centimeter (the density of the integration). Will the p-channel source/drains punch through the n tub and short to the substrate? Thus it is important to consider the whole process time and temperature sequence in determining final junction depths and if this punch-through is possible.

Device isolation, the ability to prevent unwanted communication between devices, is a particular problem in CMOS. This is partly due to the condition called *latch-up* [21]. As shown in fig. 1.24, the p-channel source, the n well, the substrate, and the n-channel source form a p-n-p-n structure. This structure can be modeled as two bipolar transistors and shunting resistors (as shown in this figure). If base drive is supplied to either bipolar base, one transistor can turn on. This will supply enough current to the base of the other transistor to turn it on. Thus, a clear current path from the V_{dd} to V_{ss} supplies is formed. This current can increase the regenerative current and burn out the device (if the current is not limited). Even if the device is not destroyed, proper function cannot be restored until supplies are removed and then reconnected.

Suppose, for example, that the npn transistor is turned on. The turn-on can result from line transients or from cosmic rays. Electrons from V_{ss} stream into the n well (the base of the pnp device). This will turn the pnp transistor on, creating latch-up. The channel stop (deep boron implant, step 16) creates a potential barrier to electrons, minimizing the likelihood of latch-up. Other techniques are also possible to suppress latch-up. A deep trench can be plasma etched into the silicon in place of the channel stop (Chapter 11), or a deep p implant into the well can reduce the well shunt resistance ($R_{\mathbf{well}}$ in fig. 1.24). Lowering substrate resistance by using epitaxially grown materials lowers $R_{\mathbf{substrate}}$. This prevents the bias voltage drop across the pnp emitter/base junction from reaching the diode turn-on voltage.

A few points should be made concerning the example frames themselves. The n-tub diffusion exists in a very limited region of space. The metal lines are thin-film wires less than 10 μm wide. These small regions must somehow be defined on the wafer. This is done by the processes of *lithography* (discussed in Chapter 8 and 9) and *pattern transfer* (Chapter 10). In the course of these processes, the silicon substrates are turned into photoplates by the application of a light-sensitive plastic called *photoresist.* Images of the desired shapes are projected on the substrate. After development, the resist remains in some areas and is gone from others. What is left is an image, in photoresist, of the desired features. The resist is called *resist* because it can hold off even powerful acid etchants. Once the resist has been defined, the pattern can be transferred to underlying materials by wet- or dry-etch processes.

For example, we can open a hole in an oxide by creating a hole in photoresist and by dipping the substrate in hydrofluoric acid. The acid attacks the oxide that is not protected by the photoresist. The resulting hole can be used as a diffusion window. After the remaining photoresist has been removed, we can place the wafer in a diffusion furnace. Dopant will reach the silicon where it is not protected by oxide. Metal lines can be produced by direct etching of metal films.

When we see such structures, the following questions immediately come to mind. How localized can these structures actually be? In other words, how narrow can we make our metal lines? How small can we make our diffusion tubs? In addition, we can ask whether or not there are interactions between regions that would make the placement of features more difficult. That is,

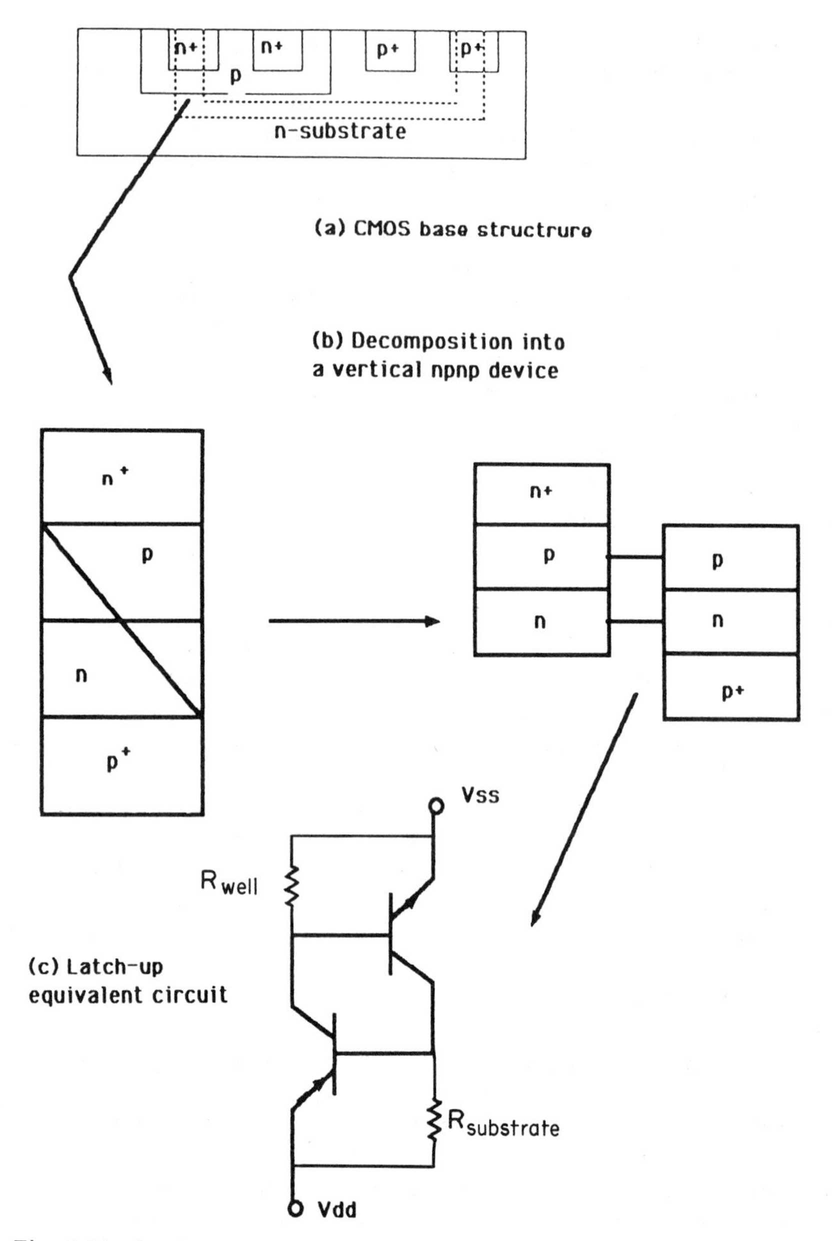

Fig. 1.24. Latch-up equivalent circuit. Plates I-IX 10 mask n-well CMOS process flow.

are two one-micron lines spaced one micron apart harder to resolve than a single, isolated one-micron line? (The answer is yes!) How close to the n-tub junction can we position a p-type source-drain? How close to the n-well junction can we place an n-channel source? The answers to all of these questions are expressed in *design rules*, that is, a group of rules in the form of statements, like: two aluminum lines can be separated no less than 4 μm; or, the n-channel drain can come no closer than within 3 μm of the n-tub junction. The setting of design rules is critical to IC technology. Design rules determine the density of integration and device yield. The materials science of factors influencing these rules is the subject of this book.

1.7 Summary

In this chapter we have illustrated an approach to process development useful in making ICs. In the first step, we try to understand the ultimate goal of the process (in this case a static RAM). We ask, what does the design engineer hope to accomplish and how can we help him or her? Next, we familiarize ourselves with the floor plan of the device to be created. This will help when we are looking through the microscope at partially completed devices. We wiil be able to decide if what we see is normal, or if there are fatal or harmless flaws. When problems arise, we will be able to help locate and eliminate them. We should have some familiarity with how the basic elements of the IC interact. This will help us in deciding which parts of the process should or could be optimized. Also, it will allow us to correctly interpret the results of on-chip test structures (like test transistors, gated diodes, etc.) Finally, we should be able to recognize design-rule violations in the chip image. The text below will address the problem of how design rules are set. Emphasis is given to the description of how the complete time and temperature history of the process impacts these design rules.

References for Chapter 1

1. S. Muroga, "VLSI Systems Design", Wiley-Interscience, New York (1982).
2. M.J. Elmansry, "Digital MOS Integrated circuits", IEEE Press, New York (1981).
3. J. Millman, C.C. Halkias, "Integrated Electronics: Analog and Digital Circuits and Systems," McGraw-Hill, New York 1972.

4. C. Mead, L. Conway, "Introduction to VLSI Systems," Addison-Wesley, New York (1980).
5. A.S. Grove, "Physics and Technology of Semiconductor Devices", Wiley, New York (1967).
6. T.J. Wallmark, H. Johnson, eds., "Field Effect Transistors," Prentice Hall, Englewood Cliffs, New Jersey (1966).
7. S. Selberherr, "Analysis and Simulations of Semiconductor Devices", Springer-Verlag, Vienna (1984).
8. K. Seeger, "Semiconductor Physics, An Introduction, 2nd Ed.", Springer-Verlag, Berlin (1982).
9. J.R. Brews, The physics of the MOS transistor in silicon, in "Integrated Circuits", Part A, D. Kahng, ed., Academic Press, New York, pp. 1-120, (1981).
10. R.H. Dennard, F. Gaensslen, H.N. Yu, V.L. Rideout, E. Bassous, A.R. LeBlanc, Design of ion-implanted MOSFETs with very small physical dimensions, IEEE J. Sol. St. Circ. SC-9, 256-268, (1974).
11. T. Enomoto, T. Ishihara, M. Yasumoto, T. Aizawa, Design, fabrication and performance of solid-state analog ICs, IEEE J. Sol. St. Circ. SC-18, 395-402, (1983).
12. R.R. Troutman, VLSI limitations from drain-induced barrier-lowering, IEEE Trans. Ed., ED-26, 461, (1979).
13. L.D. Yao, A simple theory to predict the threshold voltage of short channel IGFETs, Solid-St. Electronics, 17, 1059-1063, (1974).
14. S.M. Sze, "Physics of Semiconductor Devices, 2nd Ed.", Wiley-Inter-Science, New York, pp. 474-477, (1981).
15. A. Many, Y. Goldstein, N.B. Grover, "Semiconductor Surfaces", North Holland, Amsterdam (1965).
16. T. Ando, A.B. Fowler, F. Stern, Electronic properties of two-dimensional systems, Rev. Mod. Phys. 54 (2), 437-672 (1982).
17. A.B. Glaser, G.E. Subak-Sharpe, "Integrated Circuit Engineering-Design, Fabrication and Applications", Addison-Wesley, Reading, MA, pp. 116-130, (1977).
18. P.M. Hall, Resistance calculations for thin film patterns, Thin Solid Films, 1, 277-295 (1967).
19. R.V. Churchill, T.W. Brown, R.F. Verhey, "Complex Variable and Applications 3rd Ed.", McGraw Hill, New York (1976).
20. M. Annaratone, "Digital CMOS Circuit Design", Kluwer Academic, Boston (1986).
21. C.R. Troutman, "Latch-up in CMOS Technology", Kluwer Academic, Boston (1986).

FRAME I

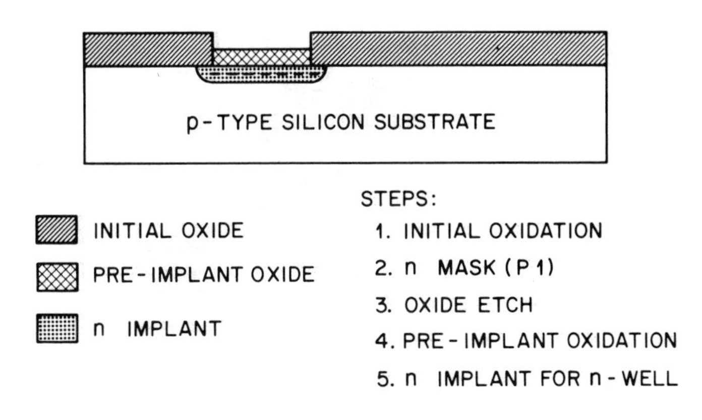

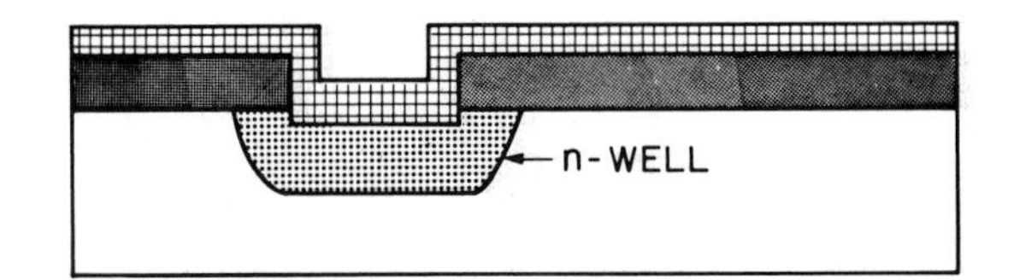

STEPS:

6. DRIVE n-IMPLANT AND GROW OXIDE

INITIAL OXIDE

n-IMPLANT

DRIVE OXIDATION (MERGED WITH PRE-IMPLANT OXIDE)

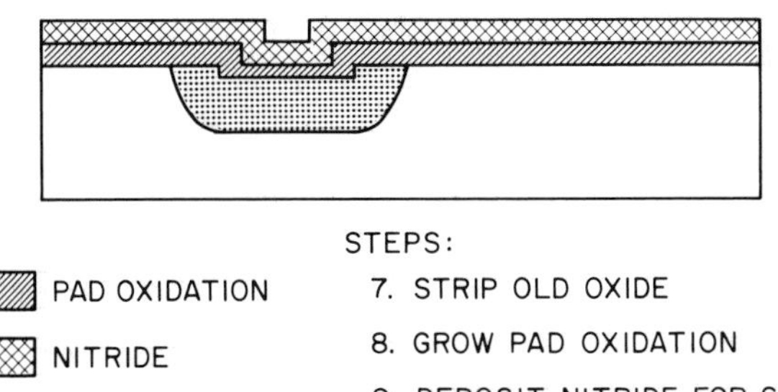

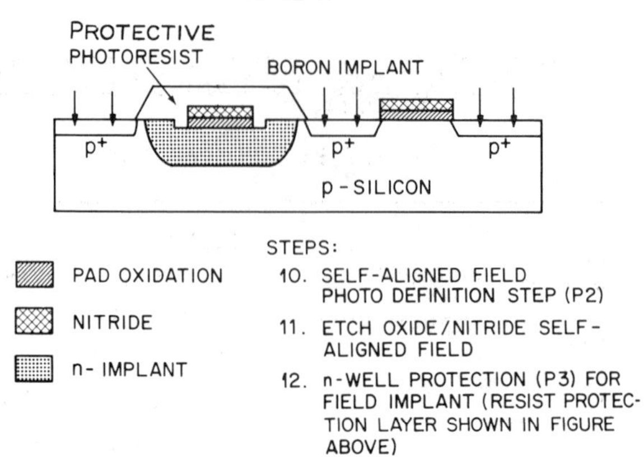
FRAME IV
PROTECTIVE
PHOTORESIST
BORON IMPLANT
p+
p+
p+
p - SILICON
PAD OXIDATION
NITRIDE
n- IMPLANT
STEPS:
10. SELF-ALIGNED FIELD PHOTO DEFINITION STEP (P2)
11. ETCH OXIDE/NITRIDE SELF-ALIGNED FIELD
12. n-WELL PROTECTION (P3) FOR FIELD IMPLANT (RESIST PROTECTION LAYER SHOWN IN FIGURE ABOVE)
13. FIELD IMPLANT (BORON)

FRAME V

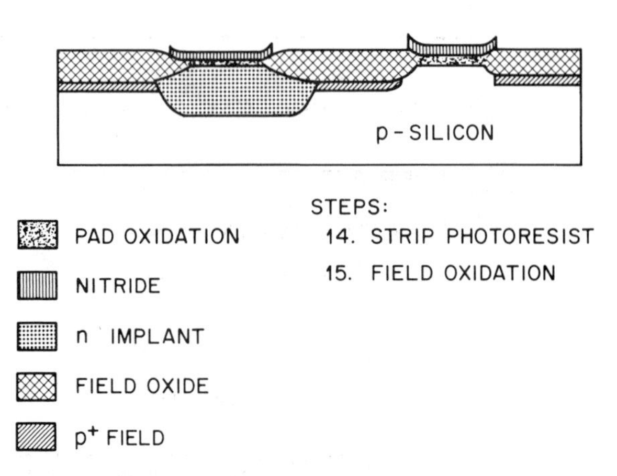
p - SILICON
PAD OXIDATION
NITRIDE
n IMPLANT
FIELD OXIDE
p+ FIELD
STEPS:
14. STRIP PHOTORESIST
15. FIELD OXIDATION

FRAME VI

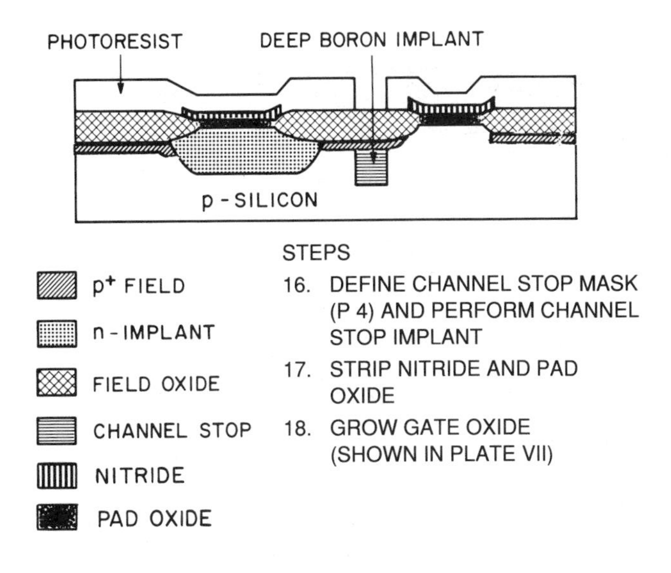

STEPS

16. DEFINE CHANNEL STOP MASK (P 4) AND PERFORM CHANNEL STOP IMPLANT
17. STRIP NITRIDE AND PAD OXIDE
18. GROW GATE OXIDE (SHOWN IN PLATE VII)

FRAME VII

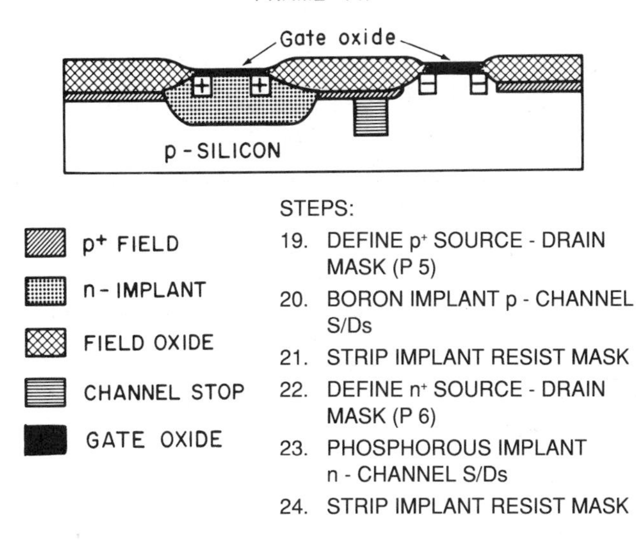

STEPS:

19. DEFINE p^+ SOURCE - DRAIN MASK (P 5)
20. BORON IMPLANT p - CHANNEL S/Ds
21. STRIP IMPLANT RESIST MASK
22. DEFINE n^+ SOURCE - DRAIN MASK (P 6)
23. PHOSPHOROUS IMPLANT n - CHANNEL S/Ds
24. STRIP IMPLANT RESIST MASK

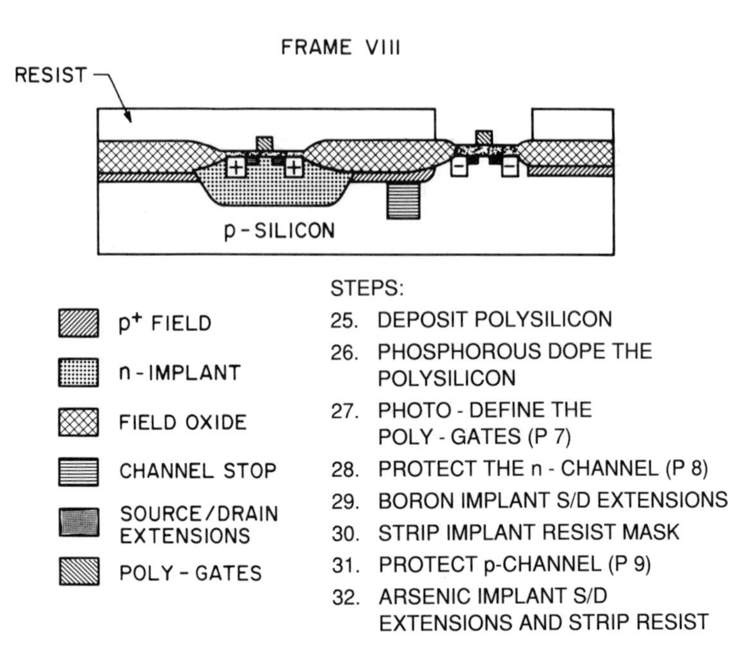

STEPS:

25. DEPOSIT POLYSILICON
26. PHOSPHOROUS DOPE THE POLYSILICON
27. PHOTO - DEFINE THE POLY - GATES (P 7)
28. PROTECT THE n - CHANNEL (P 8)
29. BORON IMPLANT S/D EXTENSIONS
30. STRIP IMPLANT RESIST MASK
31. PROTECT p-CHANNEL (P 9)
32. ARSENIC IMPLANT S/D EXTENSIONS AND STRIP RESIST

FRAME IX

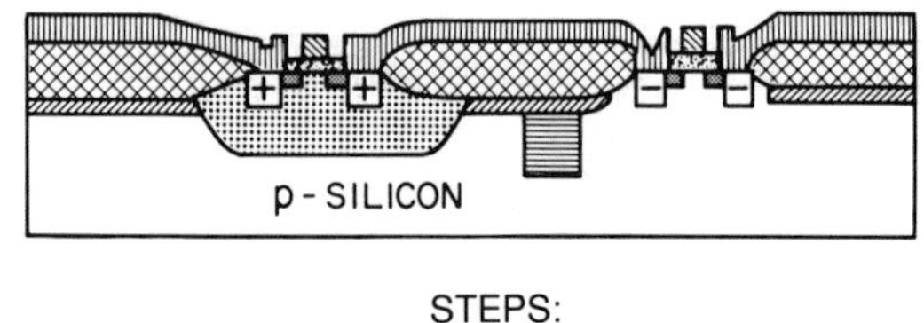

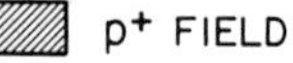

n - IMPLANT

POLY - GATES

METAL

STEPS:

33. IMPLANT ACTIVATE (FURNACE ANNEAL)
34. OPEN CONTACT WINDOWS (P 10)
35. DEPOSIT METAL
36. DEFINE METAL (P 11)

Problem Set for Chapter 1

1.1 Consider the following circuits

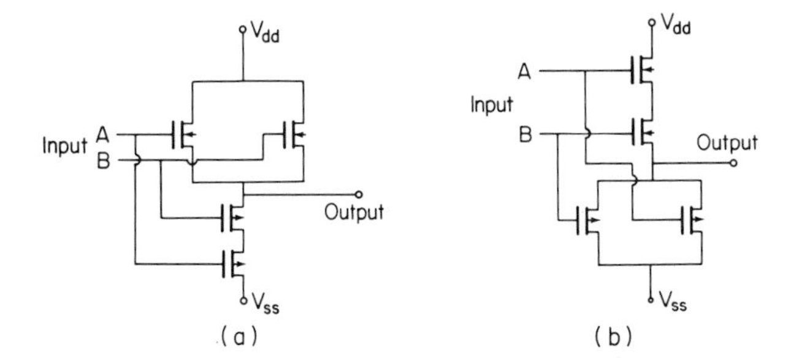

Provide:

a. Vertical and horizontal layout cross sections for these devices.
b. Label regions where device isolation is critical.
c. Truth tables for the device logic functions.

1.2. It is sometimes possible to create a doubly diffused resistor that has a relatively high sheet resistivity. Such a structure is called a *pinch-resistor.* A cross section of this device is shown below:

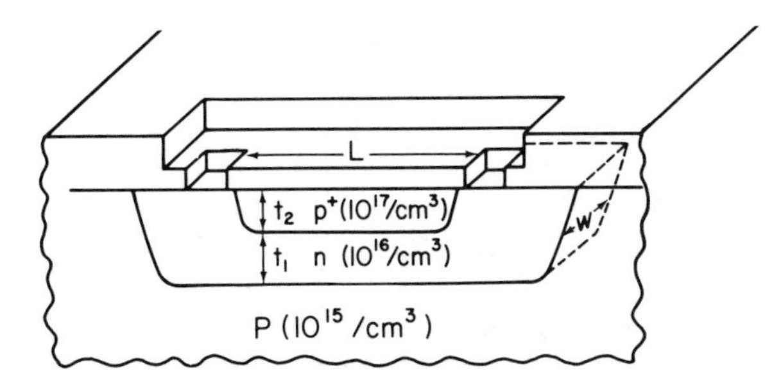

The t^s represent junction depths. Assume the widths of the p^+ and n diffusions are as shown in the figure and consider:

a. How does the p^+ diffusion influence sheet resistivity?
b. Write an expression for the sheet resistivity of the resistor in terms of t_1, t_2, L, W, and the doping concentrations assuming abrupt junctions.
c. Provide a tentative process flow for device fabrication (i.e., order of oxidations, diffusions, etc).

1.3. Ion implantation and thick field oxides are frequently used to provide device isolation and to suppress the formation of parasitic transistors. Consider the following base structures:

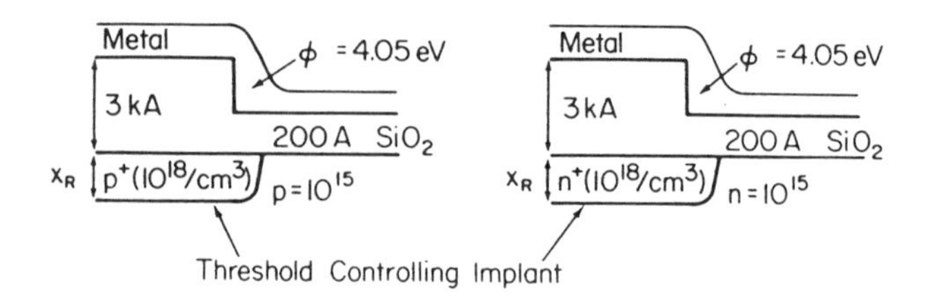

When oxide defect-related charge is zero, and the metal-semiconductor work function difference is zero:

a. What is the threshold of the parasitic transistor formed over the field oxide? (Assume the threshold control implant is uniform over the implant range).
b. What is the threshold of the gate oxide transistor?
c. How deep must X_R be to ensure that the implant change would not be fully depleted in operation.
d. What is the effect of metal-semiconductor work function difference? (Assume the metal work-function is 4.05 eV as shown.)

1.4a. Assume some ionizing radiation incident in processing introduces an oxide charge, N_{ox}, which is positive in nature and is located at the oxide/semiconductor interface. For $N_{ox} = 10^{10}/cm^2$ and $10^{12}/cm^2$, recompute the threshold calculated in problem (3) under these new assumptions.

1.4b. If the minimum observable voltage shift on a CV plotter is 1 mV, what is the minimum observable oxide defect charge, Q_{ox}?

1.5. Consider the latch-up equivalent circuit model shown in fig. 1.23. What materials parameters (i.e., doping levels, junction depths, etc.) can be altered to make latch-up more difficult?

1.6. Using the field-line trajectory estimation technique discussed in the text (see fig. 1.22) estimate the number of squares in the semi-annular end unit drawn below.

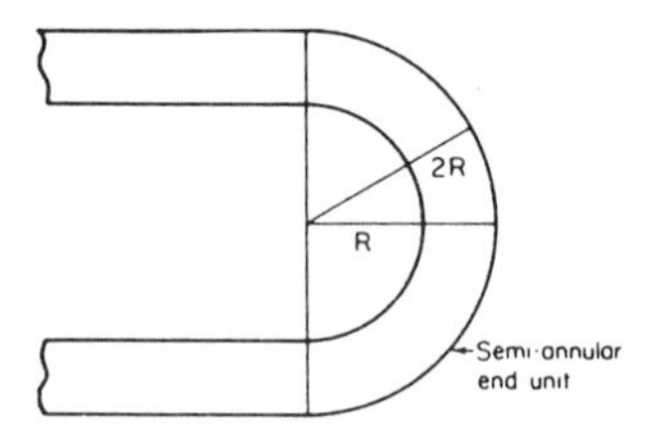

1.7. Assume a mobile carrier moves in an electric field, E, a distance l, at which point its forward momentum is redirected by a scattering process. Derive an expression for the mobility.

2 Starting Materials

2.1 Introduction

Integrated circuits are fabricated on single-crystal semiconductor substrates. These substrates are usually mirrorlike circular discs of thicknesses ranging from 1/4 to 1/2 mm. Silicon-disc diameters may range from 50 to 150 mm, with a future for 200 mm discs. The discs are called wafers. In the area of GaAs devices, 75 mm wafers are currently available and high-quality 100 mm wafers will be commercially available soon. People engaged in IC processing are rarely involved in crystal growth, but process engineers are frequently called on to help in starting material selection and modification. In this chapter the fundamentals of silicon starting materials preparation are given.

First, the current methods of preparing elemental silicon are presented. Typical residual impurities associated with these methods are listed. Next, the basics of silicon crystal structure and growth are outlined. Differences in materials properties as a function of crystal growth conditions are important. These differences are highlighted, where appropriate. Next, the process used to turn the large silicon crystal, or boule, into wafers is described. The choice of the wafer dimensions and the tolerances to which these dimensions should be maintained depend on the ultimate use of the wafer. For example, certain types of lithography require ultra-flat wafer surfaces to achieve good resolution. Just what is ultra-flat in terms of today's wafer technology? To what degree can this and other key dimensional parameters be controlled?

In addition, crystalline material has a defect structure associated with it—regions where the ideal crystal atomic repeat pattern is not precisely realized. We distinguish two types of defects: extrinsic and intrinsic. Extrinsic defects are those due to the addition of elements other than those of the host crystal. Intrinsic defects are those that are associated with the mechanical structure of the crystal. These include dislocations, stacking faults, vacancies, and interstitial atoms in the bulk crystal. Many of these defects are harmful and adversely affect the electrical performance of the devices in which they are present. Surprisingly, though, some defects can actually render harmful impurities (such as gold or other heavy metals) electrically inactive. This process is called gettering. The chapter concludes with a discussion of gettering techniques.

2.2 Growth of Single-Crystal Silicon for IC Fabrication

The silicon used in IC production comes from sand. Sand is largely composed of silicon dioxide (SiO_2). Usually, high-purity quartzite sand is selected. The sand is chemically reduced at high temperature ($\sim 2000°C$) by mixing it with carbon. The oxygen blows off as carbon monoxide, leaving bulk silicon behind. The "metallurgical" grade silicon that remains is polycrystalline and is 98-99% pure. Major impurities are: Fe, Cu, Au, Al, B, and P. Next, the metallurgical-grade silicon is reacted with anhydrous HCl to form trichlorosilane ($SiHCl_3$), which is further purified by distillation. The trichlorosilane is thermally decomposed and silicon is deposited on a thin, previously prepared, rod of silicon. The resulting polycrystalline deposit is used as a source material for the crystal growth process. While other silicon extraction schemes are possible, the most popular has been outlined here. The resulting polycrystalline material may have as little as 0.05 parts per billion (atomic) residual boron left as the dominant impurity [1].

From the point of view of silicon IC processing, there are two important crystal growth techniques. They are the Czochralski (CZ) [2] and the Float Zone (FZ) techniques [3]. The Czochralski technique is illustrated in fig. 2.1. The polycrystalline charge is placed in a quartz crucible that is sheathed by a graphite sleeve, or susceptor. An inert gas ambient (usually argon) surrounds the charge. The graphite can be resistance heated (as it is in most modern apparatus), or a radio-frequency (RF) coil can be placed around the susceptor. The susceptor would then be heated by RF induction. This, in turn, melts the charge. A single crystal seed of silicon is placed at the end of a pull rod. The crystal planes of the seed are oriented in the desired growth direction at the end of the rod. The seed is then dipped into the molten silicon. Heat is conducted out of the melt through the pull rod so that the seed remains solid when it is in contact with the melt. As the seed is slowly withdrawn from the melt, crystalline silicon grows from the end of the seed. New growth is pulled from the melt. Both the crucible and the pull rod are rotated to achieve uniform growth. The final diameter of the new-grown crystal is determined by the pull rate and by the temperature of the melt. In production, fast growth rates are economically desirable. However, this consideration must be tempered by the fact that fast growth rates frequently lead to diameter variations, inhomogeneity of crystal properties, and defects.

Modern CZ pullers hold roughly 60 kg of molten silicon. From this silicon, 100 mm and 150 mm diameter boules are drawn. The boules usually exceed 1 meter in length and are free of dislocations. Computer controllers provide $\pm$ 1 mm diameter control without human operator intervention. There is a constant customer demand to increase wafer diameter. While processing costs may increase with the number of wafers processed, these costs are relatively insensitive with respect to the wafer size–larger wafers mean less cost per individual chip.

It should also be noted that the as-grown crystal contains both oxygen and carbon [4]. This is illustrated in fig. 2.1. Oxygen enters the melt as the quartz crucible walls dissolve. As a result of convective currents in the melt, the crucible walls are continuously washed with fresh liquid silicon. The fresh silicon picks up oxygen and transports it to the growth interface. As much as 50-100 parts per million (atomic) of oxygen can be incorporated into the growing crystal in this way. Carbon may enter as a natural impurity in the melt or through vaporization of the susceptor.

The FZ process for single-crystal silicon differs from the CZ process in that the FZ process does not employ a crucible. The FZ growth apparatus is shown in fig. 2.2. Here, an RF coil is used to create a molten zone in a polycrystalline rod. The molten zone is dragged slowly away from the seed end of the rod. The molten zone is held in place by surface tension during each pass. Crystalline silicon is left behind after the zone passes through. As there is no crucible, oxygen and other impurities from the crucible are not found in FZ material. This is not always desirable. Small amounts of oxygen immobilize defects responsible for crack formation. Thus, FZ material tends to be more brittle and more breakage prone in processing than CZ. This is discussed in greater depth below.

The FZ process produces the highest purity silicon currently available. This is because typical impurities found in silicon tend to segregate in the melt. These impurities are carried along and are finally brought to rest in the crystal at the end opposite the seed. This "dirty" end of the crystal can be cut off and discarded. The float-zoning operation can be repeated again and again. Materials of resistivity near 100,000 Ω-cm have been produced in this way. Mathematically, we express the contaminant-localizing property of the melt with the aid of a segregation coefficient:

$$k_{\mathrm{s}} = \frac{C_{\mathrm{s}}}{C_{\mathrm{l}}}, \tag{2.1a}$$

where k_{s} = segregation coefficient, C_{s} = concentration of impurity at the growth interface just inside the solid portion, and C_{l} = concentration of impurity at the growth interface just inside the melt.

Boron is the most difficult impurity to remove because of its high $k_{\mathrm{s}} (\simeq 0.8)$. Despite this, boron concentration can be kept less than 0.3 parts per billion, atomic (ppba). Phosphorus can be kept in the 0.2 ppba range.

Strictly speaking, eq. 2.1a holds only for the case in which the solid is in equilibrium with the melt. In most growth situations, such an equilibrium is difficult to maintain. Usually, a solute-atom concentration gradient exists in the melt at the growth interface. This adds a diffusion-driven transport term to the arrival rate of impurity atoms on the growth surface. When this

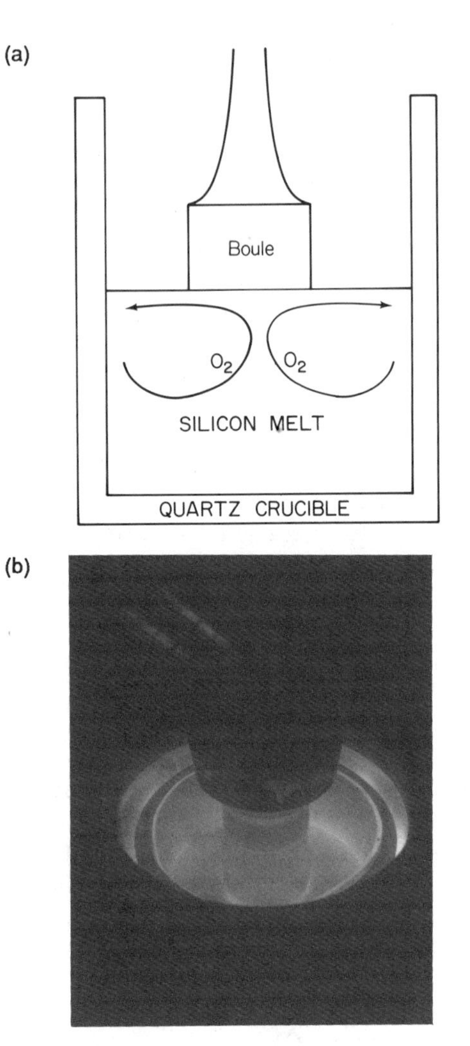

Fig. 2.1. (a) Schematic illustrating Czochralski growth. The transport of oxygen to the growth interface through convection is also illustrated. (b) Silicon crystal boule as pulled from melt.

occurs, k_s is replaced by k_s^{eff}, an effective segregation coefficient [5]:

$$k_s^{eff} = \frac{k_s}{k_s + (1 - k_s)\exp(-\frac{\delta f}{D})}, \tag{2.1b}$$

where: k_s = equilibrium segregation coefficient, δ = thickness of solute enriched volume of liquid adjacent to the growth interface, f = crystal growth rate, and D = diffusion coefficient of impurity in the liquid.

It should be noted that in a FZ boule the impurity concentration varies along the crystal axis as we move away from the seed end. This is because

Fig. 2.2. Float-zone silicon growth through "needles-eye" zone heater.

the melt zone becomes continuously enriched with impurity as it is moved to the end of the boule. To calculate the magnitude of this effect, consider the diagram in fig. 2.3a. The boule is segmented along its axis into infinitesimal volume elements of height dx. These elements are indexed as shown. Next, consider eqs. (2.2)-(2.4):

$$C_{\mathrm{s}}^{\mathrm{f}}(k) = k_{\mathrm{s}} C_{\mathrm{l}}(k+1) \tag{2.2}$$

$$\Delta(k) = C_{\mathrm{s}}^{\mathrm{i}}(k) - C_{\mathrm{s}}^{\mathrm{f}}(k) \tag{2.3}$$

$$C_{\mathrm{l}}(k) = C_{\mathrm{s}}^{\mathrm{i}}(k) + \sum_{j=1}^{k-1} \Delta(j). \tag{2.4}$$

Equation 2.2 is a redefinition for the equilibrium coefficient in the discretized model; (2.3) is the change in impurity concentration before and after the passage of the molten zone; (2.4) reflects the fact that the molten liquid contains the starting impurity concentration *and* the impurities "dragged along" by the moving melt. In these equations, k is an index parameter (not to be confused with the Boltzmann's constant or the segregation coefficient, k_{s}) the subscript s refers to the concentration of impurity in the solid, the subscript l refers to the concentration in the liquid melt; the superscript i (for initial) refers to the impurity concentration in the solid before the melt has passed through, and the superscript f (for final) is the concentration after resolidification.

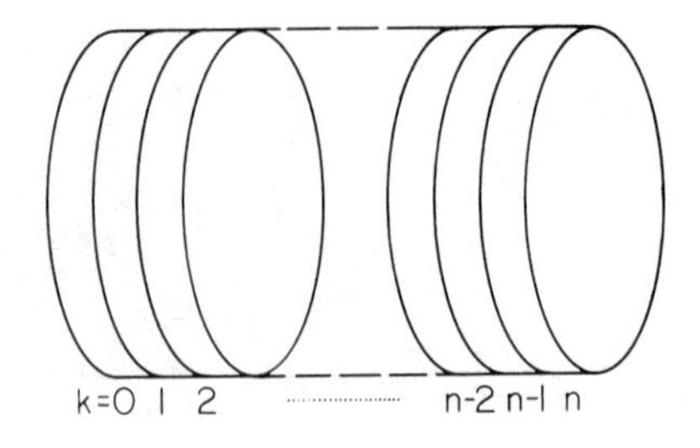

Fig. 2.3. (a) Disc model discretization for zone refining.

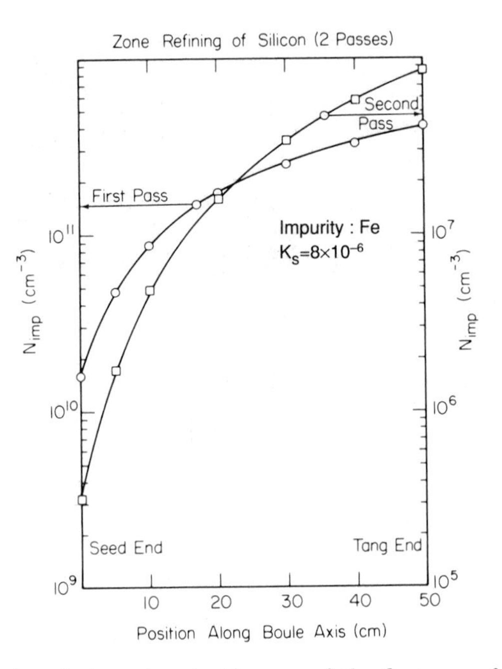

(b) The result of a single and a double pass of the float zone for an impurity of segregation coefficient k_s.

In this highly simplified model, it is assumed that: (a) the molten zone is of infinitesmal thickness, and (b) the impurity redistribution takes place only between the most recently solidified infinitesmal volume and the melt. Matter redistribution between melt and the solid ahead of it is ignored, as is mass transport in the melt itself. This creates a result which is independent of melt volume and of the speed with which the melt is transported along the boule. These restrictions are partially lifted by use of eq. 2.1b. Furthermore, it assumed that the melt impurity concentration does not approach the solid-solubility limit.

From (2.2) we see that if k_s is less than 1, there will be a reduction in impurity concentration as the molten zone moves from the k to the $k+1$

volume element. The molten zone is like a vacuum-cleaner bag, picking up impurities as it goes along. However, as the bag gets more and more loaded with impurities, as quantified by eq. (2.4), the ability of the vacuum to draw trash from successive volume elements is reduced. In fig. 2.3b, we see the results of single and double passes for iron float zoning. It is left as an exercise for the reader to simulate the results of multiple passes of a molten zone through the same boule.

The result of all of these crystallization processes and refining processes is a crystal of *electronic-grade* silicon. Silicon crystallizes in a diamond cubic-lattice structure. If we consider each lattice site as occupied by a two-atom basis, the crystal lattice is face-centered cubic (see Appendix II). This can also be viewed as two interpenetrating fcc lattices, with origins separated by a lattice vector (1/4, 1/4, 1/4). This basic structure is due to the nature of the chemical bonds between silicon atoms in the lattice: the sp^3-hybrid formation [7]. Each silicon atom is covalently bonded to four symmetrically placed silicon atoms to form a tetrahedral unit, as shown in fig. 2.4. This figure also illustrates how these tetrahedra combine to form the final crystal.

In the pure form the semiconductor crystal is practically an insulator at room temperature. The electrical conductivity of the starting material is varied by intentionally adding impurities to the bulk crystal. In the CZ process, impurity elements can be dissolved in the melt using solid additives. In the FZ process, a gaseous dopant source is usually used. Typically used dopants are phosphorous and arsenic as electron donors and boron as an acceptor (hole donor) impurity. Addition of these materials increases the mobile carrier concentration in the crystal and, thus, increases conductivity.

2.3 Wafer Preparation

After growth of the boule is completed, it is ground to create a cylinder of uniform diameter. Flat regions are also ground on the cylinder, perpendicular to key crystal planes for providing easily identifiable crystal orientation. Table 2.1 lists the flat orientations for various cuts of silicon. The wafers are sawed out of the boule with a diamond saw. The saw leaves a rough surface finish on the wafers, which must be mechanically lapped using a number of aluminum grits with successively finer diameter grit size. For optimally polished surfaces, the mechanical lapping is followed by a *chem-polish*, which employs an abrasive rubbing compound combined with a chemical etchant (usually potassium hydroxide based). This leaves a mirrorlike surface on which to build devices. The whole process is illustrated in fig. 2.5 [8]. The presence of sharp edges on the perimeter of the wafer is not mechanically desirable. It has been shown that cracks will propagate into the center of the wafer from these sharp edges. This is the result of the intense stress fields

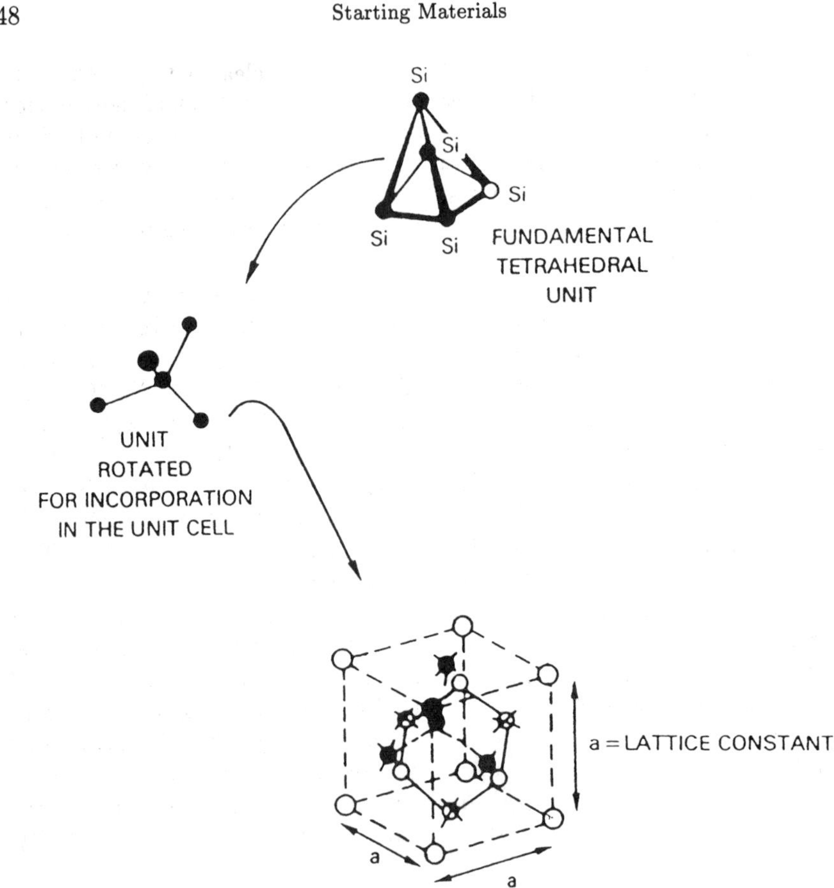

Fig. 2.4. The build-up of the diamond cubic cell from fundamental tetrahedra. Note the characteristic hexagonal ring structure.

associated with microscopic nicks that invariably appear along these edges. These cracks propagate into the center of the wafer and lead to breakage in processing. To remedy this, the edges are rounded, as shown in fig. 2.6. Edge rounding also prevents photoresist build-up along the wafer circumference during the spinning of the photoresists. Such an *edge-bead* prevents the photo mask from coming in contact with the wafer in the lithography process.

Key parameters associated with dimension specification are wafer diameter and wafer thickness. Wafer thickness can easily be controlled to within 25 μm for all wafer diameters up to 150 mm. Diameter tolerance is a function of diameter. A 75 mm wafer diameter can be controlled to (+/−)1 mm; a 100 mm wafer diameter is controllable to 2 mm. Frequently, processing equipment is made to fit one size of wafer. Wafer holders may be countersunk to provide some " grasp" on the wafer. For this reason, these dimensional

TABLE 2.1: Tentative parameters for SEMI wafer specifications

	Wafer size, mm(in)		
	80(3.150)	90(3.543)	100(3.937)
Diameter range (tolerance), mm(in)	±1 (±0.039)	±1 (±0.039)	±2 (±0.79)
Primary flat length, mm(in)	19-25 (0.748-0.984)	24-30 (0.945-1.181)	30-35 (1.181-1.378)
Primary flat location		(110)±1°	
Secondary flat length, mm(in)	10-13 (0.393-0.511)	12-15 (0.472-0.590)	16-20 (0.6-0.8)
Secondary flat location (for material identification)	(111) p-type primary flat only (100) p-type primary flat, secondary flat at 90° in either direction (111) n-type primary flat; secondary flat at 45° in either direction (100) n-type primary flat, secondary flat at 180°		
Surface orientation	On orientation for (111) and (100) ±1° Off orientation for (111) toward nearest (110) 3° ± 1° 0.5°		
Thickness μm (mils)	400±25 (15.75±1)	475±25 (18.80±1)	625±25 (24.6±1)
Orthogonal misorientation			
Bow, μm (mils)	50(2)	55(2.2)	60(2.4)
Taper (parallelism), μm (mils)	40(1.6)	45(1.8)	50(2)

Source: Semiconductor Equipment and Materials Institute (SEMI).

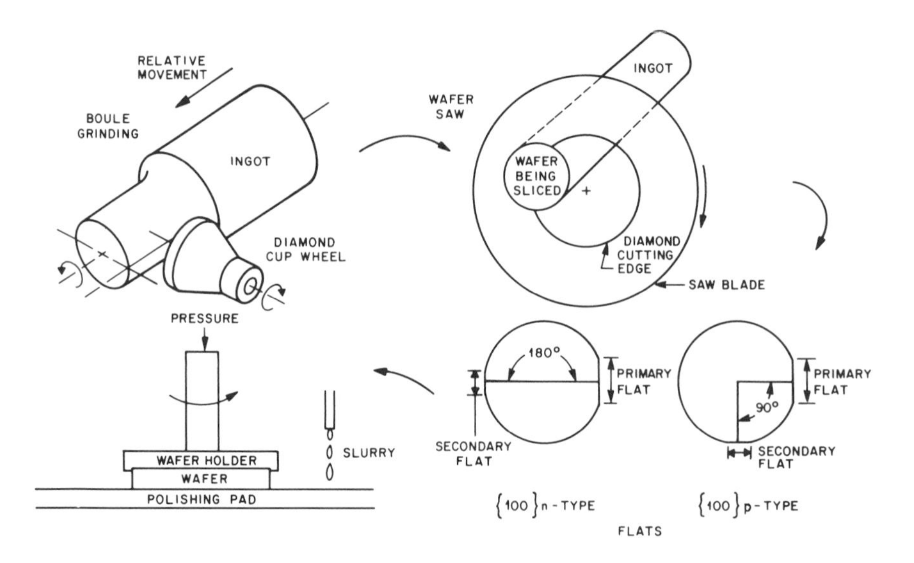

Fig. 2.5. Wafer preparation from the boule to the finished product.

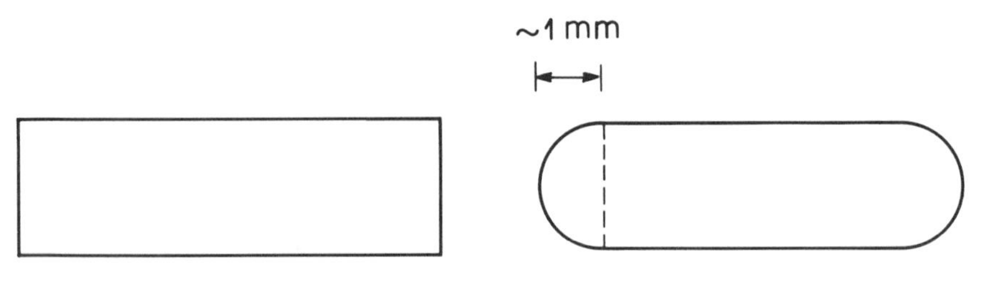

Fig. 2.6. The edge-rounding process.

tolerances are important. A complete summary of wafer tolerances in both MKS and FPS systems is shown in table 2.1.

Perhaps the most important parameter in wafer specification for large-scale integrated devices is wafer flatness. This is of critical importance for its impact on lithographic practices. In many types of lithography, an image of the desired pattern is projected on a photoresist-covered wafer. The depth of focus of high-resolution projection systems is not very great. In fact, it is about a micron for a system capable of resolving a micron. By depth of focus, we mean the distance that the plane on which the image is projected can be moved from its ideal focus to a point at which unacceptable image

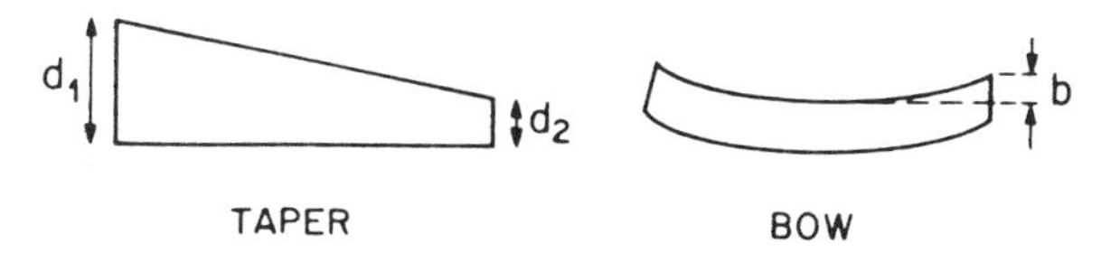

Fig. 2.7. Types of mechanical distortions appearing in prepared wafers.

blurring occurs. Typical full-wafer projection lithography systems with 2 to 3 μm resolution can tolerate (+/−)5 μm deviation of the focal plane from its ideal location. For 1 μm resolution, obviously, much better flatness is required. Current standards of flatness provide 1 μm focal plane shifts across a centimeter field. To do lithography with 1 μm resolution requires the projection field size to be limited to areas smaller than a centimeter (or out-of-plane distortion will most likely exceed the depth of focus). The image must be "stepped" over the whole wafer by serially projecting these smaller field sizes. The image must be refocused over each field. This is discussed in greater depth in Chapter 8. Wafer surfaces differ from ideal optical flats in a number of ways. The front face may not be parallel to the back face as a result of uneven pressure in the polishing process. The wafer can "bow" during processing as a result of the application of various layers on the surface and also of the introduction of mechanical defects [9]. These effects are illustrated in fig. 2.7.

2.4 Defect Structure

"Perfect" crystals, like most "perfect" objects, are creations of the mind. Real silicon wafers are a veritable zoo of defects. These defects may be [10,11]: a) point defects, which involve single atomic misplacements; b) line defects, which are interruptions in crystal perfection along a single line in the crystal; c) planar defects such as stacking faults and grain boundaries; and d) volume defects caused by clusters of vacancies and/or impurity atoms. We can also distinguish between intrinsic defects (those associated with the silicon atoms of the starting material) and extrinsic defects, which are associated with impurity atoms. All of these defects have been observed in IC silicon substrates. These defects are discussed below, as is their impact on device performance.

Extrinsic point defects typically arise from Au, Cu, and Fe heavy-metal contamination, the presence of residual dopant species such as boron, phosphorous, and aluminum; and crucible contaminants such as oxygen and carbon. Heavy metals form deep-level impurities (impurities whose energy levels lie close to silicon midgap). Such sites are extremely effective generation-recombination (GR) centers. GR centers increase device bulk leakage currents. Leakage currents are noise sources in solid-state imagers

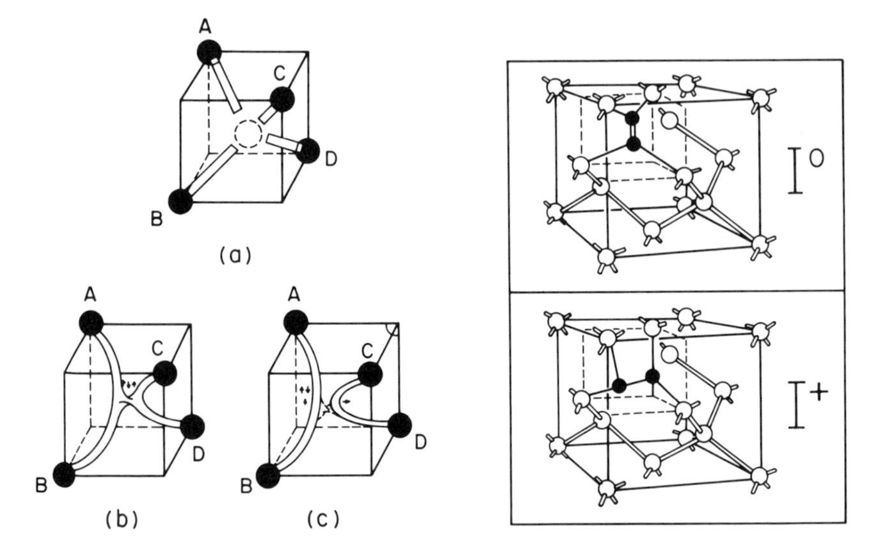

Fig. 2.8. Single-atom vacancy bond configurations that give rise to acceptor and donor states (after Watkins) (a), (b), (c): V^0_{Si}, V^+_{Si}, V^-_{Si}. Inset: Bond models of neutral and charged self-interstitials I^0, I^+ in the silicon lattice (after Seeger).

and cause dynamic random-access-memories (dRAMs) to require frequent electronic refresh.

There are two fundamental types of intrinsic point defects illustrated in fig. 2.8: vacancies and interstitials. Vacancies are sites where atoms are missing from their normal positions in the crystal lattice. The lattice relaxes somewhat about this "hole," giving rise to an extended strain field. The interstitial is an "extra" atom, stuck in a position not normally occupied by an atom in the perfect crystal. Dopant atoms may occupy interstitial sites or they may become substitutional impurities. That is, they can also occupy a site normally occupied by a silicon atom.

Equilibrium concentrations of vacancies and interstitials can be worked out using thermodynamics [12]. To do this, we must compute the free energy of formation of the defect structure. First, consider n vacancies distributed over $(N+n)$ normal atomic positions. The free energy of formation is given by

$$\Delta G = \Delta H_{\mathrm{nv}} - T\Delta S_{\mathrm{nv}}, \tag{2.5}$$

where ΔG = Gibbs free energy of formation, ΔH_{nv} = heat of formation of the n vacancies, ΔS_{nv} = change in the system entropy when the n vacancies are formed, and T = temperature. The ΔS term is really a sum of two terms:

$$\Delta S_{\mathrm{nv}} = \Delta S_{\mathrm{c}} + \Delta S_{\mathrm{v}}. \tag{2.6}$$

The ΔS_v term (frequently called the excess entropy) comes about from the change in the number of vibrational states available to the system when the vacancies are added; that is, the addition of the vacancies causes a change in the system's phonon spectrum. The ΔS_C term is called the *configurational entropy change.* It comes from the increase in the number of statistically different configurations available to the system when the n vacancies are added. To compute ΔS_C, realize that there are now n vacancy sites to be distributed over the $N + n$ possible atomic sites in the system. This gives

$$W = \frac{(N+n)!}{N!n!} \tag{2.7}$$

statistically distinct configurations (or states). This is to be compared with the single state available to the system with no vacancies. From elementary statistical thermodynamics, we have

$$\Delta S_C = \mathrm{k}\ ln(W), \tag{2.8}$$

where k is Boltzmann's constant. With the aid of Stirling's approximation (assuming $N, n >> 0$) this can be written

$$\Delta S_C = -\mathrm{k}n\ ln\left[\left(\frac{n}{n+N}\right)\right] - \mathrm{K}N\ ln\left[\left(\frac{N}{n+N}\right)\right]. \tag{2.9}$$

Since ΔH_{nv} is the heat of formation of n vacancies, $n\Delta H_v$ equals ΔH_{nv} (ΔH_v is the heat of formation of a single vacancy and is the energy required to form a vacancy in the crystal. This is equivalent to saying "the energy required to take one atom out from the bulk of the crystal and leaving it on the surface"). We can treat ΔS_{nv} similarly. Now (2.5) can be rewritten

$$\Delta G = n\Delta H_v - T\Delta S_C. \tag{2.10}$$

In equilibrium

$$\frac{\partial \Delta G}{\partial n} = 0. \tag{2.11}$$

Applying (2.11) to (2.10) yields

$$\frac{n}{(n+N)} = \exp(\Delta S_v/\mathrm{k})\ \exp\left(\frac{-\Delta H_v}{kT}\right). \tag{2.12}$$

A similar relationship holds for interstitials.

To summarize these results, we have shown that the formation of point defects is a thermally-activated process. That is, it depends exponentially on temperature and on the heat-of-formation of an individual defect. The smaller this heat-of-formation, the larger is the number of defects formed. Similarly, increasing the temperature increases the number of defects.

While line defects (dislocations) are common in metals, currently marketed starting material is usually labelled "zero-D." This means that standard techniques for revealing dislocations on starting materials (such as etch pit counts) will yield a zero dislocation count. This does not mean that dislocations cannot be introduced in processing. Thermal cycling and heavy doping do create dislocations. At temperatures as low as 700°C, SiO_2 precipitates can nucleate in bulk silicon when oxygen is present as an impurity. The stress fields associated with the accommodation of these precipitates can cause dislocations to form. This process is discussed below.

We may ask, what is the nature of the line defect? In fact, there are two basic types of line defects. They are termed *edge* and *screw* dislocations [11]. Edge dislocations can be viewed as the insertion of an extra plane of atoms into the crystal. This is shown in fig. 2.9. A segment of the extra plane is given as ABCD. The line AD is the dislocation line. Some of the valence electrons of the atoms along AD are projected into space and do not overlap valence electrons from other atoms. The normal extension of ABCD in a downward direction perpendicular to AD is not present. Application of stress perpendicular to AD causes motion of the dislocation line on the plane ADEF. ADEF is called the slip plane. When the dislocation traverses the whole crystal, the portion of the crystal above this plane will appear to have slipped an atomic plane over with respect to the portion of the crystal below ADEF. The screw dislocation is shown in fig. 2.10. This defect can be viewed as a downward displacement of one segment of the crystal (specified by the shaded volume in the figure) with respect to another. There is an apparent twisting of the crystal. No extra planes of atoms are present. What we have is a region of strain along line AB. AB is the screw dislocation line. Application of stress along AB will cause the dislocation to move along its slip plane, designated as ABCDF. If we move along the path specified by the arrow in fig. 2.10, we find that we "climb" crystal planes in a direction parallel to AB. For every circuit around the dislocation, we spiral up a distance CD.

The strength of the dislocation is an indication of the amount of crystal deformation induced by the line defect. This strength is usually specified in terms of the Burgers vector, **b**. The method of constructing Burgers vectors for the edge and screw dislocations is shown in fig. 2.11. For the edge dislocation, we create the Burgers vector by making a loop around the dislocation line. The loop is generated by picking a starting lattice site down and to the left of the dislocation on a plane perpendicular to the dislocation. Move up n lattice sites, move left n lattice sites, down n sites, and left n sites. In a perfect crystal we would have a closed circuit. Here, though, we have a discontinuity of length $|\mathbf{b}|$. For the screw case a similar clockwise algorithm is used, but a different starting point is chosen, as shown in the

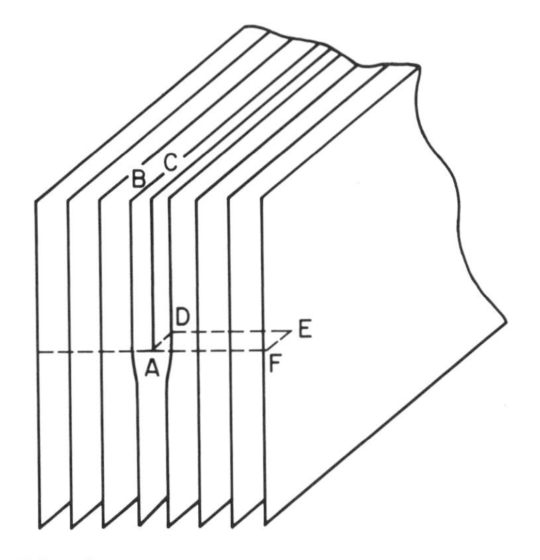

Fig. 2.9. Edge dislocation.

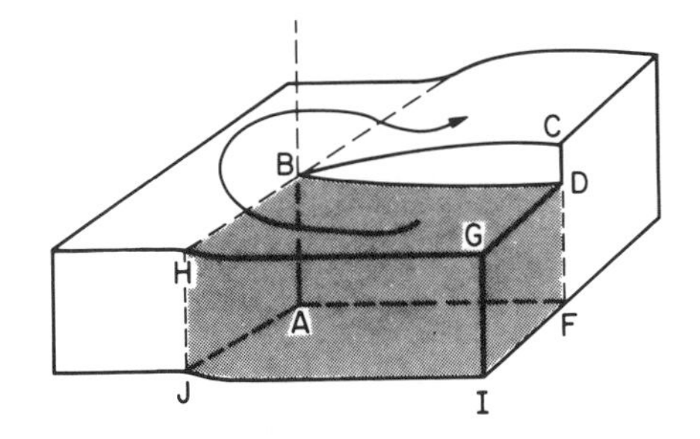

Fig. 2.10. Screw dislocation.

figure. There is always some ambiguity in choosing the direction of **b**. We use the rule that the tail of the arrow is at the termination point of the Burgers circuit. Note the difference, in the relationships of the Burger's vector and the dislocation line, between edge and screw dislocations. For edge dislocation, Burger's vector and dislocation line are perpendicular to each other. For screw dislocation, they are parallel. In real life there are mixed dislocations that have both edge and screwlike characteristics (see problem 2.3).

The self energy of a dislocation is the potential energy stored in the atoms displaced from their equilibrium positions: the energy associated with the strain field of the defect. The self energy, ζ, per unit length of the screw dislocation is:

$$\zeta = \frac{\mu \mathrm{b}^2}{4\pi} \log\left(\frac{R}{5\mathrm{b}}\right), \tag{2.13}$$

where μ = shear Modulus (Ratio of longitudinal stress to transverse strain),

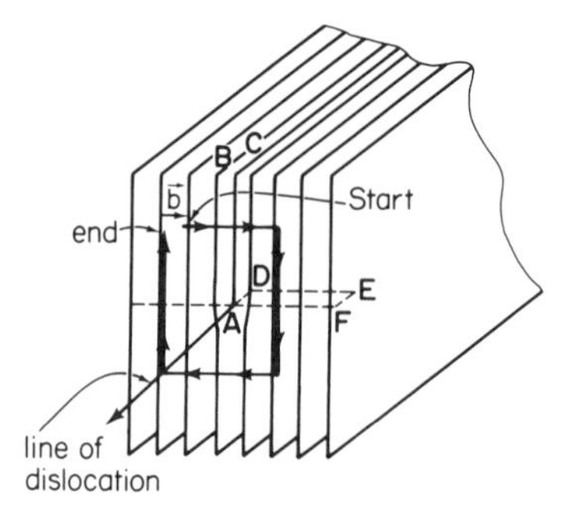

(A) Edge Dislocation

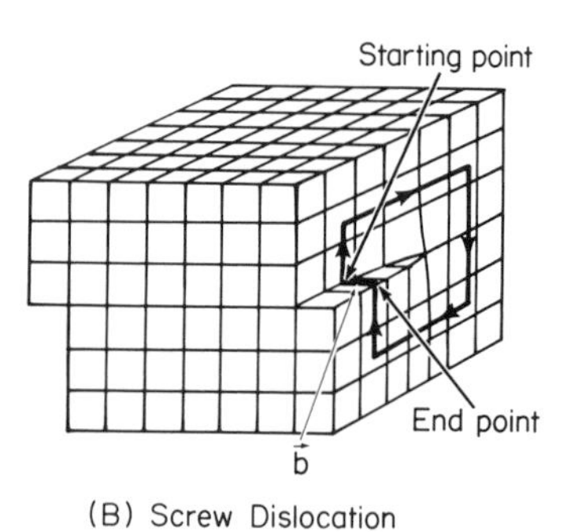

(B) Screw Dislocation

Fig. 2.11. Burgers vector for various mechanical defects in crystals.

b = magnitude of the Burgers Vector, and R = radius of the stress field surrounding the Dislocation. For an edge dislocation:

$$\zeta = \left[\frac{\mu b^2}{4\pi(1-\nu)}\right] \log\left(\frac{R}{5b}\right), \tag{2.14}$$

where ν is Poisson's ratio.

Dislocations act as acceptor impurities in silicon [6]. They can increase the mobile hole concentration in the solid. Dislocations also possess a property that is sometimes beneficial in device fabrication. They can trap heavy metal impurities and render these impurities electrically inactive. This process is called gettering. Heavy metals, such as gold, are deep-level impurities in silicon. The gettering process increases the lifetime of bulk silicon. This is discussed in greater depth in the next section.

Dislocations can be revealed by certain etches. Strained regions, or regions full of broken bonds, are more readily etched than regions of perfect crystallinity. Thus, the region in the vicinity of the dislocation is more readily etched than its surroundings. The etching process leaves pyramidal pits on the silicon surface. Typical etches for revealing dislocations are given in table 2.2. Etch pits on silicon are shown in fig. 2.12. In this figure, the geometric

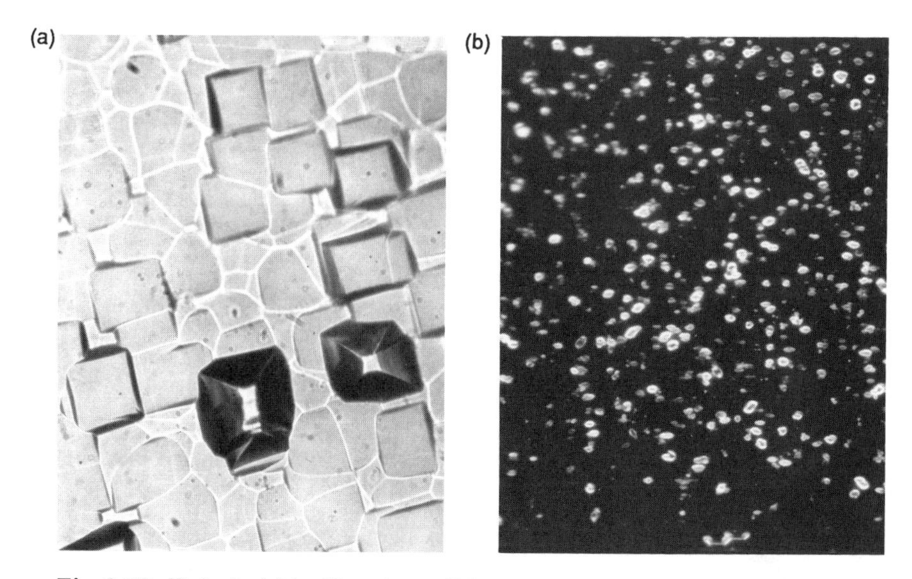

Fig. 2.12. Etch pits (a) in silicon due to dislocation contrasted with "micro defects" (b).

appearance of the pits associated with dislocations is contrasted with the ill-defined features of volume microdefects. The nature of these defects is described below.

Next, consider the possibility of planar defects in crystals. An example of such a defect in silicon is called a stacking fault (SF). To understand the stacking fault, we must view the crystal as an assemblage of tightly packed balls. For easier conceptual understanding we consider a simple fcc stacking of planes as shown in fig. 2.13. In fig. 2.13, we have labeled all the possible sites on which we can rest atoms of subsequent layers (still maintaining the close-packed arrangement). Imagine a second layer of close-packed spherical atoms fitting over the first atomic sheet. We can then place these second layer atoms into the B positions of the first layer. The third layer may fit into either the A or C depressions. This leads to two distinctly different crystal structures, illustrated in fig. 2.14. The first type of crystal is characterized by the ABABAB... pattern. The second type is a result of ABCABCABC... ordering. Silicon exhibits the ABC structure. The two-atom basis complicates the picture a bit, but this simplified picture still explains the essential feature of the flaw.

This layer structure gives rise to the possibility of a SF. The SF is a break in the ABC ordering. Part of one of the C planes can be removed, leaving an ABCAB[ ]ABC type structure. The square brackets indicated the position of the missing plane. This is called an *intrinsic stacking fault.* In addition, part of an extra plane can be inserted to give an ABC[B]ABCABC type ordering. The B in square brackets indicates the extra plane. This is termed an *extrinsic flaw.* Both flaws are illustrated in fig. 2.15. As shown in the figure, the boundary of these flaws is a dislocation. The dislocation is a closed loop of mixed character; that is, as we move around the loop,

TABLE 2.2: Etchants for defect delineation in silicon

Etch	Chemical Composition	Application
Sirtle	HF : Cr_2O_3(5 M) 1 : 1	Best applicable to {111} oriented surfaces
Dash	HF : HNO_3 : acetic acid 1 : 3 : 10	Generally applicable for both n-type and p-type substrates of {111} and {100} orientations, although the etch works best for p-type material.
Secco	HF : $K_2Cr_2O_7$(0.15 M) 2 : 1 or HF : Cr_2O_3(0.15 M) 2 : 1	Generally applicable and is particularly suitable for {100} orientations
Jenkins	HF : HNO_3 : CrO_3(5M) : Cu $(NO_3)_2$ $3H_2O$: acetic acid : H_20 60 ml:30 ml:30 ml:2 gms:60 ml:60 ml	Can be applied generally; defect-free regions are not roughened following etching

the dislocation changes from edge to screw and back again. In addition, the length of the Burger's vector for the dislocation is less than a lattice constant. Such a dislocation is termed a "partial" dislocation. It is left to the reader to show that this is the case (problem 2.3). The strain fields associated with SFs can cause bowing distortions in the wafer. Stacking faults also have gettering properties for heavy metals. Processing can induce SFs. Oxidation, in particular, releases many interstitials that can coalesce to form extrinsic stacking faults. In silicon, this coalescence tends to occur on the close-packed (111) plane. This is discussed in greater depth in Chapter 3.

SFs can grow or shrink in processing [6,13]. Consider the intrinsic fault. As seen from the above discussion, the flaw will grow if it is energetically favorable for vacancies to coalesce on the fault. For the extrinsic SF, interstitials must find it energetically favorable to coalesce homogeneously; that is to say, the system must achieve a net reduction in free energy on point defect precipitation for the SF to grow. The SF actually increases system energy through the increase in self energy of the dislocation boundary and

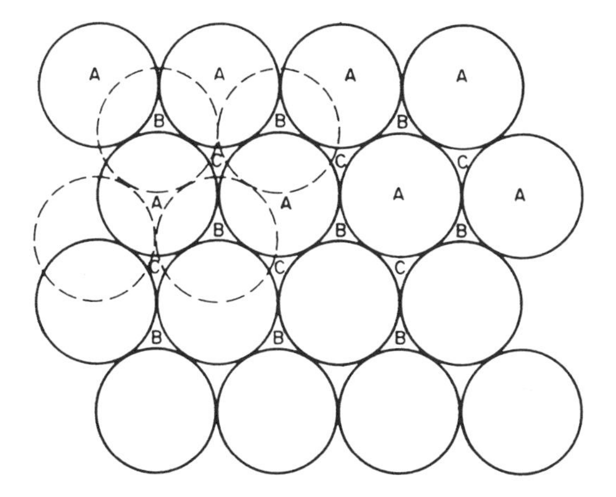

Fig. 2.13. Arrangement of close-packed spheres illustrates possible interstices for subsequent layer formation (A, B, and C points). A partially formed second layer (broken circles) is shown on a complete base layer.

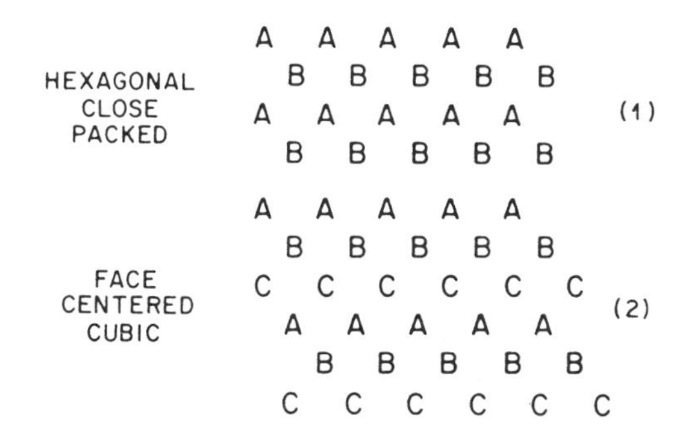

Fig. 2.14. Two possible arrangements of atomic layers in the close-packed format. Illustration (2) corresponds to silicon arrangement.

through the formation of the flaw plane surface. If, as a result of some nonequilibrium process (like oxidation or diffusion), an excess number of intrinsic point defects are created (and heterogeneous nucleation sites are provided), the system free energy will drop when these defects coalesce to form a SF plane. If the "supersaturation" level of vacancies lowers, it becomes energetically favorable for the SF to shrink.

Volume defects are also possible in silicon. Vacancies can come together to form voids; oxygen atoms can nucleate at carbon or other defect sites to form SiO_2 precipitates in the bulk [14]. Arsenic aggregates in bulk silicon are well known. Boron also aggregates in polysilicon where the impurity is present in large quantities. Such volume associations of point defects are termed microdefects [6]. These defects were previously shown in fig. 2.12. The enhancement of microdefects after oxidation is shown in fig. 2.16. Two types of microdefects are shown. The A type is relatively large and appears

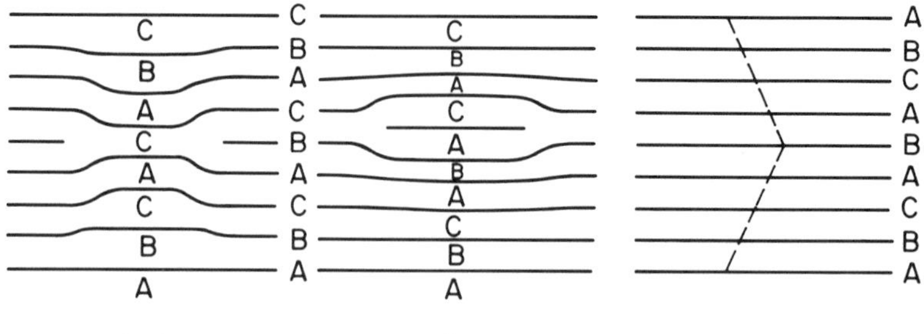

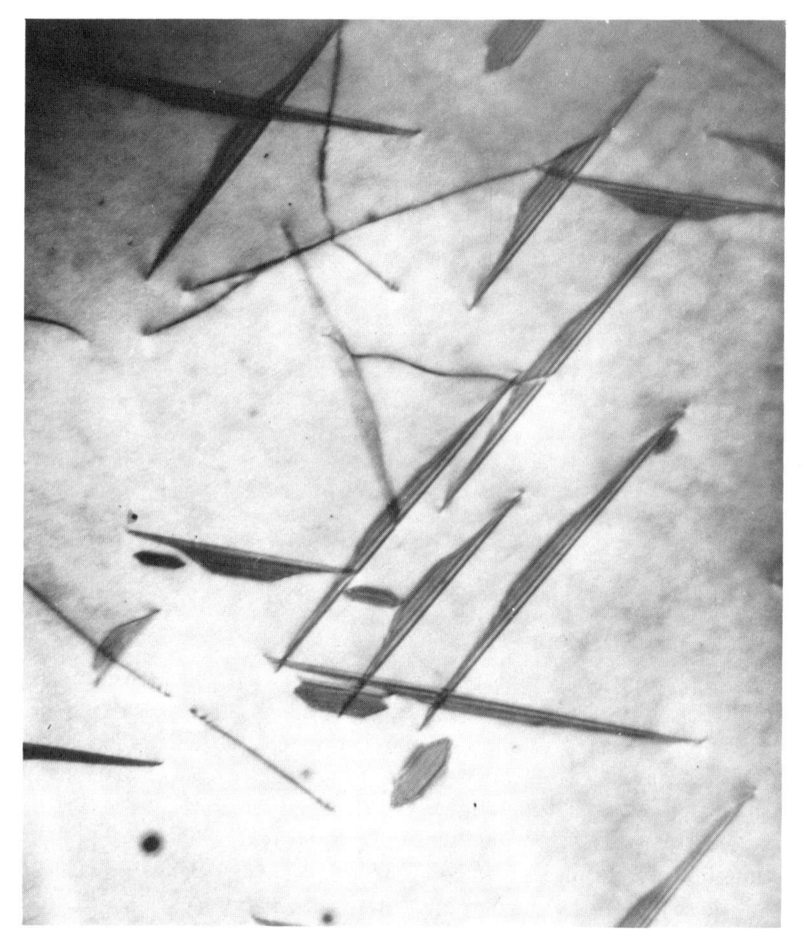

Fig. 2.15. Faults in the stacking of fcc crystals. The lines represent the edges of (111) planes. (a) Intrinsic fault, bounded by two Shockley partial dislocations. (b) Extrinsic fault, equivalent to an inserted plane. (c) Twin fault or growth fault. Stacking sequences are indicated by dashed lines and by sequence of letters. After Weertman and Weertman [11]. (d) Electron micrograph of stacking faults in silicon. Courtesy of K.V. Ravi, Crystallume, Inc.

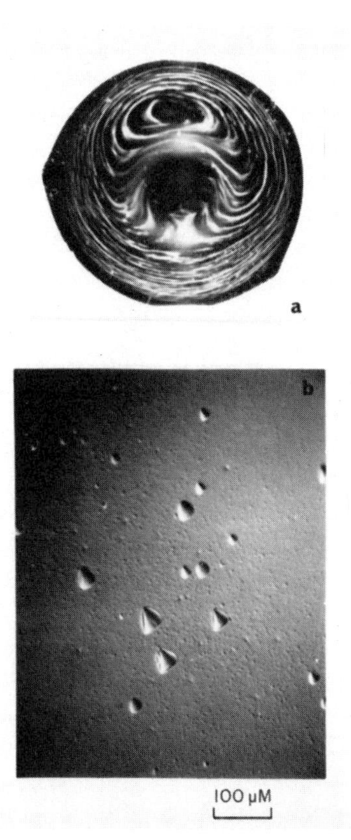

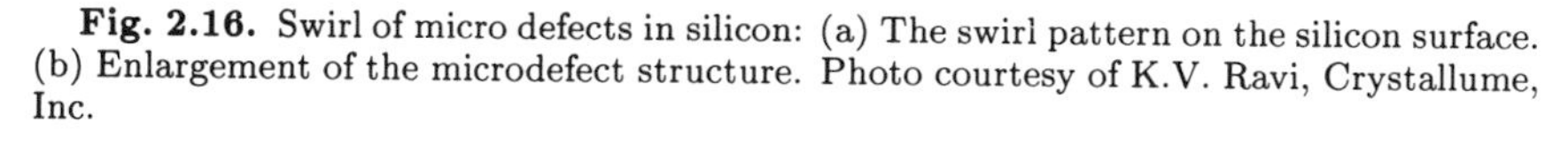

Fig. 2.16. Swirl of micro defects in silicon: (a) The swirl pattern on the silicon surface. (b) Enlargement of the microdefect structure. Photo courtesy of K.V. Ravi, Crystallume, Inc.

near the edge of the wafer. The B type is small and prevalent near the center. Both types of defect form swirls around the center of the wafer. The nature of the swirl-type defects differs as a function of the impurity concentration of the crystal. In FZ material (low oxygen content), the dominant defect is of the B type. These are clusters of silicon interstitials. These clusters can grow during the crystal growth process. Some of the larger clusters can "collapse" and form dislocation loops or other types of faults. The collapsed clusters are the A-type defect. Carbon is thought to nucleate the initial B-type cluster.

Once again, there is some moderate electrical activity to volume defects. The electrical activity of all the defects discussed so far is summarized in table 2.3. Certainly they act as GR centers. They also lower bulk carrier mobility in that they are fairly large scattering centers. However, they do have a gettering property. The oxygen precipitate is particularly interesting in this regard. Annealing techniques for manipulating oxygen precipitates to improve bulk lifetime are discussed next.

TABLE 2.3: Influence of defects on electrical properties of silicon

Defect	Electrical Activity
1. Heavy Metal - Au, Cu, etc.	1. G/R centers. Degrades mobility.
2. Oxygen	2. Donor. In high concentrations (> 1.4 $\times 10^{18}$/cc) oxygen precipitates form which, in turn, create defects within crystal that have gettering effect. Typical oxygen concentrations are 10^{17}-10^{18}/cc. Degrades mobility.
3. Dislocations[a]	3. Have gettering ability. Degrades mobility. Act as acceptor impurities.
4. Stacking faults[a]	4. Have gettering ability. Degrades mobility.

[a] Creates wafer bowing, distortion.

2.5 Anneals

As indicated above, the internal defect structure of a wafer can be both harmful and beneficial in device processing. SFs and dislocations may disturb wafer flatness and act as GR sites. However, these same sites can act as gettering centers [3,15], which neutralize more troublesome defects. For example, SFs can become decorated with heavy metals. When this decoration occurs, the trapped heavy metals become less likely to create leakage in other parts of the crystal.

There are a number of ways to achieve this gettering. Early attempts used backside damage to introduce defect complexes. The backsides of the wafers may have been "sand-blasted," or rapidly oxidized in phosphorous oxychloride ($POCl_3$). In later implementations, heavy argon implants or backside doping with phosphorous has been shown to create defects that "sink" dangerous impurities. In all cases, the intent is the same: to create defects in the wafer bulk, away from the active device areas. The damage process occurs as late as possible in the process sequence to minimize annealing out of the gettering sites and release of dangerous impurities back into the crystal bulk.

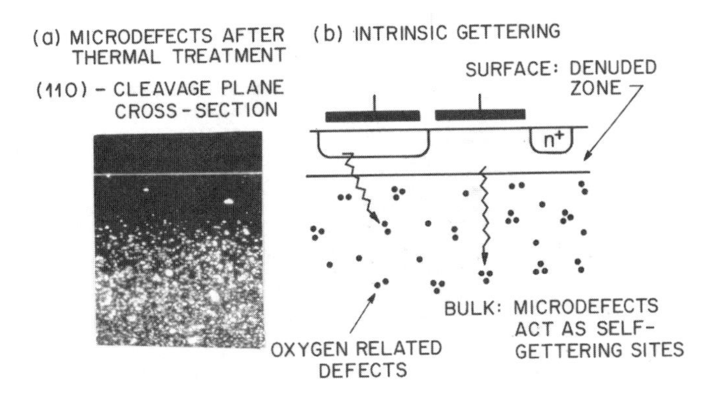

Fig. 2.17. Intrinsic gettering by oxygen related defects. Courtesy N. Thomas, Westinghouse Electric Corp.

More recently, processing engineers have been trying to manipulate the primary impurities of the starting material—oxygen and carbon—to create the gettering defects. Such processes are termed intrinsic gettering techniques. In one popular technique, the silicon is exposed to a high temperature (preferably in inert ambients) to drive off oxygen from the active device regions. At high temperatures, the oxygen is too mobile to form SiO_2 nuclei, and those precipitates that are present redissolve. This creates a denuded zone at the surface of the crystal. Wafers are then annealed at lower temperature to nucleate oxide clusters in the bulk, away from the denuded zone. It is thought that carbon acts as a nucleation site for the precipitation of the cluster. The annealing temperature is then raised to induce faster growth of the cluster. When the cluster reaches a critical radius, strain causes formation of dislocation loops and stacking faults, which act as gettering sites. The process is illustrated in fig. 2.17; the temperature behavior of silicon defects of interest in the internal gettering process is summarized in table 2.4.

The formation of the denuded zone can be regarded as a kind of doping diffusion process in reverse. Here impurity atoms leave the crystal, rather than entering it. The boundary conditions of the problem dictate that deep in the bulk the oxygen concentration becomes a constant, C_b which is higher than the equilibrium solid solubility of oxygen in silicon at that temperature. At the surface the crystal reaches some equilibrium concentration, C_s. C_s is dependent on the gas phase-solid surface segregation and the equilibrium solid solubility of oxygen in silicon. This is illustrated in fig. 2.18. To arrive at the shape of the plot of oxygen concentration versus distance from the surface, we must solve the diffusion equation for these boundary conditions [16]. The results of this calculation are given here

TABLE 2.4: Effects of heat treating oxygen in silicon crystals

Temperature	Effects of Heat Treatment
300°C–500°	Within the 300° to 500° temperature range, the O_2/O in the interstitial sites of the silicon lattice can produce donors in the form SiO_2, SiO_3, and SiO_4 complexes.
650°C	At this temperature, the subcritical oxygen complexes can be destroyed, with oxygen returning to a non-electrically active configuration.
750°C and 1000°C	Oxygen critical nuclei formed by low temperature (450°C) anneals are "ripened," and grow to form swirl defects and dislocation complexes.
1100-1200°C	Interstitial oxygen and donor complexes are diffused out through the wafer's surface, thereby creating a desirable "denuded" zone on the wafer's surface.
1350°C	At this temperature, the higher solubility of oxygen in the silicon reverses the oxygen clustering that took place at 1000°C and lower temperatures.

$$C(x,t) = C_s \text{ erfc } \left(\frac{x}{2\sqrt{Dt}}\right) + C_b \text{ erf } \left(\frac{x}{2\sqrt{Dt}}\right). \tag{2.15}$$

The expressions erfc and erf refer to the complementary error function and to the error function, respectively. The detailed discussion of the diffusion process and equations are discussed in Chapter 4. The term C_s is found to be temperature dependent

$$C_s(T) = 1.63 \times 10^{21} \exp\left(\frac{-11973}{T}\right) \text{ cm}^{-3}, \tag{2.16}$$

where T is the temperature in degrees Kelvin. D is also temperature dependent

$$D(T) = 0.17 \exp\left(\frac{-29449}{T}\right) \text{ cm}^2\text{sec}^{-1}. \tag{2.17}$$

Graphical solutions to (2.15) are given in fig. 2.18.

We can deal with the nucleation of SiO_2 particles using thermodynamics. We begin by writing the change in free energy of the system after precipitating n molecules of SiO_2 as [14]

$$\Delta G = n\Delta H_{\text{SiO}_2} - nT\Delta S_{\text{SiO}_2} + A\sigma + gV \tag{2.18}$$

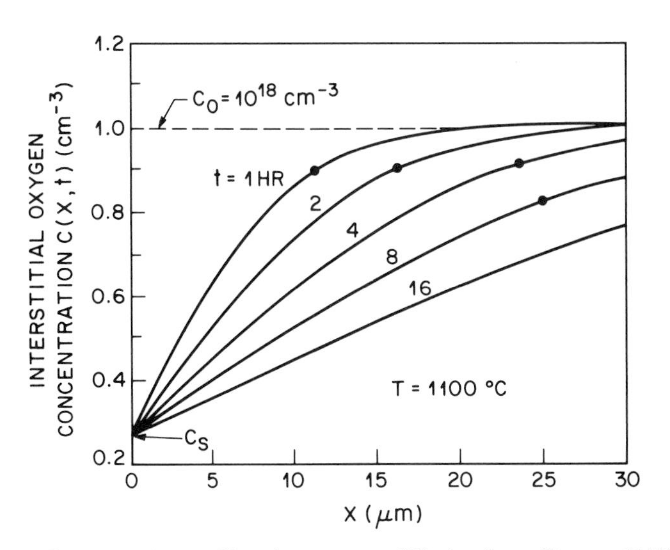

Fig. 2.18. Concentration profiles of oxygen out-diffusion from silicon at 1100°C. After Andrews [16].

where A = surface area of precipitate, V = volume of precipitate, σ = free energy of formation of the precipitate/silicon interface, and g = a constant. The first two terms on the right are familiar from the discussion of vacancy and interstitial formation given above. The last two terms represent, respectively, the heat of formation of the surface (or boundary layer between the SiO_2 and the bulk silicon) and the strain energy caused by the precipitate pushing against the host lattice. As the precipitate grows, the strain energy term gets larger. In eq. (2.18) it is assumed that the strain energy is linearly proportional to the volume of the precipitate. A and V are the area and volume of the precipitate. For simplicity, we assume a spherical precipitate. Both A and V are dependent on n. If we assume that the number of molecules of SiO_2 is proportional to V through some proportionality factor, k_1, we can write A and V in terms of n. Once ΔG is known, we can solve for the equilibrium concentration of precipitates. Since the equation for precipitate formation is

$$n(\mathrm{Si} + 2\ \mathrm{O}) \longrightarrow n\mathrm{SiO_2} \tag{2.19}$$

(assuming oxygen is present as atomic oxygen in the lattice), we have

$$[\mathrm{SiO_2}]/[\mathrm{O}]^2 = \exp\left(\frac{-\Delta G}{kT}\right), \tag{2.20}$$

where k is Boltzmann's constant and T is the temperature. The square brackets indicate concentrations.

In practice, eq. (2.18) is difficult to evaluate. First of all, precipitates usually do not form as perfect spheres. This limits our knowledge of g, since this constant depends on the shape of the precipitate. The surface energy is also difficult to ascertain experimentally. Simple inspection of this equation does tell us a bit about the nature of precipitate formation in solids. First, note that the terms of the equation contribute both positively and negatively to the total free-energy change. The basic chemical reaction is exothermic. Heat is generated when silicon and oxygen react. The entropy change also tends to raise the free energy of the system, since oxygen becomes localized in the defect. The strain and surface energy terms increase as the precipitate grows. Initially, the strain and surface energy terms dominate. Free energy increases as the reaction is initiated. These contribute to form an energy barrier to cluster formation. Once a critical number of particles has joined the assembly, the heat of reaction term dominates and further growth is thermodynamically favored (see Section 7.5). To summarize:

1. There is a critical size for the "embryo" that must be formed before the precipitate becomes thermodynamically stable and capable of growth.
2. The size of this embryo depends on the elastic properties of the solids in question, as well as on the free energies of formation of molecules and surfaces.
3. The critical size of the embryo is dependent on temperature, since ΔH and σ all depend on temperature.

Certain extrinsic impurities can act as catalysts by lowering the activation barrier for precipitation. Carbon is thought to be such an impurity. This allows a larger number of smaller-sized stable embryos to form at lower temperatures. Growth of the cluster is accomplished at higher temperatures. This growth proceeds until the cluster strain creates dislocation loops and stacking faults.

Of course, other annealing cycles are important in device work. Implantations must be annealed to electrically activate the implanted dopant. Diffusions are annealed to "drive" junctions deeper into the bulk. Metals are annealed (the terms sintered or alloyed are frequently used to describe this process) to form low-resistance contacts and silicides. These are discussed in greater depth in succeeding chapters.

2.6 Wafer Cleans

Part-per-million concentration levels of electrically—active impurities severely impact device performance. For this reason, it is of utmost importance that contamination levels be kept low during semiconductor process operations. The introduction of small amounts of contamination are unavoidable in IC processing. This is a result of the many different environments a wafer is placed in through the fabrication cycle. Processing fluids such as acids and organic solvents—even the ambient gases encountered in furnace operations, can contain harmful impurities. The impact of this contamination can be minimized by frequently cleaning the wafer surface in process. Cleans should certainly be performed prior to any furnace operation. Surface contamination would be driven into the starting material during such high-temperature operations. Special "strips" have been designed to remove photoresist and organic residuals after the photolithographic process. Pre-metal cleans have been developed to promote adhesion and to minimize the presence of undesired material sealed between the metal and the underlying substrate. In this section, we go over the types of contaminations encountered in wafer processing, and standard cleaning approaches are described.

A considerable amount of research was done in the early 1970s to characterize the nature of starting material surface contamination and to propose methods for its elimination. Kern and coworkers considered three basic types of contaminant [17,18,19]. First, they discussed materials like sodium, chlorine, fluorine, and iodine, which act as fixed or mobile charges in surface oxides used as passivation. These impurities offset MOS device thresholds from their ideal points. Such contaminants are introduced in the etch cycles encountered in various device layers. Next they considered heavy metal impurities such as gold, copper, and iron, which act as midgap impurities and which will reduce carrier lifetime. These contaminants could be introduced in the course of wafer handling and through contaminated process chemicals. Finally they considered organic contaminants from resists or from human handling of the wafers.

Kern et al. were able to show that an aqueous solution of peroxide and a base (such as 43% H_2O, 30% H_2O_2, 27% NH_4OH) oxidizes organic contaminants, which are then solvated by ammonium hydroxide. This should be followed by an aqueous solution of peroxide and an acid (such as 33% H_2O, 30% H_2O_2, 37% HCl). This solution complexes the heavy metals present. The resulting complexes are solvated in the aqueous bath.

These cleaning soaks are relatively time consuming (each bath lasts about an hour). Less time-consuming (and less effective) cleans have been developed. These employ oxidizing acids, such as HNO_3 and H_2SO_4 usually mixed with H_2O_2, followed by dilute ($\sim$10%) HF dips (to etch away on the resulting surface oxide) and a vigorous water rinse. These cleans are frequently used before less critical operations. Organic solvent rinses are sometimes used before metallization. Wafer surface preparation (grinding and polishing) frequently introduces shallow damage layers. Recent work indicates the desirability of growing a relatively thick oxide ($\sim$1000 Å) over the wafer and stripping the oxide prior to processing in order to remove this damage [20].

2.7 Characterization Techniques

Delineation of the crystal defect structure using etching techniques has been mentioned above. These defect delineation techniques are commonly known as *metallographic techniques*, standard in studying defects in all types of materials [6]. Line, area, and volume defects could be revealed quite easily in this way. However, this technique is destructive and the wafer must be discarded after examination. In addition, the electrical activity of the defects is not determined by etching. In this section, x-ray and electrical techniques for wafer characterization are described. X-ray analysis, performed on the substrates prior to device fabrication, is usually nondestructive. Capacitive pulsed techniques, such as deep-level-transient spectroscopy (DLTS), can provide a relatively complete characterization of the electrical activity of the observed defects.

2.7.1 X-Ray Characterization of Semiconductor Substrates

In this section, two types of x-ray characterization are described: the Laue backscatter method and x-ray topography [21,22]. The Laue method is useful in ascertaining the precise orientation of the crystallographic planes. In addition, the method can be used as a test for crystallinity in the sample under evaluation. X-ray topographs provide defect maps of the wafer surface.

An understanding of the reciprocal lattice concept aids in our comprehending both techniques. A brief review of crystallography and crystal structures is given in Appendix II. Here we begin our discussion with a review of the reciprocal lattice method. Consider the construction of the real-space lattice. This is done by arbitrarily selecting an origin: the (0, 0, 0) point in the crystal. Three coordinate axes are defined such that the basis elements of the crystal (i.e., the fundamental atomic or molecular units that make up the crystal) fall on these lines. The axes are not, of necessity, orthogonal. They are simply three nonparallel lines chosen so that the spacing between basis units on each is the shortest possible for the given crystal. The distance between the origin and the first basis element encountered on a given axis is called a *lattice parameter.* The crystal is thus characterized by three lattice primitive vectors: $\mathbf{a}$, $\mathbf{b}$, and $\mathbf{c}$. The magnitude of each of these primitive vectors is called a lattice parameter. The whole lattice is a collection of points ($h\mathbf{a}$, $k\mathbf{b}$, $l\mathbf{c}$), where h, k, and l are integers. A basis element occupies each lattice point.

The reciprocal lattice is defined in terms of the reciprocal lattice vectors $\mathbf{a}^*$, $\mathbf{b}^*$ and $\mathbf{c}^*$. These vectors are defined as:

$$\mathbf{a}^* = \frac{\mathbf{b} \times \mathbf{c}}{\mathbf{a} \cdot (\mathbf{b} \times \mathbf{c})} \tag{2.21a}$$

$$\mathbf{b}^* = \frac{\mathbf{c} \times \mathbf{a}}{\mathbf{a} \cdot (\mathbf{b} \times \mathbf{c})} \tag{2.21b}$$

$$\mathbf{c}^* = \frac{\mathbf{a} \times \mathbf{b}}{\mathbf{a} \cdot (\mathbf{b} \times \mathbf{c})}. \tag{2.21c}$$

The magnitude of each of these vectors is the inverse of the interplanar spacing of the planes whose surface-normal is defined by the vector cross product in the numerator of each vector. As can be seen, the reciprocal lattice vectors in eq. 2.21 are defined in terms of the real-space lattice vectors defined earlier.

The reciprocal lattice itself is formed from these lattice parameter vectors. Just as for the real-space lattice, this is done by plotting the three-dimensional network of points: $(h\mathbf{a}^*,\ k\mathbf{b}^*,\ l\mathbf{c}^*)$, where h, k, and l are integers. Also, just as for the real-space lattice, the lattice axes need not be perpendicular. The angles between the axes, though, must reflect the angles between the vector normals to the crystallographic planes they represent. With this construction in mind, we will proceed to describe the Laue and x-ray topographic techniques as they are applied to the analysis of strain and the defects in semiconductor crystals.

First, consider the Laue backscatter method. The collimated, unfiltered output of an x-ray tube is incident on the semiconductor crystal. A photographic plate is placed in front of the crystal to record the reflected x-rays. A schematic of the apparatus used is shown in fig. 2.19. X-rays will be reflected off of the crystal planes in accordance with the Bragg formula: $n\lambda = 2d \sin\theta$. Here, n is an integer, λ is the wavelength of the x-ray, d is the interplanar spacing, and θ is the diffraction angle. When this formula is satisfied, the waves of x-rays scattering off successive parallel planes all arrive in phase at a point on the film plate.

The Bragg formula was modified by Laue to express this in-phase scattering condition for a three-dimensional array of lattice points. The modification proceeds as follows. The incident and reflected waves are described by their wave-vectors: $\mathbf{K}_0$ and $\mathbf{K}$, respectively. These are vectors pointing in the propagation direction, defined by unit vectors $\mathbf{S}$ and $\mathbf{S}_0$. Their magnitudes are $1/\lambda$, where λ is the wavelength of the x-ray light. It is assumed that the reflection represents an elastic scattering process and the magnitudes of $\mathbf{K}_0$ and $\mathbf{K}$ are the same. For such a situation the Bragg equation generalizes to the three Laue equations:

$$\mathbf{a} \cdot (\mathbf{K} - \mathbf{K}_0) = h \tag{2.22a}$$

$$\mathbf{b} \cdot (\mathbf{K} - \mathbf{K}_0) = k \tag{2.22b}$$

$$\mathbf{c} \cdot (\mathbf{K} - \mathbf{K}_0) = l \tag{2.22c}$$

Satisfying all diffraction conditions represented by these equations simultaneously is extremely unlikely for a beam of fixed wavelength at an arbitrary

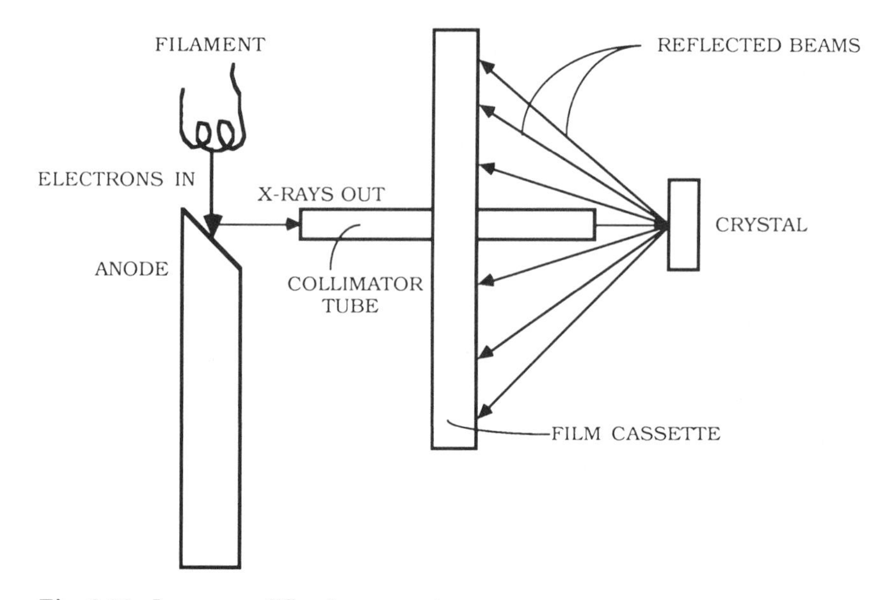

Fig. 2.19. Laue x-ray diffraction apparatus.

angle of incidence. Even when the diffraction condition is obtained, only a single set of planes will "flash out" with a reflection. In the Laue approach (fig. 2.19), the unfiltered x-ray beam is composed of many wavelengths. A typical x-ray tube output spectrum is shown in fig. 8.25. The features of this spectrum are discussed in greater depth in Chapter 8. As a result of the availability of the multitudes of wavelengths present in the incident beam, a number of crystal planes will meet the above diffraction condition simultaneously. A series of "spots" will form on the photographic plate, corresponding to the multiple images of the incident beam that satisfy eqs. 2.22. The distribution of spots on the photo is indicative of the basic symmetry of the crystal and the orientation of the crystal with respect to the incident beam. Please keep in mind that, as a result of the difficulty in satisfying eq. (2.22), it is primarily the broadband continuum that gives rise to spots, (*not* emission lines).

Which planes reflect and the angle at which that reflection appears can be visualized with the aid of the reciprocal lattice. To do this, we note that the wave vectors of the incident and reflected beams have dimensions that are reciprocal centimeters. Thus, these vectors can be plotted in reciprocal space. The incident beam is drawn on a line through the incident plane and the origin of the reciprocal space. The reflected beam is represented by a wave vector of equal magnitude to that of the incident beam. It is left as an exercise to the reader to show that the Laue equations demand that the vector difference between these two wave vectors must be a reciprocal lattice vector.

These concepts are illustrated in fig. 2.20. This approach to understanding the Laue pattern is frequently called the *Ewald sphere construction method.* The wave vector, $\mathbf{S}_0/\lambda$, is drawn incident on the origin of the reciprocal lattice (Fig. 2.20a). Its direction is defined by a line from the origin, terminating on a reciprocal lattice point representing the direction of incidence. A sphere (the Ewald sphere) is then drawn around the tail of the wave vector (point 0 in the diagram). The sphere-radius has a magnitude $1/\lambda$. All reciprocal lattice points, P, falling on the sphere give rise to allowed reflections. In the Laue case, a continuum of wavelengths is incident leading to a family of spheres. The radius of the outermost sphere is set by the shortest wavelength output from the tube (the so-called short wavelength cutoff). The innermost sphere radius is set by the longest wavelength represented by measurable scattered intensity in the tube spectrum. This is shown in fig. 2.20b). Inspection of the figure indicates that forward as well as backscattered reflections are possible. The forward reflections are the basis of the transmission Laue method. Most silicon crystals are too thick to allow significant transmission of radiation. As a result, only the Laue backscatter spots are recorded.

Typical Laue backscatter images for (100) silicon are shown in fig. 2.21, as is a slightly misaligned (100) presentation. The crystal orientation is important in a number of applications. For example, if the crystal is to be used as a seed in the CZ method, the seed face must have the desired orientation of the boule. The proper orientation is ascertained by obtaining the proper symmetry in the Laue photo. The symmetric array of spots is indicative of a well-formed crystal. In the case of a disordered solid, or glass, only short-range order is evident in the atomic arrangement. Diffraction from such a system gives rise to a characteristic ring structure (fig. 2.22). Polycrystalline solids will also give rise to x-ray scattering plots that have the appearance of spots in the process of being smeared out into rings. This is associated with the random orientation of small crystalline grains that satisfy the Laue condition simultaneously. X-ray diffraction studies can successfully be employed in studying the reactions between various materials, both single-crystalline or polycrystalline, bulk or thin films. Reactions lead to new compounds, sometimes called *intermetallics.* X-ray diffraction can easily be used to identify these reaction products as they lead to new sets of reflections at characteristic positions.

X-ray topography is an imaging technique in which contrast is achieved through changes in the crystal interplanar spacing. Such changes are usually associated in homogeneous strain or with the defect structure of the crystal. Topographs provide defect maps of the wafer surface. A typical topographic apparatus is shown in fig. 2.23. A collimated, monochromatic linear beam of x-rays irradiates the sample. The incidence angle is set to one of the Bragg-reflection angles tuned to a specific set of crystal planes. The beam is swept over the wafer. The beam and wafer are moved in such a way as

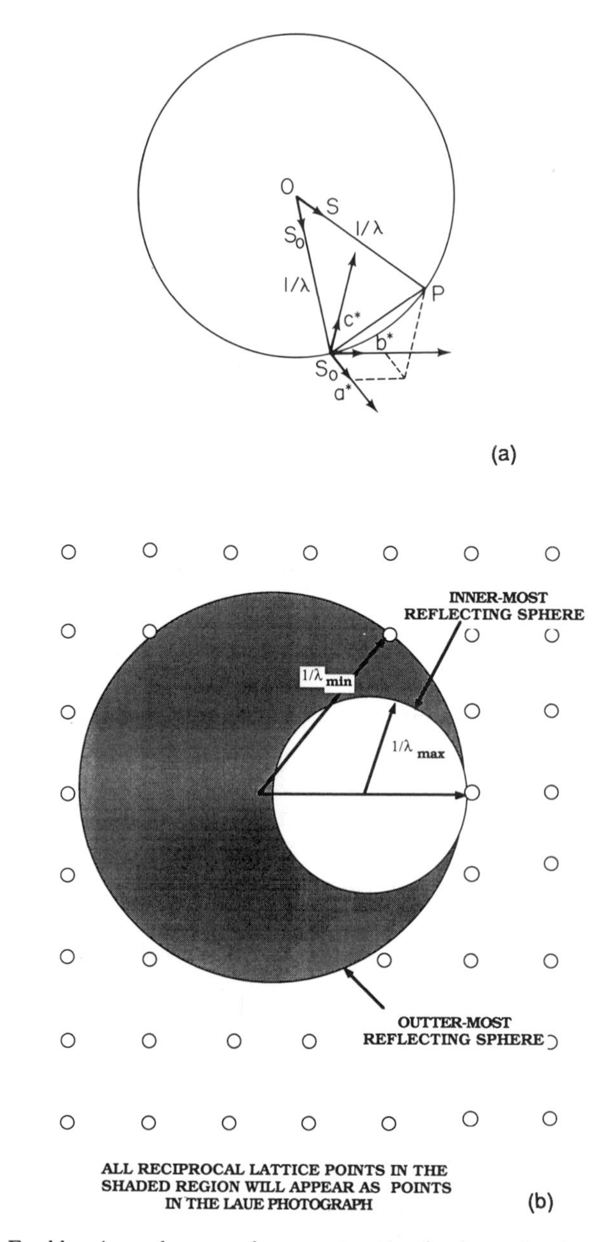

Fig. 2.20. Ewald reciprocal-space sphere construction for determination of the diffraction condition. (a) The Bragg condition for a single λ and reflecting plane at P. (b) The situation for a continuum of wavelengths incident.

to ensure that the incident angle is fixed at the Bragg angle throughout the scan. When a defect is encountered and the interplanar spacing varies, the reflected intensity goes down and the defect is identified as pointed out in the figure.

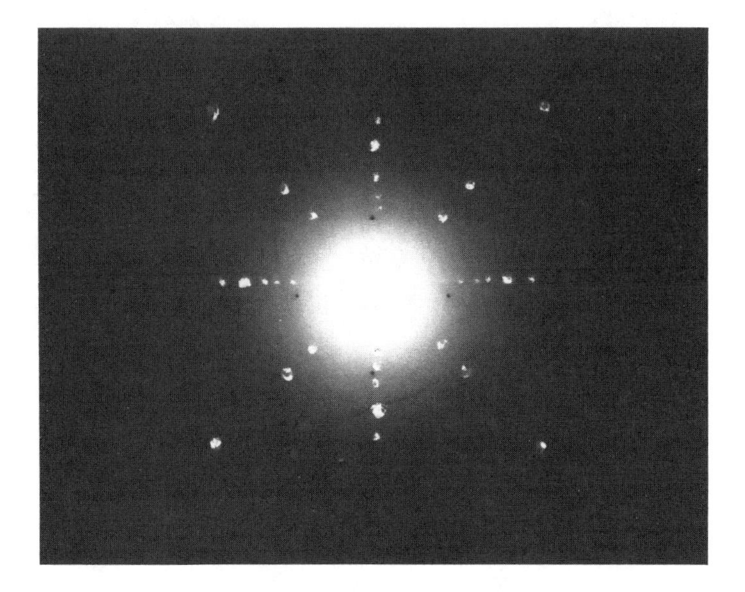

(a)

(b)

Fig. 2.21. Laue backscatter images for silicon reflection. Note the symmetric display (a) obtained when the beam is incident along a normal to a lattice plane the <100> plane in this case). In (b), we have off-axis incidence.

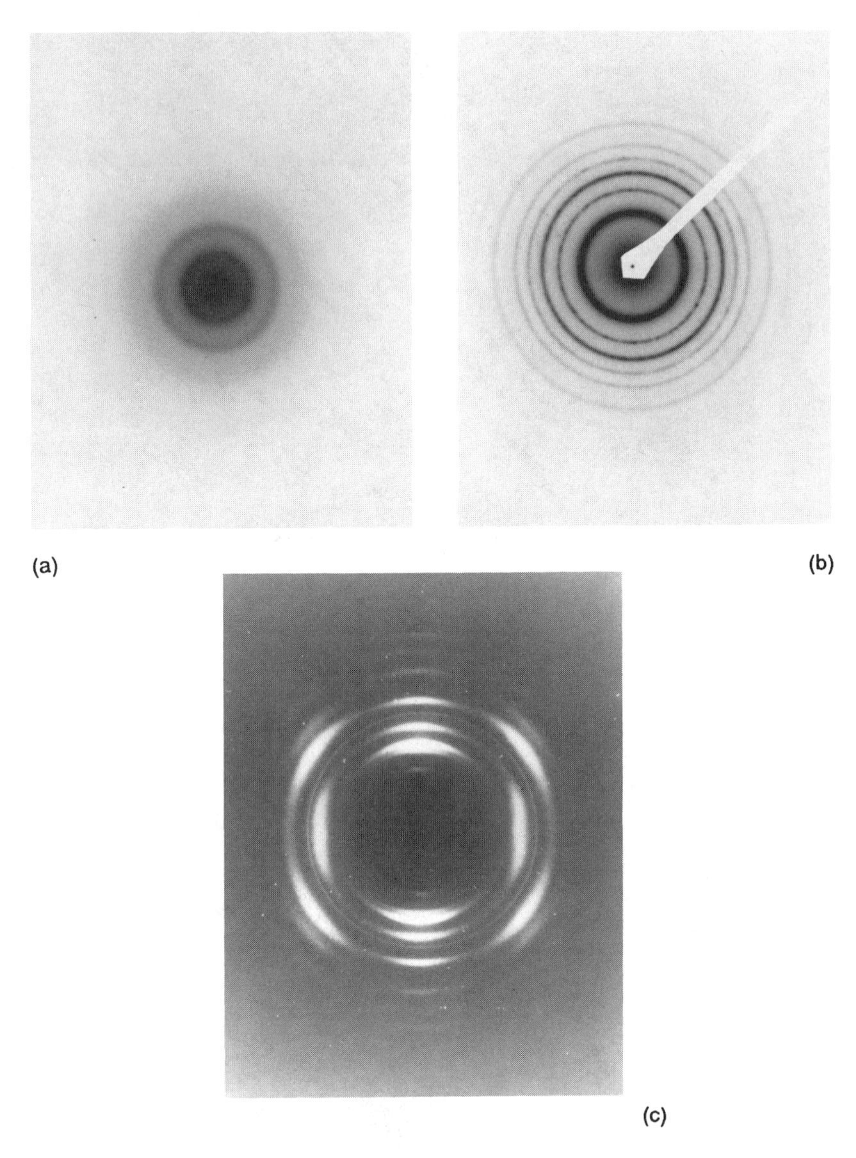

Fig. 2.22. (a) Amorphous ring pattern. (b) Spots appear in rings for randomly oriented polycrystals. (c) Oriented polycrystal growth.

Beam collimation is achieved in a number of ways. The output from the x-ray tube appears to originate (to first approximation) from a point source. If one moves far enough away from the source, the rays emerging from the source appear parallel. A slit aperture forms the line beam. Another way to achieve a collimated beam is to reflect the tube output from a collimator

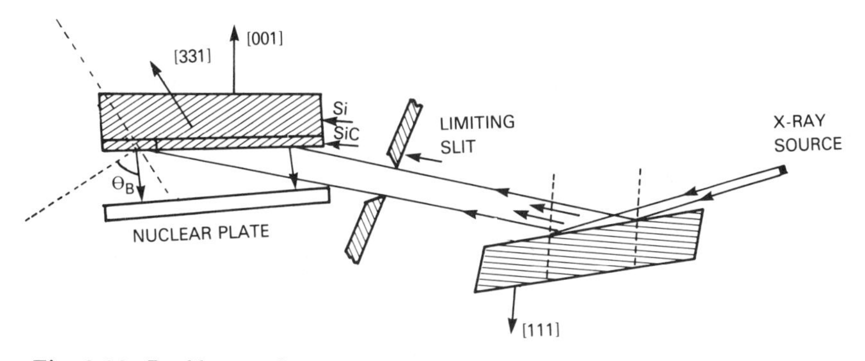

Fig. 2.23. Double crystal x-ray topographic apparatus. Courtesy of M. Fatemi.

crystal. Rays emerge from this first crystal only in the direction determined by a Bragg angle. The crystal also acts as a monochromator, reflecting x-rays of a single wavelength only. Such an arrangement is called a *double-crystal topograph.* In the case of distance collimation, filters must be used to limit the wavelength range incident on the sample. Some typical topographs of silicon and GaAs wafers are shown fig. 2.24. Line defects, dislocation tangles, stacking faults, and precipitates can be made visible (as indicated in the figure).

In addition, the technique can be used separate out one type of defect from another. For example, consider an edge dislocation (fig. 2.9). If the reflection plane for the topograph is the plane whose normal is along the dislocation line (AD in fig. 2.9), the interplanar spacing is minimally affected by the defect and the defect will not be visible. Rotating the crystal to reflect off a different plane will make the dislocation visible. Now, consider the screw-dislocation. Interplanar spacings between slip-planes are minimally affected and reflections from these planes have no topographic contrast. Next, consider reflections from planes whose surface normals are parallel and perpendicular to the Burger's vector (and, at the same time are orthoganal to the zero-contrast plane normal). Topographic reflections of these planes are affected by the dislocations. Similarly, the reflected intensities for planes whose surface normals are parallel and perpendicular to the edge of the Burger's vector (as rotated about the dislocation line) are different. The ratio of the reflected intensities for the parallel and perpendicular planes is different for the edge and screw cases. Measurement of this ratio allows a determination of the edge or screw nature of the dislocation.

The double-crystal topographic arrangement also allows for the measurement of strain in the crystal [23]. In this analytic tool, the x-ray beam is tightly collimated and the direction of beam incidence is well defined. We can determine a reflection line shape by rocking the sample crystal back and forth in the incident beam and recording the reflected intensity as a function of incident angle. This provides what is called a *rocking curve.* For an unstressed crystal, the angular width of the diffraction line will be very small. Stressing tends to bow the crystal, and a range of incidence angles show measurable reflection. Thus, the rocking curve is broadened. For a fixed angle topograph, the change in reflected intensity encountered as the beam sweeps the sample is directly related to local strain:

$$\frac{I_x}{I_0} = k\left\{\tan\theta\left[\left(\frac{\Delta d}{d}\right)+\delta_\theta\right]\right\}, \tag{2.23}$$

where: I_x = local intensity at some point, x, on the surface; I_0 = reflected intensity at the reflection maximum; k = slope of the rocking curve at its half height; $\Delta d/d$ = strain; δ_θ = component of local lattice rotation with respect to the vector—normal to the observed crystallographic plane; and θ = Bragg angle. Strain-induced changes in reflected intensity are clearly visible in fig. 2.24a.

To summarize, x-ray diffraction techniques are excellent probes of crystal properties. The Laue backscatter technique is frequently used in semiconductor crystal structure analysis to provide orientation of the crystal. In addition, it provides a coarse test of crystallinity. Topographs give defect maps of the substrate. Changing the orientation of the sample topograph gives information on the nature of the defect examined. In the double-crystal setup, a rocking curve can be generated that gives quantitative information on local strains.

2.7.2 Electrical Techniques

The electrical activity of crystallographic defects was discussed earlier in this chapter. To briefly summarize, defects can act as donor or as acceptor trapping sites in the material, and they can act as scattering centers that degrade mobility. Acceptors become negatively charged when they are occupied, and they are neutral when they are unoccupied. Donors are neutral when occupied and positively charged when they are unoccupied. An acceptor is charged (negatively) if the Fermi level is above it in the band edge versus position diagram. Donors are charged (positively) if they are above the Fermi level. We distinguish between “shallow” and “deep” levels as a function of the energy level of the traps with respect to the band edges. Deep levels lie close to midgap; shallow levels are close to the band edges.

At room temperature, for typical semiconductor doping levels, shallow levels are either always occupied or always unoccupied, depending on which band edge they are near. They primarily affect the mobile charge occupancy of the bands, influencing the position of the Fermi level in the band gap. Extrinsic donors, such as arsenic and phosphorous, make the material n type, boron makes it p type. Deep levels act as intermediate states in the band

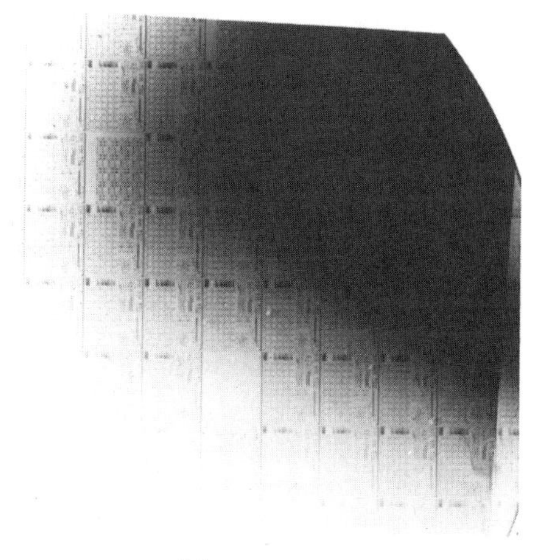

(a)

DOUBLE-CRYSTAL X-RAY TOPOGRAPHY

Defects in an LEC-Grown GaAs Wafer

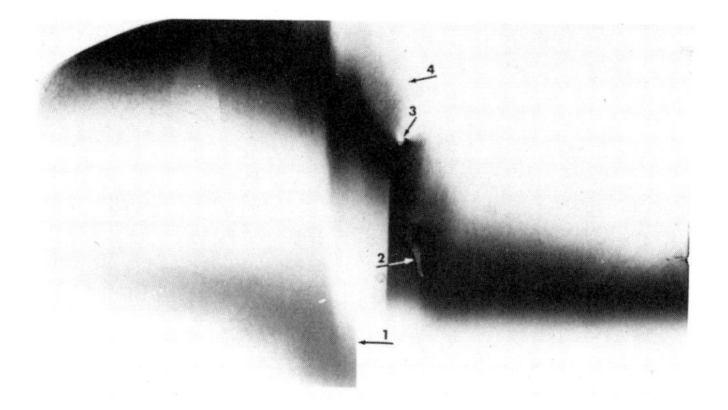

1- Tilt Boundaries 2- Low-Angle Grains 3- Inclusions 4- Dislocations

(b)

Fig. 2.24. Topographs of (a) processed silicon wafer and (b) GaAs starting materials. Courtesy of M. Fatemi.

gap and, effectively, lower the activation barrier for band-to-band transitions by providing midgap transition levels [24]. As such, they act as GR centers. When the material is thrown out of equilibrium (as a result of some injection process or through optical illumination), the band occupancies will attempt to return to their equilibrium values through the aid of the GR centers. Most intrinsic defects act, primarily, as GR centers.

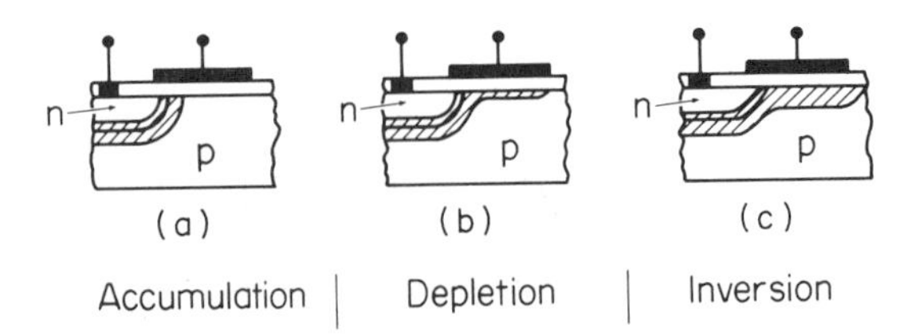

Fig. 2.25. The gated-diode structure. After ref. 2.24.

One of the most important characterization parameters associated with deep-level defects is the carrier lifetime. Consider the enhancement or depletion of the equilibrium density of mobile carriers by some technique. For example, pulsed optical illumination can give rise to a very large minority carrier density. The material is then allowed to relax to equilibrium. In the enhancement case, carriers will recombine with the aid of traps. In the depletion case, generation must occur. After the pulse, the nonequilibrium concentrations return to equilibrium exponentially in time, with a time dependence given by: $\exp\left(\frac{-t}{\tau}\right)$, where τ is called the carrier lifetime. This parameter is of prime importance in a number of devices. Lifetime helps to determine current gain in bipolar devices, the refresh time in dynamic RAMs and noise levels in nonequilibrium imagers such as charge-coupled devices.

The gated-diode is a simple electrical test structure that can give a direct measurement of carrier lifetime [24]. A gated-diode cross-section is shown in fig. 2.25. The device is used as follows: The reverse leakage current of the pn-junction diode is continuously monitored at some level of reverse bias, V_r. We define three regions of operation based on the gate bias, V_g. In the first region, the gate bias weakly accumulates the semiconductor surface underneath it. The reverse leakage is dominated by the generation current in the pn-junction diode depletion. Using the Shockley-Read GR model [24], this current is given by

$$I_{\mathrm{MJ}} = \frac{1}{2} q \frac{n_{\mathrm{i}}}{\tau_{\mathrm{MJ}}} w A_{\mathrm{J}}, \tag{2.24}$$

where I_{MJ} = leakage current in metallurgical junction of the diffused diode, q = electron charge, n_{i} = intrinsic carrier concentration, τ_{MJ} = generation lifetime in metallurgical junction, w = extent of the bulk space-charge layer, and A_{J} = junction area. The effect of the diode reverse bias is included in this equation. Increasing the reverse bias increases the diode depletion depth. This increases the reverse leakage.

In the second region of operation, the gate is biased to deplete the semiconductor beneath it. This increases the effective volume of the pn-junction depletion. The measured diode leakage increases until the region beneath the gate inverts and the gate depletion volume no longer increases with increased gate bias. The increased current is actually the sum of two currents: the gate depletion volume generation current and the generation current associated

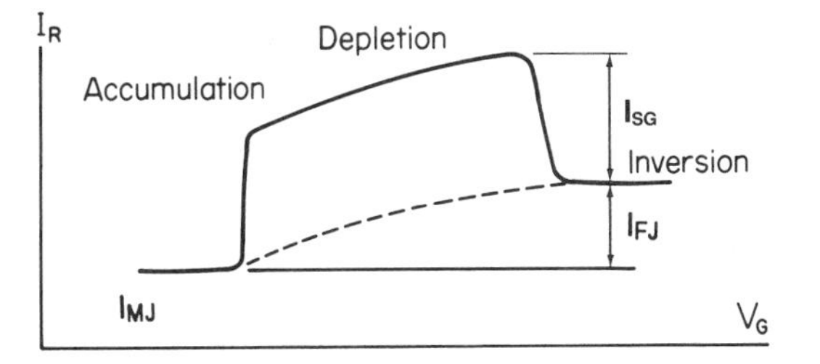

Fig. 2.26. Gated diode current-voltage plot in the three regions of operation. After ref. 2.24.

with the oxide/semiconductor surface. Again, using the Shockley-Read GR model, we have the following expressions for these currents:

$$I_{\mathrm{FJ}} = \frac{1}{2}\, q\, \frac{n_{\mathrm{i}}}{\tau_{\mathrm{FJ}}}\, x_{\max} A_{\mathrm{FJ}} \tag{2.25}$$

and

$$I_{\mathrm{SG}} = \frac{1}{2}\, q\, n_{\mathrm{i}}\, s_0\, A_{\mathrm{FJ}} \tag{2.26}$$

where I_{FJ} = field-induced junction leakage current, I_{SG} = surface leakage current, $x_{\max}$ = maximum extent of field induced junction, A_{FJ} = field induced junction area, s_0 = surface recombination velocity, and τ_{FJ} = minority carrier lifetime in the field-induced junction.

The reverse-leakage current remains relatively constant (despite some increase in the bulk depletion depth), until the surface goes into strong inversion. In this third region of operation, the mobile inversion charge saturates the surface traps and the leakage current declines by an amount equal to the surface generation current. This behavior is shown graphically in fig. 2.26. This figure also demonstrates how each of the three currents given by eqs. (2.24)-(2.26) can be measured. From this measurement, bulk lifetime and surface recombination velocity can be determined.

A more sophisticated method is required to fully characterize the defect sites present. For example, in addition to the number density of traps present, we would also like to know the energy level of the trap in the gap as well as the trap cross section. This is usually done with deep-level transient spectroscopy (DLTS) [25]. DLTS is a transient capacitance measurement that is usually done on a junction diode or on a Schottky diode. First, the diode is forward biased. This floods the semiconductor with mobile charge, causing the trapping states to charge. Next, the diode is rapidly thrown into reverse bias. Capacitance is continuously monitored as a function of time through the reverse-bias part of the measurement cycle. Electric field lines terminate on charged defect-trapping sites, as well as on charged dopant cores. Thus, the depletion layer is not the same as it would be in the absence of charged defects. This changes the measured capacitance. For example, if acceptor states are charged in n-type material, this partially neutralizes the space charge associated with the donor dopant cores. This increases the depletion depth. Thus, the charging of the deep levels decreases the capacitance. The magnitude of the change in capacitance is related to the total number of traps present. The decay rate of the trap occupancy is related to the trap emission coefficient. By measuring the trap emission coefficient as a function of temperature, and by assuming a rate-activated emission process, the trap energy can be ascertained.

2.8 Summary

In this chapter the following points were covered:

- Elemental silicon is derived by chemical reduction of sand (SiO_2).
- There are two main methods used to obtain electronic-grade crystals of silicon: float zone (FZ) and Czochralski (CZ).
- FZ silicon is the purest, having the least oxygen and carbon content.
- CZ silicon tends to have oxygen and carbon impurities (as a result of dissolution and outgassing of the crucible and susceptor) as well as heavy metals and residual boron.
- FZ silicon tends to be brittle and prone to breakage in processing as a result of the absence of impurity solution hardeners.
- In addition to extrinsic impurities, crystalline silicon can contain a variety of point, line, area, and volume defects.
- The basic defect structure of the as-grown silicon can be manipulated through anneals to improve bulk carrier lifetime.
- X-ray diffraction techniques can be used to image bulk defect structures.
- Electrical activity of defect sites is best studied through electron device structures such as gated diodes and DLTS structures.

References for Chapter 2

1. C.A. Hogarth, "Materials Used In Semiconductor Devices," Wiley-Interscience, New York, pp. 29-48 (1965).
2. W. Zulehner, D. Huber, Czochralski-grown silicon, in "Crystals—Growth: Properties and Applications," Vol. 8, J. Grabmaier, ed., Springer-Verlag, Berlin (1982).
3. W. Dietz, W. Keller, A. Muhlbauer, Float-zone grown silicon, in "Crystal—Growth: Properities and Applications," Vol. 5, J. Grabmaier, ed., Springer-Verlag, Berlin (1981).
4. W.M. Bullis, L.B. Coates, "Measurements of oxygen in silicon," Sol. St. Elect., Vol. 30(3), 69-81 (1987).
5. T. Abe, K. Kikuchi, S. Shirai, S. Muraoka, Review: Impurities in silicon single crystals—A current view, in "Semiconductor Silicon 1981," Electrochem. Soc. Proc., 81-5, H. Huff, R.J. Kriegler, Y. Takeishi, eds., pp. 54-71 (1981).
6. K.V. Ravi, "Imperfections and Impurities in Semiconductor Silicon," Wiley-Interscience, New York, 85 (1981).
7. R.E. Dickerson, H.B. Gray, G.P. Haight, "Chemical Principles," 2nd ed., W.A. Benjamin, Menlo Park, CA, pp. 507-511 (1974).
8. A.C. Bonora, Review: Wafer Preparation: Silicon, etching, polishing, in "Semiconductor Silicon 1977," H. Huff, E. Sirtle, eds., Electrochem. Soc. Proc., 77-2, 154-169 (1977).
9. V.G. Moerschel, C.W. Pearce, R.E. Reusser, A study of the effect, of oxygen content, initial bow, and furnace processing on warpage of three-inch diameter silicon wafers, in "Semiconductor Silicon 1977," H. Huff, E. Sirtle, eds., Electrochem. Soc. Proc., 77-2, 170-181 (1977).
10. M. Lanoo, J. Bourgoin, "Point Defects in Semiconductors," Vols. I and II, Springer-Verlag, Berlin (1981).
11. J. Weertman, J.R. Weertman, "Elementary Dislocation Theory," McMillan, New York (1971).
12. R.A. Swalin, "Thermodynamics of solids," 2nd ed., Wiley-Interscience, New York (1972).
13. I.R. Sander, P.S. Dobson, Phil. Mag., 20, 881 (1969).
14. H.F. Schaake, S.C. Baber, R.F. Pinizzotto, The nucleation and growth of oxide precipitates in silicon, in "Semiconductor Silicon 1981," Electrochem. Soc. Proc., 81-5, H.R. Huff, R.J. Kriegler, Y. Takeishi, eds., 273-281 (1981).
15. H.-D. Chiou, Oxygen precipitation behavior and control in silicon crystals, Sol. St. Tech., 30(3), 77-84 (1987).
16. J. Andrews, Oxygen out-diffusion model for denuded zone formation in Czochralski grown silicon with high interstitial oxygen content, in "Defects in Silicon," N.M. Bullis, L.C. Kimerling, eds., Electrochem. Soc. Proc., 83-9, 133-141 (1983).

17. W. Kern, D. Poutinen, Cleaning solutions based on hydrogen peroxide for use in silicon semiconductor technology, RCA Review, 31(2), 107-206 (1970).
18. W. Kern, Radiochemical study of surface contamination I: Adsorption of reagent components, RCA Review 31(2), 206-233 (1970).
19. W. Kern, Radio chemcial study of surface contamination II: Deposition of trace impurities on silicon and silica, RCA Review, 31(2), 234-264 (1970).
20. D.J. Mountain, K.F. Galloway, T.J. Russell, Effect of postoxidation anneal on the electrical characteristics of thin oxides, Electrochem. Soc. Jour. 13A(3), 747 (1987).
21. C.S. Barrett, T.S. Massalski, "Structure Of Metals," 3rd ed., McGraw-Hill, New York (1966).
22. B.E. Warren, "X-ray Diffraction," Addison-Wesley, Reading, MA (1969).
23. S. Qadri, D. Ma, M.C. Peckerar, High resolution x-ray determination of local strain fields in MOSFET devices, Appl. Phys. Letters, 51, 1827-1829 (1987).
24. A.S. Grove, "Physics and Technology of Semiconductor Devices," Wiley, New York, pp. 296-297 (1967).
25. D.V. Lang, Deep-level transient spectroscopy: A new method to characterize traps in semiconductors, J. Appl. Phys., 45(7), 3023-3032 (1974).

Problem Set for Chapter 2

2.1. Consider the three types of cubic lattice.

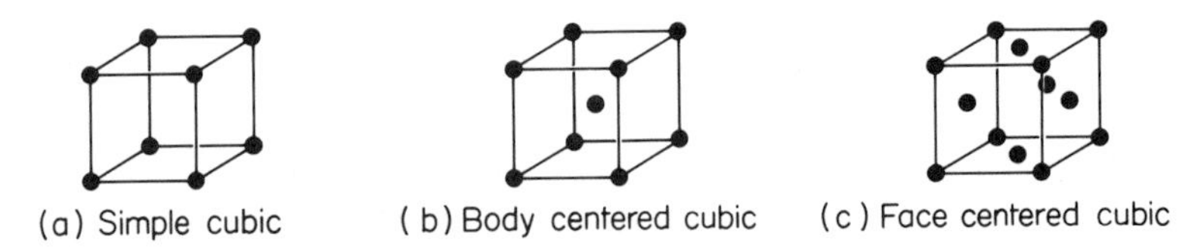

a. How many "nearest neighbor" lattice cubic points are there for each of the three lattice types?
b. Silicon is sometimes referred to as an fcc lattice. It is also referred to as a tetrahedrally bonded (fourfold coordinated) solid. Explain using a diagram of the silicon lattice for reference.

2.2. Consider the four (111) type planes intersecting the (100) plane in a cubic crystal.

a. What type of solid figure do the four (111)-planes and the (100)-plane make?
b. What is the angle between the adjacent (111) planes?
c. What is the angle between each (111) plane and the (100) plane? (Again *prove* this!)

2.3. Consider a dislocation "loop" existing on some crystal plane:

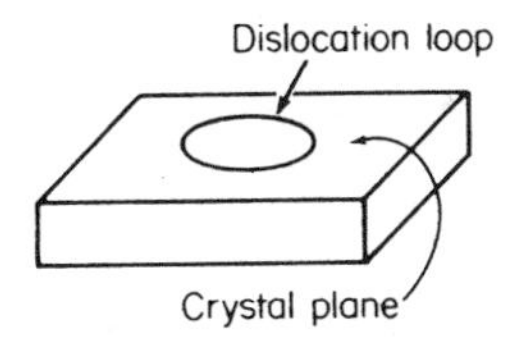

The dislocation is neither completely edge or screw in nature. It is of "mixed" character. Explain. (Assume the Burger's vector points in the direction of slip within the dislocation loops.)

2.4a. Write an expression for the total change in free energy on forming a circular stacking fault of area A, using the dislocation loop model discussed in the text.

b. Note that all the terms in the above derived expression are positive. And yet, there are conditions under which stacking faults will spontaneously form. Explain how this comes about.

2.5. Crystal surfaces and interfaces may roughen in a number of ways. Ledges may form at regular intervals, giving rise to the exposure of a high index (viscinal) plane. Jogs or kinks can appear, further roughening the surface as shown below:

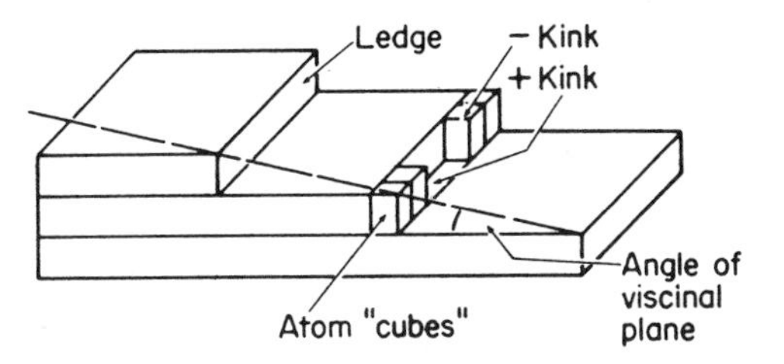

Opposing kinks are given the (arbitrary) designations + and −.

a. Give an expression that relates the spacing between the ledges and the ledge height to the viscinal plane rotation angle measured with respect to the primary plane.
b. View the individual atoms in the lattice as blocks. The atomic sites will have simple cubic coordination, and the internal energy associated with

a given site is equal to the bond energy associated with uncompensated bonds. Write a relationship between the energy of a "normal" ledge site E_1, and a kink site.

c. Assume kinks form by a rate-activated process. Show that the expression defining the ratio of kinks to normal sites is: $\frac{n_+ n_-}{n_0^2} = \exp \frac{-2E_1}{kT} = \eta^2$.

d. Assume that the number of positive kinks equals the number of negative kinks. Show that the ratio of kinked sites to the total number of ledge sites is: $\frac{n_K}{n} = \eta/(1+2\eta)$, while the ratio of the unkinked sites to the total number of ledge sites is: $\frac{n_0}{n} = 1/1+2\eta$.

e. Show the total free energy of the kinked ledge is: $\Delta G_K = -\eta kT\ell n[\eta(1+2\eta)]$.

f. Is there a condition under which free energy of the kinked ledge is zero? What would this imply? Hint: $n_+ + n_- + n_0 = n$ (total number of ledge sites); $n + n_+ + n_- =$ total number of uncompensated bonds per unit length.

2.6. Ignore, for a minute, the effect of lattice strain on oxygen precipitate free-energy. At 1100°C, the heat of formation of SiO_2 is 215 kcal/mole and the free energy is formation is 154 kcal/mole. The surface energy is 100 erg/cm^2 (note units!).

a. What is the size of a critical nucleus?

b. What effect does the omitted strain term have on the critical nucleus size?

2.7. Some dislocations can "climb" out of the slip plane by interacting with vacancies. Explain.

2.8. Demonstrate that the demand for a reflecting plane to lie on the Ewald sphere is equivalent to the demand for the satisfaction of the Bragg equations. Use fig. 2.20 and eqs. (2.22a)-(2.22c) for reference.

2.9. As a result of the enrichment (or depletion) of dopants from the melt in the Czochralski method (as determined by the segregation coefficient), the concentration of dopant changes along the crystal axis in the Czochralski process. Show that:

$$c(x) = k_s c_0 (1 - x_0)^{k_s - 1}$$

where c_0 is the dopant concentration at the seed end, and k_s is the segregation coefficient, and x_0 is the fraction of the melt solidified.

2.10. Using the float-zone disc model developed in the text, what is the impurity concentration in the zeroth disc after the first pass, assuming

a. Initial concentration 10^{12}/cc

b. $k_s = 0.8$ (Boron)
$k_s = 0.002$ (Aluminum)

2.11. Consider a Schottky diode of metal-gate area, A, on an n-type silicon substrate. The device is originally reverse-based by applying a potential v_0 (negative) to the gate. The substrate contains acceptor traps whose density is N_{ta}. N_{ta} is much less than N_d, the substrate doping. The magnitude of the device negative bias is lowered an amount ΔV for a time sufficient to saturate acceptor traps. This bias is then returned to V_0. What is the maximum change in capacitance as a result of the trap-filling pulse?

2.12. Fill in the steps leading to equation 2.12.

2.13. The criterion for the invisibility of a line defect in an x-ray topograph is usually expressed:

$$\vec{g} \cdot \vec{b} = 0$$

where $\vec{g}$ is the reciprocal lattice vector of reflecting plane and $\vec{b}$ is the Burgers vector. Explain why this is so.

3 Oxidation

3.1 Introduction

The phenomenal success of silicon as the starting material for the fabrication of integrated circuits lies in the fact that SiO_2 layers can be produced by thermal oxidation of silicon substrates and by deposition from the gas or vapor phase. The oxide films are useful in several ways:

1. They provide isolation between devices.
2. They mask against impurities, dopant diffusion, and environmental effects.
3. They provide stability against mechanical damages, e.g., scratching.
4. They act as an insulating layer between conductors.
5. They act as the gate dielectric.

Basically, the SiO_2 layers can be used in the ways described above successfully. In using such layers as gate dielectric, and for isolation purposes, the demands for control of the dielectric properties are the most severe. Dielectric properties determine transistor thresholds and parasitic leakage currents. For other applications, the requirements are less stringent.

In this chapter, thermal oxidation of a semiconductor surface is considered. Thermal oxidation of silicon in dry oxygen and in water-vapor-containing medium have been the preferred techniques. This is because oxide films produced in this way have the best properties. This chapter deals with the oxidation of silicon. In the last section of the chapter, the oxidation of compound semiconductors, such as GaAs, GaP, and InP, is discussed. For semiconductors other than silicon, the oxidations that produce acceptable oxide films have not been possible. This fact has resulted in less than the expected usefulness of such semiconductors.

3.2 Oxide Properties

Table 3.1 summarizes the most important properties of silicon dioxide. Thermally grown SiO_2 films are amorphous. They consist of a random three-dimensional network of silicon atom tetrahedrally surrounded by four oxygen atoms each [1]. The tetrahedra are joined by oxygen bridges at the corners, as shown in the two-dimensional network depicted in fig. 3.1. As is apparent,

TABLE 3.1: Properties of amorphous SiO_2

Property	Value
DC Resistivity	10^{14}-10^{16} Ω-cm at 25°C
Density	2.27 g·cm^{-3}
Dielectric constant	3.8-3.9
Dielectric strength	~ 10^7 V· cm^{-1}
Energy gap	~ 8-9 eV
Infrared absorption band	9.3 μm
Linear thermal expansion coeficient (α)	0.5×10^{-6} ppm per °C
Melting point	~ 1700°C
Refractive index	1.46
Thermal conductivity	0.014 W/(cm·°C)
Etch rate in BHF[a]	1000 Å/min for thermal oxide

[a] Buffered hydrofluoric acid etch–containing 34.6% NH_4F, 6.8% HF, 58.6% H_2O by weight.

this type of structure has open spaces. This is also reflected in the low density of the amorphous SiO_2 compared to that of crystalline SiO_2 (2.27 g·cm^{-3} versus 2.65 g·cm^{-3}). In fact, only 43% of the total measured volume is occupied by SiO_2 molecules. In spite of this open structure, SiO_2 films are excellent barriers to the diffusion of several dopants (As, B, P, Sb, Ga, etc.) and metals. Table 3.2 summarizes the diffusivities of various impurities in SiO_2. Diffusivities (in cm^2/s) can be calculated by using the Arrhenius equation $D = D_o \exp(-Q/RT)$. Let x be the diffusion depth in time t. Assuming $x^2 = 4Dt$, the values of $x/\sqrt{t}$ have been calculated for all impurities at the three most commonly used temperatures. The values are listed in table 3.2. Note the high diffusivities of H_2, O_2, H_2O, and Na in SiO_2. Of these, high diffusivities of O_2 and H_2O are very essential. During thermal oxidation (see Section 3.4), these oxidizing species diffuse rapidly through the oxide layer growing on the silicon and continue to react with silicon at the silicon-oxide interface. If the diffusivities of O_2 and H_2O in SiO_2 were low, the oxide formed at the initial stages of oxidation would have acted as a diffusion barrier and oxidation would have been self limited. High diffusivity of hydrogen in SiO_2 is also beneficial since hydrogen anneals have been shown to be very effective in reducing negative-bias instability [3] and interface trap level density [3]. For the in-depth discussion of these effects and the role of hydrogen, readers are referred to the text by Nicollian and Brews [4].

Fast diffusers like H_2O can create problems. They can diffuse in oxide at relatively low temperatures and weaken the oxide structure, affecting several oxide properties. Electron traps that are related to the presence of

TABLE 3.2: Diffusivities of various impurities in SiO_2[a]

Diffusing Species	Do (cm^2/s)	Q (eV)	$X/\sqrt{t}(\mu m/s^{1/2})$ at 800°C	900°C	1000°C
B	1.7×10^{-5}	3.37	1×10^{-6}	4.9×10^{-6}	1.8×10^{-5}
Ga	3.8×10^{5}	4.15	2.3×10^{-3}	1.6×10^{-2}	7.7×10^{-2}
P	0.186	4.03	3×10^{-6}	1×10^{-5}	9.4×10^{-5}
As	67.25	4.7[b]	1.6×10^{-6}	1.4×10^{-5}	8.4×10^{-5}
	0.037	3.7[c]	4.1×10^{-6}	4.5×10^{-5}	1.9×10^{-4}
H_2	5.65×10^{-4}	0.446	42.8	52.5	62.5
O_2	2.7×10^{-4}	1.16	0.63	1.1	1.8
H_2O	1.0×10^{-6}	0.79	0.28	0.40	0.55
Na	0.0344	1.22[d]	5.1	9.0	14.4
S	6×10^{-5}	2.6	1.2×10^{-4}	4.1×10^{-4}	1.1×10^{-3}
Au	8.5×10^{3}	3.7	3.9×10^{-3}	2.1×10^{-2}	0.09

[a] From reference 19.
[b] Using ion implantation and N_2 anneal.
[c] Using ion implantation and O_2 anneal.
[d] At temperature range of 573-1000°C.

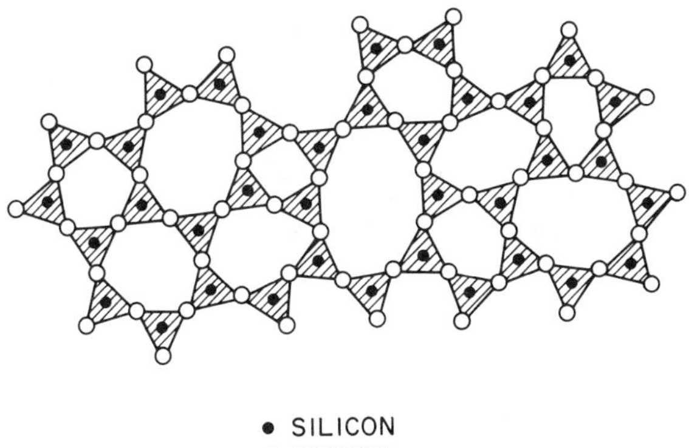

Fig. 3.1. Two-dimensional schematic representation of the amorphous structure of fused silica.

water in the oxide have been extensively studied and reported [5]. Infrared (IR) absorption characteristics change significantly with water absorption [6], making it possible to use IR measurements to detect water in the SiO_2 films. High-temperature annealing in a water-free ambient minimizes the water content in the oxide. In this case, the higher diffusivity of water molecules allows water to diffuse out of the oxide into the annealing ambient. Sodium, the other fast diffuser in SiO_2, is also undesirable. It is an electrically active impurity and therefore actively participates in controlling the properties under applied bias [4]. Sodium contamination must therefore be avoided.

SiO_2 films on silicon are very successfully used as diffusion barriers (or masks) for dopants since dopants have very low diffusivities in SiO_2 (see table 3.2). In modern practices, however, ion implantation is used to introduce dopants in silicon [7]. By controlling the energy and the dose rate of the implants, one can tailor the junction depths and the surface concentrations of the created junctions. Oxide films are used as masks to prevent ion implantation in chosen regions of the silicon. One must carefully tailor the thickness of the oxide to use it as an effective ion implantation mask. Table 3.3 gives the range and the projected standard deviation in the range (see Chapter 5) for the dopant species as a function of the ion energy [1]. For example, given a boron ion energy of 100 KeV, the range is 310.4 ± 71.0 nm, suggesting that at least a 400.0-nm thick oxide layer should be used to stop such boron ions in SiO_2. A greater thickness will be used to allow for the diffusion in the oxide during the diffusion anneal. The diffusion depth can be estimated using data in table 3.2. In addition, the amorphous layer adds a random scattering component that aids in preventing ion channeling during implantation.

Problem 3.1 *An oxidized silicon wafer is implanted with 100-keV boron and then annealed at 1000°C for 2 hours. Assume the implant range in SiO_2 at 100 keV is 310.4 ± 71.0 nm. Note the relatively large error bar on the implant range! The range will depend on the oxide density, which is determined by the process used to make the oxide. Calculate the minimum thickness of the oxide required to eliminate the possibility of boron diffusion into underlying silicon.*

Solution: Assume the diffusion begins at the *maximum* range of the implant. At 1000°C in 2 hours, boron will diffuse approximately a distance x given by

$$\begin{aligned} x &= \sqrt{4D \cdot 2 \cdot 3600} \quad \text{cm} \\ &= 169 \cdot D^{\frac{1}{2}} \quad \text{cm.} \end{aligned}$$

From table 3.2 D at 1000°C = 8.1×10^{-9} cm^2/sec. Therefore, $x = 1.52 \times 10^{-7}$ cm or 152 nm. Hence, minimum oxide thickness should be 310.4 nm + 152 nm + 71.0 nm = 533.4 nm.

TABLE 3.3: Projected range statistics for ion implantation into SiO_2

	Antimony		Arsenic		Boron		Phosphorus	
Energy	Projected Range	Projected Standard	Projected Range	Projected Standard	Projected Range	Projected Standard	Projected Range	Projected Standard
(keV)	(μm)	(μm)	(μm)	(μm)	(μm)	(μm)	(μm)	(μm)
10	0.0071	0.0020	0.0077	0.0026	0.0298	0.0143	0.0108	0.0048
20	0.0115	0.0032	0.0127	0.0043	0.0622	0.0252	0.0199	0.0084
30	0.0153	0.0042	0.0173	0.0057	0.0954	0.0342	0.0292	0.0119
40	0.0188	0.0052	0.0217	0.0072	0.1283	0.0418	0.0388	0.0152
50	0.0222	0.0061	0.0260	0.0085	0.1606	0.0483	0.0486	0.0185
60	0.0254	0.0070	0.0303	0.0099	0.1921	0.0540	0.0586	0.0216
70	0.0286	0.0078	0.0346	0.0112	0.2228	0.0590	0.0688	0.0247
80	0.0316	0.0086	0.0388	0.0125	0.2528	0.0634	0.0792	0.0276
90	0.0347	0.0094	0.0431	0.0138	0.2819	0.0674	0.0896	0.0305
100	0.0377	0.0102	0.0473	0.0151	0.3104	0.0710	0.1002	0.0333
110	0.0406	0.0110	0.0516	0.0164	0.3382	0.0743	0.1108	0.0360
120	0.0436	0.0118	0.0559	0.0176	0.3653	0.0774	0.1215	0.0387
130	0.0465	0.0126	0.0603	0.0189	0.3919	0.0801	0.1322	0.0412
140	0.0494	0.0133	0.0646	0.0201	0.4179	0.0827	0.1429	0.0437
150	0.0523	0.0141	0.0690	0.0214	0.4434	0.0851	0.1537	0.0461
160	0.0552	0.0149	0.0734	0.0226	0.4685	0.0874	0.1644	0.0485
170	0.0581	0.0156	0.0778	0.0239	0.4930	0.0895	0.1752	0.0507
180	0.0610	0.0164	0.0823	0.0251	0.5172	0.0914	0.1859	0.0529
190	0.0639	0.0171	0.0868	0.0263	0.5409	0.0933	0.1966	0.0551
200	0.0668	0.0178	0.0913	0.0275	0.5643	0.0951	0.2073	0.0571
220	0.0726	0.0193	0.1003	0.0299	0.6100	0.0983	0.2286	0.0611
240	0.0784	0.0208	0.1095	0.0323	0.6544	0.1013	0.2498	0.0649
260	0.0842	0.0222	0.1187	0.0347	0.6977	0.1040	0.2709	0.0685
280	0.0900	0.0237	0.1280	0.0370	0.7399	0.1065	0.2918	0.0719
300	0.0958	0.0251	0.1374	0.0394	0.7812	0.1087	0.3125	0.0751

3.2.1 Capacitance-Voltage ($C - V$) Characteristics

$C - V$ characteristics of the Metal-oxide-semiconductor (MOS) capacitor have provided real insight into the electronic structure of the oxide and the semiconductor. They allow us to create pictorially the effect of applied bias on the bending of the semiconductor energy bands inside the semiconductor. A comparison of the theoretically predicted and experimentally determined $C - V$ plots yields a large amount of important information, e.g., oxide thickness, dopant concentration in semiconductor, interface state charge density, oxide charge density, trap density, flat band voltage, threshold voltage, and metal workfunction. In the following, a brief description of these measurements is made. For detailed studies on this subject, references 4, 19-25 should be examined.

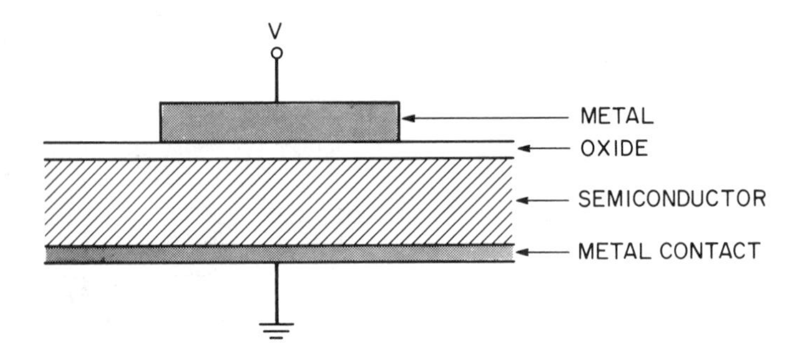

Fig. 3.2. A typical MOS capacitor cross section with metal back-contact.

Figure 3.2 shows a simple, typical MOS capacitor that is generally used to characterize the oxide. The top metal, the oxide, and the substrate metal contact are called the *gate metal*, the *gate oxide* or *insulator*, and the *back contact*, respectively. In practice, an oxide is first grown on silicon. Metal, usually aluminum, is next deposited on the oxide and then patterned into circular dots of known area. For quick characterization, metal dots can be created by evaporating the metal through a contact mask, usually made of very thin metal foil. In this case, the area of the dots will depend on the proximity of the mask to the substrate and the angle of incidence of the vaporized species. In general, the area is well approximated by the size of the pattern on the metal mask. For very accurate measurements, metal film must be patterned using photolithographic and etching techniques. A brief discussion of the $C-V$ behavior has been presented in Chapter 1. Here we present a more complete discussion, including oxide characterization aspects of $C-V$ measurements and the impact of oxide defects on these measurements.

In a typical $C-V$ measurement, a $C-V$ bridge is used. This superimposes a small AC signal at varying frequencies on a given DC voltage. The amplitude of the signal is typically 10-20 mV rms. The total applied voltage (small signal plus bias ramp) is shown in fig. 3.3. High-frequency measurements are typically made at 1 MHz. Even at these frequencies and the slowest ramp rates, equilibrium is not achieved in the deep depletion or inversion region. This generally shows up as the nonequilibrium curve shown in fig. 3.4. If, however, the ramping is stopped momentarily, when in deep depletion, equilibrium can be gained, leading to bias-independent capacitance in the inversion. The nonequilibrium behavior will occur whenever the minority carriers cannot accumulate near the silicon-oxide interface. For example, a leaky insulator will cause such nonequilibrium behavior, even under bias conditions corresponding to inversion. Also, if the capacitance is measured using a very low probing frequency (say, 1-10 Hz), the recombination-generation kinetics of the minority carriers near the surface may respond to the voltage variations causing the frequency-dependent capacitance in the inversion.

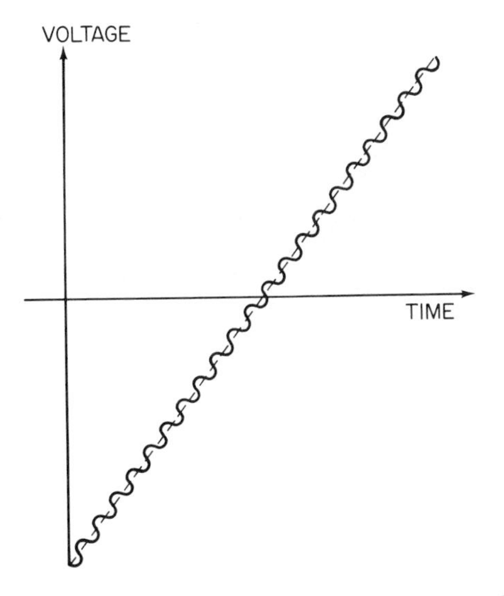

Fig. 3.3. Total applied bias on capacitor as a function of time. Note the imposed AC on the large signal ramp.

The probing frequency dependence of the $C - V$ characteristics results from the response of majority carriers (in accumulation and depletion) and minority carriers (in inversion) to the applied AC gate voltage. In accumulation and depletion, the majority carrier-charge induced at the semiconductor surface will follow the AC voltage as long as

$$\frac{1}{f} \gg \tau, \tag{3.1}$$

where f is the frequency of the applied voltage and τ is the dielectric relaxation time of the semiconductor. Dielectric relaxation time, τ is given by

$$\tau = \epsilon_{\mathrm{Si}}\rho, \tag{3.2}$$

where ρ is the resistivity of silicon. For example, for a 10 Ω-cm material, $\tau = 1.04 \times 10^{-11}$ s. This is much too small to be sensed by typical $C - V$ measurement apparatus using 1 hz to 1 MHz test signals.

Now, consider the generation of minority carriers at the semiconductor surface. They are not the result of flow or injection from the oxide or the back contact. The generation process is rather slow and the minority carrier response time τ_{R} is rather large, typically in the range of 0.01-1 s in the inversion layer at the silicon surface [4]. Thus, in inversion, for high probing frequencies, the surface charge layer (i.e., the minority charge layer) will not follow the applied AC signal. Thus, there is no change in the minority carrier charge density with variations in the gate voltage. This means then that at high frequencies the setting in of inversion does not induce any additional

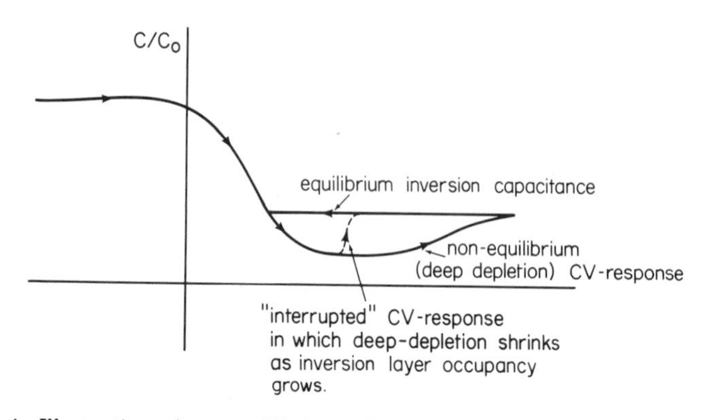

Fig. 3.4. Illustration of nonequilibrium effects on the $C - V$ response.

changes in capacitance. In fact, the capacitance practically remains equal to that prior to the onset of the inversion.

However, at low frequencies, where $1/f \geq$ the minority carrier response time (e.g., for f in the range of 1-10 Hz), the inversion layer charges can follow the AC gate signal, and the capacitor behaves like an ordinary capacitor with no variation in charge density in the surface depletion region. Thus, in inversion, the capacitance rises to the insulator capacitance.

An example of ideal $C - V$ behavior has been presented as fig. 1.18. In practice, such behavior is influenced by many factors, which include charges in the oxide and at the $Si{=}SiO_2$ interface, interface traps (these are electronic states that reside at the $Si{=}SiO_2$ interface), mobile ions such as sodium, dopant concentration and profile in silicon, and the difference between the work functions of the metal and the silicon. There is currently a generally accepted formalism for referring to the oxide defect-related charge. Q is used to refer to a Coulombic charge; N is a number density, i.e., number per square or cubic centimeter, and D refers to a number density-of-states in the semiconductor band gap. We distinguish four types of oxide charges and one density of states term associated with oxide defect structure:

- Q_{ot} (coul/cm^3): Oxide trapped charge. It is a fixed charge (either positive or negative) that is distributed through the oxide and that does not move or change charge states under bias.
- Q_{m} (coul/cm^3): Mobile oxide charge. This is the result of a mobile, charged impurity (such as sodium) that moves about the insulator under applied bias.
- Q_{f} (coul/cm^2): This is a fixed charge due to defects near the oxide/semiconductor interface. In the past these were referred to as "slow"states in that they could change charge state as a result of applied bias. This response, though, is rather sluggish.

- D_{it} (#/cm^2 - eV): Interface-state density. These are the so-called fast states that exist right at the semiconductor/oxide interface. They change state instantaneously with shifts in position of the semiconductor Fermi level at the interface.
- Q_{it} (coul/cm^2): Number of *occupied* interface states per unit of surface area.

It is necessary to elaborate in a few points relating to these defininitions. First, Q_{m} and Q_{ot} represent spatial distributions of charge. As such, they are functions of position in the oxide. If x is taken as depth into the oxide (increasing as the oxide/semiconductor interface is approached), $Q_{\text{m,ot}} = Q_{\text{m,ot}}(x)$. Charge residing near the gate has minimal effect on band bending. In the language of Chapter 1, charge residing at the oxide/gate interface causes no flat-band shift. All the image charge drawn by the oxide-defect charge onto the conductors surrounding the insulator resides on the gate itself. No band bending is produced. Charge in the oxide near the oxide/semiconductor interface has maximum effect on band bending. Hence, it creates maximum flat-band shift. All the image charge associated with these oxide defects is found in the silicon. To find the effect of such charge on flat band, we must take these distribution effects into account. In doing this, we write the flat-band shift, $\Delta\ V_{\text{fb}}$, as

$$\Delta V_{\text{fb}}^{\text{m,ot}} = \frac{Q_{\text{eff}}^{\text{m,ot}}}{C_{\text{ox}}} \tag{3.3}$$

$$Q_{\text{eff}}^{\text{m,ot}} = \frac{1}{d_{\text{ox}}} \int_0^{\text{dox}} x Q_{\text{m,ot}}(x)\ dx. \tag{3.4}$$

Next, we must realize that D_{it} is a distribution in *energies* (and not in position). The distribution referred to is shown in fig. 3.5. Occupancy of the interface state is determined by the position of the Fermi level, as shown in this figure. Once again, we distinguish donor and acceptor states. Donor states are neutral when below the Fermi level, positively charged when above it. Acceptor states are neutral when above the Fermi level, negatively charged when below it. Of course, there is *some* temperature effect to the trap occupancy, and the full Fermi-Dirac function should be applied to get a precise value for $\Delta\ V_{\text{fb}}$. The "zero-temperature"approximation described above is usually employed in most calculations. For the donor distribution shown in fig. 3.5, the occupied state area density is

$$Q_{\text{it}} = q \int_{\varepsilon_{\text{f}}}^{\varepsilon_{\text{c}}} D_{\text{it}}(E)\ dE \tag{3.5}$$

$$\Delta\ V_{\text{fb}}^{\text{it}} = \frac{Q_{\text{it}}}{C_{\text{ox}}}. \tag{3.5a}$$

As a result of these defect charges, a typical $C - V$ plot may look like curve (b), (c), or (d) in fig. 3.6, where the ideal $C - V$ plot is shown as curve (a).

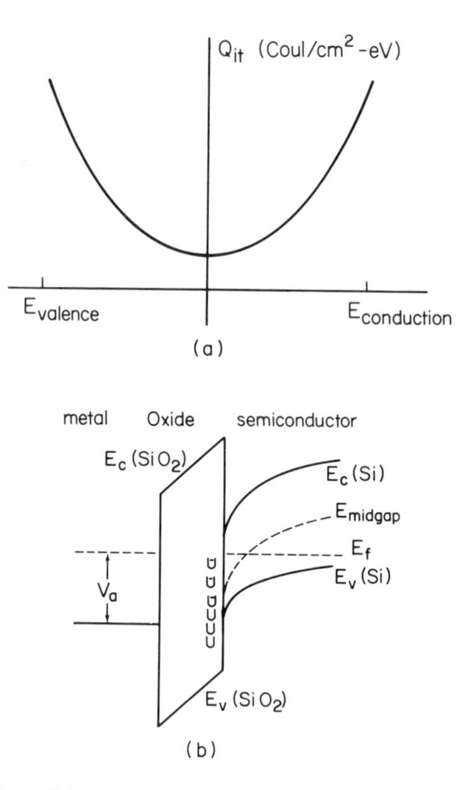

Fig. 3.5. Illustration of the interface state distribution in the band gap and its effect on band bending.

Curve (b) is a stretched-out ideal $C-V$ plot. The stretching is associated with the interface traps whose occupancy depends on the applied gate voltage. Curve (c) shows the experimental curve that has shifted parallel to the ideal curve due to a combination of one, two, or all of the following: work function difference ϕ_{ms}, Q_{m}, Q_{ot}, Q_{it}, and Q_{f}. As discussed in Chapter 1, we usually derive the "flat-band voltage" as a capacitance reference point. At flat band, the band-edge versus position diagram appears as in fig. 3.7. The mathematical expressions for evaluating C_{FB} have been given in eqns. (1.23)-(1.26). The position of the flat-band capacitance is given by

$$V_{\mathrm{FB}} = \phi_{\mathrm{ms}} - \frac{Q}{C_{\mathrm{ox}}}, \tag{3.6}$$

where

$$Q = Q_{\mathrm{m}} + Q_{\mathrm{ot}}^{\mathrm{eff}} + Q_{\mathrm{it}} + Q_{\mathrm{f}}. \tag{3.6a}$$

Sign conventions are given in Section 4 of Chapter 1. As a rule, negative charge in the oxide shifts the $C-V$ plots positively (to the right); positive oxide charge shifts the curve to the left. For ideal conditions, $V_{\mathrm{FB}} = \phi_{\mathrm{ms}}$

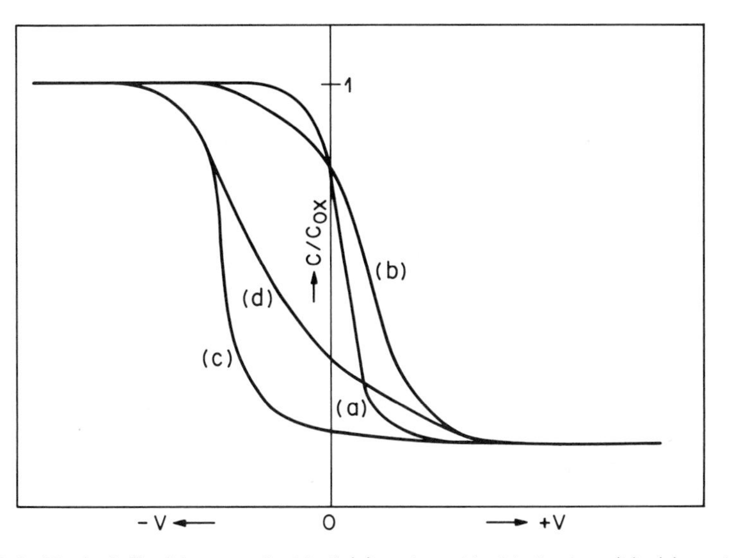

Fig. 3.6. Typical $C - V$ curves for ideal (a) and nonideal behaviors (b), (c), and (d) (see text).

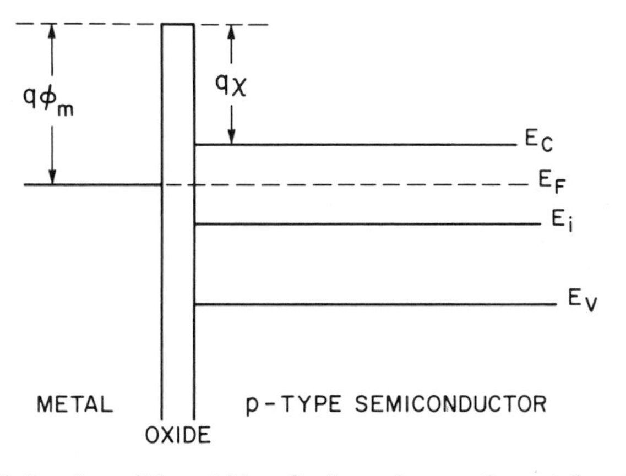

Fig. 3.7. Flatband condition at $V = O$. ϕ_m and χ are the metal work function and n-semiconductor electron affinity with respect to the vacuum level.

as reflected in the ideal $C - V$ curve (fig. 3.6a). Finally, fig. 3.6d shows the combined effect of the interface states and the flat-band shift.

Problem 3.2 *For a $\phi_{ms} = 0.7$ volts, calculate the flat band voltage for a 100-nm thick SiO_2 layer with a fixed charge density of 5×10^{10} charges per cm^2 and a bulk charge density of 2×10^{10} charges per cm^2.*

Solution: Using eq. (3.6):

$$
\begin{aligned}
V_{\mathrm{FB}} &= \phi_{\mathrm{ms}} - \frac{Q_{\mathrm{f}} + Q_{\mathrm{B}}}{C_{\mathrm{ox}}} \\
&= \phi_{\mathrm{ms}} - \frac{Q_{\mathrm{f}} + Q_{\mathrm{B}}}{\epsilon_{\mathrm{ox}}} t_{\mathrm{ox}} \\
\epsilon_{\mathrm{ox}} &= K_{\mathrm{ox}} \epsilon_0 \\
&= 3.45 \times 10^{-13} \text{ farad/cm} \\
V_{\mathrm{FB}} &= \left(0.7 - \frac{(5 \times 10^{10} + 2 \times 10^{10}) q}{3.45 \times 10^{-13}} 10^{-5} \right) \text{ volts} \\
&= 0.7 - 0.324 \\
&= 0.376V.
\end{aligned}
$$

Other important parameters include (a) the threshold voltage V_{T} of the MOS capacitor, which is defined as the voltage at which inversion sets in (see fig. 3.3) and (b) the semiconductor doping density, N_{B}, which is obtained from the minimum capacitance. The relevant equation for the latter is [19]

$$
\log_{10} N_{\mathrm{B}} = 30.38759 + 1.68278 \log_{10} C_{\mathrm{sc}} - 0.03177 (\log_{10} C_{\mathrm{sc}})^2 \tag{3.7}
$$

where

$$
C_{\mathrm{sc}} = \frac{C_{\mathrm{ox}} C_{\mathrm{min}} \text{ (HF)}}{C_{\mathrm{ox}} - C_{\mathrm{min}} \text{ (HF)}} (F \cdot \mathrm{cm}^{-2}) \tag{3.8a}
$$

and N_{B} is given in cm^{-3}.

Finally, the effect of mobile ion like Na^+ on the $C - V$ measurements is shown in fig. 3.8. Since the Na^+ is mobile, it moves with the applied voltage. The diffusion of Na^+ ions is aided by annealing the capacitor at high temperature (usually 200 - 300°C range) under applied bias. After this, the sample is cooled to room temperature and the $C - V$ characteristics are measured. The flat-band voltage shift, before and after bias-temperature aging, is a measure of the mobile ion concentration. Thus, fig. 3.8 shows, schematically, the effect of applying a positive bias-temperature aging on a MOS capacitor contaminated with mobile charge like Na^+. The curve shows a capacitor with a negligible change in the interface trap charge density and no major interfacial charge uniformity. Positive bias on the heated capacitor moves the mobile charge to the semiconductor-insulator interface, causing a shift in the $C - V$ plot. If the capacitor with a condition shown in fig. 3.8a is now heated again with gate and substrate shorted, the mobile Na^+ ions move back to the metal gate electrode, and the original $C - V$ characteristic (b) will be obtained when the measurement is made at room temperature. Thus, the mobile ions can move back and forth between the metal electrode and the semiconductor-insulator interface as the biasing conditions change. From the shift in the curve, the mobile ion charge density can be calculated using eq. (3.6).

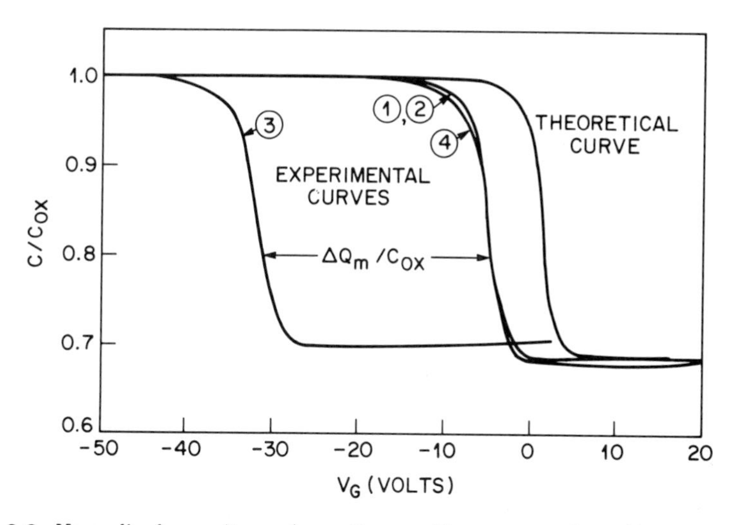

Fig. 3.8. Normalized capacitance (capacitance with respect to the oxide capacitance) as a function of the gate bias for a p-type silicon- and sodium-contaminated oxide. Theoretical curve with no contamination is compared with curves 1,2,3, and 4, which are measured as is, after heating for 5 minutes at 150°C with a $-10V$ bias, after heating for 5 minutes at 150°C with $+10V$ bias, and after final heat treatment for 5 minutes at 150°C with gate shorted to the substrate. From Snow et al.[20].

3.2.2 Current-Voltage (I − V) Characteristics

Ideally, the oxide film should truly be an insulator; it should have infinite resistivity. SiO_2 films, though, do have a finite resistivity, which ranges from 10^{14} to $10^{16}\Omega$-cm at 25°C (see table 3.1). Contamination can lower the resistivity. Thus, noticeable conduction has been observed in the oxide with electric fields as small as 1-2 $\times$ 10^6, V/cm across the oxide. There are several conduction mechanisms that can be responsible for current flow [14]. For impurity-free oxides, Frenkel-Poole emission [27] and Fowler-Nordheim tunneling [28] are the most common conduction mechanisms. In the former type, the current density is given by [25]

$$J = C_2 E \exp\left[-q\left(\phi_{\mathrm{B}} - \sqrt{\frac{qE}{\pi\epsilon_{\mathrm{ox}}}}\right)\Big/ \mathrm{k}T\right]. \tag{3.9}$$

For the Fowler-Nordheim tunneling, the current density is given by [24]

$$J = C_2 E^2 \exp\left(-\frac{E_{\mathrm{o}}}{E}\right). \tag{3.10}$$

In eqs. (3.9) and (3.10), C_1, C_2, E_o are constants, q is electronic charge, ϕ_{B} is the barrier height, E is electric field, k is the Boltzman's constant,

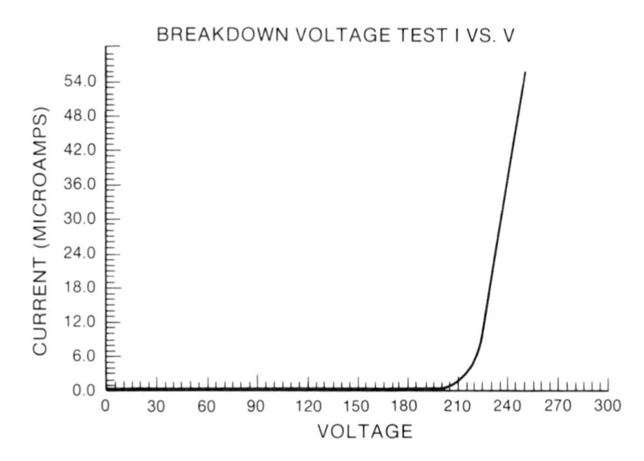

Fig. 3.9. A typical $I - V$ characteristic for Al on a SiO_2 MOS capacitor.

and T is the temperature in degrees Kelvin. Note that Frenkel-Poole-type conduction is associated with field-enhanced excitation of trapped electrons into the conduction band. It is observed only in heavily-damaged insulators. Fowler-Nordheim conduction is associated with electrons tunneling from the metal Fermi level into the oxide conduction band. More complicated tunneling mechanisms are possible. For example, charge may tunnel from the semiconductor conduction or valence band into an oxide trap. This charge may then tunnel into one of the insulator bands. It should be pointed out that the above mechanisms are not necessarily independent of each other. One may dominate over the other depending on measurement conditions like applied voltage and temperature.

Experimentally, one can determine $I - V$ or $J - E$ curves as a function of temperature. Since the Fowler-Nordheim tunneling mechanism is essentially independent of temperature [see eq. (3.10)], it is easy to distinguish from the Frenkel-Poole type, which has a temperature dependence. A typical $I - V$ plot for thin SiO_2 film with Al as the metal gate is shown in fig. 3.9.

3.2.2.1 Dielectric Strength of the Oxide

The quality of the thermally grown oxide (or any other insulator) is normally characterized by measuring the so-called dielectric or breakdown strength of the film. Normally, a large number of capacitors, fabricated directly on the oxide, are subjected to increasing DC bias across the insulator and current is measured. Capacitors that pass more current than a preset amount, usually 1-2 μA, are considered failed. The data is plotted on a log-normal paper where cumulative percent failures are plotted on the abscissa and the corresponding voltages are plotted on the ordinate. For a good oxide, practically all capacitors will fail when subjected to a field of about 10×10^6 V/cm or more across the oxide. Figure 3.10 shows two distributions: (a) for a

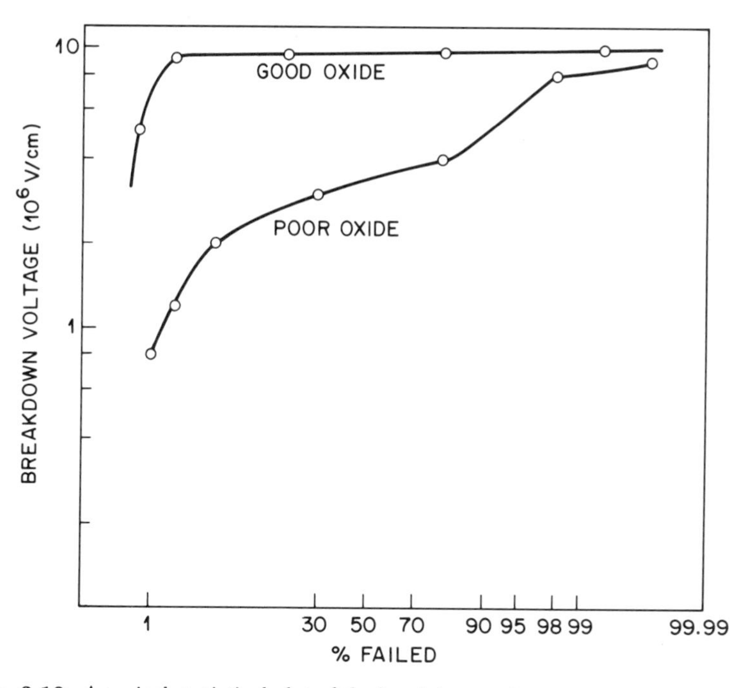

Fig. 3.10. A typical statistical plot of the breakdown voltage versus percent capacitors that failed at a given voltage.

good oxide and (b) for a poor oxide. Normally, three quantities are obtained from such plots: E_{50}, the field, at which 50% capacitors have failed; a scale parameter, σ, defined as

$$\sigma = \ln\left(\frac{E_{50}}{E_{16}}\right)$$

where E_{16} is the field at which 16% capacitors have failed; and freak population, f, which represents the early failure. For example, E_{50}, σ, and f for the two curves in fig. 3.10 are 10, 0.02, and 1; and 3.4, 0.35, and > 8, respectively. It is obvious that the highest E_{50} and the lowest σ and f are desirable.

3.2.3 Oxide Thickness Measurements

Oxide thickness can be measured by several techniques, such as the weight gain method, ellipsometry, multiple-beam interferometry, the capacitance method, the optical interference method, and by using calibrated color charts, as discussed by Nicollian and Brews [29]. The technique of ellipsometry is discussed in depth in Chapter 7.

3.2.4 Other Characterization Techniques

Among other oxide characterization techniques are the DLTS (deep-level transient spectroscopy), infrared absorption measurements, etch-rate measurements (etching performed in hydrofluoric acid solution), and chemical analyses. These techniques are helpful in identifying problems associated with poor-quality oxides that result in poor $C-V$, $I-V$, and/or dielectric breakdown characteristics.

3.3 Oxidation Kinetics

Thermal oxidation of silicon is carried out in oxygen or water vapor at temperatures generally in the range of $600-1250°$ C. The chemical reaction between the solid (s) and gas (g)

$$\mathrm{Si(s)} + O_2\mathrm{(g)} \longrightarrow \mathrm{SiO_2(s)}$$

or

$$\mathrm{Si(s)} + 2\mathrm{H_2O(g)} \longrightarrow \mathrm{SiO_2\ (s)}\ + 2\mathrm{H_2(g)}$$

occurs, leading to oxide formation at the surface. There are two stages of oxidation, as shown in fig. 3.11. At time $t = 0$, oxidation occurs at the bare silicon surface (fig. 3.11a) or at a surface that only has a native oxide on it. The thickness of the native oxide, should it be present, will lie in the range of 0.5-2.0 nm depending upon the time of exposure to atmosphere after a chemical clean. After a time $t > 0$, i.e., when an oxide layer has formed on the silicon surface, one can envisage the oxidation to continue by one of the following two possible mechanisms: a) oxidant diffuses through the growing oxide to the silicon-SiO_2 interface or b) silicon diffuses through the growing oxide to the oxide-gas interface. When the oxidant is the diffusing species, the reaction occurs at the silicon-oxide interface. When silicon is the diffusing species, the reaction will occur at the SiO_2-gas interface. The kinetics in either case will be determined by the availability of the diffusing species to the reaction interface and on the reaction at that interface.

A large number of experiments have been carried out in the past to identify the diffusing species and to determine the kinetics [4]. It is found that oxidation always occurs at the silicon-SiO_2 interface and oxidant species diffuse to this interface. These two phenomena occur in series. Specifically, diffusion of the oxidant species to the Si-SiO_2 interface precedes the reaction at the interface except at oxidation time $t = 0$. At any time greater than zero, the slower of the two phenomena will control the kinetics. In simple terms, at $t = 0$ the reaction at the Si-gas interface controls the kinetics. As the oxide thickness increases, the diffusion through the oxide starts playing a

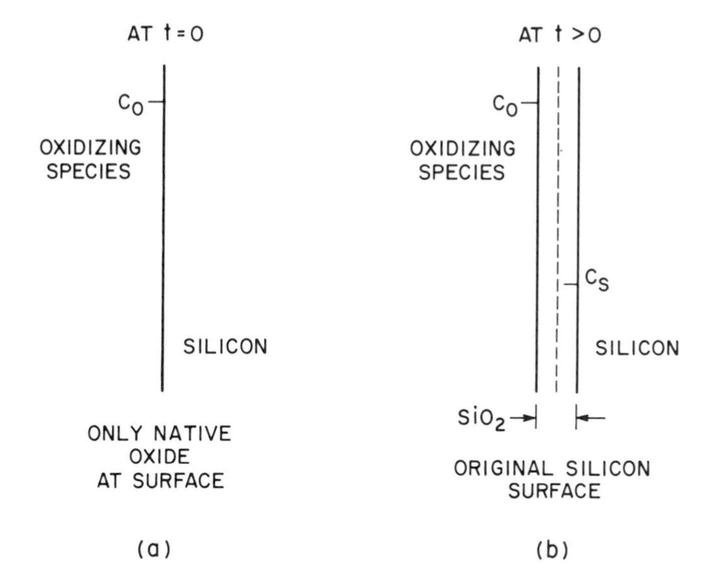

Fig. 3.11. Schematic representation of the silicon oxidation process (a) before oxidation and (b) after oxidation. C_O and C_s represent oxidant concentrations as defined in the text. Note that silicon-oxide interface moves in the silicon as oxidation proceeds.

role in determining the kinetics, since diffusion is found to be the slower process. At sufficiently large oxide thickness (approximately a few hundreds of angstroms), the diffusion through the oxide controls the kinetics completely.

Let us now use the Deal-Grove approach to derive the oxidation rates [30]. If C_o and C_s represent the concentration of the oxidant species in the silicon at the SiO_2-gas and Si=SiO_2 interfaces (see fig. 3.11b), respectively, then the flux at the SiO_2-gas interface can be approximated to be

$$F_{\mathrm{SiO_2-gas}} = \frac{D(C_o - C_s)}{x}, \tag{3.11}$$

where x is the oxide thickness and D is the diffusivity of oxygen in the SiO_2 film. Oxidizing species reacting at the Si=SiO_2 interface will react to form SiO_2. Assuming the reaction rate is proportional to the concentration of oxidizing species, the flux will be

$$F_{\mathrm{Si-SiO_2}} = kC_s. \tag{3.11a}$$

At steady state, two fluxes will be equal. This allows us to eliminate C_S from eqs. (3.11) and (3.11a) such that

$$F = F_{\mathrm{Si-SiO_2}} = F_{\mathrm{SiO_2-gas}} = \frac{DC_o}{(x + D/k)}. \tag{3.11b}$$

This steady-state flux is also given by

$$F = C_{\mathrm{I}} \frac{dx}{dt}, \tag{3.11c}$$

where dx/dt is the growth rate of the oxide thickness and C_{I} is the number of the molecules of the oxidizing species that are incorporated per unit volume of the oxide. $C_{\mathrm{I}} = 2.2 \times 10^{22}\ \mathrm{cm}^{-3}$ for molecular oxygen and $4.4 \times 10^{22}\ \mathrm{cm}^{-3}$ for steam oxidations. To understand why the C_{I} values differ by factor of two, consider the reaction equations given at the start of this section. From these equations, it is evident that a single molecule of oxygen will give rise to SiO_2, while two molecules of H_2O are required to form a single SiO_2 molecule. From eqs. (3.11b) and (3.11c) one obtains

$$\frac{dx}{dt} = \frac{D(\frac{C_{\mathrm{o}}}{C_{\mathrm{I}}})}{(x + \frac{D}{k})}. \tag{3.11d}$$

By solving equation (3.11d) for $x = x_{\mathrm{o}}$ at $t = 0$ one finds

$$x^2 + Ax = B(t + \tau_{\mathrm{o}}) \tag{3.11e}$$

where $A = 2D/k,\ B = 2DC_{\mathrm{o}}/C_{\mathrm{I}}$, and $B/A = kC_{\mathrm{o}}/C_{\mathrm{I}}$,

$$\tau_{\mathrm{o}} = \frac{(x_{\mathrm{o}}^2 + 2Dx_{\mathrm{o}}/k)C_{\mathrm{I}}}{2DC_{\mathrm{o}}},$$

x_o is the thickness of the oxide film at the start of the oxidation or when $t = 0$.

Equation (3.11e) explains the oxide growth very successfully. In this equation, A and B are oxidation constants such that B/A is a linear rate constant and B is a parabolic rate constant. As can be seen, linear oxidation kinetics describe the oxidation process when x is small. At large thicknesses, parabolic oxidation kinetics are dominant. B/A is related to the chemical surface-raction rate constant k for the reaction at the Si-SiO_2 interface [see eq. (3.11a)], the oxidant concentration, C_{o}, in the oxide in equilibrium with the partial pressure in the bulk of the oxidant gas, and the number, C_{I}, of oxidant molecules incorporated in a unit volume of the oxide. On the other hand, B has been shown to be related to the last two quantities, C_{o} and C_{I}, defined above, and the diffusion coefficient of the oxidant species in the oxide.

Table 3.4 lists the values of B and A for steam and oxygen oxidation of (100) and (111) orientation silicon. Also listed in this table are the activation energies for the linear and parabolic rate constants. Note that the values

TABLE 3.4: Oxidation constants for thermal oxidation of Si[a]

Oxidation Condition	Substrate Orientation	Temp. (°C)	A (μm)	B (μm^2/s)	B/A (μm/s)	Arrhenius-type temperature dependence[b]
Dry O_2	(111)	700	1.19	9.3×10^{-8}	7.8×10^{-8}	$B/A = 1.73 \times 10^3 \exp(-2.0\text{eV}/kT)$
0.1013 MPa		900	0.25	1.1×10^{-6}	4.4×10^{-6}	$B = 0.214 \exp(-1.23\text{eV}/kT)$
760 torr)		1100	0.082	6.6×10^{-6}	8.1×10^{-5}	
	(100)	700	2.02	9.3×10^{-8}	4.6×10^{-6}	$B/A = 1.03 \times 10^3 \exp(-2.0\text{eV}/kT)$
		900	0.42	1.1×10^{-6}	2.6×10^{-6}	$B = 0.214 \exp(-1.23\text{eV}/kT)$
		1100	0.14	6.6×10^{-6}	4.8×10^{-5}	
Wet O_2	(111)	700	8.0	8.8×10^{-6}	1.1×10^{-6}	$B/A = 4.53 \times 10^4 \exp(-2.05\text{eV}/kT)$
pyrogenic		900	0.60	4.4×10^{-5}	7.3×10^{-5}	$B = 0.107 \exp(-0.79\text{eV}/kT)$
team)		1100	0.10	1.4×10^{-4}	1.4×10^{-3}	
.0853 MPa	(100)	700	13.54	8.8×10^{-6}	6.5×10^{-7}	$B/A = 2.70 \times 10^4 \exp(-2.05\text{eV}/kT)$
640 torr)		900	1.02	4.4×10^{-5}	4.3×10^{-5}	$B = 0.107 \exp(-0.79\text{eV}/kT)$
		1100	0.17	1.4×10^{-4}	8.3×10^{-4}	

[a] From reference cited by Fair [31].
[b] Units of B/A and B are μm/s and μm^2/s, respectively.

listed in table 3.4 are for lightly doped silicon ($\sim 10^{17}$ cm^{-3} or less). Heavier doping affects the oxidation rates, as discussed in Section 3.3.2.

Most commonly, oxidation of silicon is carried out in dry oxygen, in oxygen or inert gas (such as nitrogen or argon) that is bubbled through water, or in steam. The temperature dependence of B for dry and wet oxidations yield activation energies that correspond to activation energies of diffusion of oxygen and water molecules through quartz. This provides evidence to assert that O_2 is the oxidant in dry oxidation and H_2O is the oxidant in steam oxidation (as opposed to atomic oxygen or ionic species). An excellent discussion of this subject is given in references 4 and 30.

The Deal-Grove model is unable to explain the oxidation kinetics for the early oxidation periods, specifically, for oxide thicknesses less than or equal to 30 nm. Adams et al. [32] determined oxidation kinetics in this range and pointed out that the experimental results could be fitted to any one of the six different types of thickness-time relationships. Since then, a considerable amount of work has been published [32-37] and some new models have been proposed [38-42]. Very recently, an "oxygen-diffused zone in silicon" model [43] (described below) has been proposed that mathematically explains the complete oxide thickness-time curve. The Deal-Grove model [30], discussed above, emerges as a special case of this zone model at larger thicknesses.

3.3.1 Oxygen-Diffused Zone in Silicon

This oxidation model takes into account the previously ignored diffusion of oxidizing species into the silicon substrate during the initial exposure to such species [43]. The diffusion into the substrate, in addition to interfacial oxidation, occurs only when a) diffusitivity of oxidizing species in silicon is high and b) the partial pressure (concentration) of the oxidizing species at the $Si=SiO_2$ interface is also high. The latter occurs only when the oxide layer is ultrathin. Thus, for ultrathin oxides, the diffusion resistance offered by the film is very low, leading to a very high flux of oxidizing species across the oxide, since the partial pressure of the oxidizing species at the interface is high. A homogeneous reaction occurs in the oxygen-diffused zone in silicon instead of occurring just at the interface, leading to an enhancement in the oxidation rate. As the oxide thickness increases with time, the partial pressure of the oxidizing species at the substrate-oxide interface falls rapidly (because of the increased diffusion resistance offered by the growing oxide film) to a level where all of the oxidizing species is consumed by interfacial oxidation and no diffusion occurs into the substrate. This model bears similarity to the diffusion and kinetic models for solid-gas reactions, particularly to the unreacted core shrinking model [44-45]. In this last solid-gas reaction model, the solid is considered practically impervious to the gaseous reactants and the reaction occurs at the surface of the solid and in the porous product layer [44]. Incorporation of this porous layer leads to oxide growth relationships that explain the observed oxidation kinetics and other related phenomena discussed in the following subsections.

3.3.2 Dopant Concentration Dependence of the Oxidation Kinetics

The oxidation rate of silicon is greatly enhanced when silicon is doped heavily. Figures 3.12 and 3.13 show the normalized oxidation rates for the dry and wet oxygen oxidation of silicon heavily doped with boron or phosphorus [46]. It is clear from these figures that dopants in concentrations greater than 5×10^{18} cm^{-3} enhance the oxidation rate. For phosphorus-doped silicon, Ho et al. [47] have demonstrated that only the linear oxidation rate constant (B/A) is influenced significantly by the substrate doping concentration. Figure 3.14 shows their results. Fair [31] has attempted to rationalize and explain these results on the basis of point defect (vacancies and interstitials) interactions in silicon. The results can also be explained by referring to the role of the dopants that get incorporated into the oxide and therefore affect the diffusion rate of oxidant species through the oxide. In the case of heavily boron-doped silicon, boron segregates preferentially in the oxide (see Section 3.6). Boron, like sodium, possibly weakens the bond structure of the oxide and allows faster oxidant diffusion through the oxide [4]. On the other hand, phosphorus is retained in silicon during the oxidation. Thus, oxygen diffusion through the oxide is not affected. This means that only the linear oxidation rate will be affected by the presence of phosphorus in silicon at the surface, as is observed in fig. 3.14.

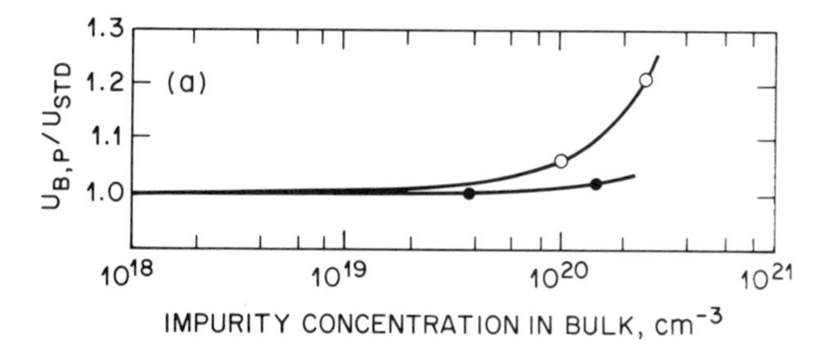

Fig. 3.12. Dependence of normalized oxidation rate (U) in dry oxygen at 1200°C on dopant concentration in the bulk Si. Open circles represent boron and black circles represent phosphorus. The value U_{STD} is the oxidation rate for a dopant concentration of 10^{16} per cm^3. From Deal and Sklar [46], Reprinted by permission of the publisher, The Electrochemical Society, Inc.

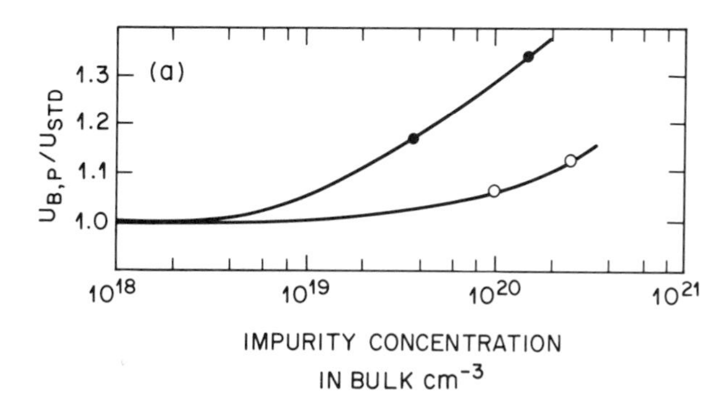

Fig. 3.13. Dependence of normalized oxidation rate (U), in wet oxygen (95°C H_2O) at 920°C, on dopant concentration in the bulk Si. Open circles represent boron and black circles represent phosphorus. From Deal and Sklar [46], Reprinted by permission of the publisher, The Electrochemical Society, Inc.

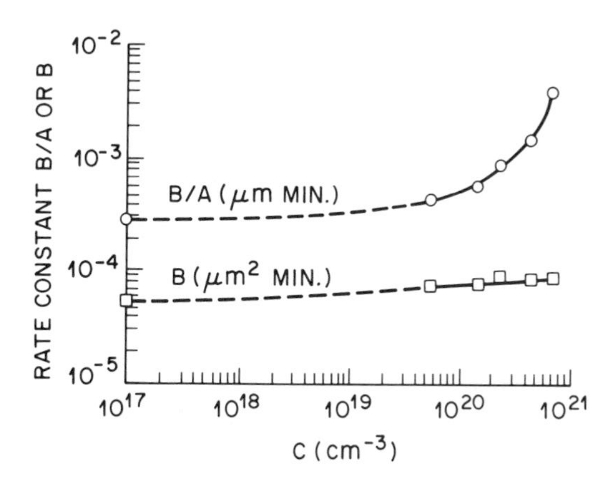

Fig. 3.14. Rate constants B/A or B versus substrate phosphorus doping concentration for oxidations carried out at 900°C. From Ho et al. [47], Reprinted by permission of the publisher, The Electrochemical Society, Inc.

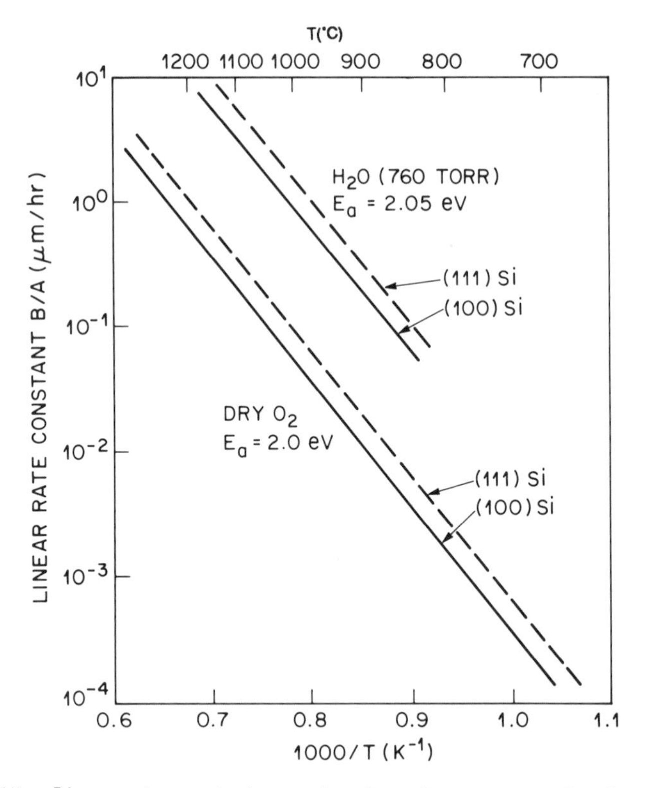

Fig. 3.15. Linear rate constant as a function of temperature for dry and steam oxidations and (100) and (111) oriented silicon surfaces. From Deal and Grove [30]. Reprinted by permission of the publisher, The Electrochemical Society, Inc.

3.3.3 Substrate Orientation Dependence of the Oxidation Kinetics

The oxidation rate of (111) silicon substrates is higher than that of (100) substrates oxidized under identical conditions of substrate doping and oxidation. The oxidation rate enhancement is due to the increase in the linear rate constant B/A, as shown in fig. 3.15. Since the parabolic rate constant B is limited by the diffusion of oxidant molecules through the growing oxide, B does not depend on the substrate orientation. Orientation dependence is discussed in considerable detail by Nicollian and Brews [4]. (111) surfaces have a higher density of silicon atoms per unit area and, therefore, provide a larger concentration of silicon atoms for reaction with oxidant species, thus increasing the reaction rate and B/A. This interpretation is supported by the data shown in table 3.2. The activation energy is the same for the oxidation of (111) or (100) substrates. Only the pre-exponential factor C_{BA} (111), however, is larger than C_{BA} (100) by a factor of approximately 1.7.

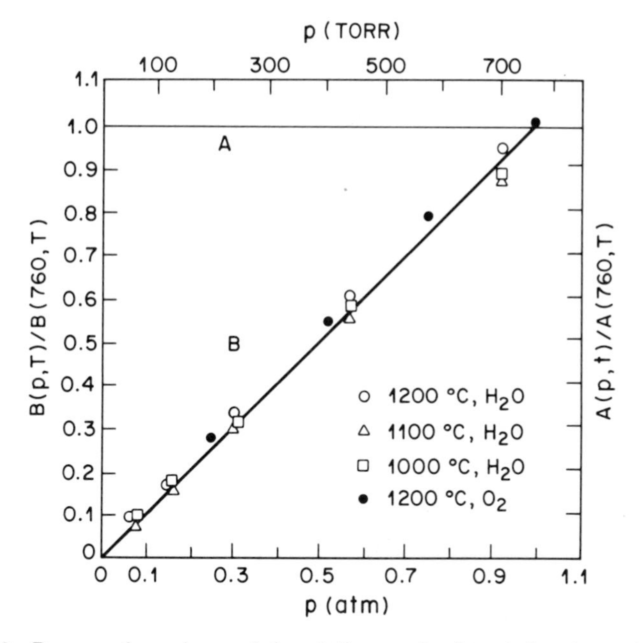

Fig. 3.16. Pressure dependence of A and B normalized to their value at 760 torr at the same temperature. From Deal and Grove [30] and Flint [48]. Reprinted by permission of the publisher, The Electrochemical Society, Inc.

3.3.4 Oxidant Pressure Dependence of the Oxidation Kinetics

The parabolic rate constant B is directly proportional to C_o. C_o represents the oxidant concentration in the oxide in equilibrium with the partial pressure P_{ox} in the bulk of the oxidant gas. C_o and P_{ox} are related

$$C_o = k_h P_{ox}, \tag{3.12}$$

where k_h is Henry's law constant [21]. Figures 3.16 and 3.17 show the pressure dependence of B at subatmospheric and above atmospheric pressures [47,48]. Since the constant A is independent of P_{ox}, it is found to be independent of pressure as expected.

The pressure dependence of oxidation rate of silicon offers means to both (a) reduce the oxidation rate for applications where very thin dielectric films of uniform thickness across the silicon substrate surface are necessary and (b) enhance the oxidation rate for applications where rather thick films are to be grown but at lower temperatures. Thus, reduced oxygen partial-pressure oxidations can be carried out for thin gate oxide growth and high-pressure oxidations can be carried out to grow thick field oxides. In both cases, the oxidations can be carried out at temperatures that are low and would not significantly affect previously diffused junction depth profiles. As

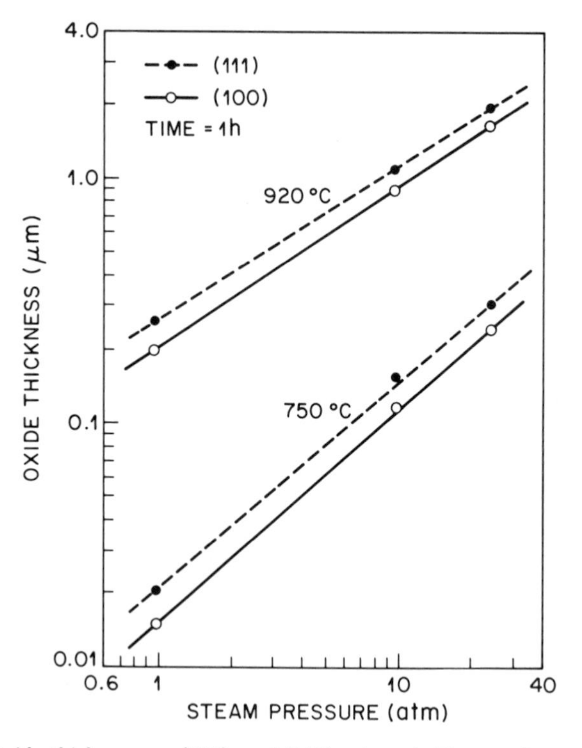

Fig. 3.17. Oxide thickness, on (100)- and (111)-oriented silicon surfaces, as a function of the pressure of steam in the furnace. From Katz et al. [50].

discussed later (Section 3.6.3), low-temperature and high-pressure oxidations could also be very useful in eliminating oxidation-induced stacking faults (OISF). OISF growth is a thermally activated process with a significantly large activation energy, permitting defect-free oxide growth at temperatures of 800°C or lower [49-51].

3.3.5 Effect of Chlorine on the Oxidation Kinetics

Presence of chlorine-containing species during silicon oxidation in dry oxygen has been found to improve the quality of the $Si{=}SiO_2$ interface. Oxides grown in chlorine-containing media were found to have little or no sodium, lower interface trap density, and superior electrical breakdown strength [52]. Also, the presence of chlorine-containing species in oxidizing media is effective in reducing OISF density and growth [53]. These advantages led to more detailed studies of the effect of chlorine on oxidation kinetics. Hess and Deal [54] showed that both parabolic and linear rate constants are increased by the additions of chlorine in the form of HCl gas. Figure 3.18 shows these results. The enhancement is perhaps associated with two factors. The first

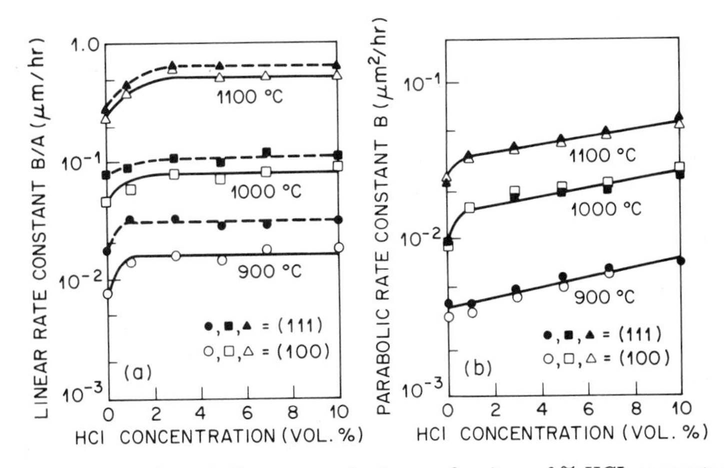

Fig. 3.18. Linear and parabolic rate constants as a funciton of % HCL concentration for (111)- and (100)-oriented silicon at different temperatures. From Hess and Deal [54], Reprinted by permission of the publisher, The Electrochemical Society, Inc.

of these is the generation of excess vacancies at the Si=SiO_2 interface due to the reaction of chlorine with silicon atoms [53]. Excess vacancies allow more silicon migration to the surface or more oxygen entrapment at the surface to occur. The second of these is chlorine incorporation into the oxide, causing the oxide lattice to expand and promote oxygen diffusion [30].

Chlorine can be introduced by using Cl_2, HCl, trichloroethylene, trichloroethane, carbon-tetrachloride, or any gaseous species that can produce chlorine without contaminating the oxide or the furnace. However, the concentration of the chlorine in the oxidizing gas should not exceed more than a few percent (the higher the oxidation temperature, the lower the chlorine concentration should be). At higher concentrations, excessive reaction with silicon produces etch pits in the silicon. The optimum concentration of HCl is about 2% by volume and depends on the oxidation temperature. The higher the temperature, the lower is the chlorine or HCl concentration [55]. Also, all chlorine-containing species react with the metals and one must be very careful or else corrosion of metallic parts of the oxidation system will result.

3.4 Oxidation Processes (Thermal)

Thermal oxidation can be carried out in any one of the following ways:

- Furnace oxidation at atmospheric, subatmospheric, or high pressures in oxygen, steam, or wet oxygen with or without chlorine-containing species can be used. Such oxidations could be assisted by the presence of plasma in the oxidizing medium.

- Rapid thermal oxidation (RTO) in similar media is also a possibility.

Alternatively, oxide films can be deposited using gas-phase chemical vapor deposition, as discussed in Chapter 7, and by anodic oxidation, as discussed in Section 3.5.

For any given method of oxidation, the silicon surface must be free of heavy metal, organic contaminants and the impurities. Silicon wafers, available from crystal and wafer producers, are generally in excellent condition and only require a chemical clean prior to oxidation. These cleans have been described at the end of Chapter 2. To avoid particulate contamination during storage of the cleaned wafers, cleaning is carried out just before oxidation. As soon as the wafers are cleaned, they are placed in the oxidation furnace.

3.4.1 Furnace Oxidations

A typical furnace setup is shown in fig. 3.19. It consists of a single-wall, high-purity, quartz tube that is separated from the furnace heating elements by a ceramic liner of high-purity alumina or silicon carbide. A double-wall quartz tube, with pure nitrogen flowing between the walls, is preferred, as this minimizes contamination (from the heating element and liner) of the oxide or the substrate during oxidation. Silicon wafers are loaded onto specially made quartz boats, which are pushed manually or mechanically into the furnace hot zone. A cross-sectional view of the arrangement is shown in fig. 3.20. Gases flow from the gas flow controllers in the back of the furnace, pass over the wafers, and exit from the open tube or capped tube outlets. Experience has shown that for the uniform oxide growth on a wafer surface and on several wafers in an oxidation lot, the gas flow in the tube should be fairly high. Flow rate should be such that viscous effects are at a minimum at all the solid surfaces in the tube so that the oxidation reaction is not limited by the availability of the oxidant species. A well-established practice is to maintain a flow rate of 1 cm/s across the tube. This translates into a volume flow rate of 11.5 liters per minute across a 150-mm internal diameter furnace tube, a size normally used for processing 100-mm wafers. Gas entering the furnace tube at this velocity will, however, expand due to high temperatures inside the furnace. Assuming the gas rises to the furnace temperature, say, of 1000°C, it will expand by nearly a factor of 4.25. Thus, the exit velocity of the gas will be approximately 4 times greater than the entering velocity, assuming that comparatively a very small amount is consumed during the oxidation of silicon. Thus, to reduce the usage of gas, velocities at the inlet is reduced. A variety of practices are followed in inserting and pulling the wafer boat in and out of the hot zone. The most preferred way is to insert the boat

Fig. 3.19. A typical oxidation furnace setup.

in an argon ambient (for temperatures $\geq 1000°C$) or in a nitrogen ambient (for temperatures $< 1000°C$). The wafers equilibrate in the inert gas for a few minutes. This allows the wafers to come to the temperature of oxidation without any oxidation during the heating cycle and also permits the surface to lose the volatile and fast diffusing surface impurities [56]. At sufficiently high temperatures ($> 1000°C$) even the native oxide will evaporate as silicon mono-oxide, leaving the silicon surface clean. In this case, silicon reduces SiO_2 by the $Si + SiO_2 \longrightarrow 2SiO$ reaction, SiO being highly volatile at such high temperatures. For this to occur, the gas, ambient must be completely oxygen free. Following this equilibration in inert gas, the desired oxidizing medium replaces the inert gas, and oxidation is allowed to proceed for the desired length of the time.

Although silicon wafers generally attain the furnace temperature quickly due to silicon's high thermal conductivity, temperature gradients can be generated in the wafer, leading to thermal stresses large enough to produce warping and crystallographic slip. This occurs at high temperatures and is generally associated with heat conduction by the heavy quartz (or silicon) boats that carry the wafers into the hot zone. Regions of the wafer in contact with the boat take longer to heat. This creates thermal gradients and stresses across the wafer. The effect is generally noticeable at temperatures $\geq 1000°C$. At such temperatures, the mass of the wafer can also contribute to warpage because wafers are generally seated in a vertical position in the

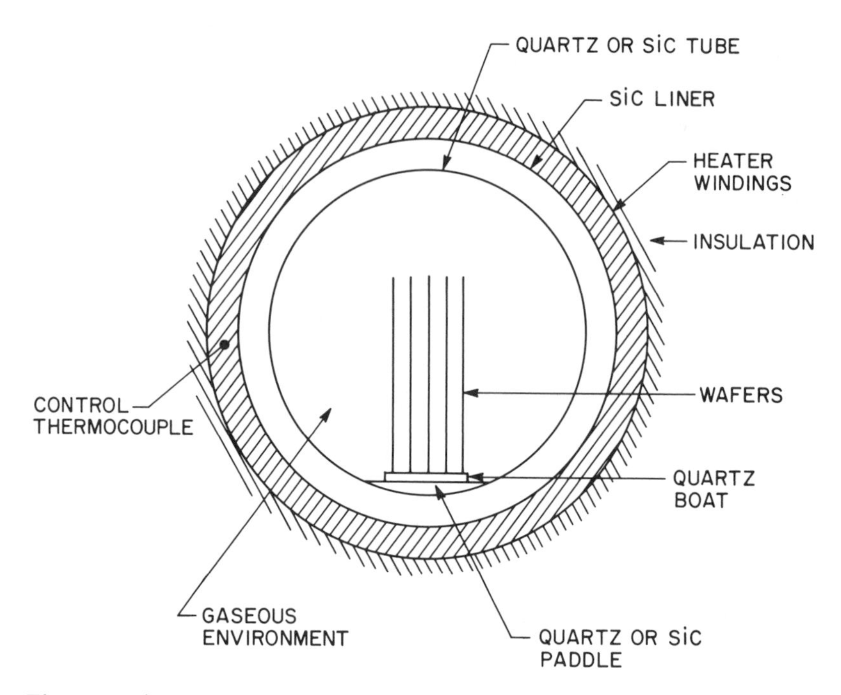

Fig. 3.20. A cross-sectional view of a typical furnace with wafers on a quartz boat.

boat and can "droop" in "process." Larger diameter and thinner wafers are more prone to such problems. To avoid these effects, boats that hold wafers at three or more spots, ramped heating, slow insertions and pull-outs, and thicker wafers are used.

Gases used for oxidation are dry oxygen, nitrogen-dry oxygen mixture, pure steam from boiling deionized water, or oxygen or argon bubbled through water maintained at a predetermined temperature. Dry oxygen is used to grow the highest quality thin oxides. In such cases, a small amount, generally 1-3% by volume, of a chlorine-containing species is added. Since chlorine readily attacks the silcon surface, the wafers are introduced into the furnace in a pure oxygen atmosphere. Chlorine, HCl, or TCE is introduced immediately afterward. The delay of a minute or two prior to the introduction of a chlorine-containing species permits chlorine-free oxidation of the surface. The oxide thus formed protects the silicon surface from direct chlorine chemical attack. For growing precisely controlled thin oxides $\leq$ 20 nm, dry oxidations at lower oxygen partial pressures, such as those created by diluting oxygen with nitrogen, are carried out [34]. Also, lowering the temperature at which oxidation takes place provides better control of the oxide thickness. In all cases, however, an argon anneal is carried out in the furnace by switching off the oxygen (or oxygen-chlorine) flow and replacing it with argon flow. The argon anneal is found to equilibrate the silicon-oxide interface and lowers the fixed charge (Q_f) density [57]. For very thin oxides ($\sim <$ 10 nm), the argon anneal should be carried out at temperatures less than 1000°C because of the silicon reduction of SiO_2 at higher temperatures that can lead to damage and defects in oxide [58].

Dry oxidations are slow, even with chlorinated oxygen. Thus, it is impractical to use dry oxygen oxidations to produce thick oxides ($\geq$ 200 nm) that are needed for field and masking applications. For thick oxide growth, steam oxidations are carried out. Two practices are followed. In one, steam is introduced into the oxidation furnace directly from a water boiler, where a constant water level is maintained through a constant water feed arrangement. In the other, water is maintained at a constant temperature of about 98°C and oxygen is bubbled through this water. Even in this case, a constant water level is maintained. In both cases, water level is maintained constant to avoid accidental drying of the bubbler. Wafers are introduced in the furnace in the steam ambient. After oxidation, steam flow is shut off and a short dry oxidation followed by argon anneal is carried out. This last dry oxidation process step results in hardening or densification of the steam oxide and also ensures a better Si-SiO_2 interface.

Figures 3.21 and 3.22 provide a few dry and wet oxygen oxidation curves that could be used to develop a tentative time-temperature oxidation schedule for a desired oxide thickness. Note that these figures refer to specific silicon wafers and oxidation conditions, as noted in the figure captions, and thus provide only an estimate for oxidations accomplished in other furnace tubes.

Oxidation rates intermediate between that of dry oxygen and steam can be achieved by controlling the water-vapor pressure in the gas that flows into the furnace. This is normally achieved by bubbling oxygen, nitrogen, or argon through water maintained at any temperature between 0 and 100°C. Water vapor pressure can be found in standard tables [59]. This technique has been successfully employed to produce thin oxides at very low temeratures (700-800°C) [60]. Figure 3.23 shows typical oxidation curves at different conditions. Excellent oxide films have been grown using this process.

High-pressure oxidation is carried out in specially made oxidation chambers or furnaces. It is much easier to carry out steam oxidations at high pressures than dry gas oxidations. A cross section of a commercial high-pressure steam oxidation system is shown in fig. 3.24 [61]. In the simplest approach, a reservoir of a known amount of water is inserted (with the wafers) in the furnace. Necessary clamps and flanges are tightened to secure the system, and the furnace is heated to the desired temperature. Steam, generated from the water, will cause a fixed and known pressure for oxidation. In spite of the fact that oxidation rates can be enhanced severalfold, thus allowing lower temperature oxidations, inherent problems associated with high-pressure generation and the safety, have made high-pressure oxidation more an academic interest than a practical process. Katz

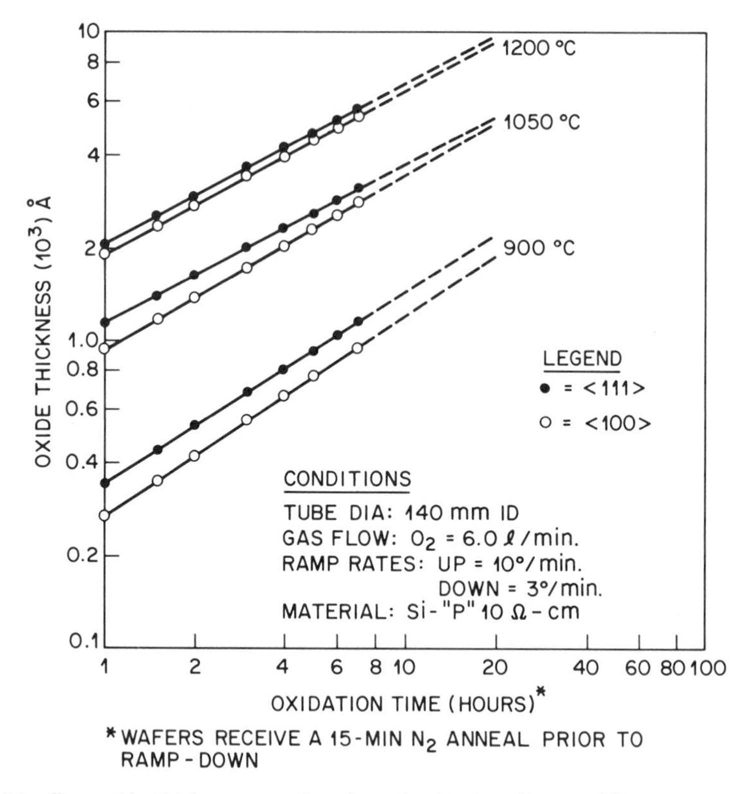

Fig. 3.21. Dry-oxide thickness as a function of oxidation time at different temperatures, for (111) and (100) wafers. Oxidation conditions are given. Note that wafers received a 15-minute nitrogen anneal prior to ramp-down to lower temperatures. From Beadle et al. [19].

and Howells [62] report that 2.026 MPa atmosphere, 725°C high-pressure oxides have higher densities, lower etch rates in hydrofluoric acid solutions, and higher refractive indexes. These effects, however, could be associated with low-temperature oxidation effects and not the high-pressure element of the method [63].

Plasma oxidation, where oxidation is carried out in an oxygen plasma excited by a high-frequency discharge of a DC electron source, offers the possibility of low-temperature oxidation with appreciable oxidation rates [63-67].

Chapter 11 discusses the rapid thermal oxidation processes in detail. RTO can be used for thin oxide growth. Excellent oxides have been grown using this method in a reasonable time without wafer distortion or diffusion spreading [68]. To summarize briefly, flash lamps or lasers can be used to provide an intense surface heating effect without significantly heating the wafer bulk. Figure 3.25 shows a plot of oxide thickness as a function of oxidation time

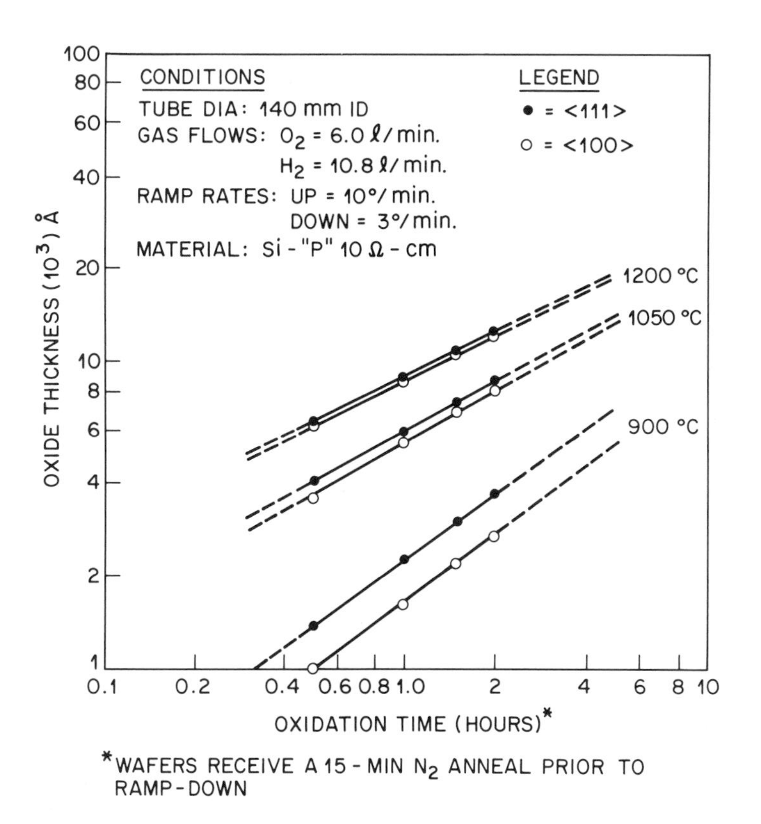

Fig. 3.22. Pyrogenic (steam) oxide thickness as a function of oxidation time at different temperatures for (111) and (100) wafers. Oxidation conditions are given. Note that wafers received a 15-minute nitrogen anneal prior to ramp-down to lower temperatures. From Beadle et al. [19].

at different temperatures. The curves are similar to those obtained for thin oxide growth by Adams et al [32], who oxidized silicon wafers in furnaces at reduced partial pressures of oxygen.

3.5 The Anodic Oxidation Process

Anodic oxidation of silicon can be carried out in a typical electrolytic environment containing oxidizing species. These species are electrically driven to a semiconductor surface by supplying a positive potential. However, such oxidations are self limiting. The grown oxide acts as a barrier against continued oxidation [69,70]. Thicknesses of oxides grown in this way do not exceed 100-200 nm. The oxide is generally porous and needs annealing to increase its density and has interface state trap densities that are usually an order of magnitude higher than thermally grown oxide. This perhaps is

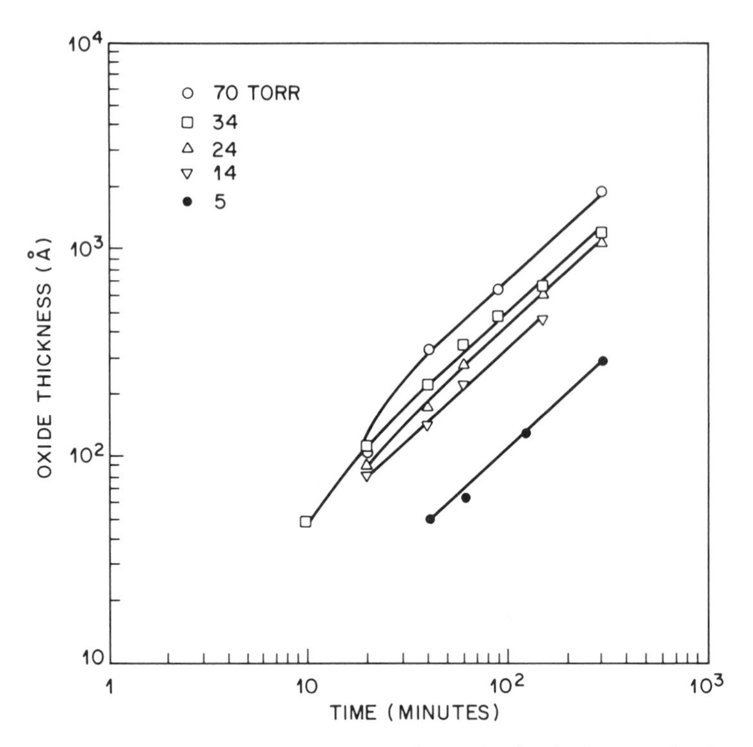

Fig. 3.23. Oxide thickness as a function of time for (100) silicon oxidized at 1000°C in various water vapor pressures (argon bubbling through water maintained at controlled temperatures to yield desired moisture level in the gas). From Beadle et al. [19].

associated with the impurities that get easily incorporated in anodic oxides. Very recently, an anodic oxidation process has been described that produces oxide films as good as those grown thermally for gate insulator applications [71]. The process employs a special galvanic cell. Ethylene glycol containing 0.04 M KNO_3 and approximately 50% (by volume) water is used as an electrolyte. Following the anodization, the oxide film is annealed in nitrogen at 500°C for 20 minutes.

In spite of the generally poor characteristics of the anodic oxide, there are some advantages in carrying out such oxidations:

- Since the oxidation is carried out at room temperature, there is no redistribution of dopant impurities.
- Anodic oxides are truly amorphous. Even the short-range order (evident in thermal oxides) is difficult to obtain.
- The process of oxidation is very simple, and thickness control is very easy and precise.
- The anodization process could be used to etch known amounts of silicon

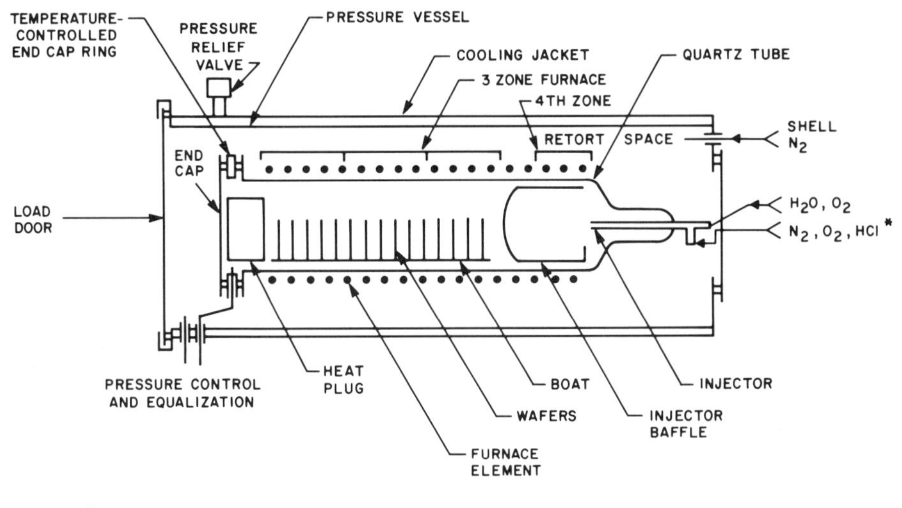

Fig. 3.24. A high-pressure steam oxidation system. From reference 61, by permission from Gasonics, Inc.

from flat surfaces. Selective etching of silicon could also be carried out by masking areas not to be etched.

- Anodization rates are very different for n- and p-type regions.

A p-type silicon oxidizes rather easily, where an n-type region may require special efforts, like light exposure, to induce anodic oxidation. This behavior allows creation of an isolation region between bipolar transistors by selectively oxidizing p regions. Ghandhi [72] explains the silicon anodization process in terms of the following reactions:

$$2H_2O \leftrightarrow 2H^+ + 2(OH^-)$$

$$Si + 2h^+ \rightarrow Si^{2+} + H_2$$

$$Si^{2+} + 2(OH^-) \rightarrow Si(OH)_2$$

$$Si(OH)_2 \rightarrow SiO_2 + H_2$$

such that the net reaction is

$$Si + 2h^+ + 2H_2O \rightarrow SiO_2 + 2H^+ + H_2.$$

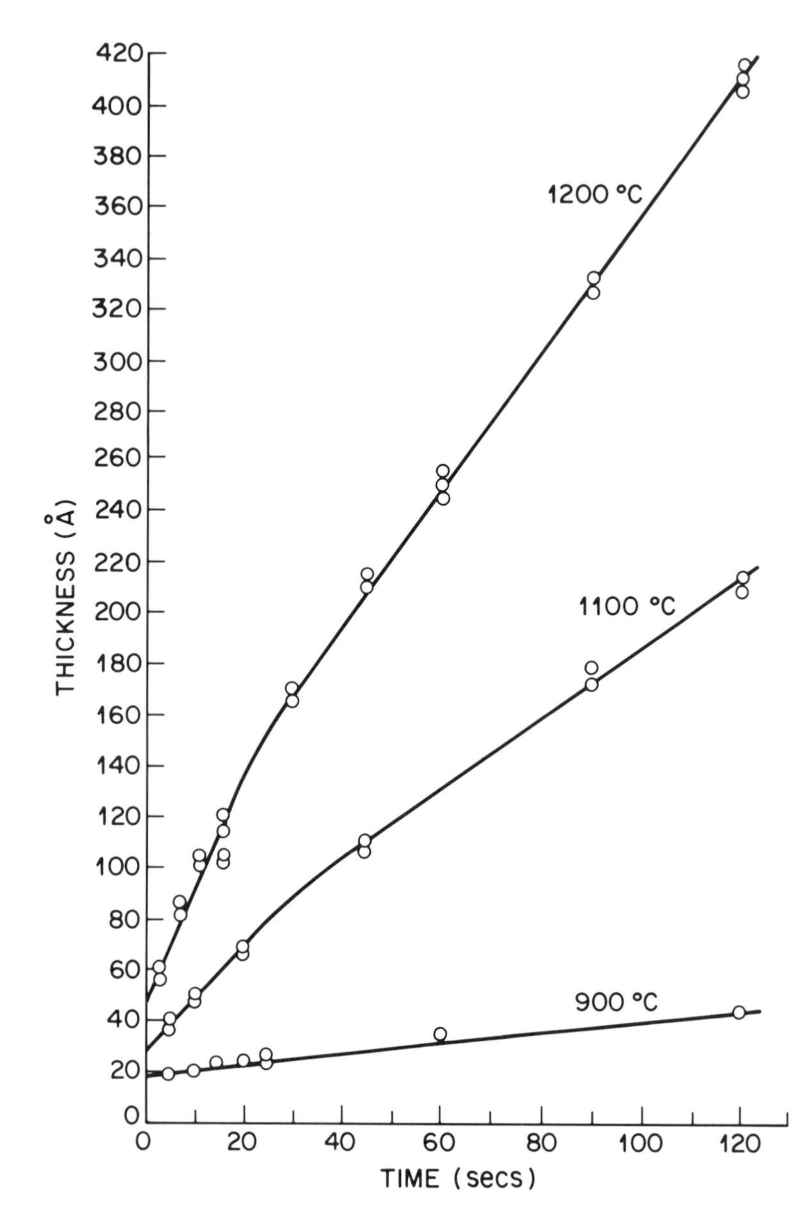

Fig. 3.25. Oxide thickness versus time for rapid thermal oxidation in dry oxygen at different temperatures.

Holes play an important role in promoting anodic oxidation. At the p-type silicon anode, it is very easy to provide holes during oxidation. For the n-type anode, a barrier to the flow of holes becomes established as soon as the voltage is applied. Thus, an external source of hole generation, such as optical illumination, becomes necessary.

3.6 Oxidation-Induced Defects

Three major types of defects are produced due to thermal oxidation: a) fixed charge and interface traps, b) strain in the silicon, and c) stacking faults in silicon. Of these, the first plays an important role in controlling the electrical properties of the MOS capacitor. Strain causes warpage in the wafer and has been associated with generation of dislocations at the edge of the oxides that are patterned on silicon substrates. Dislocations and stacking faults could become trapping centers for unwanted impurities that eventually cause leakages. Yet another effect associated with oxidation, dopant redistribution, will be discussed in Section 3.6.

3.6.1 Fixed Charges and Interface Traps

Charges at the semiconductor insulator interface lead to nonideal $C-V$ characteristics and changes in threshold voltages of MOS capacitors (see Section 3.2). Some of these charges, fixed charges and interface traps, are associated with the oxidation process.

The fixed charge, Q_f, is located in the oxide volume very close to the silicon surface. It is called *fixed*, since it cannot be charged or discharged and its density is dependent on the oxidation and annealing conditions. (100) silicon surfaces always have lower Q_f values compared to similarly oxidized (111) surfaces. The lowest value of Q_f is about 10^{10} per cm^2. Figure 3.26 shows a plot of Q_f as a function of the oxidation temperature and annealing. This triangle shaped plot, first made by Deal et al. [73], is commonly known as the Fairchild-triangle, named after the laboratory where it was first discovered. This triangle shows that lower Q_f is associated with high-temperature oxidations followed by short inert ambient anneals. In a separate experiment, Murarka [74] has shown that Q_f increases with increasing partial pressure of oxygen and Q_f is lower for chlorine-oxygen oxidations.

The origin of the fixed change Q_f is not clear. Q_f is generally associated with the dangling silicon bond (or the trivalent silicon) at the $Si{=}SiO_2$ interface. On the other hand, Nicollian and Brews [4] suggest "more probable may be the loss of an electron from a nonbridging oxygen center near the $Si{=}SiO_2$ interface to the silicon, making it a positively charged center and hence a candidate for oxide fixed charge."

Interface traps are electrically active centers with electronic states located at the $Si{=}SiO_2$ interface. These traps ionize at different potentials and result in stretched-out $C-V$ characteristics. Changes in temperature or ambient could ionize or deionize the traps, which could become positively or negatively charged or remain neutral. The individual trap levels are so close to each other in energy (i.e., their location in the band gap) that only a continuum can be inferred from experimental results. There are several

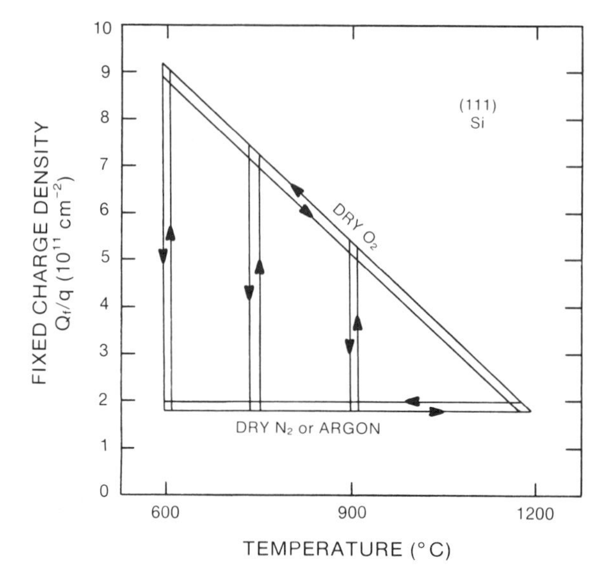

Fig. 3.26. Deal-Grove (or Fairchild) triangle showing relationship between the fixed change at the Si=SiO interface and various annealing treatments. From Beadle et al. [19].

models that try to explain the interface traps and their effect on $C-V$ characteristics [75]. Most likely the traps could be associated with chemical inhomogeneities at the $Si{=}SiO_2$ interface. This interface could be easily described as a region of great disorder — having a variety of bond structures, ions, defects, and so on. $C-V$ measurements invariably cannot isolate traps from other charges, especially Q_f. Thus, it has been very difficult to establish the desired understanding of interface traps. For an excellent discussion of the interface traps, their measurements, and models, Nicollian and Brews' text [4] is suggested.

3.6.2 Strain in Silicon

Thermal oxidation of silicon results in stress in the oxide and in the silicon due to a mismatch in thermal expansion coefficients of the growing SiO_2 film and the substrate. The linear thermal expansion coefficients of SiO_2 and Si are $5 \times 10^{-7}(°K)^{-1}$ and 3.0-4.5 $\times$ $10^{-6}(°K)^{-1}$. The room-temperature stress in the as-grown films is reported to be in the range of 2-4 $\times$ 10^9 dyn/cm^2, compressive [76-78]. At-temperature measurements of the oxide stress shows that stress decreases with temperature up to $\sim 950°C$ [77]. At higher temperatures, the stress was zero, indicating the existence of a viscous flow phenomenon at such temperatures [77].

Stress of the order of 10^9 dyn/cm^2 could easily lead to interface shear stresses that are higher than the critical stress for shear flow ($\sim 5 \times 10^7$

dyn/cm^2) in silicon [79, 80]. This indicates that dislocation generation is possible during oxidation or during post oxidation cooling. In fact, the stress in the oxide films has been known to cause warpage or bow in the wafers, especially when the oxide film is etched from the back side. Defect generation at stress centers, such as the edges of the patterned oxide on the silicon, causes enhanced oxidation. In the latter case, the effect is more pronounced when selective oxidation of silicon is carried out by protecting part of the surface by using a silicon nitride oxidation mask. This is further discussed in Section 7.6. Dislocations have been observed in the silicon at the oxide window edges [81]. Dislocation generation is aided by nucleating centers, by the heating and cooling cycles during the oxidation, and by the thermal gradients between the center and edges of the wafer.

3.6.3 Oxidation Induced Stacking Faults (OISF)

Thermal oxidation of silicon invariably leads to the generation of stacking faults in silicon. Figure 3.27 shows such stacking faults as seen by use of a) a TEM, b) an SEM, and c) an optical microscope. In the latter two cases, the silicon surface was etched in a special chemical etch to reveal the defects. All such faults are of extrinsic type. Their formation and growth is associated with silicon generation during oxidation [82]. These silicon atoms, generally interstitials in the silicon, precipitate out as an extra plane, the extrinsic stacking fault. Most of these OISF (up to 95%) on the surface of silicon have the same size. This size is found to depend on oxidation conditions. The nucleation of such faults occurs heterogeneously at the silicon surface. Contaminants, mechanical damage (such as scratches), and bulk defects are the nucleation sites. It has been demonstrated that these surface-nucleated OISF can be completely eliminated by carrying out in-situ surface cleaning in the furnace prior to oxidation [83].

Generation of OISF in silicon has been extensively studied, and Ravi [84] has discussed this subject very effectively. The important aspects of the OISF are summarized in the following: a) OISF are extrinsic stacking faults that are bound by Frank partial dislocations (see Chapter 2), with a Burgers vector $\mathbf{b} = a/3 < 111 >$, with $\mathbf{a}$ being the lattice parameter. b) The length of the stacking fault, l, is found to be given by [53].

$$l = \text{ const. } t^{\mathrm{n}} P_{\mathrm{O}_2}^{\mathrm{m}} \exp(-Q/\mathrm{k}T), \tag{3.13}$$

where t, Po_2, and T are the time, oxygen partial pressure, and temperature (in degrees Kelvin) of the oxidation, n and m are experimentally determined numerical exponents, and Q is the activation energy. The constant term is influenced by the wafer orientation, dopant concentration, and chlorine concentration in the oxidizing medium. c) Equation 3.13 is found to hold

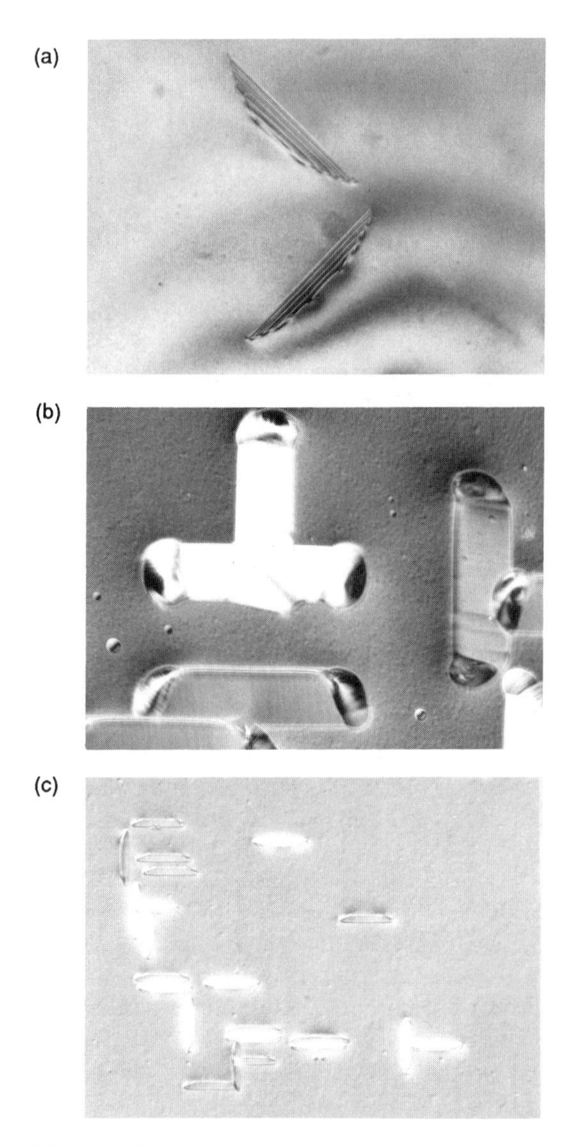

Fig. 3.27. (a) TEM, (b) SEM, and (c) optical micrographs of the oxidation induced stacking faults.

only at temperatures less than 1200°C. At higher temperatures, the length starts shrinking, and for oxidations carried out at $>$ 1240°C in oxygen or at $\geq$ 1350°C in steam, no faults are found. The upper temperature, at which no faults are seen, is a function of the oxygen partial pressure, silicon surface orientation, and also of the chlorine content in the dry oxygen. Chlorine retards (and eliminates at sufficiently high concentrations) OISF growth. These results are summarized in figs. 3.28, 3.29, and 3.30 [49, 85, 86]. d) Postoxidation inert anneal causes shrinkage of the OISF. The shrinkage phenomenon is found to be a thermally activated process with an activation energy of 5.2 eV, a value very similar to the value for the silicon self-diffusion in silicon 5.13 eV. This suggests a role for thermal vacancies in annihilating excess silicon at the OISF plane. e) Various mechanisms [82, 53, 87] have been proposed to explain growth and the retrogrowth (the shrinking phenomenon at high temperatures) of OISF. The following assumptions have been made: (i) The growth of the stacking faults at heterogeneous nucleation sites is analogous to crystalline growth at the surface kinks. Surface kinks are simple dislocation walls formed on the surface. (ii) There is an excess of silicon atoms at the $SiO_2{=}Si$ interface. (iii) The concentration of the excess silicon atoms (i.e., interstitials) in silicon depends on the partial pressure of oxygen during oxidation, the oxidation rate at which the $SiO_2{=}Si$ interface moves in silicon, and the concentration of charged (singly or doubly charged) vacancies in the adjacent bulk of the silicon. (iv) Increasing excess silicon concentration, such as during heavy phosphorus diffusion [88], will enhance OISF growth. (v) Increasing the vacancy concentration in silicon, such as at very high temperatures or due to chlorine-silicon interaction [53],

$$x\mathrm{Cl_2(g)} + \mathrm{Si\ (L)} \rightarrow \mathrm{SiCl_x(g)} + V\ \mathrm{(L)} \tag{3.14}$$

will retard growth due to the ensuing annihilation of the excess silicon interstitials

$$\mathrm{Si\ (I)} + V\ \mathrm{(L)} \rightarrow \mathrm{Si\ (L)}. \tag{3.15}$$

In these equations, (L) represents a lattice position, (g) indicates gas or vapor phase, and (I) denotes interstitial position. Very high temperature anneals in an inert ambient will also produce vacancies that migrate to OISF and reduce the fault length [89]. (vi) The orientation effect will arise because differently oriented surfaces have a different density of atoms per unit area and have a different capacity per unit area to accommodate excess silicon atoms. (vii) Murarka [53] also assumed that silicon self diffusion occurs by the formation and migration of charged vacancies. Based on these assumptions, an equation for the length of the OISF was derived [82, 53, 87]. Growth and retrogrowth phenomena were the result of the competitive and dominating generation of interstitials and vacancies.

As mentioned earlier, the growth rate of OISF is oxidation-surface orientation dependent. The growth rate is lower in (111)-oriented wafers by nearly a factor of three as compared to the growth rate in (100)-oriented wafers. There are several other orientation-dependent phenomena associated with silicon, including the oxidation-induced dopant redistribution discussed in the following section. Table 3.5 lists such phenomena and properties. Two important conclusions are, in order: a) The activation energies, obtained

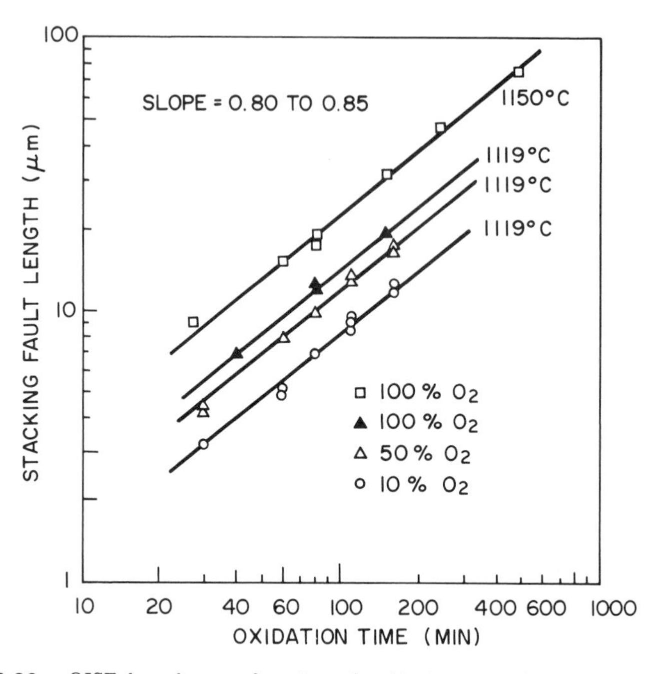

Fig. 3.28. OISF length as a function of oxidation time for oxidation carried out in varying partial pressures of oxygen. Results of oxidation at 1150°C and 1119°C are compared.

from the temperature dependence of the OISF growth, boron diffusion enhancement factor, regrowth of ion implantation damage, and the formation of the excess silicon lumps during oxidation, are the same, i.e., 2.5± 0.2 eV. This perhaps indicates an identical thermally activated process. b) The (111) surface appears to retain more silicon at the SiO_2=Si interface or in silicon near the SiO_2=Si interface. This is in agreement with the higher available bond density, absence of boron diffusion enhancement in oxidizing ambients, higher Q_f, lower OISF growth rate, higher retention of the ion implantation damage, and observation of lumps or white spots in (111) silicon. Retention of more of the (unoxidized) excess silicon will mean a lower concentration of silicon interstitials in silicon and therefore lower retrogrowth temperature. What causes the orientation dependence? For the present, there is no quantitative model to explain these differences between (111) and (100) silicon. However, the following observations are made. a) The $< 111 >$ planes are the most densely packed planes, which could retain more silicon atoms at the surface than $< 100 >$ planes. This is particularly true for diamond cubic lattice in which, perpendicular to a [111] direction, there are parallel sets of two (111) planes separated by a distance of $1/4\ a\sqrt{3}$. The two planes in the set are, however, separated by only a distance of $a/4\sqrt{3}$ where a is the lattice parameter. Thus more silicon atoms can be retained in the set

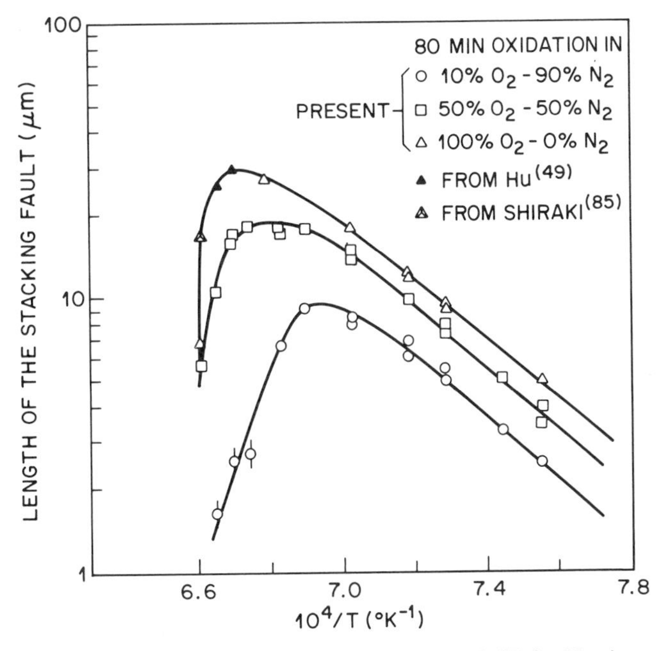

Fig. 3.29. Temperature dependence of the length of the OISF for 80-minute oxidation of silicon in different partial pressures of oxygen.

of two planes. b) Since faults grow as (111) planes, there is a probability that faults will grow parallel to the (111) surface. Such faults will either be subsequently oxidized during continued oxidation or are etched during chemical etching used to reveal faults. Since surface diffusivity is generally much higher than the bulk diffusivity and because the separation between the sets of < 111 > planes (parallel to surface) is large, the probability of fault formation parallel to surface is larger than that for fault formation parallel to the other three < 111 > planes. That more silicon atoms are obtained at the SiO_2=Si interface parallel to (111) is also supported by the fact that the epitaxial regrowth rate of the ion-implanted damaged silicon surface is lowest for (111) and highest for (100). Thus < 111 > surfaces tend to be more disordered as compared to < 100 > surfaces.

3.6.4 Oxidation-Induced Dopant Redistribution (OIDR)

In silicon integrated-circuit fabrication, the silicon is doped with dopants (As, B, P, etc.) during crystalline growth and also during device processing. When doped silicon is oxidized, a redistribution of dopant between the silicon and the growing oxide occurs. The dopant could preferentially segregate in the silicon or in the oxide, and the dopant distribution will be affected by the relative dopant diffusivities in silicon and SiO_2. Schematics of the

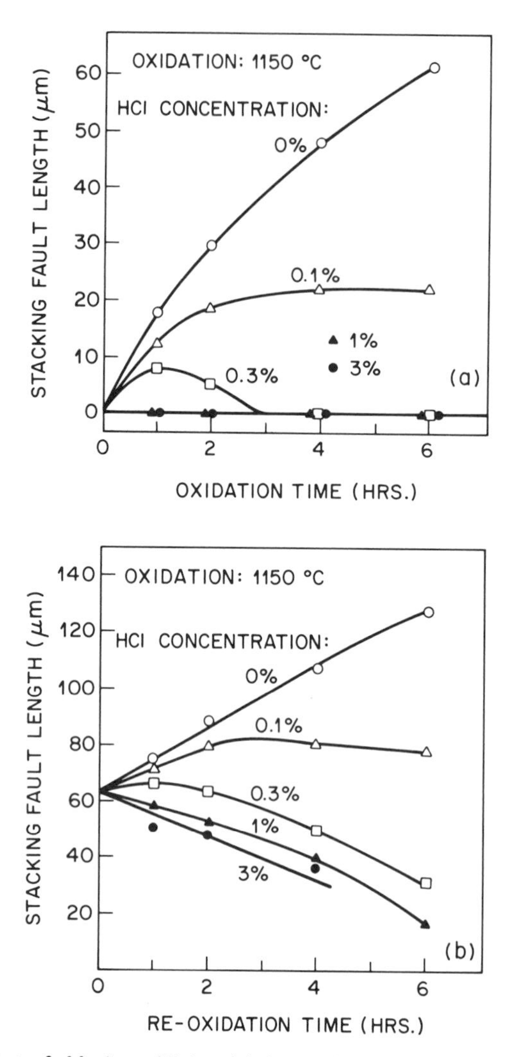

Fig. 3.30. Effect of chlorine addition (a) during first oxidation or (b) during reoxidation at 1150°C on the size of the OISF, which is plotted as a function of the oxidation time. Percent of chlorine in the oxidizing medium is shown next to the each plot. From Shiraki [85].

distribution plots are shown in fig. 3.31. Boron is lost to the oxide, decreasing its concentration in silicon at the oxide-silicon interface. Phosphorus, on the other hand, piles up in silicon at the interface. Arsenic and gallium behave like phosphorus and boron, respectively, except that they redistribute somewhat differently because of different relative diffusivities in silicon and SiO_2. Variation of dopant concentrations in silicon will affect the electrical characteristics of the devices, and an understanding of the redistribution

TABLE 3.5: Orientation-dependent phenomena in silicon

Phenomenon or property	Orientation dependence
OISF	A factor of ∼ 3 smaller in (111) than in (100) silicon. For wafers a few degrees off (100), the length decreases as a function of misorientation. No P_{O_2} dependence of the OISF length in (111). Activation energy of growth (∼2.3 eV) is independent of orientation.
Boron diffusion in oxidizing conditions	B diffusivity in (100) silicon is enhanced in in presence of oxygen, but is unchanged in (111) silicon. Enhancement factor activation energy ∼ 2.6 eV.
Surface-state fixed charge (Q_f)	Decreases in the order (111) > (110) > (100).
Linear oxidation-rate constant	Greater for (111) than that for (100).
Energy of formation of the silicon-oxygen bonds (calculated)	Lower for (111) than that for (100).
Ion implantation damage	Higher residual damage in (111) than that in (100). Activation energy for regrowth (∼2.4 eV) is independent of orientation.
Lattice strain (oxidation)	Near the specimen surface is larger in (100) than in (111). It extends deeper in (100) than in (111) wafers.
Available silicon bond density	Higher for (111) than that for (100).
Retrogrowth of OISF	Occurs at lower temperature for (111) than that for (100) wafers.
Structural difference at the Si=SiO_2 interface	(111) dry-oxidized samples had spots at the interface. (100) surfaces did not show these spots.

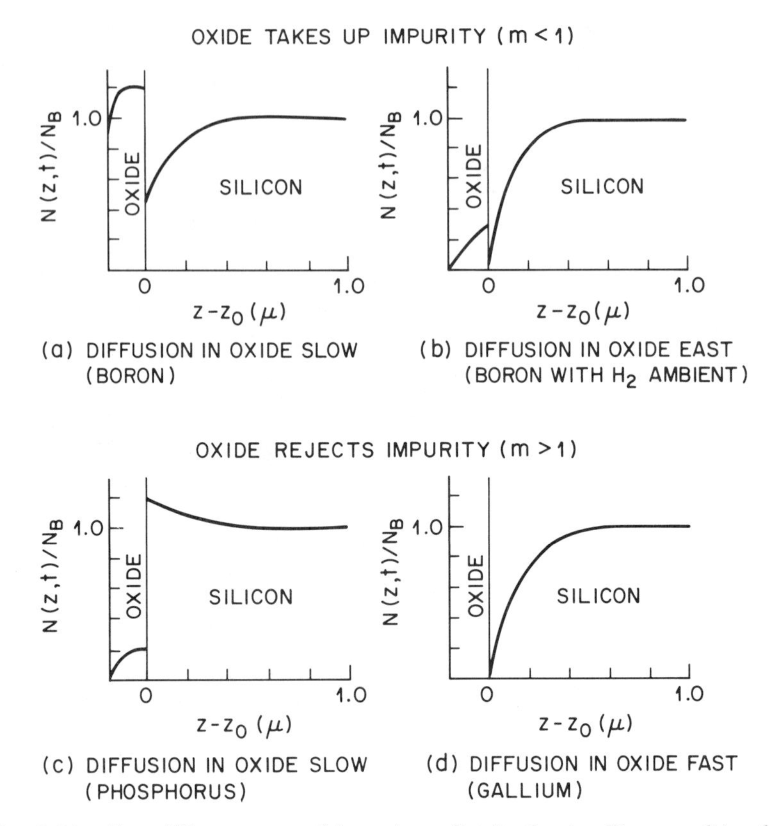

Fig. 3.31. Four different cases of impurity redistribution in silicon resulting from thermal oxidation. From Grove [21].

is important so that necessary corrections can be made. The model for dopant diffusion in silicon during oxidation can be best described by a double coordinate system [90], as shown in fig. 3.32. Here, the moving Si=SiO_2 boundary is defined by the X coordinate, movement in the silicon is given by the Y coordinate and the moving oxide-gas interface is described by the Z coordinate. Thus, the impurity concentrations in the silicon and in the oxide are given by $C(Y,t)$ and $C(Z,t)$. At any time t, the silicon-silicon oxide interface will be given by $X = 0$, $Y = Y_o(t)$, and $Z = Z_o(t)$, where Y_o is the thickness of silicon transformed into oxide of thickness Z_o during this time, such that

$$Y_o(t) = \alpha Z_o(t). \tag{3.16}$$

Here α is the ratio of the thickness of silicon consumed during oxidation to the oxide thickness. It can be easily calculated from the densities and molecular weights of silicon and silicon dioxide and assuming 100% dense oxide formation. The calculated value of α is 0.45.

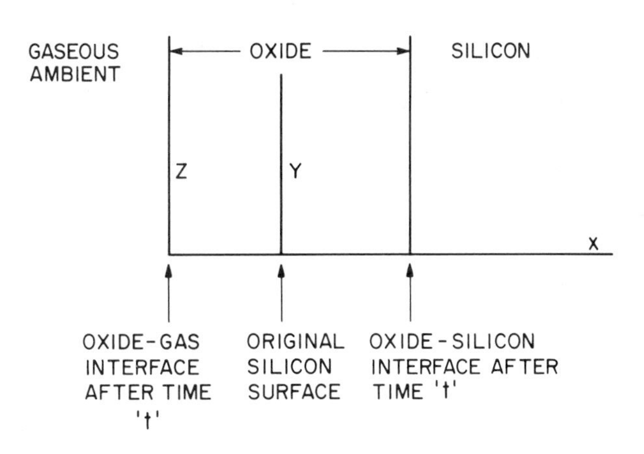

Fig. 3.32. Mathematical model for diffusion in silicon in oxidizing ambients.

The dopant distribution will be given by Fick's second diffusion equation

$$\frac{dc}{dt} = \frac{d}{dZ}(D_{ox}\frac{dc}{dZ}),\ 0 < Z < Z_o \text{ in the oxide} \tag{3.17}$$

and

$$\frac{dc}{dt} = \frac{d}{dY}(D\frac{dc}{dY}),\ Y > Y_o \text{ in the silicon,} \tag{3.18}$$

where D_{ox} and D are the dopant diffusion coefficients in oxide and silicon, respectively.

Dopants could preferentially segregate in SiO_2 or silicon. Thermodynamically, this preferential segregation is determined by a quantity called the *segregation coefficient*, m, which is the ratio of the concentration of dopant in silicon to that in the oxide. Mathematically,

$$m = \frac{c(Y_o, t)}{c(Z_o, t)}. \tag{3.19}$$

Using the continuity equation for the impurity flux across the $Si{=}SiO_2$ interface and neglecting the diffusion in the oxide, one obtains an equation

$$c\frac{dZ_o}{dt}\left(\alpha - \frac{1}{m}\right) = D_{ox}\frac{dc}{dZ} - D\frac{dc}{dY}. \tag{3.20}$$

Equation (3.20) forms one of the boundary conditions needed to solve the diffusion equations (3.17) and (3.18). The latter equation is more often solved, as we are interested mostly in diffusion in silicon. In eq. (3.20), dZ_o/dt is the oxidation rate of silicon that can be obtained from eq. (3.11) by substituting Z_o for x.

TABLE 3.6: Comparison of diffusion and segregation coefficient values[a]

Diffusion source	Temperature range (°C)	Drive-in ambient	m	Reference
Out diffusion of B-doped Si	1200-1250	Steam or dry O_2	0.32	90
B_2O_3	1120-1230	Dry O_2	0.1	91
BN	1100-1250	Wet O_2	0.1	95
BBr_3	1100	Dry O_2	0.25	96-98
B-doped glass	1000-1280	Dry N_2	0.06	99
Ion implanted	1000-1200	Steam	$4.29 \times 10^3\ e^{-1.135eV/kT}$ (0.1 - 0.66)	93
Ion implanted	1100	Dry O_2	1.0	92
Ion implanted	1050-1250	Dry O_2	$9.82\ e^{-0.29eV/kT}$ (0.75 - 1.0)	92
Ion implanted	1000-1200	Dry O_2	$33.3\ e^{-0.52eV/kT}$ (0.27 - 0.51)	94

[a] From reference 92.

Thus, the problem of calculating dopant concentration profiles in silicon after a diffusion drive-in in an oxidizing ambient (i.e., oxidation) reduces to solving eq. (3.18) for the boundary condition (3.20). Generally, the solutions are obtained for different conditions using numerical methods and computers. For boron, which diffuses very slowly in SiO_2, the boundary condition (3.20) could be simplified by setting D_o equal to zero. For solutions, see references 90, 91, and 92.

The segregation coefficient, m, has been determined or estimated for various dopants. Values ranging from 10^{-3} to greater than 10^3 have been reported. For donor atoms (As, P, Sb), Grove et al. [90] report a value of about 10. Murarka [92] has summarized the values of m for boron. As shown in table 3.6, m values of 0.1 to 1 have been reported and a few authors [72, 91-93] report m to be temperature dependent. Fair [31] has rationalized these data by pointing out the role of small amounts of water vapor and organic contaminants in oxygen in determining the segregation behavior. Faster oxidation rates, in the presence of water, decrease m for boron, indicating the presence of more boron in oxide and rather more open structure of boron-SiO_2-glass when oxidation occurs in the presence of steam.

3.7 Effect of Radiation on SiO_2 Properties

Exposure of SiO_2 to radiation sources can lead to radiation damage in SiO_2. The type and extent of the damage are determined by the type, energy, and flux of the radiation and on the temperature at which the exposure occurs. Undesirable electrical activity can be produced in SiO_2, leading to undesirable device characteristics. This undesirable electrical activity could be result of a electron-hole pair production, atomic defects (mono- and di-interstitials or vacancies, interstitial or vacancy impurity complexes, defect clusters, and ionized species) produced in the film. This could be a result of interaction with photons (photoelectric effects, compton scattering, or pair production in the order of increasing energy), charged particles (Rutherford scattering or nuclear interactions in case of heavy particles), and neutrons (elastic or inelastic scattering or transmutation reactions). Since SiO_2 films are thin, penetrating radiation may cause similar damage in silicon.

The importance of radiation damage arises due to instabilities arising in the devices due to radiation exposure during processing or during use in outside environments where special applications in such areas as those in the space and nuclear industries could cause exposure to large fluxes of radiation. Radiation exposure during processing results from sputtering; dry plasma or reactive ion etching; e-beam evaporation; the use of lithographic radiation sources such as UV, visible, e-beams, ion beams, or x-rays; laser annealing; ion implanation; etc. Space and nuclear exposures could include all types of radiation including γ-rays, neutrons, and other energetic particles. Processing-induced radiation damage can generally be repaired through annealing at different temperatures in different gas ambients. Hydrogen is the most effective ambient [2, 4]. For protection against environmental damage in actual use, the devices have to be specially designed and processed to minimize the effect of radiation. Studies that investigate radiation effects invariably use MOS capacitors and examine $C - V$ characteristics to determine the effects of radiation exposure, annealing, and process variables that may affect the effect of radiation exposure [4].

3.8 Oxidation of Compound Semiconductors

Oxidizing semiconductors like GaAs, GaP, InP, and GaAlAs leads to complexities associated with the compound nature of the material. The complexities are compounded by the formation of the volatile oxides of arsenic or phosphorus and the resulting uncertainties about the oxide-semiconductor interface, which in turn controls the oxidized material's electrical behavior. Various attempts to form a native oxide layer, similar to SiO_2 on silicon, have been made without any significant success.

Thermal oxidation [100-107], anodization [108-111], and plasma anodization [112, 113], processes have been relatively more successful techniques. Several attempts have been made to thermally oxidize GaAs in oxygen, water vapor, or air [100-104]. Most of these oxidations [99-102] were carried out at higher temperatures ($\geq$ 600°C), resulting in polycrystalline oxide that are mostly $\beta - Ga_2O_3$ with small amounts of $GaAsO_4$ or even free aresenic. Arsenic oxides are generally not seen in bulk oxide films. Navratil [104] reported the results of oxidation in dry oxygen at temperatures as low as 400°C. Murarka [105] carried out thermal oxidation in air at 350, 450, and 500°C. In these studies, the oxides grown at lower temperatures (at or below 500°C) were amorphous, and Murarka reported good mechanical stability, uniformity, and electrical behavior. Oxides grown above 500°C had increasing amounts of crystallinity. At 500°C, a linear oxidation growth law was obtained. Figure 3.33 shows a log-log plot of the oxide grown in air at 500°C. At lower temperatures, the kinetics is not well established, although Navratil described a parabolic growth law for the oxide at 400 and 450°C.

Takagi et al. [106] used arsenic trioxide (As_2O_3) vapors to oxidize GaAs in an attempt to eliminate the loss of arsenic during oxidation. It was also hoped that this procedure would lead to oxides containing no free arsenic. Oxide thus grown had considerable amounts of As_2O_3 as one of the bulk constituents. However, free arsenic was also found in such films [106]. The presence of metallic arsenic in thermally grown oxides has been detrimental to oxide electrical properties. In addition, all thermal oxides retain their amorphous nature during subsequent anneals only at temperatures less than ~500°C. At higher temperatures, continued loss of arsenic (from the film) and crystallization to $\beta - Ga_2O_3$ occurs, making those films unsuitable for MOS applications.

A Ga-As-O condensed phase-diagram (fig. 3.34) has been constructed by Thurmond et al. [108] to predict the equilibrium phases expected under certain conditions used to oxidize GaAs. This is done by following a line that connects the oxidant to GaAs and examining the intersections of this line and the phase field boundaries it crosses. Figure 3.35 shows three compositions of the oxide for oxidation under different conditions of oxygen pressure and temperature (whereas only one composition is predicted for all As_2O_3 oxidations) [107]. These predictions have been verified qualitatively in experiments.

Anodic oxidation of GaAs has also been studied extensively [109-113]. Oxides thus grown are found to be more reliable than thermal oxides. However, these oxides also lack stability at temperatures higher than 500°C. Anodic oxidation on GaAs has been carried out using H_2O_2 and H_3PO_4 solutions with or without glycol [109-114]. The oxide growth rate depends on the pH of the electrolyte, applied bias, temperature, and, most importantly, on the resistivity of the substrate. The pH dependence [110] is shown in fig. 3.36. The curve shows a considerable decrease in oxidation rate for pH values ~ 3.5. These values approximate the pH of a 30% H_2O_2 solution with no acid or base added to it. The minimum at this pH is thus associated with a

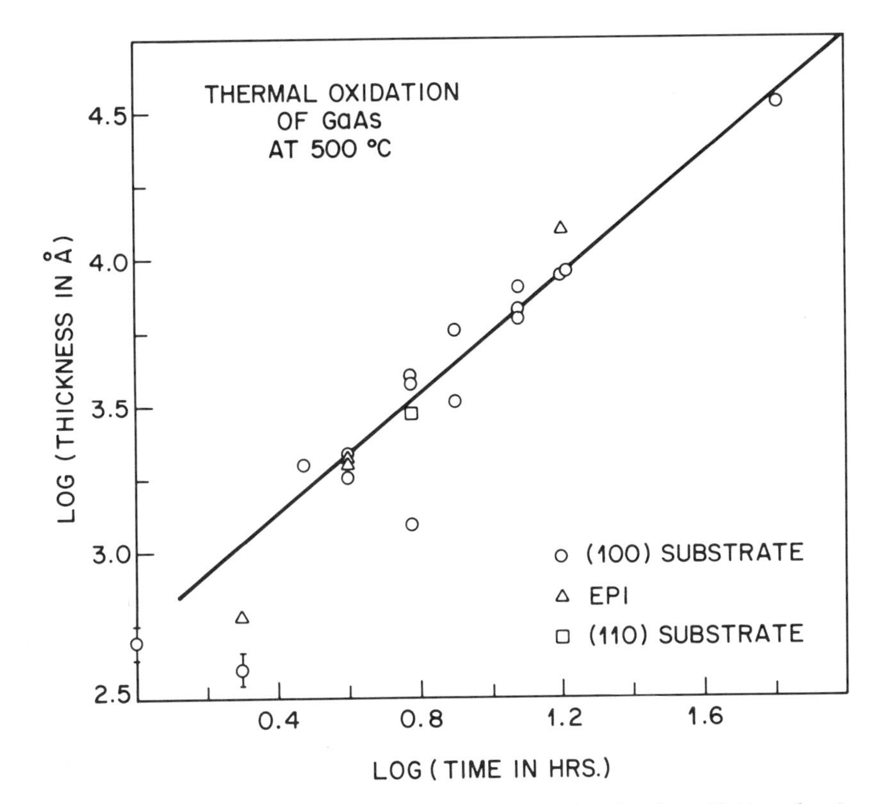

Fig. 3.33. Log-log plot of the oxide thickness versus time for the oxidation of various n-GaAs substrates in dry oxygen at 500° C.

conductivity minimum of the solution due to a lack of added ions (NH_4OH to increase pH and H_3PO_4 to decrease it). Figure 3.37 shows the bias and temperature dependence of the oxide growth for a ~0.01Ω-cm n substrate. Figure 3.37a (top of the figure) shows thickness as a function of time for two voltages. Oxide thickness builds up in the first few seconds. Anodization virtually stops at longer times because of the insulating oxide layer formed on the surface. Increasing the applied bias increases the growth rate and initial anodization cut-off time. Figure 3.37b shows the voltage dependence of the oxide thickness with a slope of 1.84 nm/V at room temperature. Thus, increased anodization rates can be achieved at higher voltages. However, at very high voltages, the anodization results in poor oxide quality—nonuniform deposition and poor electrical breakdown characteristics. Even increases in the refractive index of the oxide, granularity, and blotchiness on the substrate (after removal of the oxide) have been reported [109]. Figure 3.37c shows the temperature dependence of the anodization. Higher tem-

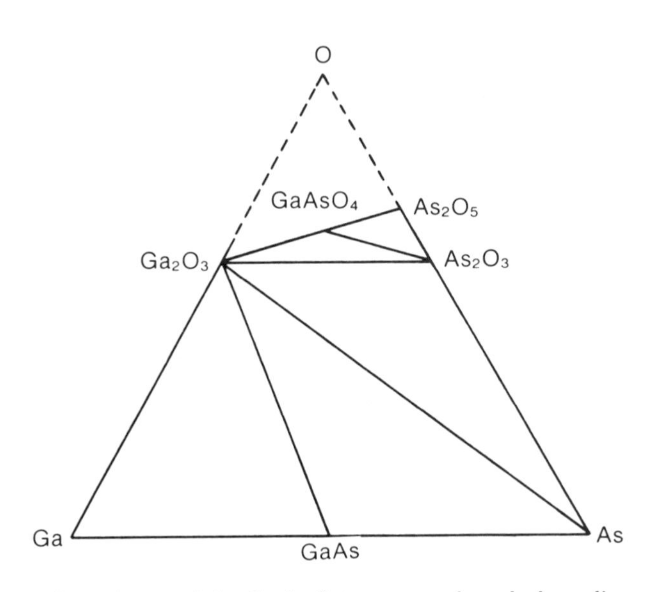

Fig. 3.34. An estimate of the Ga-As-O ternary condensed phase diagram for temperatures below the melting point of As_2O_3 (278° C, arsenolite). A portion of this phase diagram, the oxygen-rich region above the line Ga_2O_3-$GaAsO_4$-As_2O_5 and bounded at the sides by the dashed lines, has not been studied and is an excluded part of the Ga-As-O ternary diagram. From Thurmond et al. [108]. Reprinted by permission of the publisher, The Electrochemical Society, Inc.

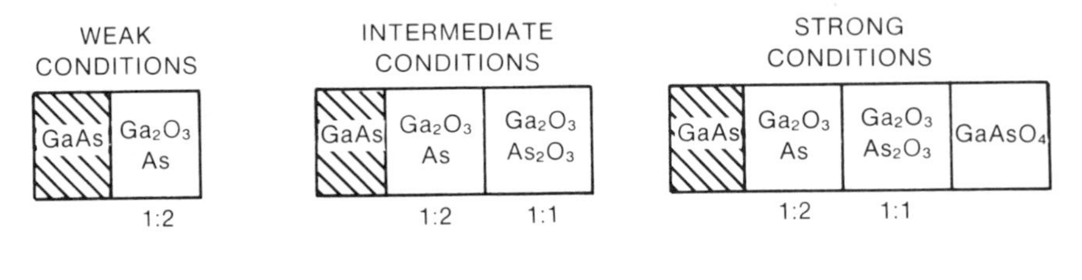

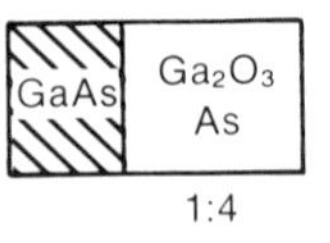

Fig. 3.35. Schematic diagrams of the phases formed in the oxidation of GaAs by oxygen and As_2O_3 under various oxidizing conditions according to the Ga-As-O phase diagram of fig. 3.34. The predicted molar ratio of the upper to lower phase constituents is given beneath the oxide phase. From Thermond et al. [108]. Reprinted by permission of the publisher, The Electrochemical Society, Inc.

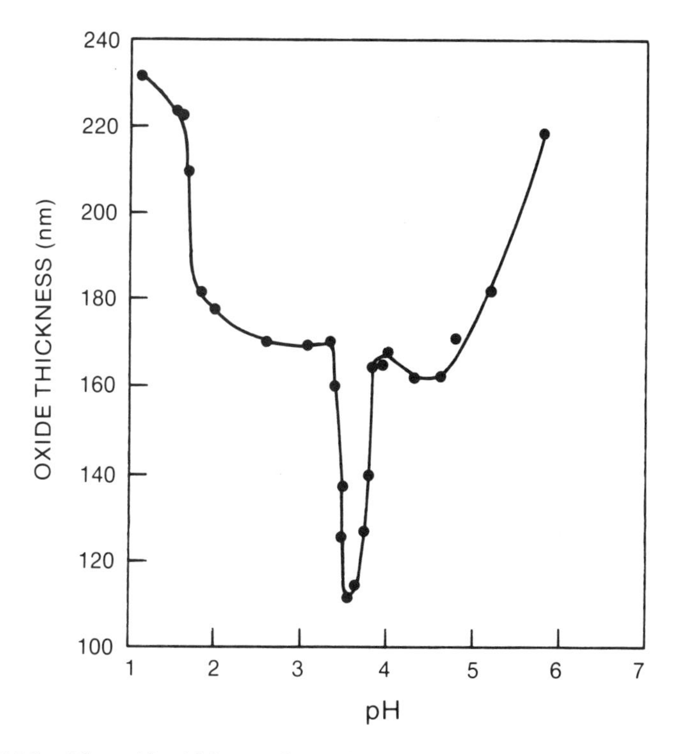

Fig. 3.36. The oxide thickness obtained in 10-minute anodizations at 100 V as a function of the pH of the electrolyte (H_2O_2). At higher pH values (~6.5) the oxide became nonuniform in thickness. From Logan et al. [109]. Reprinted by permission of the publisher, The Electrochemical Society, Inc.

perature aids the anodization process due to increased ionic mobilities and (perhaps) decreased resistance of the electrolyte and the oxide.

Figure 3.38 shows the oxide thickness as a function of the resistivity of the substrate for a given anodization condition [110]. Anodization of the substrates with resistivity greater than 10 Ω-cm was not possible. This result is significant in the application of anodic oxidation to produce oxides on a variety of substrates, i.e., the oxidation rate is a function of the substrate resistivity. For low-resistivity substrates with high-resistivity epitaxially grown layers, the oxide thicknesses on the front and back surfaces will be different. Invariably, anodization of the substrate (back) surface will occur first until saturation, followed by the oxidation of the epi (front) surface.

Schwartz [115] defines the concept of a working potential drop across the growing oxide layer (V_{ox}) as:

$$V_{ox} = (V_a - V_b - V_s) \quad (f[r(t)]), \tag{3.21}$$

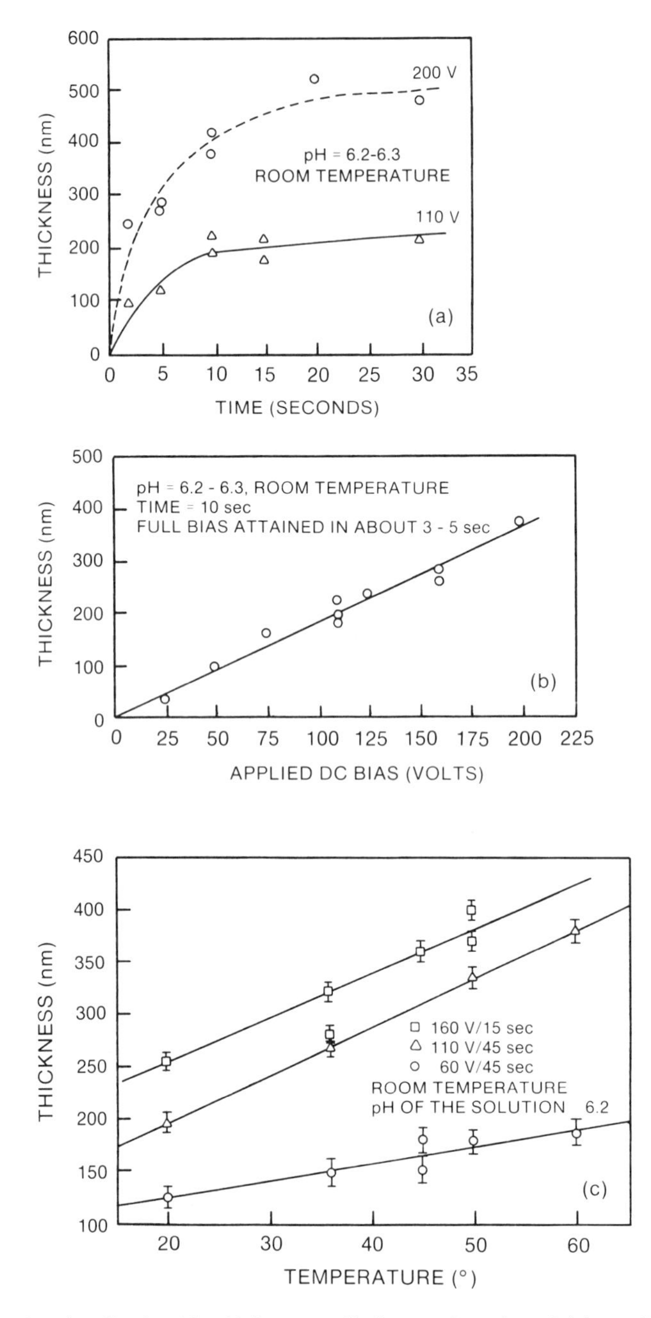

Fig. 3.37. Anodized oxide thickness on GaAs as a function of (a) anodization time, (b) DC bias, and (c) electrolyte temperature.

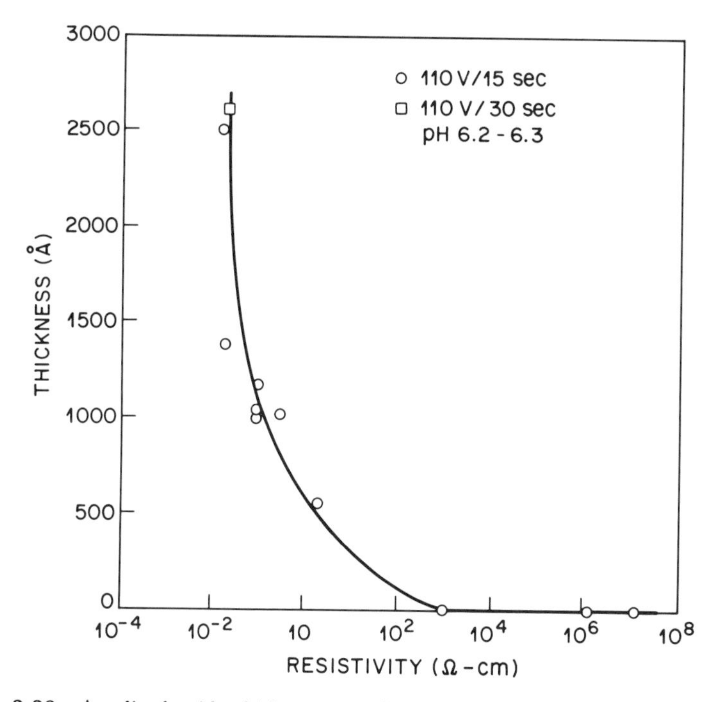

Fig. 3.38. Anodized oxide thickness as a function of the GaAs substrate resistivity for room temperature anodizations.

where V_a is the applied bias, V_b is the breakdown voltage of the semiconductor, V_s is electrolyte double-layer voltage, and $f[r(t)]$ is a resistance function dependent on the oxide thickness and resistivity dependences (figs. 3.37 and 3.38). The higher the applied bias is, the larger the voltage drop across the oxide becomes. Similarly, the lower the substrate resistivity is, the lower is V_b. This results in more voltage drop across the oxide. The more the voltage across the oxide is, the higher the anodization rate becomes, as has been seen.

Similar studies of anodizing (or thermal oxidizing) other III-V compounds have been reported. In all cases, uniform anodic (thermal) oxides can be grown in a usable thickness range of a few hundred angstroms to about 5000 Å. The oxides, however, lack reliability due to electrical interface instabilities and crystallization at higher temperatures ($\geq$500°C). Inability to grow good native oxides has led to the use of deposited oxides in GaAs, GaAlAs, and InP technology.

3.9 Summary

In this chapter we have examined the oxidation of semiconductor materials such as silicon and GaAs. We reviewed

- The properties and applications of SiO_2 including its use as an MOS capacitor.
- Characterization, especially using $C - V$ methods, of SiO_2.
- Kinetics of oxidation and oxidation mechanisms.
- Effect of dopant concentration, substrate orientation, and chlorine on oxidation kinetics.
- Experimental methods (thermal and anodic) of oxidizing semiconductors.
- Oxidation-induced defects and dopant redistribution.

References for Chapter 3

1. J.M. Stevels, in "Non-crystalline Solids," V.D. Frechette, ed., Wiley, NY, Chapter 17 (1960).
2. A.K. Sinha, H.J. Levinstein, L.P. Adda, E.N. Fuls, E.I. Povilonis, Solid-St. Electronics 21, 531 (1978).
3. P. Balk, Extended Abstracts, Electronics Div. Electrochemical Society 14 (1), 237 (1965).
4. E.H. Nicollian, J.R. Brews, "MOS Physics and Technology," Wiley, NY, Chapter 15 (1982).
5. D.J. DiMaria, D.R. Young, R.F. Dekeersmaecker, W.R. Hunter, C.M. Serrano, J. Appl. Phys. 49, 5441 (1978).
6. J.M. Aitken, D.R. Young, J. Appl. Phys. 47, 1196 (1976).
7. D.J. DiMaria, J.M. Aitken, D.R. Young, J. Appl. Phys. 47, 2740 (1976).
8. R.A. Gdula, J. Electrochem. Soc. 123, 42 (1976).
9. A. Vshirokowa, E. Suzuki, M. Warashina, Jpn. J. Appl. Phys. 12, 398 (1973).
10. V.J. Kapoor, F.J. Feigl, S.R. Butler, J. Appl. Phys. 48, 739 (1977).
11. S.R. Butler, F.J. Feigl, Y. Ota, D.J. DiMaria, in "Thermal and Photo-Stimulated currents in Insulators," D.M. Smyth, ed., Electrochem. Soc. Princeton, NJ, p. 149 (1976).
12. J.M. Aitken, D.R. Young, K. Pan, J. Appl. Phys. 49, 3386 (1978).
13. E.H. Nicollian, C.N. Berglund, P.F. Schmidt, J.M. Andrews, J. Appl. Phys. 42, 5654 (1971).
14. T. Bill, G. Hetherington, K.H. Jack, Phys. Chem. Glasses 3, 141 (1962).
15. T. Drury, J.P. Roberts, Phys. Chem. Glasses 4, 79 (1963).

16. A.E. Owen, R.W. Douglas, J. Soc. Glass Technol. 43, 159 (1959).
17. K.H. Beckman, N.J. Harrick, J. Electro. Chem. Soc. 118, 614 (1971).
18. W.-K. Chu, J.W. Mayer, M.-A. Nicolet, "Backscattering Spectrometry," Academic Press, NY (1978).
19. W.E. Beadle, J.C.C. Tsai, R.D. Plummer, eds., "Quick Reference Manual for Silicon Integrated Circuits," Wiley, NY (1985).
20. E.H. Snow, A.S. Grove, B.E. Deal, C.T. Sah, J. Appl. Phys. 36, 1664 (1965).
21. A.S. Grove, "Physics and Technology of Semiconductor Devices," Wiley, NY (1967).
22. B.G. Streetman, "Solid State Electronic Devices," 2nd ed., Prentice-Hall, Engelwood Cliffs, NJ (1980).
23. P.Richman, "MOS Field-Effect Transistors and Integrated Circuits," Wiley, NY (1973).
24. R.F. Pierret, "Field Effect Devices," Modular Series On Solid State Devices, Vol. IV, R.F. Pirret, G.W. Neudeck, eds., Addison-Wesley, Reading, MA (1983).
25. S.M. Sze, "Physics of Semiconductor Devices," 2nd ed. Wiley, NY (1981).
26. S.M. Sze, ed., "VLSI Technology," McGraw-Hill, NY (1983).
27. J. Frenkel, Tech. Phys. USSR 5, 685 (1938); and Phys. Rev. 54, 647 (1938).
28. R.H. Fowler, L. Nordheim, Proc. Roy. Soc. (London) 119 (1928).
29. E.L. Nicollian, J.R. Brews, "MOS Physics and Technology," Wiley, NY, p. 712 (1982).
30. B.E. Deal, A.S. Grove, J. Appl. Phys. 36, 3770 (1965).
31. R.B. Fair, in "Silicon Integrated Circuits," Part B, Applied Solid State Science Suppl. 2, D. Kahng, ed., Academic Press, NY, p. 1 (1981).
32. A.C. Adams, T.E. Smith, C.C. Chang, J. Electrochem. Soc. 127, 1787 (1980).
33. E.A. Irene, J. Electrochem. Soc. 125, 1708 (1978).
34. A.M. Goodman, J.M. Breece, J. Electrochem. Soc. 117, 982 (1970).
35. F.P. Fehlner, J. Electrochem. Soc. 119, 1723 (1972).
36. S.I. Raider, L.E. Forget, J. Vol-Sci: Technol. 12, 305 (1975).
37. J. Nulman, J.P. Krusius, A. Gat, IEEE Electron Device Letters EDL-5, 205 (1985).
38. V. Murali, S.P. Murarka, J. Appl. Phys. 60, 4327 (1986).
39. A. Fargeix, G. Ghibaudo, J. Appl. Phys. 56, 589 (1984).
40. S.M. Hu, J. Appl. Phys. 55, 4095 (1984).
41. H.Z. Massoud, J.D. Plummer, E.A. Irene, J. Electrochem. Soc. 132, 2693 (1985).
42. R.H. Doremus, A. Szewczyk, Paper presented at the Spring MRS Meeting, Boston (1986).
43. V. Murali, S.P. Murarka, J. Appl. Phys. 60, 2106 (1986).

44. S. Yagi, D. Kunii, "Fifth Symposium (International) on Combustion," Reinhold, NY, p. 231 (1955).
45. M. Ishida, C.Y. Yen, Inst. Chem. Eng. Journal 14, 311 (1968).
46. B.E. Deal, M. Sklar, J. Electrochem. Soc. 112, 430 (1965).
47. C.P. Ho, J.D. Plummer, B.E. Deal, J.D. Meindl, J. Electrochem. Soc. 125, 665 (1978).
48. P.S. Flint, "Extended Abstracts for May 1962 Electrochemical Society Meeting," Abstract 94, pp. 222-223 (1962).
49. S.M. Hu, Appl. Phys. Lett. 27, 165 (1975).
50. L.E. Katz, B.F. Howells, L.P. Adda, T. Thompson, D. Carlson, Solid State Technol. 21, 87 (1981).
51. S.P. Murarka, G. Quintana, J. Appl. Phys. 48, 46 (1977).
52. B.R. Singh, P. Balk, J. Electrochem. Soc. 125, 453 (1978).
53. S.P. Muraraka, Phys. Rev. 1316, 2849 (1977); and Phys. Rev. 1321, 692 (1980).
54. D.W. Hess, B.E. Deal, J. Electrochem. Soc. 124, 735 (1977).
55. R.J. Kriegler, "Semiconductor Silicon 1973," H.R. Huff, R.R. Burgess, eds., The Electrochem. Soc., Pennington, NJ, p. 363 (1973).
56. S.P. Murarka, H.J. Levinstein, R.B. Marcus, R.S. Wagner, J. Appl. Phys. 58, 4001 (1977).
57. B.E. Deal, M. Sklar, A.S. Grove, E.H. Snow, J. Electrochem. Soc. 114, 226 (1967).
58. H. Kauget, M.S. Thesis, Rensselaer Polytechnic Inst., Troy, NY (1985).
59. "Handbook of Chemistry and Physics, 60th, ed.," Chemical Rubber Co., Cleveland, OH (1979-80).
60. W.W. Weick, S.P. Murarka, unpublished work, Bell Laboratories, Murray Hill, NJ.
61. High-Pressure oxidation furnace system by Gasonics, Rockaway, NJ.
62. L.E. Katz, B.F. Howells, J. Electrochem. Soc. 126, 1822 (1979).
63. L.E. Katz, in "VLSI Technology," 1st ed., S.M. Sze McGraw-Hill, NY, p. 153 (1983).
64. J.R. Ligenza, J. Appl. Phys. 36, 2703 (1965).
65. V.Q. Ho, T. Sugano, IEEE Trans. Electron Devices ED-27, 1436 (1980).
66. J.R. Ligenza, M. KLuhn, Solid State Technol. 13, 33 (1970).
67. A.K. Ray, A. Reisman, J. Electrochem. Soc. 128, 2466 (1981).
68. V. Murali, S.P. Murarka, "Proceedings of First VLSI Conference," Electrochemical Soc. Pennington, NJ 1987); also see ref. 36.
69. D.R. Tuner, J. Electrochem. Soc. 105, 402 (1958).
70. A. Uhlir, Bell Syst. Techn. J. 35, 333 (1958).
71. G. Mende, J. Wende, Thin Solid Films 142, 21 (1986).
72. S.K. Ghandhi, "VLSI Fabrication Principles-Silicon and Gallium Arsenide," Wiley, NY, p. 401 (1983).
73. B.E. Deal, M. Sklar, A.S. Grove, E.H. Snow, J. Electrochem. Soc. 114, 266 (1967).

74. S.P. Murarka, Appl. Phys. Lett. 34, 587 (1979).
75. E.H. Nicollian, J.R. Brews, "MOS Physics and Technology," Wiley, NY, p. 826 (1982.
76. R.J. Jaccodine, W.A. Schlegel, J. Appl. Phys. 37, 2429 (1966).
77. E.P. EerNisse, Appl. Phys. Lett. 35, 8 (1979).
78. S.P. Murarka, J. Appl. Phys. 54, 2069 (1983).
79. J.R. Patel, A.R. Chaudhuri, J. Appl. Phys. 33, 2223 (1962).
80. S.M. Hu, J. Appl. Phys. 40, 4413 (1969).
81. B.O. Kolbesen, H.P. Strunk, in "VLSI Electronics-Microstructure Science," vol. 12, N.G. Einspruch, H. Huff eds., Academic Press, Orlando, FL, p. 144 (1985).
82. S.M. Hu, J. Appl. Phys. 45, 1567 (1974).
83. S.P. Murarka, H.J. Levinstein, R.B. Marcus, R.S. Wagner, J. Appl. Phys. 48, 4001 (1977).
84. K.V. Ravi, "Imperfections and Impurities in Semiconductor Silicon," Wiley, NY (1981).
85. J. Shiraki, in "Semiconductor Silicon 1977," H.R. Huff, E.R. Sirtl, eds., The Electrochem. Soc. Princeton, NJ, p. 546 (1977); see also Jpn. J. Appl. Phys. 15, 1 and 83 (1976).
86. S.P. Murarka, J. Appl. Phys. 48, 5020 (1977).
87. A.M. Lin, D.A. Antoniadis, R.W. Dutton, W.A. Tiller, J. Electrochem. Soc. 128, 1121 (1981).
88. C.L. Claeys, G.J. Declerck, R.J. Van Over Straeten, in "Semiconductor Characterization Techniques," P.A. Barnes, G.A. Rozgony, eds., The Electrochem. Soc. Princeton, NJ, p. 366 (1978).
89. H. Shimizu, A. Yoshinaka, Y. Sugita, Jap. J. Appl. Phys. 17, 767 (1978); see also S. Kishino, S. Isomae, M. Tamura, M. Maki, J. Appl. Phys. 49, 3255 (1978).
90. A.S. Grove, O. Leistiko, Jr., and C.T. Sah, J. Appl. Phys. 35, 2695 (1964).
91. T. Kato, U. Nishi, Jpn. J. Appl. Phys. 3, 377 (1964).
92. S.P. Murarka, Phys. Rev. 1312, 2502 (1975).
93. J.L. Prience, F.N. Schwettmann, J. Electrochem. Soc. 121, 705 (1974).
94. R.K. Jain, R. Van Overstraeten, J. Appl. Phys. 44, 2437 (1973).
95. J.S.T. Huang, L.C. Welliver, J. Electrochem. Soc. 117, 1577 (1970).
96. G. Masetti, S. Solmi, G. Soncini, Solid State Commun. 12, 1299 (1973).
97. G. Masetti, P. Negrine, S. Soluie, Alta Frequenza, XLII, 626 (1973).
98. G. Masetti, P. Negrini, S. Solnic, G. Soncini, Alta, Frequenza XLII, 346 (1973).
99. M.L. Barry, P. Olofsen, J. Electrochem. Soc. 116, 854 (1969).
100. H.J. Minden, J. Electrochem. Soc. 104, 733 (1962).
101. M. Rubenstein, J. Electrochem. Soc. 113, 540 (1966).
102. D.H. Phillips, W.W. Grannemann, L.E. Coerver, G.J. Kuhlmann, J. Electrochem. Soc. 120, 1087 (1973).

103. K.H. Zaininger, A.G. Reversz, J. Phys. Paris 25, 208 (1964).
104. K. Navratil, Czech. J. Phys. 1318, 266 (1968).
105. S.P. Murarka, Appl. Phys. Lett. 26, 180 (1975).
106. H. Takagi, G. Kano, I. Teramoto, J. Electrochem. Soc. 125, 579 (1978).
107. G.P. Schartz, J.E. Griffiths, D. DiStefano, G.J. Gualtieri, B. Schwartz, Appl. Phys. Lett. 34, 742 (1979).
108. C.D. Thurmond, G.P. Schwartz, G.W. Kammlott, S. Schwartz, J. Electrochem. Soc. 127, 1366 (1980).
109. R.A. Logan, B. Schwartz, W.J. Sundburg, J. Electrochem. Soc. 120, 1385 (1973).
110. S.P. Murarka, unpublished work (1973).
111. H. Hasegawa, K.E. Forward, H.L. Hartnagel, J. Electrochem. Soc. 123, 567 (1976).
112. B. Schwartz, F. Ermanis, M.H. Brastad, J. Electrochem Soc. 123, 1089 (1976).
113. T. Sugano, Y. Mori, J. Electrochem. Soc. 121, 113 (1974).
114. O.A. Weinreich, J. Appl. Phys. 37, 2924 (1966).
115. S. Schwartz, CRC Crit. Rev. in Solid State Sciences 5, 609 (1975).

Problem Set for Chapter 3

3.1 Our present understanding of the mechanism for the oxidation of silicon is based on oxidation studies on flat silicon wafers. These studies have shown that oxidation behavior can be represented by the expression (1)

$$x^2 + Ax = B(t + \tau), \tag{1}$$

where x is the oxide thickness, t is the oxidation time, and τ is an initial oxidation period in which oxide growth occurs in a manner that is not well understood. For short oxidation times it has been found that oxidation follows a linear rate law

$$x = \frac{B}{A}(t + \tau), \tag{2}$$

where $B/A \approx kC_o/C_I$, and for subsequent oxidation, follows a parabolic rate law

$$x^2 = B(t + \tau), \tag{3}$$

where $B = 2DC_o/C_I$. The rate constant for the silicon/SiO2 interface reaction $Si + O_2 = SiO_2$ is represented by k, C_I is the number of oxidant molecules incorporated in one unit volume of oxide, and D is the diffusion constant of oxygen within the oxide.

There are three oxidation regimes in which the above description has not been satisfactory: a) the initial oxidation period τ, b) oxidation of a silicon surface at the edge of a wall of previously grown thick oxide and c) oxidation of polycrystalline silicon.

To investigate the second problem, (100) oriented p-type silicon wafers were patterned and ion etched to produce a structure illustrated in fig. 3.39. With such a structure, the affect of the deviation from planarity of an oxidizing surface on the oxidation kinetics can be observed.

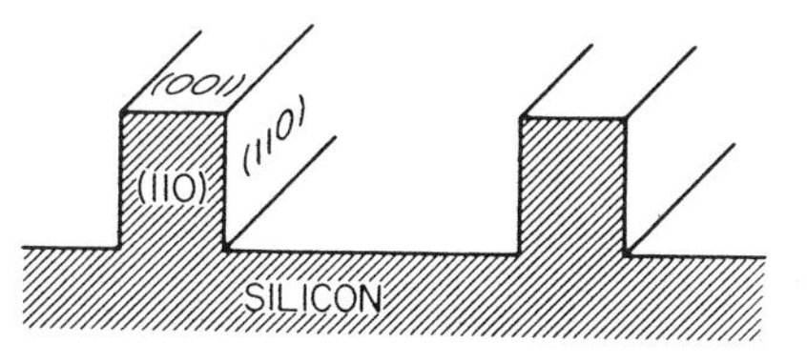

Fig. 3.39

Oxidations were performed at four different conditions: 900° for 3.5 hours, 950° for 2 hours, 1050° for 42 minutes, and 1100° for 25 minutes, such that the films are of nominally the same thickness. A typical result is shown in fig. 3.40, and important regions are also identified in the drawing. Based on these experimenal results, explain:

a. Why is G/F always greater than 1, and why is it greater at lower temperature (i.e., 1.36 at 900° versus 1.11 at 950°)?

b. Why is the oxidation rate severely depressed at an inside corner (ratio L/F), particularly for the 950° oxidation?

c. Suppose that a conducting layer is deposited conformly on top of the oxide, and the thickness is inferred from the planar oxidation data. What adverse electric behavior will happen for this trench type structure?

3.2 Figure 3.41 shows a typical oxide thickness versus growth time plot in wet ambient at 1000°C for a specific furnace. If it is requested to grow 3000 Å oxide, how much time is required? After this oxide is grown, it is found that 5000 Å, rather than 3000 Å is needed. How much extra time is required?

3.3 You would like to make an MOS capacitor on silicon with 1Ω-cm resistivity and boron as dopant. The capacitor should have an oxide capacitance of 1000 picofarad and an inversion threshold voltage of +1.6 volts. Make and explain reasonable assumptions that you will make in order to design this capacitor.

3.4 Identical (100) silicon wafers are oxidized in two different laboratories as per the process sequences described below. Assuming the Arrhenius temperature dependences given in table 3.4 and that Deal-Grove mechanism

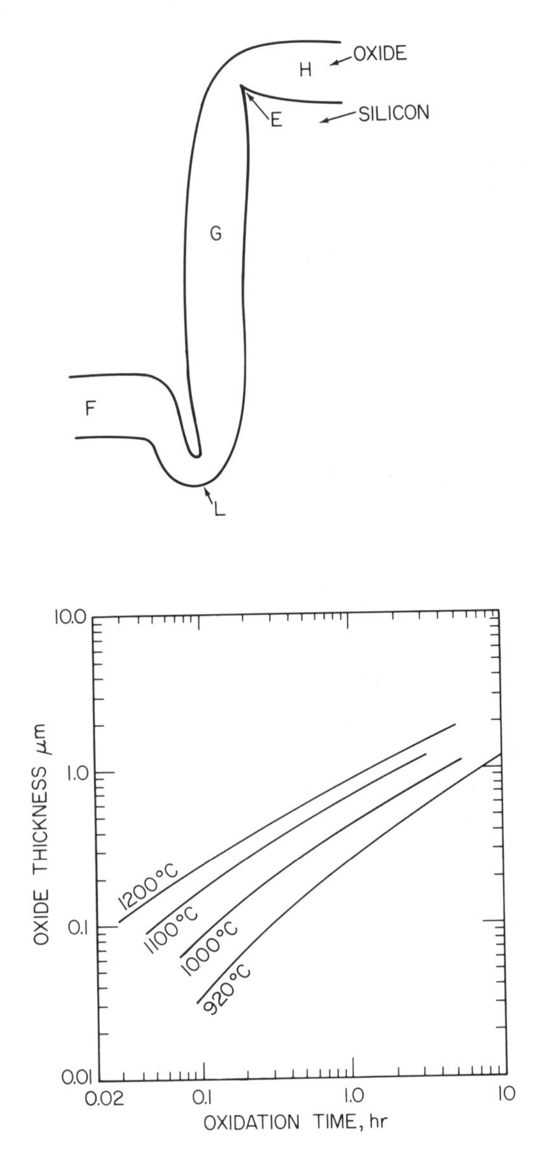

Fig. 3.40.

Fig. 3.41.

is valid predict the oxide thickness in two cases. Assume that cleaned silicon wafer, prior to oxidation, has 2 nm of the native oxide present on surface.

3.5 To reduce the height of the masking oxide from the active silicon surface, silicon nitride is deposited on silicon, patterned, and then oxidized. Silicon nitride acts as a mask and thus does not allow silicon to oxidize. However, for a patterned nitride on silicon, the oxidant can diffuse along the

Laboratory A	Laboratory B
Oxidation in dry O_2 for 2 h at 1050°C	Oxidation in steam at 1050°C for 1 h
+	+
Oxidation in steam for 1 h at 1050°C	Oxidation in dry O_2 for 2 h at 1050°C

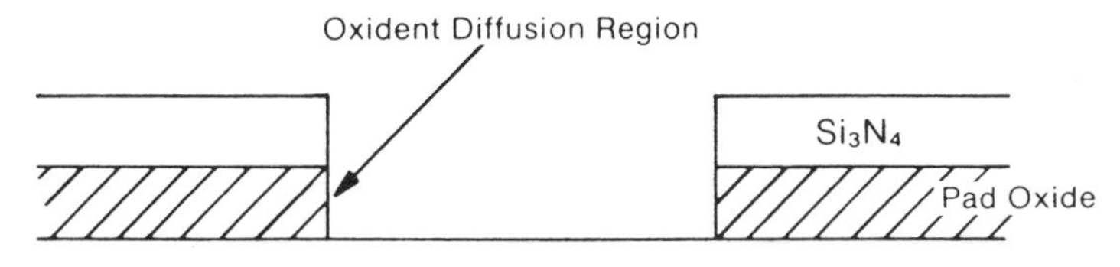

Fig. 3.42.

edge of the pattern and oxidize silicon under this edge, as shown in fig. 3.42. Calculate the length p to which the oxide penetrates under the nitride surface and lifts the nitride. Also calculate the angle. Assume oxidation is carried out in steam at 1100°C for 1 h and that no oxidant molecules can penetrate Si_3N_4. If you are making assumptions, state them clearly.

3.6 A silicon wafer has several $p-n$ junctions already formed. The wafer must be oxidized to form 2000 Å oxide. However, the diffusion junction depth of boron in silicon could not be allowed to increase by more than 0.3 nm. Using the information given in this chapter suggest, an oxidation treatment that can accomplish this task.

3.7 Assume SiO_2 layer of thickness X_0 has a uniform positive charge density $+\rho$ cm^{-3}:

a. Derive an expression for the flat-band voltage (V_{FB}) including work function effects. Sketch V_{FB} versus oxide thickness qualitatively.
b. If the ion density is 10^{17} cm^{-3}, what is V_{FB} for 1000 Å SiO_2 on n type Si with $N_D = 10^{15}$ cm^{-3} with Au contact?

3.8 MOS structures are fabricated with SiO_2 film thicknesses of 1000, 2000, and 3000 Å and corresponding V_{FB}'s are -2.25, -3.75, and -5.25 V, respectively. Obtain ϕ_{ms} and Q_{ss}. If the substrate is p-type with $N_A = 10^{14}$ cm^{-3}, what do you think the metal plate is?

3.9 A MOS transistor can be shown in a simple sketch in fig. 3.43:
Here, I_{ds} is the drain-to-source current, V_{ds} is the drain to source voltage, V_{gs} is the gate voltage with respect to source, and V_{th} is the threshold voltage required to cause inversion. Derive a simple mathe-

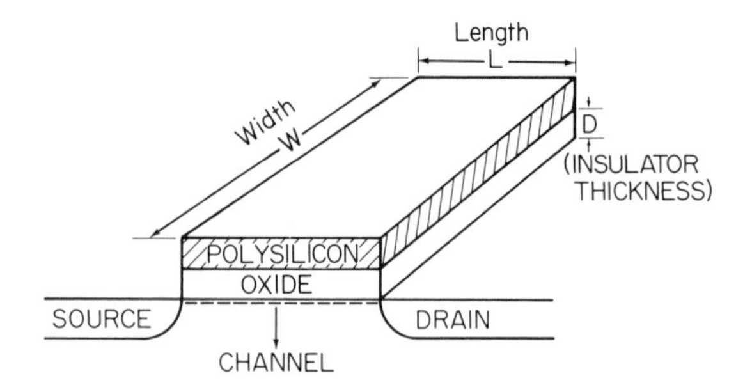

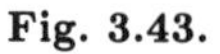
Fig. 3.43.

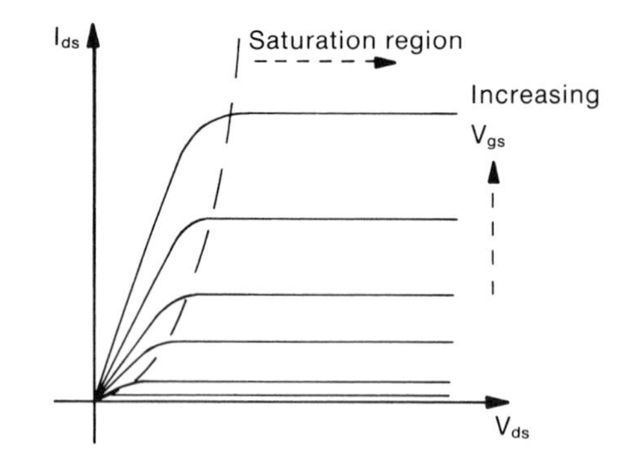

Fig. 3.44.

threshold voltage required to cause inversion. Derive a simple mathematical expression to explain the set of curves shown in fig. 3.44 for a MOS transistor.

4 Diffusion

4.1 Introduction

The properties of semiconductors are significantly affected by the presence of impurities. Physical, chemical, and optical properties are all of influence by the impact of impurities on the band structure of the semiconductors. Thus, the parts-per-billion (ppb) to parts-per-million (ppm) concentration levels of donor or acceptor impurities change the resistivity of the semiconductor crystal by several orders of magnitude. One can introduce these impurities into the semiconductor either during the growth of the crystal or by diffusion from a source into the wafer. The diffusion-controlled introduction of impurities such as phosphorus, arsenic, boron, and gallium into silicon and germanium, and of zinc, beryllium, silicon, and tellurium into gallium arsenide and indium phosphide, has led to the successful fabrication of solid-state devices and circuits. Current trends have led to the increased use of ion implantation as a method of introducing the dopant into the solid. However, the subsequent redistribution of impurities during high-temperature process cycles is determined by diffusion.

Diffusion of one material into another occurs if mixing leads to the lowering of free energy of the system. In gases, diffusion leads to a complete homogeneous mixture. In condensed phases, i.e., liquids and solids, mixing is a) limited to the solubility of one phase into another and b) generally a precursor to compound formation. In solids, diffusion occurs very slowly and is characterized by high activation energy. The activation barrier is determined by the interaction between defects in the solid and the diffusing atoms. A vast amount of literature is available on diffusion. References 1-20 list some of the books on this subject. Of these, the *Mathematics of Diffusion* by J. Crank [1] is perhaps the best source for a mathematical treatment of the subject. *Diffusion Data* [21] is a compilation of most of the data on diffusion.

In this chapter, diffusion in semiconductors and related materials, such as thin oxide and nitride films, is examined. The mathematics of diffusion and diffusion equations are given in Section 4.2. Methods to determine concentration profiles and diffusion coefficients are discussed in Section 4.3. Primary techniques for diffusing dopants are developed in Section 4.4. Atomistic diffusion mechanisms are discussed in Section 4.5, where the importance of the charge state of the local defect structure for determining the diffusion coefficient is emphasized.

4.2 Diffusion Equations

Like the conduction of heat where heat transfer is the result of a temperature gradient, the diffusion of matter is the result of a concentration gradient. Fick's first law [22] describes the relationship between the flux, J, per unit area, $\partial c/\partial x$ the concentration gradient in the x direction, where c is the concentration of the diffusing species given in units of per unit volume, and a quantity D called the *diffusion coefficient* or *diffusivity* is given in units of area/time

$$J = -D\frac{\partial c(x,\ t)}{\partial x}. \tag{4.1}$$

Here J gives the diffusing flux in the x direction, c is considered a function of x and time t, and D is considered independent of position and time. The equation is valid in the absence of external forces such as those caused by thermal and electrical gradients. Equation (4.1) simply describes that rate of transfer of diffusing species across a unit area perpendicular to the x direction and is proportional to the concentration gradient in that direction. The negative sign indicates that the transfer occurs from the higher concentration side to the lower concentration side. Generally x, t, c, D, and J are given in units of cm, second, cm^{-3}, cm^2/sec, and cm^{-2}, respectively.

An equivalent three dimensional equation can be written as

$$\vec{\mathbf{J}} = -D\nabla\vec{\mathbf{C}}, \tag{4.2}$$

where $\vec{\mathbf{J}}$ and $\vec{\mathbf{C}}$ are now vector quantitites and D is considered to be isotropic in all directions. For an anisotropic medium, the flux component in the $x, y,$ or z direction is a sum of the flux contributions due to the concentration gradients in all three directions. Thus

$$J_x = -D_{xx}\frac{\partial c}{\partial x} - D_{xy}\frac{\partial c}{\partial y} - D_{xz}\frac{\partial c}{\partial z} \tag{4.3a}$$

$$J_y = -D_{yx}\frac{\partial c}{\partial x} - D_{yy}\frac{\partial c}{\partial y} - D_{yz}\frac{\partial c}{\partial z} \tag{4.3b}$$

$$J_z = -D_{zx}\frac{\partial c}{\partial x} - D_{zy}\frac{\partial c}{\partial y} - D_{zz}\frac{\partial c}{\partial z}. \tag{4.3c}$$

Here J_x, J_y, and J_z represent the magnitudes of the vector $\vec{\mathbf{J}}$ in the x, y, and z directions. The first subscript of D represents the flux direction under consideration, and the second subscript of D indicates the direction of the concentration gradient contributing to the flux under consideration. For our purposes the anisotropy of the diffusion need not be discussed because all semiconductors under consideration are cubic. Thus for most practical applications isotropy of the medium is assumed, and Fick's law of diffusion for one dimension (4.1) is applicable.

Fick's second law of diffusion can be derived by considering the diffusion across an element of volume and using the law of conservation of matter. The rate at which the concentration changes in this volume must equal the local decrease of the diffusion flux. In other words, the accumulation rate of matter in the element of volume under consideration must equal the difference in the rate at which matter diffuses into this volume and the rate at which the matter diffuses out of this volume. These considerations lead to the equation

$$\frac{\partial c(x,\ t)}{\partial t} = \frac{\partial}{\partial x}[D\frac{\partial c(x,\ t)}{\partial x}] \tag{4.4}$$

and (for concentration-independent D) to the equation

$$\frac{\partial c(x,\ t)}{\partial t} = D\frac{\partial^2 c(x,\ t)}{\partial x^2}. \tag{4.5}$$

Equations (4.4) and (4.5) are one-dimensional equations defining Fick's second law of diffusion for concentration-dependent and concentration-independent diffusion coefficients, respectively.

4.2.1 Solving the Diffusion Equations

To determine the diffusion behavior of a given material into any other material (host), one must carefully determine the concentration of the diffusing material as a function of its depth in the host material. Equation (4.4) or (4.5) is then solved using the boundary conditions that have been defined by the experiment. *Mathematics of Diffusion* by Crank [1] gives a variety of solutions fitting various boundary conditions. For most practical cases of diffusion in semiconducting materials, two different boundary conditions are employed. In the first case, the diffusing substance is deposited within a certain restricted region at $t = 0$ and allowed to diffuse throughout the host medium. This diffusing substance could be in the form of a thin film, or an instantaneously deposited layer of a dopant such as those resulting from PBr_3 or $POCl_3$ predepositions (see Section 4.4.2) or ion implantation. Thus, in this case, the total amount of matter is constant, i.e.,

$$\int_0^\infty c(x,\ t)\ dx = M. \tag{4.6}$$

The other boundary condition is

$$c(\infty,\ t) = 0, \tag{4.7}$$

suggesting that the host is an infinite medium in the x direction. Also the condition at $t = 0$ is

$$c(x,\ o) = 0. \tag{4.8}$$

The solution to eq. (4.5) that obeys boundary conditions (4.6) through (4.8) is simply the Gaussian distribution

$$c(x,t) = \frac{M}{(\pi Dt)^{1/2}} \exp(-\frac{x^2}{4Dt}). \tag{4.9}$$

Note that boundary condition (4.7) is applicable in all those cases where the diffusion host's dimension in the diffusion direction (in this case x) is significantly larger than the diffusion depth. This is true in most cases, diffusion depths being on the order of microns compared to the host dimensions of hundreds of microns. Also note that after a diffusion time, t, the surface concentration at $x = 0$ is given by c_s = c(0,t), i.e.,

$$c_s = \frac{M}{(\pi Dt)^{1/2}}. \tag{4.10}$$

In the second case, the diffusion occurs from a constantly maintained surface concentration. For example, the diffusion source could be a gas such as phosphine or diborane maintained at a constant pressure, a deposited oxide such as heavily doped phoshorus glass, or a BN disc in contact with the semiconductor. This implies that the surface concentration, c_s, is constant, i.e.,

$$c(o,t) = c_s = \text{constant}. \tag{4.11}$$

Once again

$$c(x,0) = 0, \text{ and} \tag{4.12}$$

$$c(\infty,t) = 0. \tag{4.13}$$

The solution to eq. (4.5) that satisfies these boundary conditions is the complementary error function distribution

$$c(x,t) = c_s \operatorname{erfc}(\frac{x}{2\sqrt{Dt}}), \tag{4.14}$$

The complementary error function is related to the error function as shown,

$$\operatorname{erfc}(z) = 1 - \operatorname{erf}(z). \tag{4.15}$$

The error function is defined as

$$\operatorname{erf}(z) = \frac{2}{\sqrt{\pi}} \int_0^z \exp(-\eta^2)\, d\eta. \tag{4.16}$$

TABLE 4.1: Values of the error function and error function complement

x	erf x	erfc x	x	erf x	erfc x
0	0	1.0000	0.70	0.6778	0.3222
0.05	0.0564	0.9436	0.75	0.7112	0.2888
0.10	0.1125	0.8875	0.80	0.7421	0.2579
0.15	0.1680	0.8320	0.90	0.7969	0.2031
0.20	0.2227	0.7773	1.0	0.8427	0.1573
0.25	0.2763	0.7237	1.1	0.8802	0.1198
0.30	0.3286	0.6714	1.2	0.9103	0.0897
0.35	0.3794	0.6206	1.4	0.9523	0.0477
0.40	0.4284	0.5716	1.6	0.9763	0.0237
0.45	0.4755	0.5245	1.8	0.9891	0.0109
0.50	0.5205	0.4795	2.0	0.9953	0.0047
0.55	0.5633	0.4367	2.4	0.9993	0.0007
0.60	0.6039	0.3961	2.8	0.9999	0.0001
0.65	0.6420	0.3580			

Table 4.1 lists the values of the error function and complementary error function as a function of the argument z [23].

The amount of material M' diffused (from a constant concentration source) in a given time t can be calculated by first calculating the rate J_R at which material diffuses across the surface $x = 0$. J_R is given by

$$J_{\mathrm{R}} = \left(D \frac{\partial c}{\partial x} \right)_{x=0}. \tag{4.17}$$

Using eq. (4.14) for c and solving eq. (4.17) will give

$$J_{\mathrm{R}} = \frac{Dc_{\mathrm{s}}}{(\pi Dt)^{1/2}}. \tag{4.18}$$

Now

$$M' = \int_0^t J_{\mathrm{R}}\, dt,$$

$$M' = 2c_{\mathrm{s}}\left(\frac{Dt}{\pi}\right)^{1/2}. \tag{4.19}$$

Compare eq. (4.19) with (4.10). In eq. (4.19) M', the amount of matter diffused in, is a function of time and c_s is constant, whereas in eq. (4.10)

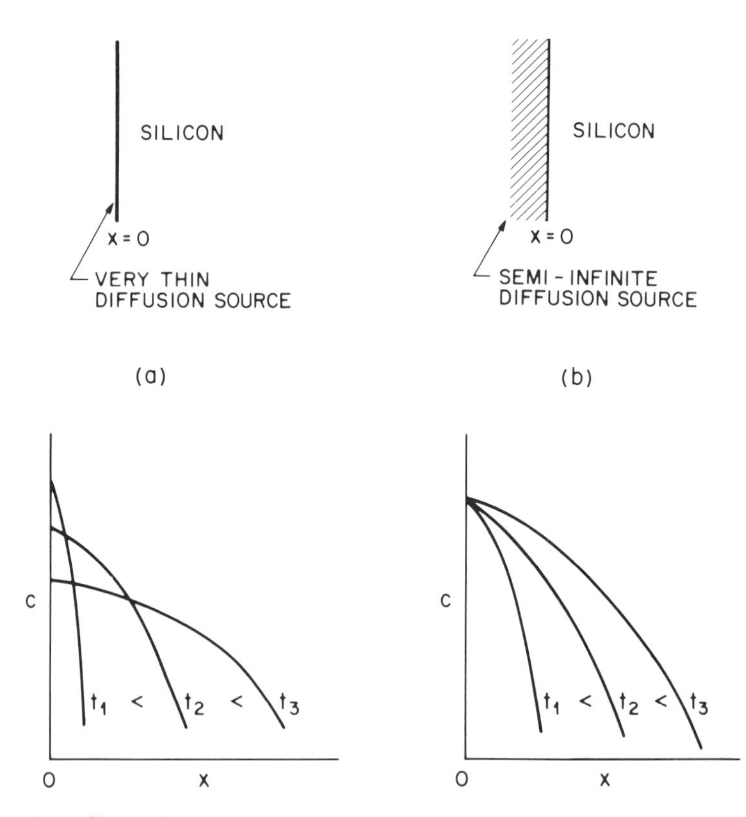

Fig. 4.1. Shows the typical concentration versus depth curves for diffusion studies using (a) a thin film source, resulting in the use of the Gaussian solution [eq. (4.9)] of the diffusion equation and (b) a semi-infinite source, resulting in the use of error function solution [eq. (4.14)]. The t indicates diffusion anneal times in the increasing order.

the total amount of matter M is constant and independent of time and c_s is a function of time. Figure 4.1 shows c versus x plots for the cases (a) and (b) corresponding to eqs. (4.9) and (4.14), respectively. Integrating the area under the curves gives M and M', and relationships in eqs. (4.10) and (4.19) then become obvious.

4.2.2 Diffusion Depth and Junction Depth

Diffusion depth x_D is simply defined as the depth to which the diffusing species have diffused in a given time, t, at the given temperature T°K. Theoretically x_D should be infinite, since x_D will be determined by the condition $c = 0$ at $t > 0$. For most practical considerations x_D is defined as

$$x_D^2 = 4Dt. \tag{4.20}$$

Use of this equation to determine a typical diffusion depth simply indicates that c/c_s has a predetermined value: For Gaussian solution (4.9) $c/c_s = 1/e = 0.37$, and for the error function solution (4.14) $c/c_s = 0.1573$.

Junction depth x_j refers to the diffusion depth at which the concentration of the diffusing species equals a concentration in the host material. For example, silicon may have a base uniform doping of phosphorus c_n in the crystal. If boron is diffusing, x_j will be the depth at which boron concentration c_p equals c_n using eq. (4.9)

$$c = c_p = c_n, \text{ and } x = x_j \text{ so that}$$

$$x_j^2 = 4Dt\ell n\frac{c_s}{c_n}, \tag{4.21}$$

where D is the diffusion coefficient of boron in phosphorus-doped silicon and t is the diffusion time. A similar expression for x_j can be obtained using the error function solution of eq. (4.14).

Problem 4.1: *In a diffusion experiment, phosphorus, from an infinitesimal fixed deposit, was diffused in silicon (with boron concentration of 10^{15} per cm^3) at 1000° C for 2 hours, resulting in a junction depth of 0.537 μm. Estimate the diffusion coefficient and the total amount of the phosphorus initially deposited on the surface.*

If, on the other hand, a constant surface concentration of 5×10^{20} per cm^3 was maintained during the above diffusion anneal, calculate the total amount of phosphorus diffused in silicon.

Solution: We can estimate D by using eq. (4.20)

$$\begin{aligned} x^2 &= 4Dt \\ D &= \frac{x^2}{4t} = \frac{(0.537 \times 10^{-4})^2}{4 \times 2 \times 3600} \\ &= 1.0 \times 10^{-13} \text{ cm}^2/\text{s}. \end{aligned}$$

Using the Gaussian solution of eq. (4.9) to the diffusion equation, we can write the amount of material initially deposited as

$$\begin{aligned} M &= c(n,\ t)\ \sqrt{\pi Dt}\ \exp\left(\frac{x^2}{4Dt}\right) \\ &= 10^{15}\ \sqrt{\pi \times 1 \times 10^{-13} \times 2 \times 3600}\ \exp(1) \\ &= 4.756 \times 10^{10}\ e\ =\ \underline{1.29 \times 10^{11}} \text{ per cm}^2. \end{aligned}$$

For the constant surface concentration case, the total amount of material difference is given by eq. (4.19), i.e.,

$$M' = 2\ c_s \left(\frac{Dt}{\pi}\right)^{1/2}$$
$$= 2 \times 5 \times 10^{20} \left(\frac{1 \times 10^{-13} \times 2 \times 3600}{3.1416}\right)^{1/2}$$
$$= 1.5 \times 10^{16} \text{ per cm}^2.$$

4.2.3 Concentration-Dependent Diffusion Coefficient

Equation (4.5) and the solution of this equation described in Sections 4.2.1 and 4.2.2 are applicable when D does not depend on the concentration. This is generally true for dilute solutions, i.e., at low concentrations. In most cases the diffusing species are diffusing into the host free of such species. There are situations where the diffusing species diffuse from a higher concentration source to a lower concentration host, i.e. $c(x, o) \neq 0$. In such cases, i.e., at higher concentrations, the diffusion coefficient is concentration dependent, and eq. (4.5) is not valid. Instead eq. (4.4) should be used. Solution of eq. (4.4) is not trivial. Crank [1], in his Chapters IX and XII, has considered various approaches in solving eq. (4.4). These approaches, called Boltzmann-Matano analyses, are described below.

4.2.3.1 Boltzmann-Matano Analysis

For D, a function of concentration c only, a new variable is defined as:

$$\eta = \frac{x}{2t^{1/2}} \tag{4.22}$$

so that

$$\frac{\partial c}{\partial x} = \frac{1}{2t^{1/2}}\frac{dc}{d\eta}, \tag{4.23}$$

and

$$\frac{\partial c}{\partial t} = -\frac{x}{4t^{3/2}}\frac{dc}{d\eta}. \tag{4.24}$$

Eq. (4.4) is then transformed into

$$\frac{dc}{d\eta} = -\frac{1}{2\eta}\frac{d}{d\eta}(D\frac{dc}{d\eta}). \tag{4.25}$$

Eq. (4.25) is an ordinary differential equation and is valid for diffusion in an infinite or a semi-infinite host with the conditions

$$c = c_1 \text{ for } x < 0,$$
$$\text{i.e., } \eta = -\infty \text{ at } t = 0,$$

and

$$c = c_2 \text{ for } x > 0, \tag{4.26}$$
$$\text{i.e., } \eta = \infty \text{ at } t = 0.$$

One can obtain $D(c)$ from eq. (4.25)

$$D(c) = \frac{-2 \int_{co}^{c} \eta dc}{dc/d\eta} \tag{4.27}$$

by plotting c as a function of η for a specific time and numerically obtaining the slope $dc/d\eta$ and the integral $\int_{co}^{c} \eta dc$ for each c value. Since an experimental c versus x profile can be plotted for a given diffusion annealing time t, one can treat t as constant and convert eq. (4.27) to

$$D(c) = \frac{-\int_{co}^{c} x dc}{2t(dc/dx)}. \tag{4.28}$$

Once again from the c versus x curves one can calculate the integral and the slope at concentration c and obtain $D(c)$ at this value of c. A typical Boltzmann-Matano type plot is shown in fig. 4.2.

The Boltzmann-Matano analysis has frequently been used in the studies of diffusion in alloy systems. Applications in semiconductor systems are not so common. The assumption that two variables x and t can be replaced by η must be checked by obtaining a large number of concentration profiles for various diffusion times, with boundary conditions, temperature, and pressure being kept constant. The concentration versus η profiles should be identical, and the x versus $t^{1/2}$ plot for a given c should be a straight line.

4.2.3.2 Assumed Concentration Dependences

One can assume D to be given as

$$D = D_a f(c), \tag{4.29}$$

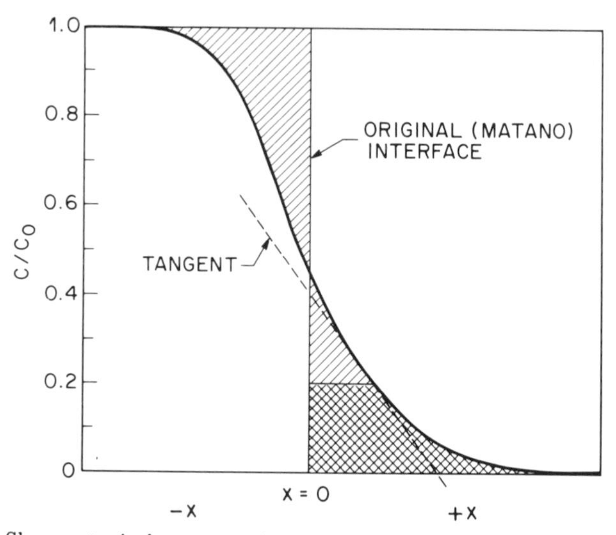

Fig. 4.2. Shows a typical concentration versus depth plot for a binary diffusion couple leading to the Boltzmann-Matano type analysis. Diffusion coefficient can be calculated by drawing a tangent and determining the slope and area of the curve at a given value of c/co, as shown at $c/co = 0.2$.

where $f(c)$ is an arbitrary function of c, and D_a is the concentration-independent diffusion coefficient or a diffusion coefficient at a designated concentration. Crank [1] has examined various possible solutions, and the range of applicability for a given solution. The most commonly selected function is $f(c) = c/c_o$, with boundary conditions

$$c = c_o \text{ at } x = 0 \text{ for } t > 0 \text{ and}$$
$$c = o \text{ at } x > 0 \text{ for } t = 0. \tag{4.30}$$

For a given set of experimental concentration profiles one can use trial and fit or iterative methods to obtain an optimum $f(c)$ and hence $D(c)$. We shall discuss concentration-dependent diffusivities again when considering diffusion mechanisms (see Section 4.6.1).

4.2.4 Temperature-Dependent Diffusion Coefficient

Diffusion of one species into another strongly depends on the temperature, since the phenomenon is a thermally activated process (see Section 4.5). Diffusion could be a result of a single thermally activated process, or it can be the result of a number of processes each with its own activation energy. For a uniquely defined single-step activated process, one can write the temperature dependence of D in the Arrhenius form

$$D = D_o \exp(-\frac{Q}{kT}), \tag{4.31}$$

where D_o is the preexponential factor in the units of D (length)^2/time, the most popularly used unit being cm^2/s; Q is the activation energy in electron volts (eV); k is Boltzmann's constant in eV per degree Kelvin; and T is given in degrees Kelvin. D_o is related to the lattice jump distance, atomic jump frequency, the frequency factor, and the entropy of diffusion. Q is related to the specifc atomic migration path and the enthalpies associated with jumps along that path; that is, Q is determined by the energy expended in getting an atom to move from an interstitial site to another interstitial site or from a lattice site to a vacancy, etc. A number of available pathways are discussed in Section 4.5.

4.3 Determination of Diffusion Coefficient—Concentration Profiles

Determination of the diffusion coefficient involves careful determination of concentration versus depth profiles of the samples that have been heat treated at various temperatures and under various conditions. Classically, the determination of the diffusion profiles has been easier for diffusion depths of several hundred micrometers. This technique involves determination of the solute concentration in thin (1 to 2 μm) layers removed from the specimen surface by mechanical (grinding or turning on a precision lathe) or by electrochemical means (e.g., anodic oxidation). Radiotracer technique, in which a radioactive isotope of the diffusing species is used for easy identification and analysis, has provided the most accurate diffusion profiles and coefficients. In semiconductors the diffusion depths are very shallow and soon the useful depths will be in the range of 10-200 nm. This is a challenge for the analytical techniques since the total numbers of atoms in such a small volume will be in the range of $10^9 - 10^{16}$. Analysis by sectioning of the diffused layer will reduce this number by another order of magnitude. In addition, as shown in fig. 4.3, the concentration gradient is very high, and concentration drops by a factor of $10^5 - 10^6$ in a distance of 10-200 nm. Finally, an accurate measurement of the diffusion coefficient requires very accurate control of the diffusion anneal temperature T. A small error of ΔT in T will introduce an error ΔD in D given by

$$\frac{\Delta D}{D} = (\frac{Q}{RT^2})\Delta T. \tag{4.32}$$

An uncertainty of 5°K in temperature (i.e., ΔT = 5°K) will cause an error of 14% or 18% in D for a Q of 4 or 5 eV, respectively, at a diffusion temperature of 1273°K. Thus, a temperature control of better than 5°K is essential.

The following profiling techniques could be used:

1. Spreading resistance measurements

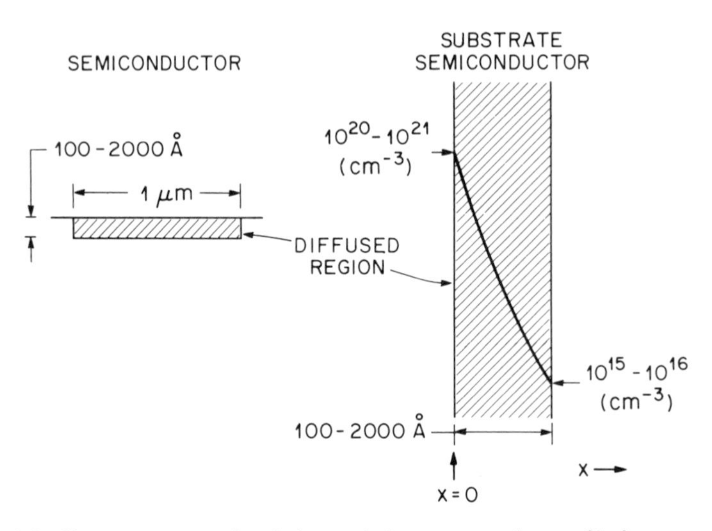

Fig. 4.3. Shows a cross-sectional view and the concentration profile for a very shallow and narrow junction in a semiconductor.

2. Sheet resistance measurements
3. $C - V$ method
4. Radio-tracer studies
5. Auger Analysis coupled with ion-milling
6. Secondary Ion Mass Spectrometry (SIMS)
7. Rutherford Backscattering Analysis (RBS)
8. Transmission Electron Microscopy with or without EDAX

Each of these techniques are discussed below.

4.3.1 Spreading Resistance Measurements

Consider a contact made between the semiconductor surface and the tip of a small conducting probe. When a voltage is applied to the probe, the current spreads around the contact area, as shown in fig. 4.4. Due to this current spreading, approximately 80% of the total potential drop along the surface occurs at a distance of about 5 times the cylindrical or hemispherical contact radius. The resistance between two probes associated with current spreading is given by [24]

$$R_{\text{sp}} = \frac{\rho}{A} \ , \tag{4.33}$$

where subscript sp represents spreading, ρ is the average semiconductor resistivity, and A is the probe diameter.

In deriving eq. (4.33), the following assumptions have been made: a) The semiconductor has much higher resistivity than the probe material, b) the

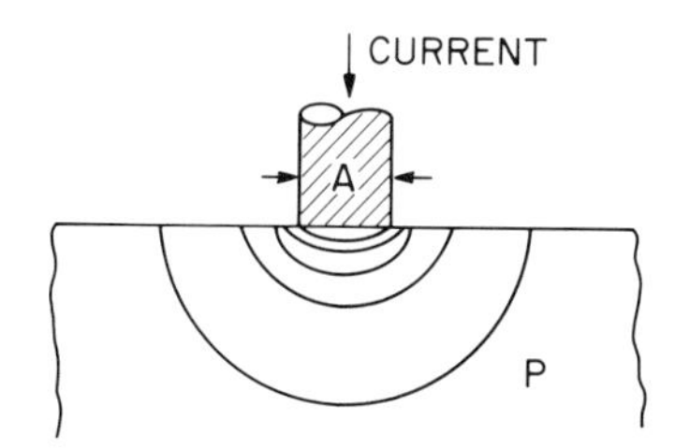

Fig. 4.4. Schematic drawing of potential distribution, in a semi-infinite medium, associated with the current constriction through a small contact.

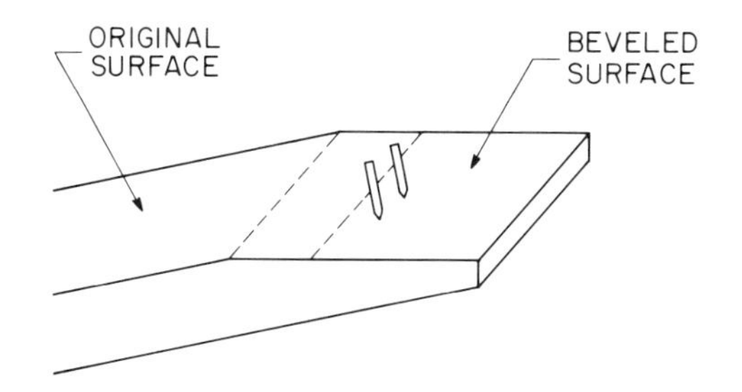

Fig. 4.5. Beveled surface with spreading resistance measuring probes.

semiconductor surface area is large compared to the area of contact with the probe, c) there is no damage to the surface, d) ohmic behavior is applicable, and e) spacing between probes (typically in the range of 10 – 1000 μms) is such that the bulk resistance between the probes is small compared to the resistance associated with charge injection into the substrate away from the volume between probes. Thus R_{sp} can be taken as the measured resistance between two probes on the semiconductor surface. In practice it is difficult to separate the small bulk resistance and the resistance associated with contact barrier effects from the spreading resistance.

The spreading resistance is very sensitive to local dopant concentration variations. Because of this, the measurements using two probes have been used very successfully in determining spreading resistance depth profiles. These profiles can then be translated into the dopant concentration depth profile. The technique is independent of the type of dopants or the methods used to introduce them. To carry out a depth profile, the doped sample surface is sectioned at a very shallow angle of about half a degree. This beveled edge exposes the diffused layer where probes can be placed and moved along the depth. Figure 4.5 shows a typical arrangement. From the measured value of R_{sp}, ρ is determined as a function of depth. Finally, ρ is then converted into dopant concentration as described in 4.3.2. Figure 4.6 compares a spreading resistance generated concentration profile with that generated using the incremental sheet resistance measuring technique. As is apparent, the two techniques yield similar profiles.

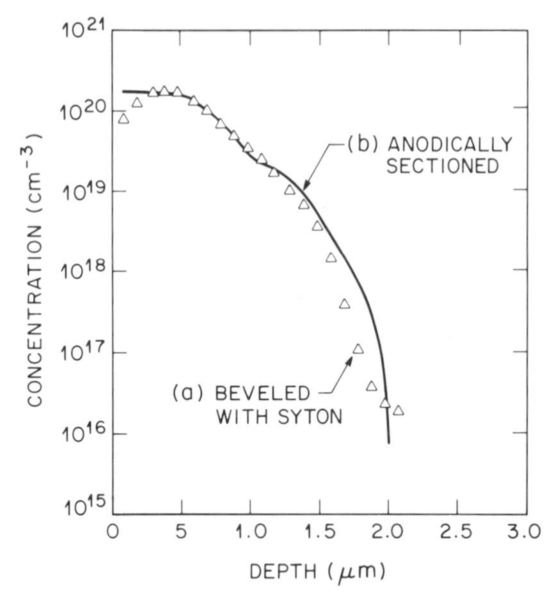

Fig. 4.6. Phosphorus diffused concentration profiles obtained by (a) spreading resistance measurements on the beveled specimen and (b) incremental sheet resistance measurement. From D'Avonzo et al. [26]. Reprinted by permission from the publisher, The Electrochemical Society, Inc.

The spreading resistance technique, although apparently sensitive to concentration variation in layers with a total thickness of 100 nm or less [25], is mostly used for comparing an unknown specimen with a known. The difficulty in making absolute determinations of concentration is related to the assumptions made earlier. For a precise, independent depth profile measurements, painstaking care should be taken in preparing the beveled surface to assure ohmic contact formation. Probe separation must be precisely measured, and the constant current source and voltage meters be in good calibration.

4.3.2 Sheet Resistance Measurements

For a plane sheet of length ℓ, thickness x, and width w, the electrical resistance is given as

$$R = \rho \frac{\ell}{xw}, \tag{4.34}$$

where R is given in ohms, ρ (the resistivity of the material) is given in ohm-cm and ℓ, w, and x are each given in cm. When the plane sheet is a square $\ell = w$, we obtain a quantity R_s, called sheet resistance, given by

$$R_s = \frac{\rho}{x}, \tag{4.35}$$

where R_s is now given in units of ohms per square (usually written as $\Omega/\square$). It is clear that R_s is independent of the size of the square.

As discussed in Chapter 1, sheet resistance, in $\Omega/\square$, is very commonly used in calculating resistance of the conductors in circuits and chips. The entire length of the conductor is divided into a number of squares. This number is then simply multiplied by the metallic film sheet resistance. For example, a 1-cm long 1-μm wide conductor will have 10^4 squares. Thus, the total resistance of such a conductor will be $10^4\ R_s$. R_s can be calculated using eq. (4.35) and ρ and x.

The sheet resistance, R_s, for a diffused layer as derived in Chapter 1 is given as

$$\frac{1}{R_s} = q\int_o^{x_j} \mu(x)c(x)\,dx. \tag{4.36}$$

Note that mobility has been considered to be concentration dependent, especially at high dopant concentrations.

An effective conductivity σ is now defined as

$$\sigma = \frac{1}{R_s x_j} \tag{4.37}$$

such that

$$\sigma = \frac{q}{x_j}\int_o^{x_j} \mu(x)c(x)\,dx. \tag{4.38}$$

Equations (4.37) and (4.38) provide a means of generating diffusion profiles $c(x)$ versus x and of calculating the diffusion coefficients. Irvin [27] has numerically computed eq. (4.38) for various diffusion profiles, substrate doping levels, and erfc or Gaussian distributions. For a known substrate dopant concentration, the surface dopant concentration of an assumed concentration profile has been evaluated in terms of the quantity σ. Figure 4.7 shows a typical set of Irvin curves for erfc diffusion profile and bulk substrate concentration (c_B) in the range of 10^{14} to 10^{20} per cm^3. Compilations of such curves are given in references 28 and 29.

The concentration profile is determined experimentally by measuring the sheet resistance of the diffused layer using a four-point probe technique, and the junction depth is measured by chemically staining a beveled edge and using an interference-fringe method [30]. The most common four-point probe arrangement is shown in fig. 4.8a [29]. For this arrangement

$$R_s = \frac{V}{I}G_f, \tag{4.39}$$

where G_f is the geometrical correction factor that depends on the sample dimensions (a and d for a rectangular sample and a for a circular sample are shown in fig. 4.8b) and the probe spacing s. Table 4.2 lists the values of G_f

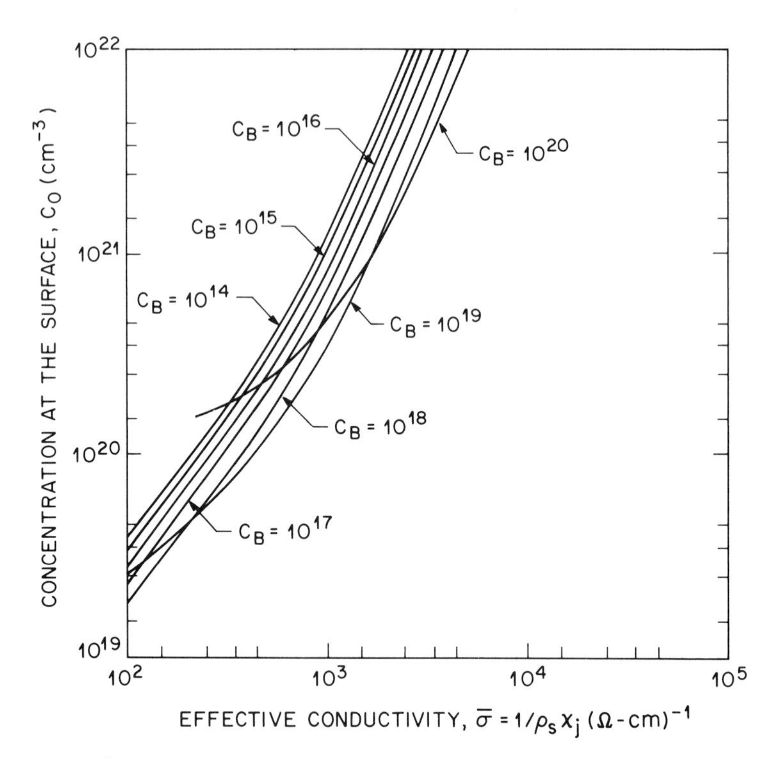

Fig. 4.7. Effective conductivity of diffused layers in n-Si. Error function solution for surface concentration in the range of $10^{22} \geq \text{Co} \geq 10^{19}$.

as a function of d/s for a circular or a rectangular sample. For a semi-infinite sample $G_f = 4.53$ so that

$$R_s = 4.53\frac{V}{I}. \tag{4.40}$$

Generally, a constant current of 4.53 mA is maintained between the outer probes and the voltage drop, between two inner probes, is read in mV to give R_s directly in $\Omega/\square$. The correction factor G_f, as given in table 4.2, is applicable only for a total diffusion depth that is considerably less than the probe spacing. Also diffusion should have occurred only on the measuring side of the specimen, i.e., no continuity or electrical conduction should occur between the surface being evaluated and other surface, such as the sides and the back of the wafer. For very shallow diffusion depths, this type of R_s measurement could be difficult to make due to associated contributions from the surface generation effects, contact resistance of the probes, and the use of higher voltages or currents.

Junction depth can be measured by beveling the junction region on a shallow angle chuck (usually $\sim 1°$). A special staining solution that selectively stains the doped region is then used to produce the desired contrast. Finally, the sample is examined under a microscope using interference-fringe

TABLE 4.2: Geometrical correction factor G_f (for sample thickness or junction depth $< s/2$)

d/s	Circle	$a/d = 1$	$a/d = 2$	$a/d = 3$	$a/d \geq 4$
1.00				0.9988	0.9994
1.25				1.2467	1.2248
1.50			1.4788	1.4893	1.4893
1.75			1.7196	1.7238	1.7238
2.00			1.9454	1.9475	1.9475
2.50			2.3532	2.3541	2.3541
3.00	2.2662	2.4575	2.7000	2.7005	2.7005
4.00	2.9289	3.1137	3.2246	3.2248	3.2248
5.00	3.3625	3.5098	3.5749	3.5750	3.5750
7.50	3.9273	4.0095	4.0361	4.0362	4.0362
10.00	4.1716	4.2209	4.2357	4.2357	4.2357
15.00	4.3646	4.3882	4.3947	4.3947	4.3947
20.00	4.4364	4.4516	4.4553	4.4553	4.4553
40.00	4.5076	4.5120	4.5129	4.5129	4.5129
inf	4.5324	4.5324	4.5324	4.5324	4.5324

techniques of Tolansky [30]. From the number of fringes, the junction depth is directly calculated. For a sodium light source (wavelength at 589.593 and 588.996 nm), the distance between fringes is approximately 0.29 μm. For junctions shallower than 0.29 μm, other techniques must be employed.

One can construct an experimental diffusion profile by a) serial etching the diffused layer, b) measuring the sheet resistance and junction depths, and c) using Irvin's curves for an erfc or a Gaussian profile. Serial etching can be carried out in a chemical etch or anodically. Figure 4.9 shows an experimental and computer-generated diffusion profile for boron diffusion in silicon. The best fit to experimental data was obtained by iteration using the diffusion coefficient and the segregation coefficient (see Chapter 3 for diffusion during an oxidizing anneal).

4.3.3 Capacitance-Voltage Method

For a one-sided abrupt $p - n$ junction, the measurement of the reverse-bias junction capacitance C_{pn} can provide useful information about the dopant distribution when used in the expression

$$c(w) = \frac{2}{q\epsilon_s} \cdot \frac{1}{d(1/C_{pn})^2/dV}. \tag{4.41}$$

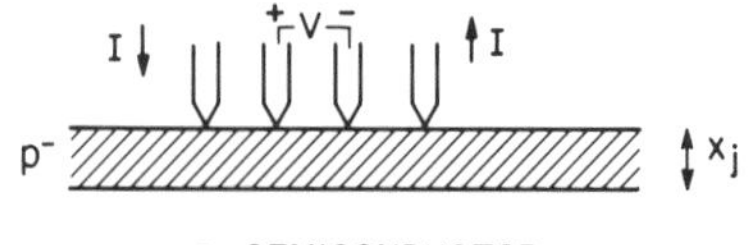

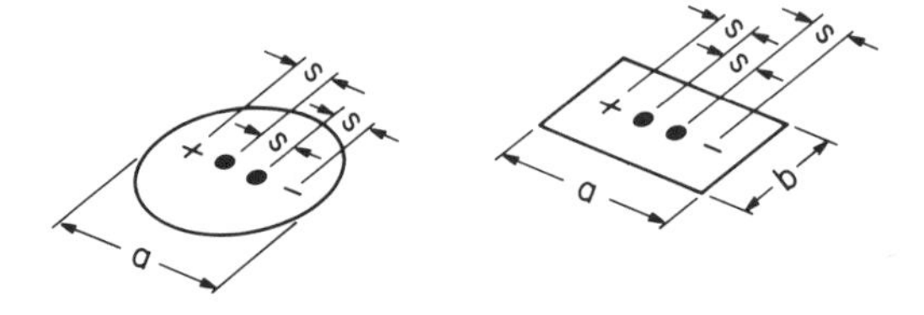

(b)

Fig. 4.8. (a) A four-point sheet resistance measurement system. A fixed current is forced through outer probes and voltage is measured between two inner probes. (b) Shows sample dimensions and probe spacings for circular and rectangular samples.

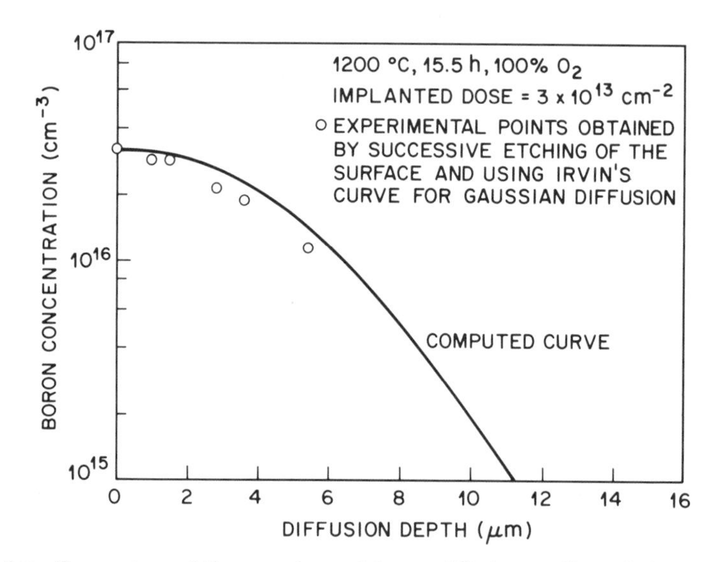

Fig. 4.9. Comparison of the experimental boron diffusion profile and the computed profile for boron diffusion at 1200°C/15.5h in 100% oxygen and assuming $m = 1$.

Thus measurements of C_{pn} as a function of reverse-bias voltage can be used to calculate the concentration $c(w)$ at the space charge layer edge. In eq. 4.41

w is the width of the depletion layer. MOS capacitors, $p-n$ junctions, and Schottky barriers have been used to obtain $1/C^2 - V$ plots that lead to dopant concentration profiles [31-35].

Wu et al. [32] have examined the limitations of this technique in determining impurity profiles. Validity of the depletion approximation was considered, and it was concluded that the conventional $C-V$ method is limited to distances of a few extrinsic Debye Lengths [see eq. (1.24)] from the depletion layer edge at zero bias. Thus, although $C-V$ measurements are easy to carry out and nondestructive, at best they provide a good approximation.

4.3.4 Radio-Tracer Studies

The most direct way of determining the impurity concentration profiles of diffusion samples is to use a radioactive isotope (the tracer) as the diffusing species [36]. Table 4.3 lists the radioisotopes that could be employed for dopant diffusion studies in semiconductors. Selection of the isotope depends on its producibility, decay half-life, diffusion anneals, health hazards and disposability, and on the availability of analytical equipment for detection and counting of it.

The technique is simple in principle, but, due to health hazard and radioactive contamination problems, the experimentation requires very careful planning and complete control at each step, from making the diffusion samples to the annealing, the sectioning, the counting, and the final disposal. In a typical diffusion profile generating sequence, a radioactive tracer will be deposited by electroplating or evaporation on the polished oxide-free surface of the semiconductor. Samples will then be encapsulated in glass ampoules and annealed. They will then be sectioned to divide the diffused zone into thin sections so that both the diffusion depth perpendicular to the sample surface and the concentration of the radiotracer in the thin section can be determined. Thin layers of diffused material are removed from the surface by using a lathe, by grinding, or by slicing with a microtome. The latter is a sectioning instrument with a precisely positioned steel blade to remove thin layers from softer material. In all cases layers are removed in a fashion that the advancing surface is parallel to the original surface and the thickness of the diffused layer is determined in the direction normal to this surface. Some other methods of sectioning are chemical etching, anodization, and sputtering. In all cases, the removed material is collected and counted for its tracer content. Section depth is determined on the basis of weight loss, predetermined chemical etch rate, anodization rate, or sputtering rate. In this way a concentration versus depth profile is determined in the most direct manner.

Alternatively, the diffusion sections are removed and the residual activity from the face of the crystal is monitored [38]. The residual activity, I_n, left

TABLE 4.3: Radioisotopes of dopants [37]

Semiconductor	Radioisotope	Half-life	Useful Radiation	
			Type	Energy (MeV)
Silicon	Silicon-31	2.62h	γ	1.270
	Silicon-32	650y	β^-	0.217
	Phosphorus-32	14.3d	β^-	1.710
	Arsenic-76	26.5h	γ	0.559
	Antimony-122	2.8d	γ	0.561,0.687
	Antimony-124	60.3d	γ	0.603,1.692
	Gallium-72	14.1h	γ	0.630,0.835
	Indium-114M	50.0d	γ	0.192,0.556
Germanium				
	Germanium-77	11.3h	γ	0.264,0.212
III-V	Silicon-31	2.62h	γ	1.270
	Silicon-32	650y	β^-	0.217
	Germanium-77	11.3h	γ	0.264,0.212
	Sulphur-35	88d	β^-	0.167
	Selenium-75	120.4d	γ	0.265,0.136
	Tellurium-121	17d	γ	0.573,0.508
	Tellurium-123M	117d	γ	0.159
	Zinc-65	243.6d	γ	1.116,0.511
	Cadmium-115	2.25d	γ	0.530,0.490
	Cadmium-115M	43.0d	γ	0.935,1.295
	Mercury-203	46.57d	γ	0.279
	Tin-117	14d	β	0.317
	Tin-125	9.4d	γ	1.068,0.811

in the diffusion sample after removal of the $\mathrm{n^{th}}$ layer at a depth x_n from the original surface, is given by

$$I_n = \int_{x_n}^{\infty} c(x,t) \cdot \exp[-\mu(x - x_n)]\, dx. \tag{4.42}$$

The general solution of this equation is

$$\mu I_n - \frac{dI_n}{dx_n} = A\, c(x_n). \tag{4.43}$$

In eqs. (4.42) and (4.43), μ is the linear absorption coefficient (in $\mathrm{cm^{-1}}$) of the monitored radiation in the diffusion sample, and A is a constant.

Eq. (4.43) is independent of the functional form of $c(x)$. The left-hand side of this equation can be evaluated by plotting I_n versus x_n and then fitting the curve with the help of a computer. The residual analysis method is most accurate when μ is so large that most of the emitted radiation is absorbed in a thin section on top. In such a case, eq. (4.43) reduces to

$$\mu I_n \approx A\ c(x_n). \tag{4.44a}$$

Such applications use low-energy β emitters such as those from silicon-32 or nickel-63. On the other hand, for high-energy γ radiation sources, μ is so small in most materials that the first term in the left-hand side of eq. (4.43) can be neglected, leading to

$$-\frac{dI_n}{dx_n} = A\ c(x_n). \tag{4.44b}$$

The residual activity method is a very convenient method, but its accuracy could be compromised because of required controls on sectioning and counting, and a lack of accurate knowledge about μ.

The radiotracer method, although most direct, is usually limited by the detectability of the isotope's radiation. However, removal of extremely thin sections in cases where the total diffusion depth may not exceed 250 nm, will be very difficult. In such cases, an integral activity method may be useful [39, 40]. In this method, the radiations emanating from both the front and the back surfaces of a thin sample are monitored as a function of the diffusion time. If $I_1(t)$ and $I_2(t)$ are the activities from the front and back surfaces and are separated by a distance ℓ (i.e., sample thickness $= \ell$), it can be shown that

$$\ell n\left(\frac{I_1 - I_2}{I_1 + I_2}\right) = \ell n K - \frac{\pi^2 Dt}{\ell^2}, \tag{4.45}$$

where t is the diffusion time and K is the term containing all absorption terms. Thus, a plot of the left hand side of eq. (4.45) as a function of t, yields D directly. The requirement for successful use of this technique is that ℓ should be small, on the order of μ^{-1}.

4.3.5 Auger Analysis Coupled with Ion-Milling

Auger analysis is a surface analytical tool for investigating a few monolayers of a surface [41]. This is due to the short range of low-energy Auger electrons in most solids. Auger electrons are emitted from the high energy electronic shells of the atom. The process is a result of a chain of events: a) ionization of the atom due to the release of an electron from a lower shell (e.g., a K-shell-electron), b) the release of a photon associated with the trapping of an

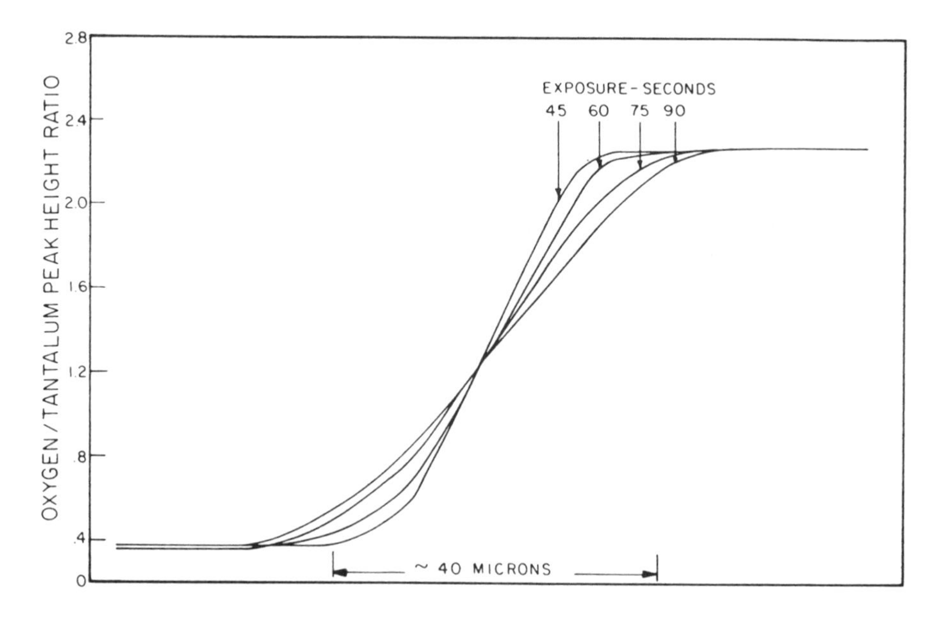

Fig. 4.10. Shows typical concentration profile obtained using Auger analysis coupled with ion milling. In this case, diffusion of oxygen in Tantalum is shown as a function of the oxygen exposure time. From Hudson [42].

electron from an outer shell orbital to fill the vacancy in the lower shell, and finally c) absorption of the emitted photon energy taken up by one of the electrons in the outermost shells causing its ejection from the atom as an Auger electron. The Auger energy is very small (20-2000 eV) and uniquely associated with a given segment of transitions; therefore, identification of the surface atom by an emitted Auger electron is possible. The number of Auger electrons collected is proportional to the number of emitting atoms at the surface analyzed. Quantitative Auger analysis, coupled with ion milling to remove material from the surface, is used to determine a depth profile.

In ion milling, collimated energetic inert gas ions (Ar^+, Ne^+, or Kr^+) are directed to the surface of the specimen to sputter off surface atoms. This leads to a crater formation. Auger spectra are taken continuously during ion-milling processing. Concentration of the species on the surface is then obtained from the Auger peak-to-peak height monitored as a function of the ion-milling time, which is then converted into the depth scale using the known ion-milling rate to obtain a depth profile. Figure 4.10 shows a typical depth profile obtained by this technique [42].

Depth calibration and resolution are difficult to ascertain because of inherent uncertainties associated with ion milling. The degree of the surface roughness, the effect of ion milling on this roughness, interference from the absorbed ion-milling species, crater size, and incident electron beam size, peak shifts associated with chemical bonding, and charging during the ion-milling process make Auger analysis coupled with ion milling, a comparative tool at best. When used with calibration standards, this depth-profiling technique could be used to obtain very shallow depth profiles, e.g., the kind required for VLSI or ULSI diffusions.

4.3.6 Secondary Ion Mass Spectroscopy (SIMS)

In SIMS analysis, the ions sputtered off the surface of a sample are analyzed using a mass spectrometer [43]. A primary beam of Cs^+, 0_2^+, or 0^- is used to ion mill material from the surface, and mass analysis is then carried out on the ions that have been sputtered off. The depth profile is generated in a fashion very similar to that of Auger analysis coupled with ion milling. The ion-milling rate determines the depth. The geometrical shape and texture of the ion milled crater play a crucial role in the depth determination. By using the same ion energy and current, various materials can be ion milled at different rates. Thus, an internal standard consisting of a similar specimen becomes essential for proper calibration. The concentration of the species is determined by the ion current. Thus, accuracy and sensitivity are determined by the ability to separate ions of different masses and to minimize interferences. Production of species of identical masses invariably leads to interferences that are difficult to deal with. Thus, separation of $^{31}P^+$ from $^{31}SiH^+$ peak or of $^{59}Co^+$ from $^{59}Si_2^+$(^{29}Si- ^{30}Si dimer) becomes virtually impossible. In addition, impurity redistribution during ion milling leads to significant errors.

Careful SIMS profiling is becoming very popular in spite of these difficulties. For accurate profiling, the use of calibration standards is generally very effective. Its lower detection limits and its ability to analyze hydrogen, boron, oxygen, phosphorus, arsenic, and many others also make this technique very applicable. One of the serious limitations of SIMS is its spatial (horizontal surface) resolution, which is defined by the primary beam diameter. To improve the detection limit, a large beam diameter is desirable. This, however, is undesirable for smaller size samples, such as lines and spaces, and windows in VLSI and ULSI devices. Finer primary beams (up to redundant ≤ 100 nm in diameter) can be used, but the secondary ion yield is very low, and accurate determination of the ion current has not been possible with current mass analyzers.

A typical SIMS diffusion profiles for B and Sn diffusions [44] in silicon are shown in fig. 4.11. Note that SIMS is the only technique that can yield direct boron profiles reasonably well. In fig. 4.11, the total boron concentration (both electrically active and inactive) is given because SIMS analysis yields total concentration only. Similar profiles can be generated for other diffusions. To separate the active and inactive contributions, electrical techniques used to determine ionized (i.e., active) dopant concentrations

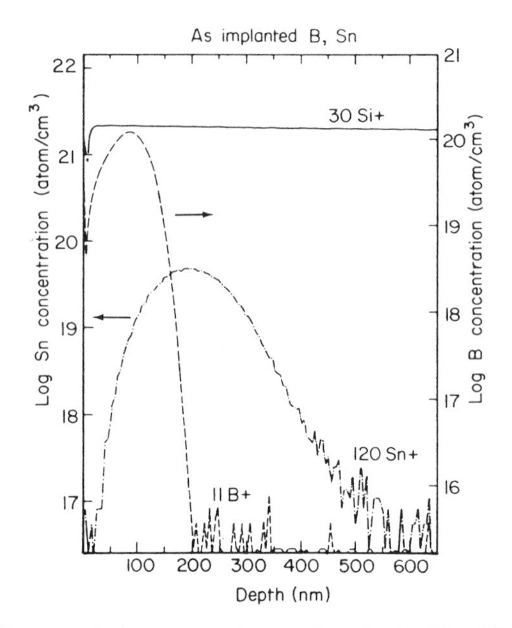

Fig. 4.11. Shows typical concentration profiles obtained by SIMS coupled with ion milling. From Dietrich [44].

should be used to check profiles generated not only generated by SIMS but also by Auger, radiotracer, RBS, or TEM methods.

4.3.7 Rutherford Backscattering (RBS)

In this technique [45], a beam of monoenergetic and collimated helium ions (He^+) impinges perpendicularly on a surface. Some of these ions are backscattered. The number of the backscattered ions is monitored as a function of energy. A typical RBS spectrum (from reference 45, figs. 5.10 and 5.11) of arsenic-diffused silicon is shown in fig. 4.12. The backscattered ion count is plotted as a function of the backscattered ion energy, and this data can be easily converted into a concentration versus depth profile as shown in fig. 4.13.

In fig. 4.12, RBS spectra of as-implanted and diffused samples are shown. Arsenic counts at the high-energy end are due to scattering at or near the surface of the sample where energy loss is small; this leads to higher backscattered energy in the detected helium ions. The backscattered energy decreases because it is being scattered from arsenic atoms at deeper locations within silicon. Thus the tail of the arsenic spectrum in fig. 4.12a indicates diffusion. Note that the tail is missing in the spectrum of the undiffused

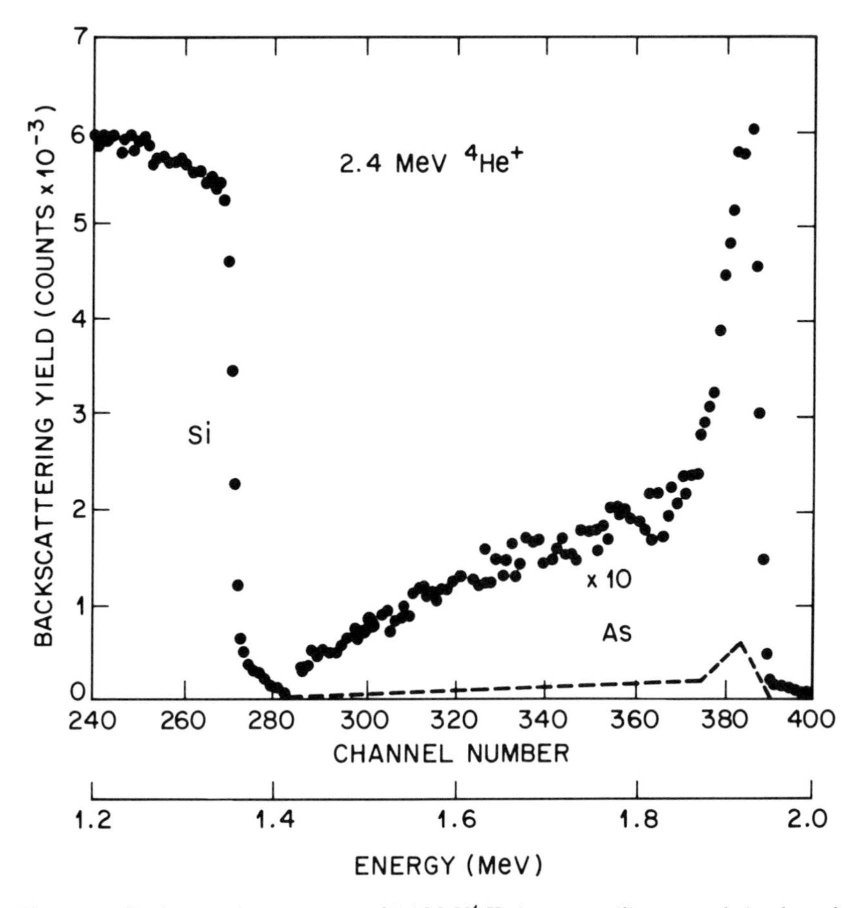

Fig. 4.12. Backscattering spectrum of 2.4 MeV^4 He ions on a silicon sample implanted with a nominal dose of 3.4 x 10^6 arsenic per cm^2 and then heat treated to produce a depth diffusion profile of As. The data points for the arsenic signal are scaled by a factor of 10 above the original signal height (dashed line). From Chu, Mayer, and Nicolet [45].

sample (fig. 4.12b). Figure 4.13 is the concentration versus depth profile generated from the data in fig. 4.12.

The RBS technique provides a very fast and nondestructive way of obtaining diffusion profiles. It does not require a calibration standard, and depth resolution is limited only by the energy resolution of its ion detectors. A depth resolution of 10 nm is now common. With specialized detectors, the resolution has been improved to the 1-2 nm range [46]. Since this is a nondestructive technique, same sample can be used to determine depth profiles as a function of diffusion time.

The sensitivity of RBS is limited, compared to the SIMS or radiotracer methods. Typical RBS sensitivity is $\sim 10^{18}$ atoms/cm^3. The heavier the element, the better the detection limit. Lighter elements like B, C, N, and O are difficult to analyze, although the use of lower mass substrates

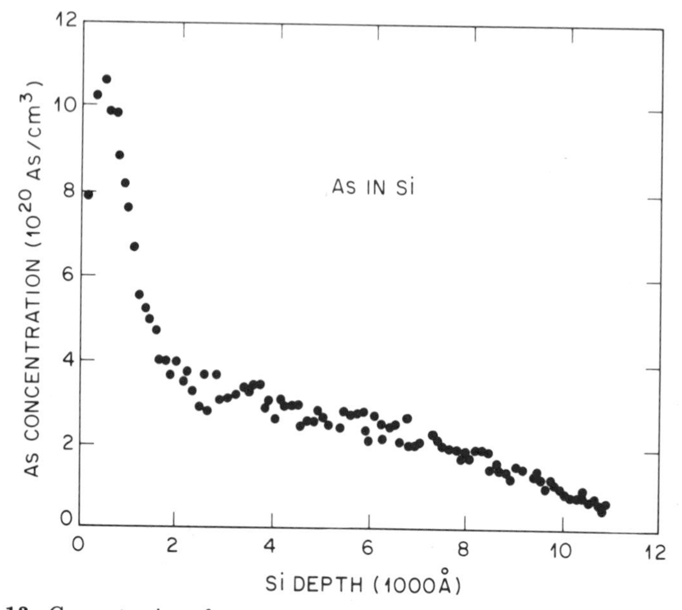

Fig. 4.13. Concentration of arsenic in silicon versus depth obtained from the data of fig. 4.12. From Chu, Mayer, and Nicolet [45].

and of channeling to minimize substrate contribution is being employed to enable detection of these light elements. In addition, elements of very similar masses like P and Si cannot be distinguished from each other, and because RBS utilizes a larger area ion beam, its application is limited to large area samples. Because of the roughness of the surface, scratches, particulates, and neighboring materials on the surface will interfere and lead to incorrect spectras. Thus, to ensure the accuracy of the diffusion profiles, surface cleanliness and smoothness must be insured.

4.3.8 Transmission Electron Microscopy (TEM)

Cross-sectional TEM, often called XTEM, coupled with a junction staining technique, provides the most accurate measurement of junction depths. A detailed description of the sample preparation and the staining technique is given by Marcus and Sheng [47]. Figure 4.14 shows the junction under the field oxide in a MOSFET. Revelation of this $n^+ - p$ junction was made possible by the staining solution 0.5% HF in HNO_3 acid. The solution etches n^+ material at a much higher etch rate than it etches p material or undoped silicon. A similar approach used to delineate $p^+ - n$ junctions has not been very successful. For a successful delineation, an oxide film on the silicon surface should be preserved to provide a depth to fiducial mark. Most p^+ preferential etches etch oxides at a high rate, making delineation

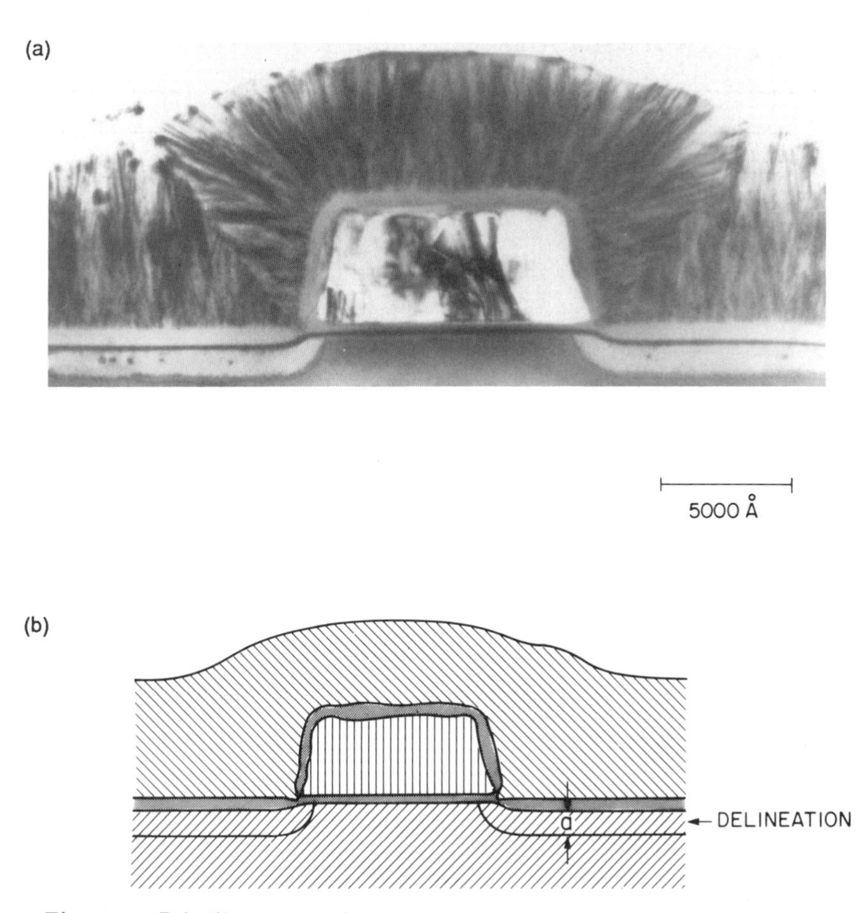

Fig. 4.14. Polysilicon gate and source and drain regions are revealed in (a) the TEM cross section after n^+p junction delineations and (b) schematic drawing of (a). The junction 'a' can easily be measured. The junction was formed by arsenic implantation and diffusion anneal. From Marcus and Sheng [47].

obscure. Very accurate sheet resistance measurements and TEM junction measurements can be used, then, to extract depth profiles, as described in Section 4.3.2.

4.4 Diffusion Methods

All diffusion methods are two step processes. In the first step a source of impurities is provided. The second step is the annealing to allow the impurities to diffuse to the desired depths. In semiconductor doping, the second step electrically activates the dopants in semiconductors. Ion implantation is, at present, the most commonly used technique to introduce dopants. This technique is discussed in the following chapter. Other techniques use

TABLE 4.4: Diffusion sources for the dopants

	Sources for			
Form	Silicon		GaAs	
	n	p	n	p
Solid				
	Polysilicon[a] glass[a]	Polysilicon[a] glass[a]		
	SOG[b]	SOG[b]		
	P-impregnated discs	BN B_2O_3		
Liquid				
	PBr_3 $POCl_3$	BBr_3 $BC1_3$		
Gaseous				
	AsH_3 PH_3	H_2B_6	S,Te,Se[c]	Zn[c]
Ion implantation				
	As,P,Sb	B	S,Te,Se	Zn,Be

[a] Deposited by CVD technique with heavy dopant concentration in the film.
[b] Spin-on-glasses, see text.
[c] Usually solid form, heated to vaporize and produce gaseous diffusion source.

diffusion from solid, liquid, or gas sources. Table 4.4 lists these sources for the doping of silicon and GaAs [48].

4.4.1 Solid Sources

Films containing large concentrations of the diffusing species and deposited on the semiconductor and solid material, in specially formed shapes, in contact with the semiconductor have been used as diffusion sources. Heavily doped polysilicon or SiO_2 films deposited at relatively low temperatures using CVD techniques (Chapter 7) and spin-on-glasses fall into the first category. Preformed BN discs fall into the second category. Some phosphorus impregnated discs are also available for this purpose [49].

Polysilicon films—heavily doped with phosphorus, boron, or arsenic—can be deposited on patterned or unpatterned substrates. Dopants are introduced in polysilicon during deposition, and concentrations in the range of $10^{20} - 10^{22}$ atoms per cm^3 can be attained. Because of the chemical vapor deposition process, a doped polysilicon film can be deposited with excellent step coverage, even over steep surface topographies. Following deposition, the film can be patterned, if so desired, and then annealed to allow dopant diffusion into the substrate. This technique has been successfully used in the so-called poly-plug process in the dynamic random access memories [50]. The polysilicon is left in these regions to reduce topography and to act as a diffusion barrier between the silicon substrate and the top aluminum. In this particular example, which is the most commonly used one, phosphorus doped polysilicon is used. However, phosphorus is introduced using a PBr_3 source in a furnace after the polysilicon deposition. One can similarly use boron- or arsenic-doped polysilicon.

Heavily doped CVD glass films [51] can also be used as a diffusion source in a manner similar to that of polysilicon; all glasses, however, are insulators, and it will be necessary to remove them from the silicon surface to allow for electrical contact formation. Spin-on-glasses contain inorganic or organic compounds of boron, phosphorus, or arsenic, e.g., a B_2O_3-SiO_2 or P_2O_5-SiO_2 mixture, triphenyl phosphate, or arsenosiloxane in a carrier. The compound is spun on the wafer, leaving a reasonably flat surface. A preheat treatment in an oxidizing medium is carried out to evaporate the carrier and to decompose the organic compound into a desired glass. Subsequent high temperatures are used for diffusion into the substrate [48]. After the diffusion anneal, glass is etched off in an HF solution. The major drawbacks of all these spin-on-film type sources are the surface damage, the nonuniform doping, and the rather less than acceptable reproducibility. Some of these problems are associated with the application, cracking, dopant density fluctuations in the liquid, degasing during the predeposition or diffusion anneal, sensitivity to environment, and removal of the film after anneal.

Examples of other solid sources are BN and phosphorus-impregnated discs. In a typical diffusion run, the discs are activated by annealing them at high temperatures (900-1200°C) in an oxdizing medium. This leads to formation of a volatile oxide (B_2O_3 or P_2O_5) on the surface. The activated discs are then placed in a quartz boat, facing the silicon surface to be doped. Thus, two silicon surfaces, placed at equal distance from the BN disc, can be doped by annealing in nitrogen at 900-1200°C. Surface boron oxide evaporates and dopes the silicon surface. The surface of silicon, however, is coated with a layer of boron glass, $[(B_2O_3)_x(SiO_2)_y]$, which is difficult to etch. To remove this layer, a technique of annealing the diffused wafer in steam followed by dissolution in dilute HF solution has been developed. Annealing in steam has been called *boron glass cracking* and is also used after BBr_3 or $BC\ell_3$ diffusions. Exposure to steam cracks the boron glass-film. This ensures the penetration of hydrofluoric acid between the film and the silicon and successful removal of the glass. Thus a typical BN diffusion will follow these steps:

1. Activation of BN discs in oxygen at 900-1200°C
2. Diffusion anneal with BN discs facing silicon wafers
3. Boron glass cracking, and
4. Dissolution of boron glass in HF medium.

Deeper diffusions could be carried out by a longer time or by high-temperature annealing in the presence of BN discs [52] or prior to the removal of the boron glass [53]. In spite of problems associated with the BN-disc doping process, the use of high-purity pyrolytic BN discs [52] has made doping rather easier and contaminaton free [53].

4.4.2 Liquid Sources

The most common method for introducing phosphorus into silicon is the use of liquid phosphorus tribromide (PBr_3) or phosphorus oxychloride ($POC\ell_3$). In this method very high purity (usually 99.999% or purer) liquid is placed into a quartz vial (A) with a gas inlet and outlet, as shown in fig. 4.15. The vapor pressure of the liquid determines the amount of PBr_3 or $POC\ell_3$ carried into the diffusion furnace; therefore it is necessary to control the temperature of the liquid source. This is usually done by placing the quartz vial in a thermoelectric freezer (B); nitrogen is bubbled through the liquid at a predetermined rate, and PBr_3- or $POC\ell_3$-impregnated gas is carried into the mainstream of gas (C) that flows into the furnace. The mainstream of gas carries a small amount of oxygen for effective doping. Figure 4.16 shows the effect of oxygen in the main gas stream on the sheet resistance of the doped silicon wafer. The sheet resistance is used to determine the amount of dopant that has diffused into the silicon substrate. The minimum point in fig. 4.16 signifies a maximum doping condition.

The role of oxygen can be qualitatively explained on the basis of relative amounts of P_2O_5 and SiO_2 in the phosphorus glass layer $[(P_2O_5)_x\ (SiO_2)_y]$ that forms on the silicon surface. When there is no oxygen in the gas phase, PBr_3 does not oxidize; consequently there is no P_2O_5 available to form a phosphorus glass layer on silicon, and the native SiO_2 on silicon then reduces the phosphorus flux from the gas phase into silicon. As the oxygen concentration increases, the P_2O_5 concentration in the phosphorus glass layer increases, since most of the oxygen is consumed in oxidizing gaseous PBr_3 and relatively little or none is available for SiO_2 formation at the

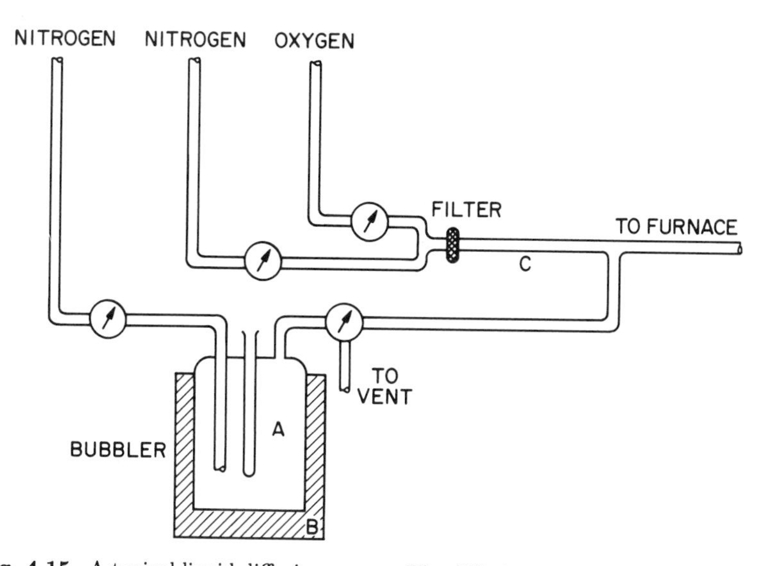

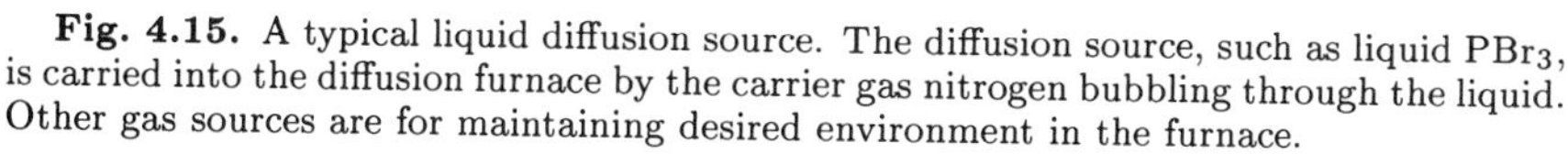
Fig. 4.15. A typical liquid diffusion source. The diffusion source, such as liquid PBr_3, is carried into the diffusion furnace by the carrier gas nitrogen bubbling through the liquid. Other gas sources are for maintaining desired environment in the furnace.

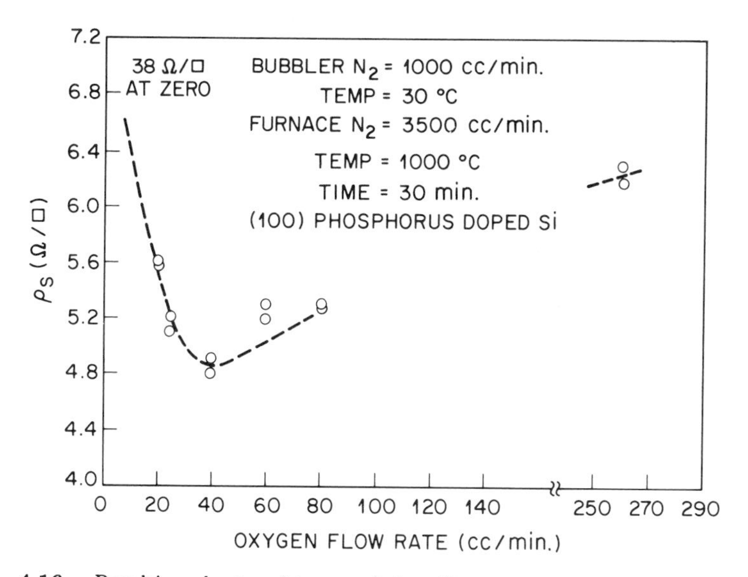

Fig. 4.16. Resulting sheet resistance of the silicon wafer after the defined PBr_3 treatment as a function of the oxygen flow rate in the furnace gas.

silicon surface. This continues until all gaseous PBr_3 is consumed, signifying maximum phosphorus (P_2O_5) concentration in the phosphorus glass layer, maximum doping in the silicon, and the lowest sheet resistance. At oxygen concentrations higher than this, more SiO_2 forms, leading to reduction in

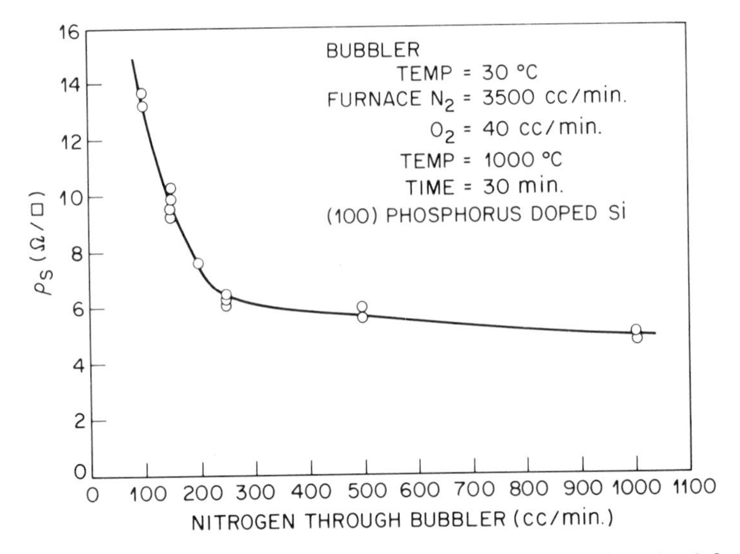

Fig. 4.17. Resulting sheet resistance of the silicon wafer after the defined PBr_3 treatment as a function of the nitrogen flow rate through the PBr_3 bubbler.

the phosphorus concentration in the glass and to increased sheet resistance of the doped surfaces.

Figures 4.17 and 4.18 show that by increasing the flow rate of PBr_3 (either by increasing the nitrogen flow rate through the bubbler or by increasing the bubbler temperature so that the PBr_3 vapor pressure is increased) into the furnace, more phosphorus can be effectively introduced into the silicon, leading to a lowering of the sheet resistance. The doping can also be enhanced by increasing the diffusion temperature or time.

Phosphorus doping by this method has been very reproducible and, therefore, has been a reliable method for doping silicon or polysilicon. The same doping method has also been used in producing heavily doped gettering layers on the backside of wafers. The term *gettering* has been used for a technique of removing electrically active impurities, such as gold and iron, away from the active regions of the devices. Such impurities are fast-diffusing species. They quickly diffuse to gettering layers and get trapped due to solubility enhancement, ion pairing with phosphorus, and in misfit dislocations. Heavily doped layers produce dislocations, strained layers, and a large excess of phosphorus, all of which trap impurities that cause leakages. The process of reducing leakage by this method of trapping impurities has been called gettering. Gettering is generally employed as the last high-temperature step in the processing sequence.

Boron doping has also been carried out in a similar fashion using a liquid BBr_3 or $BC\ell_3$ source. As mentioned in the previous case, the use of BN as a doping source and BBr_3 or $BC\ell_3$ also necessitates a boron-glass cracking treatment for the removal of the glass formed during the diffusion process.

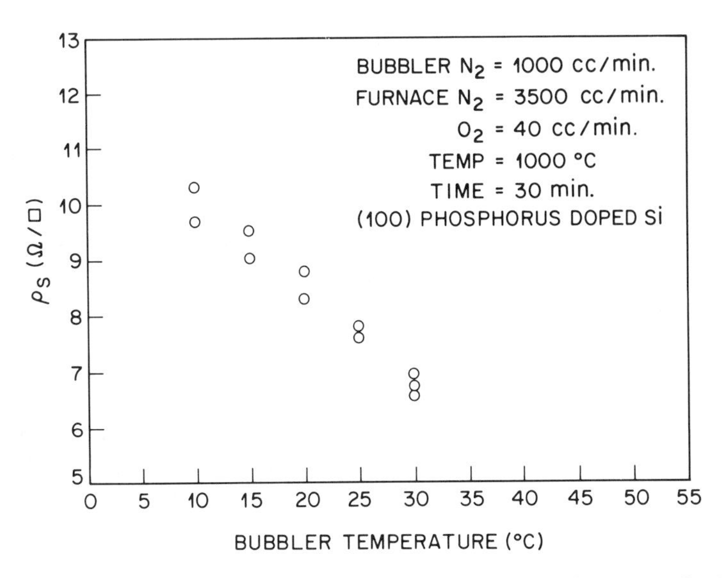

Fig. 4.18. Resulting sheet resistance of the silicon wafer after the defined PBr_3 treatment as a function of the PBr_3 bubbler temperature.

4.4.3 Gaseous Sources

For doping silicon, phosphine (PH_3) and arsine (AsH_3) have frequently been used as n-type gaseous sources. Diborane (B_2H_6) has been used as a source for p-type doping. They are flown directly into the diffusion furnace, usually mixed with large amounts of a diluent gas such as nitrogen, hydrogen, or argon; thus, instead of the bubbler (see fig. 4.15), a gas bottle is connected to the diffusion system. In all cases, small amounts of oxygen are introduced (as in the case of the liquid sources) to enable the formation of a surface layer from which dopants diffuse into silicon. In the case of arsenic, formation of an arsenic-free oxide of a thickness greater than a few nanometers is avoided, since a thicker oxide could become a barrier to continued arsenic diffusion.

All the gases, PH_3, AsH_3, and B_2H_6, are highly toxic and therefore dangerous. For this reason their use in diffusion furnaces has not become popular.

4.4.4 Ion Implantation

Ion implantation has become a very popular technique for doping semiconductors. In ion implantation, ionic species of the dopant are accelerated, collimated, and bombarded into the semiconductor. The method provides extremely good control over doping profiles. The process is presented in detail in the following chapter (Chapter 5).

4.4.5 Diffusion in Compound Semiconductors

Heat treating compound semiconductors like GaAs, InP, GaP, and CdTe, at temperatures where a measurable diffusion can occur, requires special attention. This is because the high vapor pressure of one of the components (e.g., As, P, and Te) of the semiconductor leads to a loss of stoichiometry from near the surface regions. Most reliable diffusions have therefore been carried out in closed tube furnaces with or without superimposed high pressures of As, P, or Te. In this case, the semiconductors and a dopant source like Zn are put into a quartz ampoule that is then sealed. Generally, solid As or an alloy of Zn with As or GaAs is also sealed inside to provide necessary As vapor pressure as well. The sealed tube is then placed in the furnace at the desired temperature for the desired time. After the anneal, the tube is cooled and the seal broken, and the diffused semiconductor is taken out.

With the use of ion implantation, problems associated with dopant diffusion in GaAs are generally taken care of; thus ion implantation has become a popular method in this case as well. For more details on doping compound semiconductors, especially GaAs, see reference (54).

4.5 Diffusion Mechanisms

Why does the diffusion occur? Thermodynamically, mixing of pure materials leads to a lowering of the free energy. This can be easily understood by examining the equation for the entropy of mixing. In the simplest terms, entropy, is a measure of the amount of disorder in the system: the higher the entropy, the lower the free energy and the stabler the system for which free energy is minimum. The entropy of mixing, ΔS_{m}, for a two component (A, B) system is given by

$$\Delta S_{\mathrm{m}} = -N_o \mathrm{k}[x_A \ell n x_A + x_B \ell n x_B], \tag{4.46}$$

where x_{A} and x_{B} are the atom fractions of A and B in the mixture. N_o is the total number of atoms and k is the Boltzmann constant. Let us examine eq. 4.46 more carefully. When either x_A or x_B is very small, the other is close to unity; so, although one of the terms is near zero, the other in eq. 4.46 becomes very large. By mixing a very small amount of a foreign species into a theoretically pure material, the entropy of the system is increased and the free energy is decreased tremendously. Thus, there is a natural tendency to mix and thus to create impure materials. The amount of one material that will dissolve into another to form a mixture with the lowest free energy depends on several factors. The dissolution limit for solids, called the *solid solubility*, is uniquely determined by the component materials and the

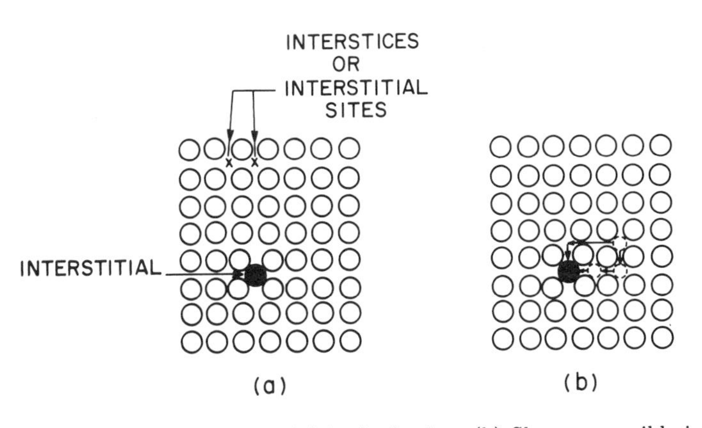

Fig. 4.19. (a) Shows an interstitial in the lattice. (b) Shows a possible interstitial migration through interstitial sights only.

temperature. The definition implies that the crystal structure and properties of the major constituent are preserved. In a semiconductor, this mixing leads to significant changes in electrical properties, although crystal structure and most other properties are not affected.

The thermodynamic considerations relating to mixing lead to diffusion; and yet, solids are rigid and the kinetics of the mixing processes are not obvious. In a perfectly crystalline solid, foreign species can diffuse from interstitial site to interstitial site (see fig. 4.19). This, however, causes gross atomic displacement in the host lattice. Such motion would generally be unlikely. However, there are no perfectly crystalline materials. As shown in Chapter 2, there is a defect structure intrinsic to the crystal. Vacancies, interstitials, and other point defects exist that provide more energetically favored migration channels.

In addition, atoms in a lattice are not fixed to a definite point in space. They oscillate about their equilibrium positions, and this oscillation amplitude is temperature dependent; thus we have imperfect material all the time. Occasionally the vibrations get so violent that atomic or molecular exchange between neighboring sites can occur, thus causing diffusion in solids. In addition, these thermal vibrations cause an occasional loss of an atom or molecule from a crystalline site to the surface, thus leaving vacancies behind. In another circumstance, the thermal vibrations may cause the atom or molecule to move into the interstices of the crystal, resulting in a vacancy on the original position occupied by the atom or molecule and an interstitial. In the first case, we have only one defect created—the vacancy as shown in fig. 4.20a. In the second case, a vacancy-interstitial pair is created, as shown in fig. 4.21. In the last case, it is possible that during crystal growth, where thermodynamic equilibrium is not obtained, some excess number of atoms or molecules are retained in the interstices, as shown in fig. 4.19. Such interstitials do not have associated vacancies. Vacancies can also be

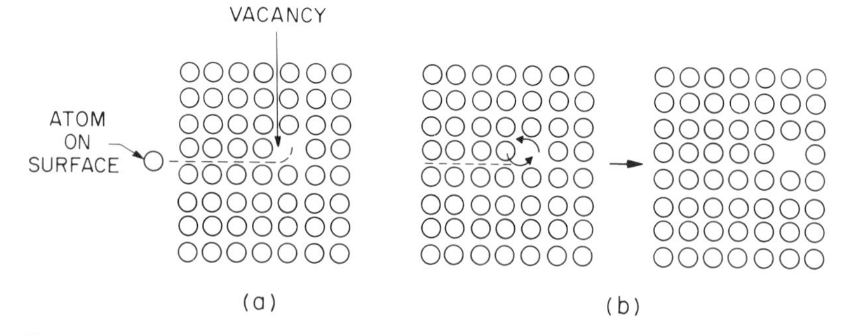

Fig. 4.20. (a) Shows formation of a vacancy in the lattice by movement of an atom to the surface. (b) Shows vacancy motion by exchanging places with a neighboring atom.

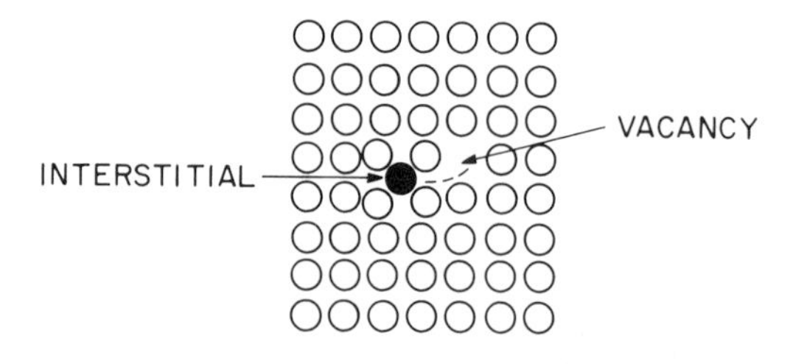

Fig. 4.21. Shows an atom from lattice site pushed into an interstitial site, thus creating an interstitial and vacancy pair.

created during crystal growth. These defects do not exist outside the crystal. They are called *point defects* and in physical terms are treated as atoms or molecules of foreign materials. Their creation, like mixing with foreign material, leads to the lowering of the free energy of the system. These point defects lower the activation barrier for diffusion.

In addition to the point defects, there are so-called line and surface defects that cause diffusion at a rate much higher than caused by point defects. They are dislocations, grain boundaries, and surfaces (see Chapter 2). These defects provide regions in the crystal where diffusion jumps are significantly easier. Semiconductor crystals are free of dislocations, grain boundaries, and internal surfaces, so these defects do not contribute in diffusion. Diffusion in nonepitaxial films, however, is dominated by dislocations, grain boundaries, and surfaces.

4.5.1 Interstitial Diffusion

Figure 4.19b shows a diffusion process, as indicated by arrows, in which an interstitial atom jumps from one interstitial site to another interstitial site. During such a jump, activation energy will be required to push the interstitial through the neighboring atoms on site. For a matrix in which the interstitial concentration is very small, the random walk theory of diffusion [19] gives the diffusion coefficient D_i, at a temperature T°K, as

$$D_i = D_{oi} \exp^{-(\Delta H_{mi}/kT)}, \tag{4.47}$$

where

$$D_{oi} = \gamma a_o^2 \nu \exp(\frac{\Delta S_{mi}}{k}) \tag{4.48}$$

and γ is a geometrical factor related to the crystal structure of the matrix in which diffusion is occurring, a_o is the lattice parameter, ν is taken as the vibration frequency of the atom on its equilibrium site, ΔS_{mi} is called the entropy of migration of an interstitial, ΔH_{mi} is the activation energy or enthalpy of migration for an interstitial, and k is Boltzmann's constant. Note that ΔS_{mi} and ΔH_{mi} are given in units of per molecule or per atom.

Generally, interstitial diffusion is associated with smaller diffusing atoms since the interstitial volume is small compared to the volume associated with a lattice position. In the diamond cubic lattice, like those of Ge, Si, or GaAs, the lattice is comparatively more accommodating, interstitial volume is large, and interstitial diffusion of dopants like boron in Si is postulated. The role of self-silicon interstitial in enhancing the diffusion of dopants during oxidation is discussed in Section 4.6.

4.5.2 Vacancy Diffusion

Figure 4.20b shows the vacancy mechanism of diffusion. An atom on a regular lattice site exchanges place with the vacancy, resulting in a migration, as shown by the arrows. For diffusion to occur by the vacancy mechanism, it is necessary that a vacancy be the nearest neighbor of the diffusing atom. Since vacancies on the lattice site have to be created, it requires a certain amount of energy to form a vacancy. Note the difference between the vacancy on the interstitial site and the vacancy on the regular lattice position. Interstitial sites are nearly all vacant, and no energy is required to form them except in those cases where a very high concentration of atoms of one kind or another have occupied most of these interstitial sites. We are not concerned about such cases. In addition, an activation energy will be required to cause a jump of the atom from its position into the neighboring vacant site. Thus, for the vacancy mechanism of diffusion, two energies are necessary, the formation energy ΔH_{fv} and the migration energy ΔH_{mv}. The diffusion coefficient at temperature T is given as

$$D_v = D_{ov} \exp[\frac{-(\Delta H_{fv} + \Delta H_{mv})}{kT}], \tag{4.49}$$

where

$$D_{ov} = a_o^2 \nu \exp(\frac{\Delta S_{mv} + \Delta S_{fv}}{k}), \tag{4.50}$$

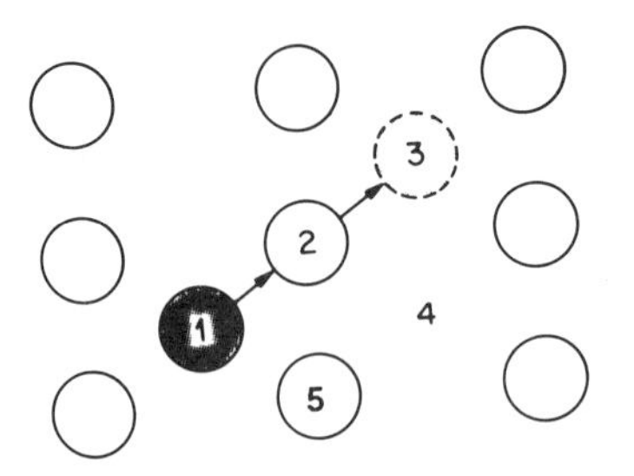

Fig. 4.22. Interstitialcy mechanism of diffusion is illustrated. Interstitial (dark) atom on interstitial site (1) pushes the atom on a regular site (2) to the second interstitial site (3). Such diffusion requires lower dilation of the lattice surrounding the diffusing interstitial than if the interstitial had to diffuse between regularly occupied sites (2) and (5) to the interstitial site (4).

and ΔS_{mv} and ΔS_{fv} now represent the entropies of vacancy migration and formation, respectively.

The vacancy mechanism of diffusion is very common in metals. In silicon, this mechanism is dominant in cases of P, As, Sb, and Ga diffusions.

4.5.3 Interstitialcy Diffusion

Diffusion can also occur by a mechanism in which an interstial atom and a lattice atom both participate. This is shown in fig. 4.22. An interstial atom pushes its nearest neighbor lattice atom into another interstial site and itself occupies the lattice position vacated by the lattice atom in this position. If such exchanges continue, effective diffusion results. This mechanism is called the *interstitialcy mechanism.* Such a mechanism is more likely to occur with interstitial atoms that are comparatively large and would have difficulty in pushing through nearest neighbors into adjacent interstitial site. This is because lattice distortions occurring during diffusion by interstitialcy mechanisms will be small compared to those in diffusion by pure interstitial mechanism. In effect, the total diffusion coefficient will be a function of both the interstitial and vacancy diffusion coefficients.

The interstitialcy mechanism has been suggested to explain the oxidation enhanced dopant diffusion in silicon (see Section 4.6). In this case, oxidation produces a large flux of silicon self interstitials, which then exchange places with the dopant on substitutional (or lattice) sites. If the exchange between an impurity atom (I_{s}) a on substitution site and the silicon self interstitial (Si_{I}) can be described by the expression

$$I_{\mathrm{s}} + Si_{\mathrm{I}} \overset{\mathrm{K}}{\rightleftharpoons} I_{\mathrm{I}} + Si_{\mathrm{s}}, \tag{4.51}$$

where I_{I} and Si_{s} denote impurity on the interstitial site and silicon on the lattice site, respectively, and K is the thermodynamic equilibrium constant.

Then it has been shown [55] that

$$D_T = D_v + K c_{SI} D_I. \tag{4.52}$$

Here c_{SI} and D_I are the self-interstitial concentration and diffusion coefficient of the impurity by pure interstitial mechanism, respectively. D_T and D_v are the total and the vacancy diffusion coefficients.

It is clear from the eq. (4.52) that when the self-interstitial concentration approaches zero, D_T will be D_v. Similarly, for the pure interstitial mechanism, D_v will approach zero and D_T will only be determined by D_I.

In compound semiconductors, defects on both anion and cation sites (vacancies or interstitials) play an equally important role. The diffusion mechanism changes with the dopant, anion, or cation concentration, which can be varied by use of external sources. For III-V semiconductors, the diffusion of acceptor impurities (such as zinc) has been postulated to occur by an interstitialcy mechanism in which the dopant interstitial changes sites with the vacancy on the anion site (e.g., the gallium site in gallium arsenide). Interstitials thus play a very important role in these semiconductors. On the other hand, for donor impurities (such as sulphur), diffusion into GaAs is postulated to occur by vacancies on arsenic sites.

4.5.4 Grain Boundary Diffusion

The boundary between the grains provides a very fast diffusion path for diffusing species; thus when diffusing species diffuse into a polycrystalline material, their flux is divided in two. Some of the material diffuses into the grains (or lattice) and their flux (J_L) is determined by the cross-sectional area of the grain, A_L, the concentration gradient, and the lattice diffusion coefficient, D_L. On the other hand, the flux through the grain boundary (J_{gb}) will be determined by the grain boundary crosssectional area, A_{gb}, the concentration gradient, and the grain-boundary diffusion coefficient, D_{gb}; thus, in a simple calculation (with the same imposed concentration gradient), assuming no exchange of diffusant occurs between grain boundary and the grain, one obtains

$$\frac{J_{gb}}{J_L} = \frac{D_{gb}}{D_L}\frac{A_{gb}}{A_L}. \tag{4.53}$$

Generally the grain boundary area is of the order of a few percent of the grain area unless the grain size is around 1 nm. On the other hand, the D_{gb}/D_L ratio depends on temperature and could vary from as large as 10^6 to values below unity. Large D_{gb}/D_L values are found at low temperatures where grain boundary diffusion will dominate. At higher temperatures, when D_{gb}/D_L is small, the contribution of grain boundaries to diffusion will be negligible.

A more accurate expression for the J_{gb}/J_L is obtained when diffusion from the grain boundaries into the adjoining grains is permitted. Such an expression is derived by considering a situation in which the concentrations on the front and back surfaces of the slab of the polycrystalline material are maintained constant and a steady state is assumed such that all concentration gradients parallel to the plane of the slab are eliminated. In such a case the diffusion fluxes through the slab per unit area are related by

$$\frac{J_{gb}}{J_L} = \frac{2\delta D_{gb}}{D_L} \tag{4.53a}$$

where δ is the grain boundary width.

For diffusion along grain boundaries and dislocations (that behave like grain boundaries), no defect formation energy is required; thus the diffusion activation energy is simply given by the energy of migration. As mentioned earlier for diffusion in semiconductors that are monocrystalline, there is no contribution from grain boundaries and dislocations, except for diffusion in polycrystalline films like polysilicon and metal films, where the diffusion contribution from these defects is significant at the process temperatures of interest ($< 1000°$C). This is especially true for as-deposited films for which grain size is generally very small.

4.5.5 Diffusion via Ionized Point Defects

In Sections 4.5.1 through 4.5.3, only vacancies and interstitials that are electrically neutral were considered. These defects can ionize and, therefore, can act as donors and acceptors themselves. A defect in the lattice causes lattice distortion around it, leading to generation of energy states in the forbidden gap of the semiconductor. The activation energy for diffusion associated with each defect "channel" is a function of the charge state of the channel defect. Fair [55] has presented the estimated vacancy energy levels in the silicon band gap at 0°K, as shown in fig. 4.23. Thus vacancy can ionize according to one or more of the following equations:

$$V^x \rightleftharpoons V^- + \mathrm{h}^+, \tag{4.54a}$$

$$V^- \rightleftharpoons V^= + h^+, \text{ or} \tag{4.54b}$$

$$V^x \rightleftharpoons V^+ + e^-, \tag{4.54c}$$

$$V^+ \rightleftharpoons V^{++} + e^-. \tag{4.54d}$$

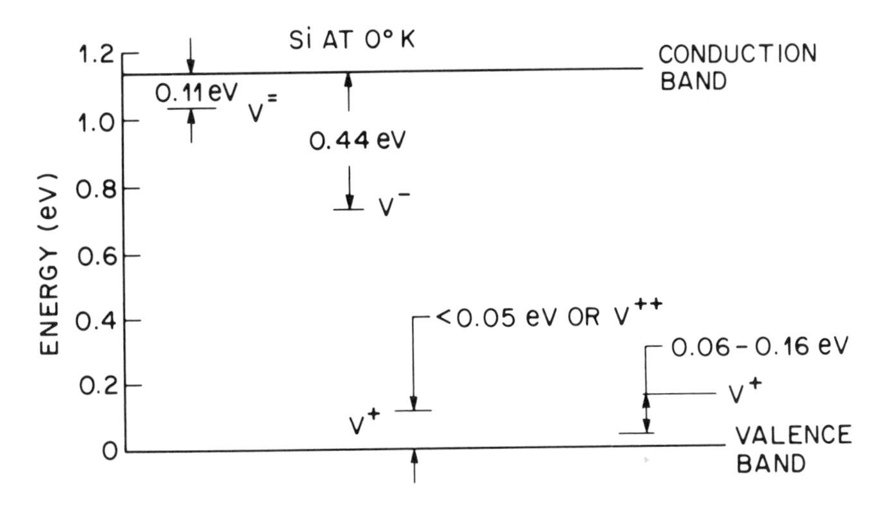

Fig. 4.23. Estimated vacancy energy levels in the silicon bandgap at 0°K. From Fair [55].

Here superscripts $x, -, +, =$, and $++$ indicate the neutral, singly charged negative, singly charged positive, doubly charged negative, and doubly charged positive states of the defect, respectively.

Similarly, an interstitial atom can ionize by accepting or donating an electron:

$$I^x \rightleftharpoons I^- + h^+, \text{or} \tag{4.55a}$$

$$I^x \rightleftharpoons I^+ + e^-. \tag{4.55b}$$

The concentration $[V^r]$ of a vacancy in the rth charge state can be obtained using Fermi-Dirac statistics and is given as [56]

$$[V^r] = \frac{[V]_\text{t}}{1 + 1/2\ \exp(\frac{E_\text{A}^\text{r} - E_\text{F}}{\text{k}T})}, \tag{4.56}$$

where $[V]_\text{t}$ is the total vacancy concentration given by

$$[V]_\text{t} = [V^x] + [V^-] + [V^+] + [V^=] + [V^{++}] + ... \tag{4.57}$$

E_A^r is the energy of the vacancy level in the band gap and E_F is the Fermi energy of the crystal under consideration. The charged vacancy concentration will depend on the electron or hole concentration reaction defined by a thermodynamic equilibrium constant K^-

$$\text{K}^- = \frac{[V^-]\,p}{[V^x]}, \tag{4.58}$$

where p gives the hole concentration; thus

$$[V^-] = \frac{K^-[V^x]}{p}. \tag{4.59}$$

If we have a doped and an intrinsic semiconductor, $[V^-]$ will be different in two crystals, and the ratio will be given by

$$\frac{[V^-]_e}{[V^-]_i} = \exp\left(\frac{E_{Fe} - E_{Fi}}{kT}\right). \tag{4.60}$$

Here subscripts e and i specify extrinsic and intrinsic semiconductors. The right hand side of eq. (4.60) also equals n_e/n_i where n_e and n_i are the respective electron concentrations in extrinsic and intrinsic semiconductors. The same conclusion can be obtained by the use of eq. (4.59); thus

$$\frac{[V^-]_e}{[V^-]_i} = \frac{n_e}{n_i}. \tag{4.61a}$$

For a positively charged vacancy it can be similarly shown that

$$\frac{[V^+]_e}{[V^+]_i} = \frac{n_i}{n_e}. \tag{4.61b}$$

For n-type semiconductors, therefore, $E_{Fe} - E_{fi}$ is positive, and $[V^-]_e/[V^-]_i$ will be expected to be greater than one, whereas for p-type semiconductors $[V^-]_e/[V^-]_i$ will be lower than one since $E_{Fe} - E_{Fi}$ is negative. For Ge with 10^{18} per cm^3 donor or acceptor concentration, $[V^-]_e/[V^-]_i$ at 300°C is found to be 8 or 0.12, respectively [57].

Defects in all charged states will contribute to diffusion. The total self-diffusion coefficient in an intrinsic semiconductor will then be given as

$$\frac{1}{f} D_i = \frac{[V^x]}{N_s} D_v^x + \frac{[V^-]_i}{N_s} D_v^- + \frac{[V^=]_i}{N_s} D_v^= + \frac{[V^+]_i}{N_s} D_v^+ + \frac{[V^{++}]_i}{N_s} D_v^{++} \tag{4.62}$$

Here f is the correlation factor and D_v^r is the diffusivity of the vacancy in the rth charged state. N_s is the number of the lattice sites in the crystal. Eq. (4.62) can be written as

$$D_i = \frac{f}{N_s} \sum [V^r]_i D_v^r, \tag{4.63}$$

where summation is carried out over all charged states—neutral, negative, and positive. For self diffusion in an extrinsic crystal $[V^r]_i$ must be replaced with $[V^r]_e$ so that

$$D_e = \frac{f}{N_s} \sum [V^r]_e D_v^e. \tag{4.64}$$

Now the silicon diffusivity is related to the vacancy diffusivity by

$$D_{\mathrm{Si}}^{r} = \frac{f}{N_{\mathrm{s}}}[V^{r}]_{\mathrm{i}} D_{\mathrm{v}}^{r}, \tag{4.65}$$

and eq. (4.64) can be transformed into the following form

$$D_{e} = \sum D_{\mathrm{Si}}^{r} \frac{[V^{r}]_{\mathrm{e}}}{[V^{r}]_{\mathrm{i}}}. \tag{4.66}$$

When only the charged states shown in eq. (4.54) or (4.62) are considered, eq. (4.66) can be rewritten as

$$D_{\mathrm{e}} = D_{\mathrm{Si}}^{x} + D_{\mathrm{Si}}^{-}\left(\frac{n}{n_{\mathrm{i}}}\right) + D_{Si}^{=}\left(\frac{n}{n_{\mathrm{i}}}\right)^{2} + D_{\mathrm{Si}}^{+}\left(\frac{n_{\mathrm{i}}}{n}\right) + D_{\mathrm{Si}}^{++}\left(\frac{n_{\mathrm{i}}}{n}\right)^{2}, \tag{4.67}$$

where transformations described in eq. (4.61) have been used. Equations of type (4.56) through (4.67) can be written for interstitial or interstitialcy diffusion involving charged interstitials or both charged interstitials and vacancies.

4.5.6 Dopant Diffusion in Semiconductors

Diffusion of foreign atoms into a host crystal is, in general, associated with activation energies that are lower than the self-diffusion values. Table 4.5 lists various D_{o} and Q values compiled by Fair [55] for dopant diffusion in intrinsic silicon. Here superscripts x, $+$, $-$, or $=$ represent the defect charge state of the dominant diffuser for which D_{o}^{r} and Q_{r} values were determined. As can be seen for each dominant diffuser defect, the silicon self-diffusion activation energy is larger. For example, in the case of diffusion occurring through neutral defects (vacancies), the silicon self-diffusion activation energy is 3.89 eV, which is higher than that for B, P, or As (3.46, 3.66, or 3.44 eV, respectively).

The lowering of the diffusion activation energy for dopants is the result of the dopant-defect interactions. These interactions are associated with the ionization states and the position of the defect level in the semiconductor band gap, coulombic interaction between ionized impurity and defects, and the short-range strain term due to overlap between the closed shells of impurities and host atoms [55, 58]. Even for metals, where the free electron clouds screens the charge on the impurity or defect and the coulombic interaction is treated differently [59, 60], the coulombic and the strain terms lead to an impurity diffusion activation energy that is very different from the self-diffusion value. Fair [55] has presented a discussion on the energetics of impurity diffusion.

TABLE 4.5: Dopant diffusivities in intrinsic silicon [55]

Diffuser	Defect responsible for diffusion	D_o (cm^2/s)	Q (eV)
Si	V^x	0.015	3.89
	V^-	16	4.54
	$V^=$	10	5.1
	V^+	1180	5.09
As	V^x	0.066	3.44
	V^-	12.0	4.05
B	V^x	0.037	3.46
	V^+	0.76	3.46
Ga	V^x	0.374	3.39
	V^+	28.5	3.92
P	V^x	3.85	3.66
	V^-	4.44	4.00
	$V^=$	44.2	4.37
Sb	V^x	0.214	3.65
	V^-	15.0	4.08
N	V^x	0.05	3.65

The lowering of the activation energy for dopant diffusion generally implies higher diffusivity of the dopant, especially at lower temperatures. At higher temperatures, the relative values of the pre-exponential factors will determine its relative magnitude. In case of arsenic diffusion in intrinsic silicon, for example, diffusion via singly ionized vacancies will be dominant at higher temperatures, whereas diffusion via neutral vacancies will dominate at lower temperatures (see table 4.5). Using the pre-expontential factors, D_o, and activation energies Q from this table, one can calculate that the crossover

TABLE 4.6: Diffusivities of other impurities in silicon

Diffuser	D_o (cm^2/s)	Q (eV)	References
Ge	6.25×10^5	5.28	61
C	1.9	3.1	61
Pt	150-170	2.22-2.15	62
Na	1.6×10^{-3}	0.76	18
Cu(800-1100°C)	0.04	1.0	18
(300-700°C)	4.7×10^{-3}	0.43	18
Ag	2×10^{-3}	1.6	18
Au(800-1200°C)	1.1×10^{-3}	1.12	18
Au(interstitial)	2.4×10^{-4}	0.39	18
(substitutional)	2.8×10^{-3} (700-1300°C)	2.04	
Fe	6.2×10^{-3}	0.87	18
Co	9.2×10^4	2.8	63
S	0.92	2.2	61
O_2	0.19	2.54	64
H_2	9.4×10^{-3}	0.48	18
He	0.11	1.26	61

will occur at 1085°C. However, in most cases this crossover does not occur for diffusion in solid silicon, which melts at 1410°C.

Table 4.6 lists the diffusivity of several other impurities in silicon. Most of these diffusers are interstitial diffusers accompanying very low activation energies and extremely high diffusivities. High diffusivities of metals lead to considerable problems. They quickly diffuse (from undesired contamination sources) into silicon and provide levels in the band gap and thus cause leakages. A specific example is that of gold, which acts as donor as well as acceptor in silicon [65]. Such impurities should be avoided during silicon integrated fabrication.

Table 4.7 lists the diffusivities of Ga, As, dopants, and other impurities in GaAs and InP. Note that the activation energies of the anion and cation self diffusions are very different. Also, as in the case of the dopant or impurity diffusion into silicon, the dopant or impurity diffusion activation energies in GaAs and InP are lower compared to the self-diffusion activation energies, indicating an interaction between diffusion ions or atoms, and the defects.

TABLE 4.7: The diffusivities of Ga, As, and other elements in GaAs[a] and InP

	GaAs		InP	
Diffuser	D_o (cm^2/s	Q (eV)	D_o (cm^2/s)	Q (eV)
As	0.7	3.2	–	–
Ga	10^7	5.6	–	–
In	–	–	10^5	3.85
P	–	–	7×10^{10}	5.65
Ag	25	2.27	3.6×10^{-4}	0.59
Au	29	2.64	1.32×10^{-5}	0.48
Cu	0.03	0.53	3.8×10^{-3}	0.69
Cr	4300	3.4	–	–
Li	0.53	1.0	–	–
Mg	0.026	2.7	–	–
Mn	0.65	2.49	–	–
O	0.002	1.1	–	–
S	0.0185	2.6	–	–
Se	3000	4.16	–	–
Sn	0.038	2.7	–	–
Cd	Concentration dependent		1.8	1.9
Zn	Concentration dependent		Concentration dependent	
Te	$D(1000°=10^{-13}$ cm^2/s		–	–
	$D(1100°=2 \times 10^{-12}$ cm^2/s		–	–

[a] From reference [54]

4.6 Special Diffusion Effects

In the above section, diffusion in the intrinsic semiconductor was considered. The effects of the already present dopants and impurities and the external factors such as the annealing environment will be introduced in this section. Also included will be a brief discussion of the enhanced diffusivity observed during silicide formation on silicon.

4.6.1 Concentration-Dependent Diffusion

Diffusion of arsenic, phosphorus, and boron in silicon is significantly influenced by the concentration of dopants (and hence of the charge carriers electron) or hole. Fair [55] summarizes the observations:

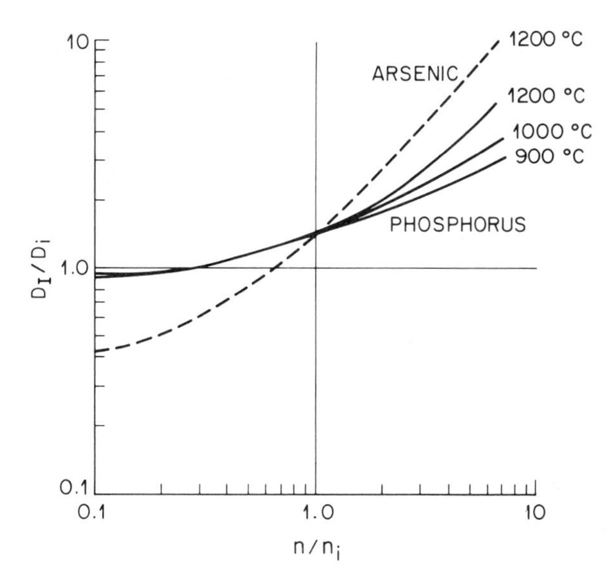

Fig. 4.24. Plots of normalized diffusivity for phosphorus and arsenic versus n/n_i. From Fair [55].

1. For dopant concentrations above $\sim 1.5 \times 10^{20}$ cm^{-3}, the total diffused phosphorus or arsenic concentration depends on n^3.
2. D_p or D_{As} is concentration dependent. Both are enhanced at concentrations above $\sim 1 \times 10^{20}$ cm^{-3} and in the temperature range of 800 - 1200°C. Figure 4.24 shows the plots of the normalized donor diffusivities (with respect to the diffusivities in the intrinsic silicon) as a function of n/n_i [55]. Assuming donors are ionizid almost completely, n represents the total donor concentration. Increased donor concentration therefore increases diffusivities, especially at $n/n_i \approx 1$ or greater than 1. In the case of As and P, the enhancement is observed even at $n/n_i = 1$. The effect is related to electrical interactions between the ionized donors and the electrons.
3. Boron diffusivities show higher concentration dependence, as is shown in fig. 4.25. The concentration dependence is seen even at concentrations for which p/n_i is less than 1, but unlike in fig. 4.24, the extrinsic to intrinsic diffusivity ratio does not show enhancement at $p/n_i = 1$, indicating a negligible electrical interaction between ionized boron atoms and holes.
4. At very high dopant concentrations (usually above the solubility limit in the silicon), the dopant diffusivity is retarded.

For a complete discussion of these results, reference 55 should be reviewed. In the following, a few expressions are derived to explain the observed results. It was shown in Section 4.5.5 that the silicon self diffusion is related to the electron concentration in the silicon [see eq. (4.67)]. A similar expression can be written for the impurity (dopant) diffusion coefficient D_{I} [55].

$$D_{\mathrm{I}} = hD_{\mathrm{i}}^{x} + D_{\mathrm{i}}^{-}(\frac{n}{n_{\mathrm{i}}}) + D_{\mathrm{i}}^{2-}(\frac{n}{n_{\mathrm{i}}})^{2} + D_{\mathrm{i}}^{+}(\frac{n_{\mathrm{i}}}{n}) + D_{\mathrm{i}}^{2+}(\frac{n_{\mathrm{i}}}{n})^{2}, \tag{4.68}$$

where the superscript on D indicates the charged state of the vacancy under consideration and the subscript i represents intrinsic condition; h is a factor that corrects for the electric field effect on diffusion of the impurity (see Section 4.6.2). It is clear from eq. (4.68) that changing the electron concentration will affect D_{I} and an enhancement in D_{I} will be expected.

For phosphorus and arsenic diffusion in silicon, the contribution from positively ionized vacancies can be neglected, giving

$$D_{\mathrm{PAs}} = hD_{\mathrm{i}}^{x} + D_{\mathrm{i}}^{-}(\frac{n}{n_{\mathrm{i}}}) + D_{\mathrm{i}}^{2-}(\frac{n}{n_{\mathrm{i}}})^{2}. \tag{4.69}$$

For arsenic diffusion, arsenic-vacancy complex formation occurs at high arsenic concentrations. In this case, two arsenic atoms are bound coulombically to a $V^{=}$ vacancy, i.e.,

$$2\mathrm{As}^{+} + V^{=} \rightleftharpoons (V\mathrm{As}_{2})^{x}. \tag{4.70}$$

Such complexes diffuse slowly and exist in significant concentration only at high concentrations ($\gtrsim 1.5 \times 10^{20}\mathrm{cm}^{-3}$). These complexes break down at lower concentrations and $(V\mathrm{As})^{x}$ pairs become dominating diffusion species.

$$(V\mathrm{As}_{2})^{x} \rightleftharpoons (V\mathrm{As})^{x} + \mathrm{As}^{+} + e^{-}. \tag{4.71}$$

For most practical cases V^{-} vacancy plays an important role in determining diffusion of arsenic by producing pairs according to the equation

$$V^{-} + \mathrm{As}^{+} \rightleftharpoons (V\mathrm{As})^{x}.$$

For arsenic diffusion, therefore, the first two terms of eq. (4.69) are most important, except when the arsenic concentrations are so high that $V^{=}$ form complexes with ionized arsenic. For phosphorus, on the other hand, all three terms are important, although the contribution of $V^{=}$ dominates the diffusion mechanism [55].

In the extrinsic region, boron diffusion is dominated by singly charged positive vacancies (donors), giving

$$D_{\mathrm{B}} = D_{\mathrm{i}}^{+}\left(\frac{n_{\mathrm{i}}}{n}\right). \tag{4.72}$$

Note that $np = n_{\mathrm{i}}^{2}$. Eq. (4.72) can then be rewritten as

$$D_{\mathrm{B}} = D_{\mathrm{i}}^{+}\left(\frac{p}{n_{\mathrm{i}}}\right). \tag{4.73}$$

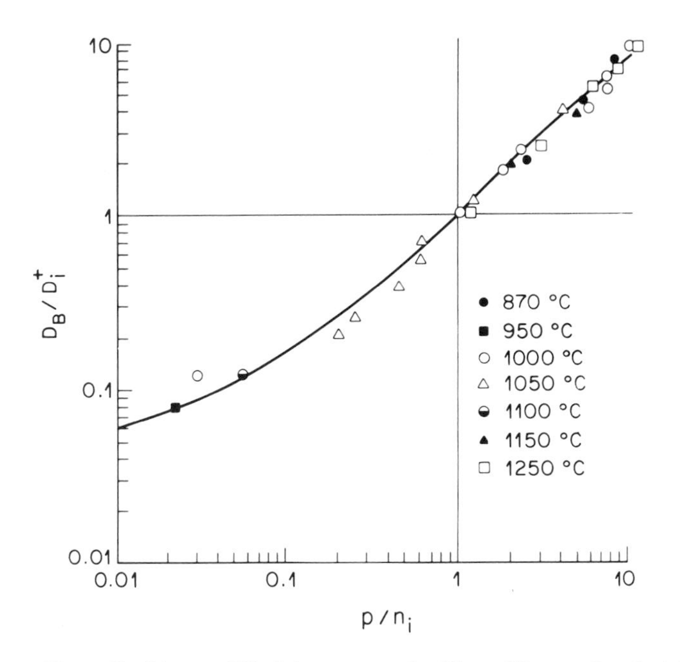

Fig. 4.25. Normalized boron diffusivity versus p/n. The solid curve is calculated with the vacancy-donor level at E_v (valency band) + 0.05 eV. From Fair [55].

Equations (4.69) and (4.73) show that the dopant diffusivities will increase with increased electron and hole concentrations. Figures 4.24 and 4.25 confirm these predictions. Fair and Tsai [66] proposed an impurity-vacancy pair diffusion model and dissociation model [cf. eq. (4.70) and (4.71)] to explain the typical phosphorus and electron concentration profiles shown in fig. 4.26. The curve represents a diffusion condition at very high phosphorus concentrations. Near the surface, the total phosphorus and electron concentrations practically remain unchanged. Beyond this region a kink occurs in the profile at an electron concentration of n_e, which is found to correspond to a Fermi level of 0.11 eV below the conduction band. Finally, a deeper diffusion tail is observed. The results of fig. 4.26 are explained as follows [66]:

Near the surface phosphorus ions interact with $V^{=}$ resulting in pairs:

$$\mathrm{P}^+ + \mathrm{V}^= \rightleftharpoons (\mathrm{P}V)^-. \tag{4.74}$$

As phosphorus moves deeper into the bulk of the silicon, its concentration falls to lower levels and the $(\mathrm{P}V)^-$ pairs start dissociating. This was proposed to occur by $E_\mathrm{F} = E_\mathrm{c} - 0.11$ eV. This produces a change in the mechanism of diffusion. The dissociation occurs according to

$$(\mathrm{P}V)^- \rightleftharpoons \mathrm{P}^+ + V^- + e^- \tag{4.75}$$

and now diffusion is dominated by V^-.

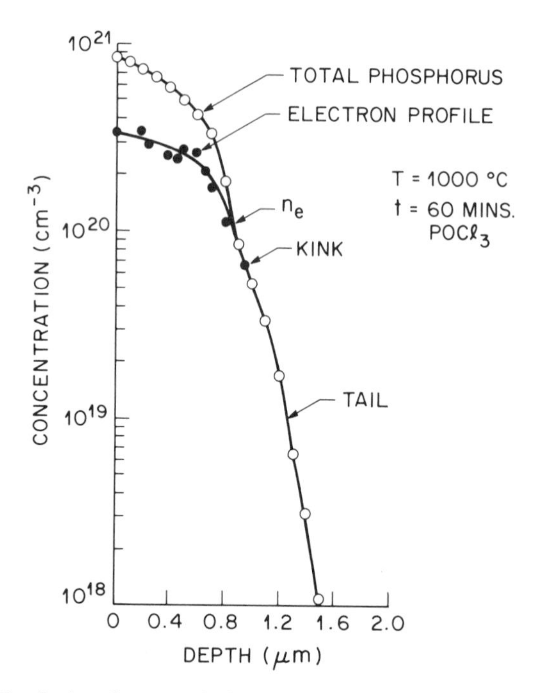

Fig. 4.26. Total phosphorus and electron concentration profiles obtained by SIMS and differential conductivity measurements. From Reference 66, reprinted by permission of the publishers, The Electrochemical Society, Inc.

Near the surface, (4.74) holds and one can write

$$[(PV)^-] = K[P^+][V^=]. \qquad (4.76)$$

It can be shown that $[V^=]$ is proportional to n_s^2 and $[P^+] = n_s$ for complete ionization of donors, where n_s is the electron concentration in the near surface region, so that

$$[(PV)^-] \alpha\ n_s^3. \qquad (4.77)$$

The near surface electron concentration is therefore determined by $[(PV)^-]$, which is assumed to remain constant throughout the high-concentration region near the surface (see fig. 4.26). In deeper regions, where V^- is formed [eq. (4.71)] due to dissociation of $(PV)^-$, the diffusion enhancement occurs due to the interactions of V^- with P^+ ions.

4.6.1.1 Concentration-Dependent Diffusion in Compound Semiconductors

Dopant concentration-dependent diffusion has also been reported for several compound semiconductors. The diffusion of zinc in GaAs, for example, has been extensively investigated as a function of the zinc concentration

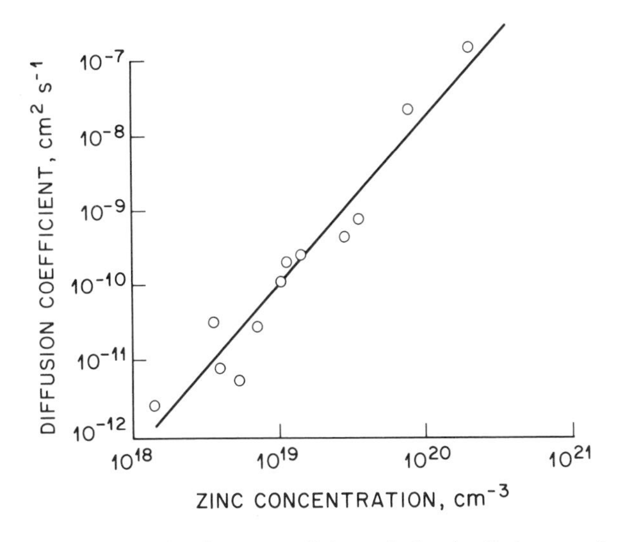

Fig. 4.27. Variation of diffusion coefficient of zinc in GaAs as a function of zinc concentration for diffusion at 1000°C at a dissociation pressure of approximately 10^{-4} atom. From Tuck [9].

in GaAs. Figure 4.27 shows the diffusion coefficient of Zn in Zn-doped GaAs at 1000°C. It can be seen that $D_{\mathrm{Zn}}\alpha[\mathrm{Zn}]^2$. The following interstitial-substitutional mechanism explains the results. In this mechanism, the zinc interstitial exchanges place with a vacancy on the gallium site, thus changing itself into a substitutional zinc. Diffusion occurs by the continued exchanges between interstitial and substitutional sites. This reaction can be described as

$$\mathrm{Zn}_{\mathrm{i}}^{+} + V_{\mathrm{Ga}}^{x} \rightleftharpoons \mathrm{Zn}_{\mathrm{Ga}}^{-} + 2h^{+}, \tag{4.78}$$

where it is assumed that zinc on an interstitial site is a donor, and zinc on a substitutional site is an acceptor. Assuming that the semiconductor is p type, one can write $[\mathrm{Zn}_{\mathrm{Ga}}^{-}] \approx p$ and $[\mathrm{Zn}_{\mathrm{i}}^{+}] \ll [\mathrm{Zn}_{\mathrm{Ga}}^{-}]$. Also, if the vacancy concentration is considered to be its concentration at equilibrium, it can be shown theoretically that $D_{\mathrm{Zn}}\alpha[\mathrm{Zn}_{\mathrm{Ga}}^{-}]^2$. This type of agreement between experiment and the model supports the interstitial-substitutional exchange mechanism model.

Diffusion in other III-V semiconductors occurs by a mechanism similar to that suggested for GaAs. In other compound semiconductors, complex mechanisms that are different for different concentration ranges have been proposed.

4.6.2 Retarded Dopant Diffusion

At very high concentrations of dopants in silicon, the diffusivities of phosphorus, boron, and arsenic have been found to be lower than those expected

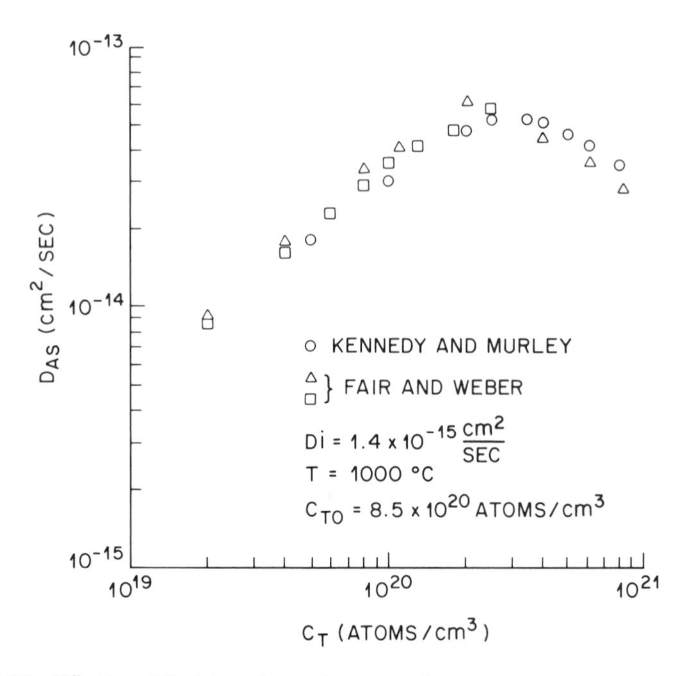

Fig. 4.28. Effective diffusivity of arsenic versus C_T for diffusions into p-type silicon at 1000°C. From Fair [55].

from the extrapolation of the values for moderate concentrations. The phenomenon of retarded diffusion occurs when the dopant concentration exceeds its solid solubility in silicon. This behavior is especially observed after a high temperature PBr_3, $POC\ell_3$, or BBr_3 diffusion where dopants at concentrations of $10^{21}cm^{-3}$ or higher can be easily introduced into silicon. Figure 4.28, which is reconstructed from the data of Kennedy and Murley [68] and Fair and Weber [69] (see also ref. 55), shows this effect for arsenic diffusion in p-type silicon at 1000°C.

High concentrations of dopants are known to introduce a large density of misfit dislocations [70-75]. The dislocations are generated to relieve the stress associated with the large concentrations of dissolved impurities. Prussin [76] relates the maximum stress σ_{max} with the dopant concentration by the equation

$$\sigma_{max} = \beta E c_s(1 - \eta), \tag{4.79}$$

where β is the lattice contraction coefficient, E is Young's modulus, c_s is the surface concentration, and η is Poisson's ratio. For a phosphorus concentration of $10^{21}cm^{-3}$ in (111) silicon with $E = 1.19 \times 10^{12}$ dyn/cm^2, $\eta = 0.29$, and $\beta = 2.4 \times 10^{-24}cm^3$ per atom [77]) one obtains $\sigma_{max} = 3.2 \times 10^{11}$ dyn/cm^2. Such stresses would affect the generation and migration of defects through the thermodynamic relation [19]

$$(\frac{\partial \Delta G}{\partial p})_T = \Delta V,$$

where ΔG is the sum of the free energies of formation and migration of defects, P is the pressure, and ΔV is called the activation volume. To obtain the effect of the pressure effect on diffusivity, one can differentiate ℓnD with respect to pressure and obtain

$$[\frac{\partial \ell n(D/a^2\nu)}{\partial p}]_T = -\frac{1}{RT}(\frac{\partial \Delta G_{\mathrm{v}}}{\partial T})_p - \frac{1}{RT}(\frac{\partial \Delta G_{\mathrm{m}}}{\partial T})_p, \tag{4.80}$$

where ΔG_{v} and ΔG_{m} are the free energies of formation and migration of the defect. Equation (4.80) can be rewritten in terms of activation volumes

$$\left[\frac{\partial \ell n(D/a^2\nu)}{\partial p}\right]_p = -\frac{\Delta V_{\mathrm{a}}}{RT}, \tag{4.81}$$

where $\Delta V_{\mathrm{a}} = \Delta V_{\mathrm{v}} + \Delta V_{\mathrm{m}}$, is the sum of the activation volumes associated with the formation of vacancies and their migration. Equation (4.81) thus suggests that the effect of pressure or stress will be to retard the diffusivity.

Problem 4.2: *Activation volume ΔV_{a} of the vacancies is 6 $cm^3/mole$. Calculate the change in the diffusion coefficient when the applied stress is 1×10^9 dyn/cm^2.*

Solution: From eq. 4.81

$$\left(\frac{d\ell nD/a^2\nu}{dP}\right)_T = -\frac{\Delta V_{\mathrm{a}}}{RT}.$$

Therefore

$$\frac{\ell nD_2}{a^2\nu} - \frac{\ell nD_1}{a^2\nu} = -\frac{\Delta V_{\mathrm{a}}}{RT}(P_2 - P_1),$$

when D_2 and D_1 are the diffusion coefficients at stresses P_2 and P_1. For $P_1 = 1$ atm $= 1.01\times10^6$ dyn/cm^2, and $P_2 = 1 \times 10^9$ dyn/cm^2, $P_2 - P_1 \approx P_2$. Therefore,

$$\ell n\frac{D_2}{D_1} = -\frac{\Delta V_{\mathrm{a}} P_2}{RT}$$

$$\Delta V_a P_2 = 10 \times 10^9 \text{ ergs/mole} = 10^{10} \text{ ergs/mole}$$

$$= 239 \text{ cal/mole}.$$

Therefore $\ell n D_2/D_1 = \frac{239}{1.987 \times 1273} = -0.0945$. Therefore $D_2 = 0.91\ D_1$; thus the diffusion coefficient is lowered by a factor of 0.91.

Stress on a semiconductor crystal also affects its band gap. The change ΔE_{g} in the band gap has been given as [78]

$$\Delta E_{\mathrm{g}} = -1.08 \times 10^{-11} \sigma \text{ (eV)}. \tag{4.82}$$

This would mean that the concentration of ionized vacancies, which were shown to be responsible for diffusion behavior, will be affected. For stresses that are positive [as given by eq. (4.79)], ΔE_{g} will be negative and thus the ionized vacancy concentration will be lowered, leading to decreased diffusivity [66]. This is because the ionized vacancy concentration in a stressed crystal $[V^-]_{\mathrm{s}}$ and in an unstressed crystal $[V^-]$ are related to ΔE_{g} by [66]

$$\ell n \left\{ \frac{[V^-]_{\mathrm{s}}}{[V^-]} \right\} = \frac{\Delta E_{\mathrm{g}}}{2\mathrm{k}T}. \tag{4.83}$$

A similar relation holds for doubly charged vacancies:

$$\ell n \left\{ \frac{[V^=]_{\mathrm{s}}}{[V^=]} \right\} = \frac{\Delta E_{\mathrm{g}}}{\mathrm{k}T}. \tag{4.84}$$

Thus for a stress of 3×10^9 dyn/cm^2, $\Delta E_{\mathrm{g}} = -0.0324$ eV. For this value of ΔE_{g} and $T = 1000°$C, $[V^-]$ and $[V^=]$ will be reduced by approximately 14% and 26%, respectively. The diffusivities will therefore be reduced by appropriate factors in both ranges of ionized vacancies. The decrease in vacancy concentration (or diffusivity) is related to the dopant concentration by eq. (4.79).

4.6.3 The Effect of an Electric Field on Diffusion

In eqs. (4.68) and (4.69), a factor, h, was used to correct for the electric field effect on diffusion. The electric field, $E(x,t)$, is the result of a variation in the Fermi potential $\phi(x,t)$ due to the impurity gradient inside the semiconductor. $\phi(x,t)$ is related to the quantity $1/q(E_{\mathrm{F}} - E_{\mathrm{c}})$, where E_{c} is the conduction band energy and E_{F} is the Fermi level that is a function of the impurity concentration. $E(x,t)$ is then given by

$$E(x,t) = -\frac{d\phi(x,t)}{dx} \tag{4.85}$$

or

$$E(x,t) = \frac{1}{q}\frac{d}{dx}(E_c - E_F). \tag{4.86}$$

Following Hu's approach [79-80], Fair [55] shows that the total diffusion flux will be a sum of the fluxes due to a chemical gradient and an electric field. This leads to a net diffusivity that is h times the diffusivity in absence of the electric field. h is given as

$$h = 1 + \frac{c}{2n_i}[(\frac{c}{2n_i})^2 + 1]^{-1/2}, \tag{4.87}$$

where c and n_i are the impurity concentration and the intrinsic electron concentration, respectively. Experimental findings show that the electrical field factor could be as large as 2 when $c/n_i > 2$. Equation (4.87) also predicts h to approach a value of 2 at $c/n_i >> 2$.

4.6.4 Oxidation-Enhanced Diffusion

Diffusion of boron [81-84], phosphorus [85-88], and arsenic [88] in silicon is enhanced during oxidation of silicon. The enhancement is related to the fact that the oxidation of silicon leaves approximately one silicon atom in a thousand unoxidized. The unoxidized silicon atom could diffuse into the growing oxide or into the silicon below the $Si{=}SiO_2$ interface. The silicon atoms, which diffuse in silicon, become silicon interstitials and actively participate in the dopant diffusion mechanism. Hu [88] suggested that in the presence of a large excess of silicon self interstitials, dopant diffusion occurs by an interstitialcy mechanism. The effective diffusion coefficient is given by [89, 55] [of eq. (4.52)]

$$D_e = D_s + D_I K_I C_I, \tag{4.88}$$

where D_e is the enhanced diffusion coefficient (under oxidizing medium), D_s is the substitutional diffusion coeficient representing the vacancy mechanism, D_I is the interstitial diffusion coefficient representing the self-interstitial-substitutional impurity exchange mechanism, C_I is the self-interstitial concentration, and K_I is the equilibrium constant of the silicon self interstitial. Si_I, and the substitutional impurity atom, I_s, exchange reaction effectively leads to the movement of the impurity atom to the interstitial site, I_I

$$Si_I + I_s \overset{K_I}{\rightleftharpoons} I_I. \tag{4.89}$$

It is obvious from eq. (4.89) that the enhancement in diffusion $(D_e - D_s)$ is dependent on K_I and C_I, both of which will depend on the temperature and the oxidation conditions. For oxidations carried out at low oxygen partial pressures, the enhancement is considerably smaller [83]. At least for boron diffusion, the enhancement is dependent on the orientation of the crystal, being maximum for (100). No enhancement is reported for diffusion in (111)-oriented substrates. The effect is explained on the basis of the ability of the surface kinks to capture the self interstitials generated during oxidation. Surface kinks are surface atoms that do not have neighboring atoms on one or more sides. Hu [88] points out that the surface kink density is highest for (111) and lowest for (100) surfaces when comparing (111), (110), and (100) surfaces. The oxidation rate of (111) surfaces is also high compared to other surfaces, increasing the probability of the trapping of the silicon self interstitials by the advancing SiO_2=Si interface. Both the larger kink density and the increased oxidation rate will therefore decrease the silicon self-interstitial generation. This in turn will then lead to reduced or no enhancement of diffusion.

In addition to the diffusion enhancement, oxidation causes redistribution of impurities at the Si=$Si0_2$ interface, as discussed in Chapter 3. The redistribution, which is controlled by the segregation coefficient m (Chapter 3), and the relative diffusivities of dopant in silicon and SiO_2 determine the diffusion profile. Four different situations are defined by Grove [90], as shown in fig. 3.31. In this figure, special cases of diffusion that result in various diffusion profile types are shown. For example, fig. 3.31a could represent boron diffusion in an oxidizing ambient. For boron the segregation coefficient m is less than one, which means that boron prefers to stay in SiO_2 rather than in Si. This leads to a depletion of boron in silicon near the surface of silicon. Occasionally this leads to change in the electrical characteristics of the Si-$Si0_2$ MOS capacitor and low-energy boron ion implantation is done to minimize the effect of boron loss. This implantation is usually called a *threshold adjust implant.* The redistribution effects are, however, negligible with n-type silicon.

4.6.5 Reaction-Enhanced Diffusion

Metals react with the silicon substrate to form compounds called *silicides.* It is found that when near-noble metals, like Pd, react with a doped silicon substrate, the dopant diffusion in silicon is enhanced [90]. The enhancement is associated with a) the production of point defects during the metal-silicon interaction, which leads to the silicide formation, and b) the rejection of dopants by the growing silicide layer. At low temperatures, metal atoms diffuse in the silicon, breaking bonds and creating point defects. If these point defects are fast-diffusing self interstitials, an interstitialcy mechanism, similar to that invoked to explain oxidation-enhanced diffusion, may be operating to cause the enhanced diffusion. The enhanced diffusion during silicide formation could be put to practical use in forming very shallow junctions with silicide contacts in one anneal at temperatures in the range of 200-

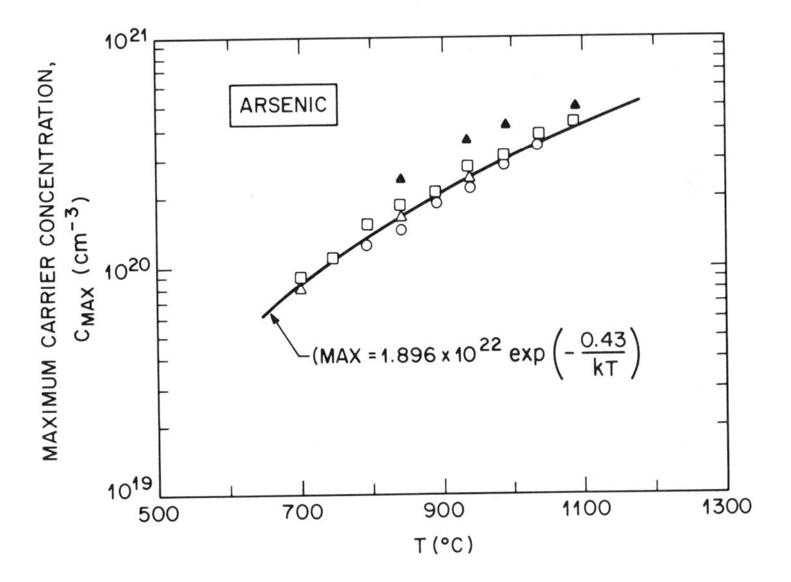

Fig. 4.29. Maximum carrier concentration of arsenic-implanted silicon versus annealing diffusion temperature. From Reference 91, reprinted by permission of the Electrochemical Society, Inc.

500°C. Very shallow depth dopant implantation is carried out followed by metal deposition. The semiconductor surface is cleaned insitu in the metal deposition chamber prior to the deposition of metal. After the metal deposition, annealing at the desired temperature will give junction with silicide contact on surface.

4.6.6 Electrical Activity of Diffused Atoms and Carrier Concentrations

Dopants ionize to produce electrons or holes that contribute to electrical activity in semiconductors. Donors, like P, donate electrons to the conduction band, whereas acceptors, like B, trap or accept electrons to create holes in the valence band. For substitutional impurities in Si, like As, B, Ga, P, and Sb, electrical activity is associated with their presence on substitutional sites. Dopants present in any other form, like interstitials and precipitates, are regarded as electrically inactive; thus, when a dopant is diffused into the semiconductor, all of the diffused atoms may not be electrically active. The maximum carrier concentration is related to the annealing temperature, which also determines the solid solubility of dopants in the semiconductor. Figures 4.29 and 4.30 show the maximum carrier concentrations for arsenic and boron, respectively, as a function of temperature. Thus electrical activity is increased by annealing at higher temperatures. For boron, various curves represent different techniques used to determine the concentrations.

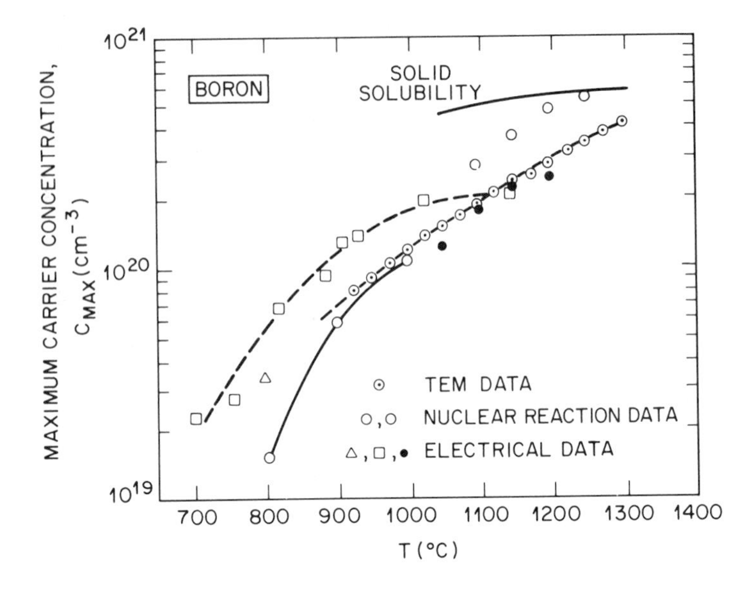

Fig. 4.30. Maximum carrier concentration of boron-implanted silicon versus annealing diffusion temperature. From Reference 92.

It is clear from these curves that diffusion of more arsenic or boron will not lead to increased electrical activity. Excess diffused dopants will possibly stay inactive as interstitials, aggregates, or a precipitate in silicon.

For impurities other than group V or group III elements (of the periodic table), the electrical activity varies for each impurity. Oxygen in silicon, therefore, stays inactive when present as interstitials. Oxygen when present in complex forms (like O_4), however, behaves like a donor. Gold diffuses as an interstitial and also as a substitutional impurity; thus it can be considered as an interstitial-substitutional diffuser. Au in silicon has both donor and acceptor levels and acts as a donor or an acceptor, depending on the substrate type and the diffusion conditions. Most transition elements behave like gold. Alkali metals, with the exception of lithium, are interstitial diffusers and are generally electrically inactive.

Figure 4.31 shows a plot of the carrier concentration as a function of zinc and sulphur concentrations in GaAs. There is a one-to-one relationship between carrier and zinc concentrations, except for very high zinc doping. For sulphur, the electron concentration is about 50% to 60% of the sulphur concentration. It is also found that the electrical activity of donors decreases when GaAs crystals have an arsenic deficiency, i.e., more vacancies on the arsenic sites. This behavior of sulphur has been attributed to the presence of sulphur in two different configurations in GaAs: a) approximately 50% of sulphur is present in the form of substitutional donors on arsenic sites and b) the other 50% of sulphur is present in the form of a (S-V_{As}) com-

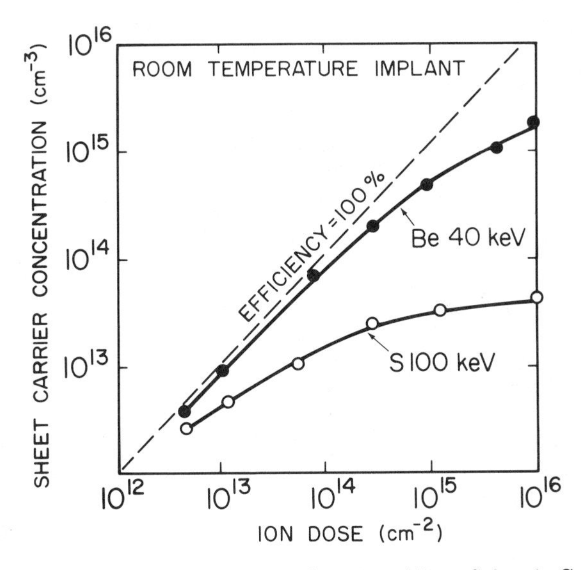

Fig. 4.31. Sheet carrier concentration as a function of Be or S dose in GaAs, indicating lower concentration with S than with Be. Courtesy of Poate, see also Ref. [91].

plex that is electrically inactive [91]. An increase in the V_{As} concentration inceases the (S-V_{As}) complexes, thereby reducing the substitutional sulphur concentration and consequently the donor electrical activity.

4.7 Diffusion in Dielectric Films

SiO_2 and Si_3N_4 films are most commonly employed in silicon integrated circuits and also in III-V compound semiconductor based circuits. SiO_2 is used as the gate dielectric, the field oxide to define the desired active regions and to prevent the diffusion of dopants into the underlying silicon, and as the interlevel insulator. Si_3N_4 is generally used as a barrier to sodium diffusion and is thus deposited as a final dielectric coating to act as an encapsulant.

Table 3.2 lists the diffusivity parameters (D_o and Q) for various other impurities in SiO_2. For ion-implanted dopants the range and the standard deviation in the range (see Chapter 5) in the amorphous SiO_2 film are given in table 3.3. Table 4.8 gives the ion-implantation ranges and the standard deviation in the range for dopants in Si_3N_4. Note, although accurate dopant diffusion data in Si_3N_4 are not available, Si_3N_4 is known to be an excellent diffusion barrier. A few reported results suggest the use of a thin Si_3N_4 layer on top of the thin SiO_2 gate oxide to prevent boron diffusion through the gate dielectric in cases where boron doped polysilicon is used as the gate level metal.

TABLE 4.8: Projected range statistics for ion implantation into Si_3N_4

	Antimony		Arsenic		Boron		Phosphorus	
Energy (keV)	Projected Range (μm)	Projected Standard Deviation (μm)	Projected Range (μm)	Projected Standard Deviation (μm)	Projected Range (μm)	Projected Standard Deviation (μm)	Projected Range (μm)	Projected Standard Deviation (μm)
10	0.0056	0.0015	0.0060	0.0020	0.0230	0.0111	0.0084	0.0037
20	0.0090	0.0024	0.0099	0.0033	0.0480	0.0196	0.0154	0.0065
30	0.0119	0.0033	0.0135	0.0045	0.0736	0.0267	0.0226	0.0092
40	0.0147	0.0040	0.0169	0.0056	0.0990	0.0326	0.0300	0.0118
50	0.0173	0.0047	0.0202	0.0066	0.1239	0.0377	0.0376	0.0143
60	0.0197	0.0054	0.0235	0.0077	0.1482	0.0422	0.0453	0.0168
70	0.0222	0.0061	0.0268	0.0087	0.1719	0.0461	0.0532	0.0192
80	0.0246	0.0067	0.0301	0.0097	0.1950	0.0496	0.0612	0.0215
90	0.0269	0.0074	0.0334	0.0108	0.2176	0.0527	0.0693	0.0237
100	0.0292	0.0080	0.0367	0.0118	0.2396	0.0555	0.0774	0.0259
110	0.0315	0.0086	0.0400	0.0127	0.2610	0.0581	0.0856	0.0280
120	0.0338	0.0092	0.0433	0.0137	0.2820	0.0605	0.0939	0.0301
130	0.0360	0.0098	0.0467	0.0147	0.3025	0.0627	0.1022	0.0321
140	0.0383	0.0104	0.0500	0.0157	0.3226	0.0647	0.1105	0.0340
150	0.0405	0.0110	0.0534	0.0167	0.3424	0.0666	0.1188	0.0358
160	0.0428	0.0116	0.0568	0.0176	0.3617	0.0684	0.1271	0.0377
170	0.0450	0.0122	0.0603	0.0186	0.3807	0.0700	0.1354	0.0394
180	0.0472	0.0128	0.0637	0.0195	0.3994	0.0716	0.1437	0.0411
190	0.0495	0.0134	0.0672	0.0205	0.4178	0.0731	0.1520	0.0428
200	0.0517	0.0139	0.0706	0.0214	0.4358	0.0744	0.1602	0.0444
220	0.0562	0.0151	0.0776	0.0233	0.4712	0.0770	0.1767	0.0475
240	0.0606	0.0162	0.0847	0.0252	0.5056	0.0793	0.1931	0.0505
260	0.0651	0.0174	0.0918	0.0270	0.5390	0.0815	0.2094	0.0533
280	0.0696	0.0185	0.0990	0.0289	0.5717	0.0834	0.2255	0.0559
300	0.0741	0.0196	0.1063	0.0307	0.6037	0.0852	0.2415	0.0584

4.8 Summary

In this chapter diffusion in semiconductors was discussed. We examined:

- Fundamental diffusion equations and their solutions
- Experimental procedures for diffusing dopants into semiconductors
- Experimental methods to determine concentrations versus time profiles and hence the diffusion coefficient
- Diffusion mechanisms—interstitial, vacancy, interstitially, and others, including the role of ionized point defects and their association with the diffusion
- Special diffusion effects associates with high dopant concentration, internal electric field, oxidation, and reaction with metals.

Principles and phenomena discussed in this chapter are applicable to most solid-state diffusion processes. However, this treatment has focused on semiconductor applications.

References for Chapter 4

1. J. Crank, "The Mathematics of Diffusion," Oxford University Press, London (1970).
2. W. Jost, "Diffusion in Solids, Liquids, Gases," Academic Press, NY (1952).
3. R.M. Barrer, "Diffusion in and Through Solids," Cambridge University Press, Cambridge (1941).
4. C. Wells, "Atom Movements," ASM, Cleveland (1951).
5. D. Shaw, ed., "Atomic Diffusion in Semiconductors," Plenum Press, NY (1973).
6. A. Gangulee, P.S. Ho, K.N. Tu, eds., "Low Temperature Diffusion and Applications to Thin Films," Elsevier Sequoia, Switzerland (1975).
7. A.S. Nowick, J.J. Burton, eds., "Diffusion in Solids: Recent Developments," Academic Press, NY (1975).
8. G.E. Murch, A.S. Nowick, eds., "Diffusion in Crystalline Solids," Academic Press, NY (1984).
9. B. Tuck, "Introduction to Diffusion in Semiconductors," Peter Peregrinus, Herts, England (1974).
10. J.C.C. Tsai, Diffusion in "VLSI Technology," S.M. Sze, ed., McGraw-Hill, NY (1983).
11. C.P. Flynn, "Point Defects and Diffusion," Clarendon Press, Oxford (1972).
12. J.H. Albany, ed., "Defects and Radiation Effects in Semiconductors," 1978 Conf. Ser. No. 46, The Inst. of Physics, London, (1979).
13. S.Z. Bokshtein, "Diffusion and Structure of Metals," Oxonion Press, New Delhi, India (1985).
14. V.M. Falćhenko, L.N. Larikov, V.R. Ryabov, "Diffusion Processes in Solid-Phase Welding of Materials," Oxonian Press, New Delhi, India (1984).
15. J.R. Manning, "Diffusion Kinetics for Atoms in Crystals," Van Nostrand, Princeton, NJ (1968).
16. R.B. Fair, in "Impurity Doping Processes in Silicon," F.F.Y. Wang, ed., North-Holland, Amsterdam (1981).
17. F. Beniére, C.R.A. Catlow, eds., "Mass Transport in Solids," Plenum, London (1983).
18. B.L. Sharma, "Diffusion in Semiconductors," Trans. Tech. Publications, Clausthal, Germany (1970).

19. P.G. Shewmon, "Diffusion in Solids,"McGraw Hill, NY (1963).
20. F.F.Y. Wang, ed., "Impurity Doping Processes in Silicon,"Vol. 2, North-Holland, NY (1981).
21. "Diffusion Data,"Trans. Tech. Publications, Rockport, MA (1967).
22. A. Fick, Ann. Phys. Lpz. 170, 59 (1955).
23. See "Tables of the Probability Function,"Work Project Association, NY (1941).
24. R. Holm, "Electrical Contacts Theory and Applications,"Springer-Verlag, NY (1967).
25. J.R. Ehrstein, "Semiconductor Measurement Technology,"Proc. of Spreading Resistance Symposium, NBS Publication #400-10 (Dec. 1974).
26. D.C. D'Avonzo, R.D. Rung, A. Gat, R.W. Dutton, J. Electrochem. Soc. 125, 1170 (1977).
27. J.C. Irvin, Bell System Tech. Journal 41, 387 (1962).
28. H.F. Wolf, "Silicon Semiconductor Data,"(Pergamon Press, NY (1969).
29. W.E. Beadle, J.C.C. Tsai, and R.D. Plummer, "Quick Reference Manual for Silicon Integrated Circuit Technology,"Wiley-Interscience, NY, 1985.
30. S. Tolansky, "Multiple-Beam Interferometry of Surfaces and Films," Clarendon Press, Oxford, (1948).
31. J. Hilibrand, R.D. Gold, RCA Rev. 21, 245 (1960).
32. C.P. Wu, E.C. Douglas, C.W. Mueller, IEEE Trans. Elec. Dev. ED-22, 319 (1975).
33. R.A. Moline, J. Appl. Phys. 42, 3553 (1971).
34. Y. Zohta, Solid State Electron. 11, 124 (1973).
35. W.C. Niehaus, W. VanGelder, T.D. Jones, P. Langer, in "Silicon Device Processing,"NBS Special Publication 337, 256 (1970).
36. S.J. Rothman, in "Diffusion in Crystalline Solids,"G.-E. Murch, A.S. Nowick, ed., Academic Press, Orlando, p. 1 (1984).
37. C. Meixner, "Tables of Gamma–Ray Energies for Activation Analysis,"Verlag Karl Thiemig, Munich (1970).
38. P.L. Gruzin, Dok. Akad. Nauk. SSSR 86, 289 (1952).
39. A.A. Zhukovitskii, S.N. Kryukov, V.A. Geodakyan, Symp. 34, Moscow Steel Inst., Moscow (1955), (English Transl.: AEC Transl. 3100, Part II, p. 3.).
40. D. Gupta, D.R. Campbell, Philos. Mag. A42, 513 (1980).
41. C.C. Chang, in "Characterization of Solid Surfaces,"P.F. Kane, G.R. Larrabee, eds., Plenum Press, NY, p. 509 (1974).
42. J.B. Hudson, RPI, Troy, NY, private communication.
43. J.A. McHugh, in "Methods of Surface Analysis,"A.W. Czanderna, ed., Elsevier, NY, p. 223 (1975).
44. H. Dietrich, Naval Res. Lab., Washington, DC, private communication.

45. W.-K. Chu, J.W. Mayer, M.-A. Nicolet, "Backscattering Spectrometry," Academic Press, NY (1978).
46. For examples see L.C. Feldman, J.W. Mayer, S.T. Picraux, "Materials Analysis by Ion Channeling Submicron Crystallography," Academic Press, NY (1982).
47. R.B. Marcus, T.T. Sheng, "Transmission Electron Microscopy of Silicon VLSI Circuits and Structures," Wiley, NY (1983).
48. S. Kirtley, S. Gupta, private communication.
49. N. Jones, D.M. Metz, J. Stach, R.E. Tressler, J. Electrochem. Soc. 123, 1565 (1976).
50. J.T. Clemens, U.S. Patent # 4291322 (1981).
51. W. Kern, R.S. Rosler, J. Vac. Sci. Technol. 14, 1082 (1977).
52. S.P. Murarka, unpublished results.
53. W.A. Waggner, unpublished results.
54. S.K. Ghandhi, "VLSI Fabrication Principles Silicon and Gallium Arsenide," Wiley, NY (1983).
55. R.B. Fair, in "Silicon Integrated Circuits," Part B in Applied Solid Science Series Supplement 2, D. Kahng, ed., Academic Press, NY, p. 1 (1981).
56. R.A. Longini, R.F. Greene, Phys. Rev. 102, 992 (1956).
57. R.A. Swalin, "Thermodynamics of Solids," first ed., Wiley, NY, p. 265 (1962).
58. R.A. Swalin, "Atomic Diffusion in Semiconductors," D. Shaw, Plenum, NY, p. 32 (1973).
59. D. Lazarus, Phys. Rev. 93, 973 (1954).
60. A.D. LeClaire, Phil. Mag. 1, 141 (1962); and Phil. Mag. 10, 641 (1964).
61. D.L. Kendall, D.B. Vries, in "Semiconductor Silicon 1969," R.R. Haberect, E.L. Kern, eds., The Electrochem. Soc., Inc., NY, p. 414 (1969).
62. R.B. Bailey, T.G. Mills, in "Semiconductor Silicon 1969," R.R. Haberect, E.L. Kern, eds., The Electrochem. Soc., Inc., NY p. 481 (1969).
63. W. Wurker, K. Roy, J. Hesse, Mat. Res. Bult. (USA) 9, 971 (1974).
64. M. Stavola, J.R. Patel, L.C. Kimerling, P.E. Freeland, Appl. Phys. Lett. 42, 73 (1983).
65. W.M. Bullis, Solid-State Electronics 9, 143 (1966).
66. R.B. Fair, J.C.C. Tsai, J. Electrochem. Soc. 122, 1689 (1975).
67. R.B. Fair, J. Appl. Phys. 50, 860 (1979).
68. D.P. Kennedy, P.C. Murley, Proc. IEEE 59, 335 (1971).
69. R.B. Fair, G.R. Weber, J. Appl. Phys. 44, 273 (1973).
70. J.E. Lawrence, J. Electrochem. Soc. 115, 860 (1968).
71. M.C. Duffy, F. Barson, J.M. Fairfield, G.H. Schwuttke, J. Electrochem. Soc. 115, 84 (1968).
72. M.L. Joshi, F. Wilhelm, J. Electrochem. Soc. 112, 185 (1965).

73. M. Watanabe, H. Muraoka, T. Yonezawa, Jpn. J. Appl. Phys. 44, 269 (1975).
74. R.B. Fair, J. Electrochem. Soc. 125, 923 (1978).
75. G. Rozgonyi, P.M. Petroff, M.H. Read, J. Electrochem. Soc. 122, 1725 (1975).
76. S. Prussin, J. Appl. Phys. 32, 1876 (1961).
77. A. Fukuhara, Y. Takano, Acta. Cryst. Sect. A. 33, 137 (1977).
78. J.J. Wortmann, J.R. Hauser, R.M. Burger, J. Appl. Phys. 35, 2122 (1964).
79. S.M. Hu, S. Schmidt, J. Appl. Phys. 39, 4272 (1968).
80. S.M. Hu, J. Appl. Phys. 43, 2015 (1972).
81. L.E. Katz, Nat. Bur. Stand. (U.S.), Spec. Publ. 337, 192 (1970).
82. W.G. Allen, K.V. Anand, Solid-State Electronics 14, 397 (1971).
83. S.P. Murarka, Phys. Rev. 12, 2502 (1975); for a review of older data see this reference.
84. D.A. Antoniadis, A.G. Gonzalez, R.W. Dutton, J. Electrochem. Soc. 125, 813 (1978).
85. G. Masetti, S. Solmi, G. Soncini, Solid-State Electronics 16, 1419 (1973).
86. R. Francis, P.S. Dobson, J. Appl. Phys. 50, 280 (1979).
87. D.A. Antoniadis, A.M. Lin, R.W. Dutton, Appl. Phys. Lett. 33, 1030 (1978).
88. S.M. Hu, J. Appl. Phys. 45, 1567 (1974).
89. P. Baruch, Inst. Phys. Conf. Ser. 31 126 (1977).
90. M. Wittmer, K.N. Tu, Phys. Rev. B29, 2010 (1984).
91. E. Guerrero, H. Potzl, R. Tielert, M. Grasserbauer, G. Stingeder, J. Electrochem. Soc. 129, 1826 (1982).
92. H. Ryssel, K. Muller, K. Haberger, R. Henkelmann, F. Jahael, J. Appl. Phys. 22, 35 (1980).
93. F. Sette, S.J. Pearton, J.M. Poate, J.E. Rowe, J. Stohr, Phys. Rev. Lett. 56, 2637 (1986).

Problem Set for Chapter 4

4.1 Derive eq. (4.4) from Fick's first law and the requirement that matter is conserved. (Hint: Consider diffusion across an element of volume and calculate the rate at which concentration changes in this volume).

4.2 Typically grain-boundary diffusion activation energies are nearly half the activation energy for the diffusion in the grain or crystal. However, at very high temperatures diffusion in the bulk dominates. To see this, plot ln

D_T as a function of $1/T$, where $D_T(cm^2/s)$ is given by

$$D_T = 0.895 \exp - (2.00 \text{ eV}/kT) + 2.3 \times 10^{-5} \exp(-1.15 \text{ eV}/kT).$$

Explain the resulting curve. Based on this computation, describe how diffusion behavior in a polysilicon film differs from that of bulk silicon.

4.3 The diffusion coefficient of Si in Al is given by the Arrhenius relationship, $D_L = 0.25 \exp(-1.3 \text{ eV/kT}) \text{ cm}^2/\text{s}$, for the diffusion in the grains; and, $D_b = 2.5 \times 10^{-5} \exp(-0.7 \text{ eV/kT}) \text{ cm}^2/\text{s}$, for the diffusion in the grain boundaries. A 1 μm thick polycrystalline Al film is deposited on silicon substrate. Assuming that Al-Si interface is no barrier to diffusion, calculate the time taken for silicon to diffuse through the Al film at 200, 300, 400, 450, and 500°C. Can you now explain why Si-Al interactions are so important at 450°C? How can we minimize these interactions?

4.4 In an experiment phosphorous is diffused into a p-type silicon wafer with bulk resistivity of 0.01 Ω-cm, which corresponds to a doping of ~ 1 × 10^{19} boron per cm^3. After diffusion, sheet resistance is measured on the surface of the wafer and after successful removal of 200 nm layers at a time. The independently measured junction depth is 2 μm. Construct a diffusion depth profile if the sheet resistance measurements are 2.08, 3.27, 4.81, 10.20, 26.04, 52.63, and 69.44 Ω/□ after consecutive removal of the layers. Note, the first measurement reflects measurement on the surface prior to layer removal. (Hint: Use fig. 4.7.)

4.5 Predeposition is carried out on a silicon wafer at 975°C for 30 minutes in the presence of an excess of phosphorus. The substrate is boron doped to a concentration of 10^{17} atoms/cm^3.

a. Find C_s for this procedure.
b. Calculate $\sqrt{Dt}$ for this procedure.
c. Calculate the dopant concentration in the silicon at the following depths: x = 0.2 μm, and x = 0.5 μm.
d. Determine the depth at which the metallurgical p-n junction occurs.
e. Find the total amount of dopant incorporated into the wafer per unit area.
f. Assuming uniform density of the dopant atoms in the silicon after the diffusion, calculate the conductivity of the diffused layer when μ_n = 1450 cm^2/V-s.

4.6 Czocharalski grown silicon wafers generally contain oxygen in excess of their solid solubility limit. On annealing in an inert ambient at high temperatures, oxygen diffuses out of the silicon. The starting bulk concentration of oxygen in the wafer is 20 ppma and the oxygen solid solubility at 1100°C is 5×10^{16} per cm^3. Plot an oxygen concentration profile in silicon after an 1100°C/48h/inert ambient anneal. The diffusivity of oxygen in silicon is given by:

$$D = 0.17 \exp\left(-2.54 \frac{\text{eV}}{\text{k}T}\right).$$

Comment on the role of the oxygen concentration reduction in the near surface region in defect formation.

4.7 Comment on the often-made statement that vacancies and interstitials are thermodynamic defects, whereas dislocations are not.

4.8 Often, interstitial and vacancies are formed in pairs and their concentrations are related. Derive an equation, using the law of mass action, relating these concentrations.

4.9 One possible way to make silicon a conductor is to reduce its band-gap energy to zero by applying tensile stress. Calculate this stress. Can such stresses be realized without affecting the basic crystal structure?

5 Ion Implantation

5.1 Introduction

Ion implantation is the process of introducing energetic ions into a substrate. The depth to which the ions penetrate depends on the ion energy, mass, charge, angle of incidence, and the substrate material. The process can be used for implanting any species that can be ionized in the gas phase in an evacuated chamber. The process is unique in separating and then selecting the desired ion species by the use of analyzer electromagnets. It is, therefore, very useful in introducing dopants such as boron, phosphorus, or arsenic into silicon and silicon, zinc, or telurium into GaAs. The advantages of choosing ion implantation over other doping techniques (as described in Chapter 4) are several, the most important of which is the precise control in the number in ions introduced. The most important disadvantage of the use of ion implantation is the associated damage induced into the crystal.

In addition, ion implantation offers a means of introducing dopants not possible by any other means. For example, by a proper choice of the ion energy and dose, a base can be implanted through the emitter to create a bipolar transistor. Threshold voltage of a metal-oxide-semiconductor field effect transistor (MOSFET) can be tailored by the use of dopant implants in the near surface region of the semiconductor. Graded doping profiles can be created to suit a device design and performance.

There are other applications of this technique [1,2]. These applications include modification of materials by ion implantation. Generally such modifications are confined to the near-surface regions of the materials since the change induced by ion implantation is limited to the penetration depth that can be achieved. Recently very-high-energy and high-flux (or high-ion-current) implanters have become available so that species like oxygen or nitrogen can be implanted into silicon to produce SiO_2 or Si_3N_4.

In this chapter, the fundamentals of ion implantation are presented in Section 5.2, followed by the available options in Section 5.3, and equipment and technology in Section 5.4. Sections 5.5 and 5.6 discuss new applications, e.g., ion-beam mixing and materials modification and dielectric film formation, respectively.

5.2 Fundamentals

Energetic ions interact with solids at all energies. The outcome of this interaction is, however, not the same in all cases and depends very strongly on the ion energy and the proximity of the ion to the target atom and less so on the mass and charge of the ion and of the atoms that constitute the solid substrate. At very low energies the interaction causes electronic excitation associated with the absorption of energy in the range of a few eV to a few hundreds of eV. The outermost electrons are affected at very low energies. As the energy increases, inner electrons interact. As the energy of the ions becomes greater than a few hundreds of eV, the screened coulombic collisions between the moving ion and the substrate nuclei start playing a role in slowing down the incoming ion. Such nuclear interactions become important at incoming ion energies in the range of a few keV (kilo electron volts) to about a few hundred of keV. At higher energies the interactions between the electrons of the moving ion and those in the solid become important. For ion energy in the range of a few MeV, the interaction between the incoming particle and the substrate nuclei once again occurs by coulombic repulsion. At such high enegies the velocities of the incident particles are so great that one may, to a very good approximation, neglect the screening of the nucleus by its electrons. Such is the case of Rutherford backscattering of MeV range α particles (doubly ionized helium ions). For our ion implantation application, this high energy range is not of any interest and therefore will not be considered. Finally mass of the ion, relative to the substrate, also plays an important role in determining the above ranges for interactions of different types. For ions, lighter with respect to the substrate atom, electronic stopping becomes important at lower energies. For example, for boron traveling in silicon, electron stopping is dominant even at ~10 keV, whereas for P and As, the nuclear collision mechanism is important up to ~130 and 700 keV, respectively [3].

In the above, various types of ion-target interactions were classified based on the energy of the incoming ion. Close proximity of the ion to the target atom was assumed. If, however, the distance between the moving ion and the target atom increases, the severity of interaction (or the energy transfer to the target atom) decreases rapidly with distance. A high-energy ion, at large distances from the target atom, may only excite outermost electrons. As the distance of closest approach between the moving ion and target nuclei decreases, different types of interactions will occur effectively, leading to the results similar to those described with increasing energy. Thus to derive a meaningful result, the distance between the target atom and moving ion and the ion energy both must be considered, together with the masses and charges of the interacting species.

Figure 5.1 shows schematically the effects of ion-solid interactions and various processes. As the energy of the ion, being implanted into a solid, increases, the stopping mechanism changes. For ions with very low energies (say, ~10 eV) all of the ion energy is used up in exciting the outermost electrons of the surface layers of the target, leading to a deposition of the layer of the implanting species. At energies of the order of 1000 eV, incoming ions lose energy by exciting innermost electrons of the target atoms and may

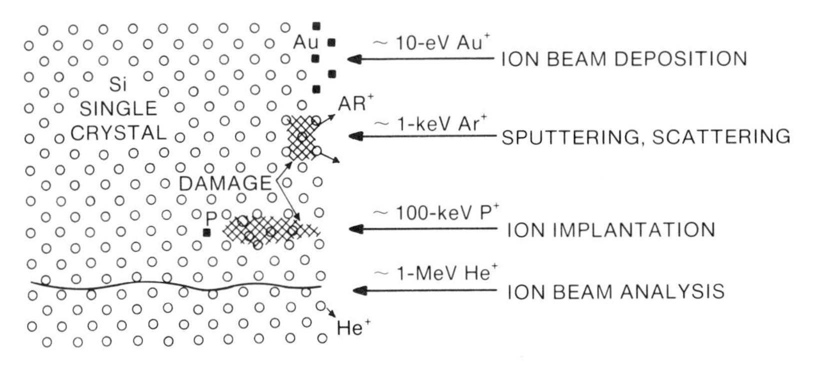

Fig. 5.1. Schematic illustrating the interactions of ion beams with a single-crystal solid. Directed beams of ~10 eV are used for film deposition and epitaxial formation. Ion beams of ~1 keV are employed in sputtering application; 100-keV ions are used in ion implantation. Both of these processes damage and disorder the crystal. Higher energy light ions are used by channeling analysis.

even incur collisions with target nuclei, causing sputtering or scattering away from the surface.

At higher ion energies, ion penetration into the substrate increases, leading to ion implantation. The interaction with the target and electrons leads to dislodging of target nuclei from their equilibrium position resulting in so-called damage. At very high energies, where slowing down of the moving ion is the result of interaction between the electrons of the moving ion and those in the target, the ions travel very large distances without causing any significant damage in the target. Only a very small fraction of the incoming ions make direct collision with target nuclei, causing scattering of the incoming particle and dislocation of the target nuclei. Note that sputtering, scattering, and implantation cause damage and disorder in the target as shown by shaded regions.

For our applications, ion implantation is of interest. Ion energies in the range of a few keV to a few hundred keV are generally used. At such energies, ions entering the target lose energy mainly by two mechanisms, the nuclear collisions and electronic interactions, as discussed in the preceding paragraph. As the ion travels through the solid, its trajectory changes several times prior to the ion coming to a stop. The changes in the trajectory occur due to scattering associated with nuclear and electronic interactions. Depending upon the ion energy, the scattered target nuclei will produce their own trajectories prior to coming to a stop. Thus a cascading effect occurs, leading to a disordered region surrounding the trajectory of the incoming ion, as shown in fig. 5.2a. The heavy zig-zag path between the surface of the solid and the stopped ion represents the incoming ion trajectory. The side branches are the trajectories of the dislocated target atoms. The shaded area around this tree represents the damage consisting of vacancies and interstitials (point defects) and an amorphous region.

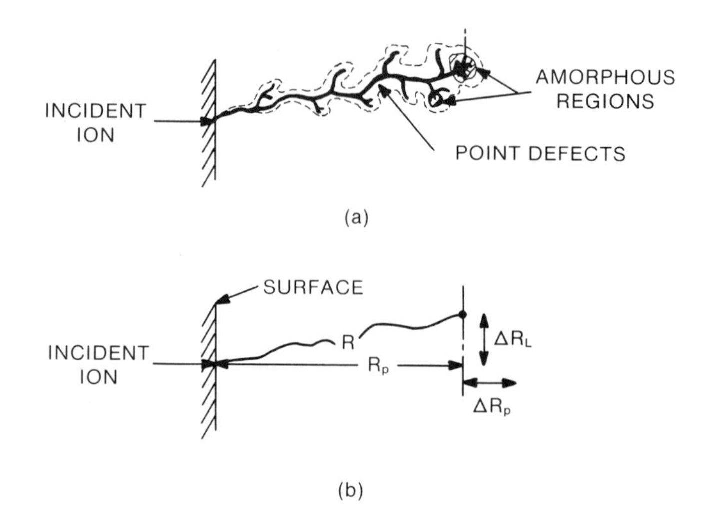

Fig. 5.2. (a) A "tree"of disorder for a typical implanted ion. (b) A schematic of the ion range R, projected range R, uncertainty in R_p or projected straggle ΔR_p, and the lateral straggle ΔR_L.

In fig. 5.2b, the range R, the projected range R_p, the projected straggle ΔR_p which is the standard deviation of the Gaussian distribution (in the incoming direction of the ion) and lateral straggle ΔR_L are defined.

In most implantation, the incident-beam direction is at angle (say θ) with the surface normal. In such a case the mean penetration depth x measured along the surface normal, and the depth straggling Δx along this normal, are related to R_p and ΔR_p (measured along the incident beam direction) by the following relationship [4]:

$$x = R_p \cos\theta \tag{5.1a}$$

and

$$\Delta x^2 = R_p^2 \cos^2\theta + \frac{1}{2} R_L^2 \sin^2\theta. \tag{5.1b}$$

The range R is the total distance traversed by the ion. R_p is the projected distance of travel in the direction of incidence. ΔR_p is also defined along this direction. If the energy loss per unit path length is defined as $-dE/dx$, then range R at energy E_0 is given as

$$R(E_0) = \int_0^{E_0} \frac{dE}{(-dE/dx)}, \tag{5.1c}$$

where E_0 is the energy of the ion prior to entering the target and the negative sign indicates the loss of energy.

5.2.1 *Range*

The calculation of range from theoretical considerations is complicated, and only the essential features are presented here. The ion energy loss per unit length is considered to be a sum of two terms: the nuclear stopping term and the electronic stopping term. The stopping power can be defined as energy loss per unit length, i.e., $(-dE/dx)$, where E represents the energy, x is the distance, and the negative sign makes the stopping power a positive quantity, since dE/dx is a negative term, as energy of the ion decreases with distance traveled. We can define nuclear stopping power S_n as $(-dE/dx)_\mathrm{n}$ and electronic stopping power S_e as $(-dE/dx)_\mathrm{e}$. Total stopping power then is the sum of S_n and S_e. Lindhard, Scharff, and Schiott (now called LSS) calculated both the nuclear stopping and electronic stopping term. New dimensionless parameter for range and energy were defined as [5]

$$\rho = R\pi a^2 N \frac{M_1 M_2}{(M_1 + M_2)^2} \tag{5.2}$$

$$\epsilon = \frac{Ea}{e^2}\left[\frac{M_2}{Z_1 Z_2 (M_1 + M_2)}\right], \tag{5.3}$$

where

$$a = 0.8853 a_0 \left(Z_1^{2/3} + Z_2^{2/3}\right)^{-1/2} \tag{5.4}$$

and a_0 is the Bohr radius equal to 0.529 Å, N is the number of atoms per unit volume of the target, Z and M are atomic number and mass, e is the electronic charge, and subscripts 1 and 2 refer to the incident ion and the target, respectively. In terms of the new parameters S_n and S_e become $(-d\epsilon/d\rho)_\mathrm{n}$ and $(-d\epsilon/d\rho)_\mathrm{e}$ respectively.

LSS derived the electronic stopping power as

$$S_e = \left(-\frac{d\epsilon}{d\rho}\right)_\mathrm{e} = k_e \epsilon^{1/2}, \tag{5.5}$$

where

$$k_\mathrm{e} = \zeta\left[\frac{0.0793 Z_1^{1/2} Z_2^{1/2} (M_1 + M_2)^{3/2}}{(Z_1^{2/3} + Z_2^{2/3})^{3/4} M_1^{3/2} M_2^{1/2}}\right] \tag{5.6}$$

and ζ is approximately equal to $Z_1^{1/6}$ and therefore lies between 1 and 2.

LSS also derived a relationship between nuclear stopping terms $S_\mathrm{n} = \left(-\frac{d\epsilon}{d\rho}\right)_\mathrm{n}$ and $\epsilon^{1/2}$. Figure 5.3 shows both the stopping terms as a function of $\epsilon^{1/2}$. Note there are a series of curves (one for each k_e value) for the

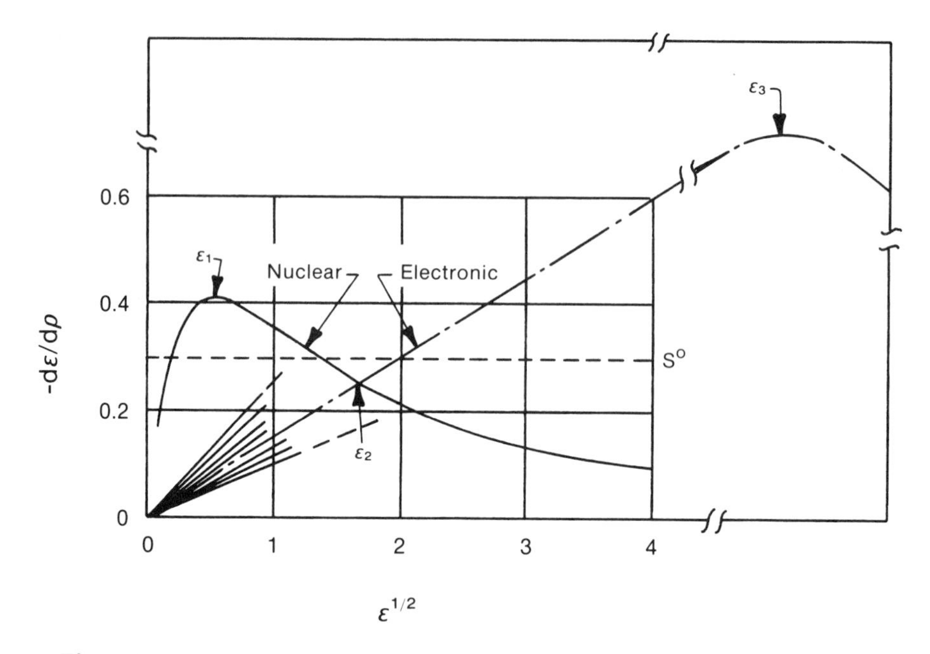

Fig. 5.3. Theoretical nuclear and electronic stopping-power curves, expressed in terms of the reduced variables p and ϵ. (Based on Lindhard et al. [5]). For electronic stopping, a family of lines (one for each combination of projectile and target) is obtained; the majority of cases fall within the limits shown. The dot-dash line represents the electronic stopping for k = 0.15. The horizontal line labeled S° represents the constant- stopping-power approximation suggested by Nielson [6]. From Mayer et al. [7].

electronic stopping power. It is clear from the figure 5.3 that nuclear stopping power dominates at low energies, whereas electronic stopping power becomes important at high energies.

Figure 5.4 shows the calculated values of the nuclear and electronic stopping powers as a function of the energy for dopants commonly used in silicon. Although the nuclear and electronic contribution to the stopping power varies with the ion energy, the total stopping power is nearly independent of energy.

The range R at an energy E can be calculated from the known relationship between ρ and ϵ. An approximate relationship has, therefore, been given [6,7] as

$$R(\mathring{A}) = \frac{60E(\text{keV})}{g} \frac{M_2}{Z_2} \frac{M_1 + M_2}{M_1} \frac{\left(Z_1^{2/3} + Z_2^{2/3}\right)^{1/2}}{Z_1}, \qquad (5.7)$$

where energy E is given in keV and range R is in angstroms, and g is the density of the target in g/cm^3.

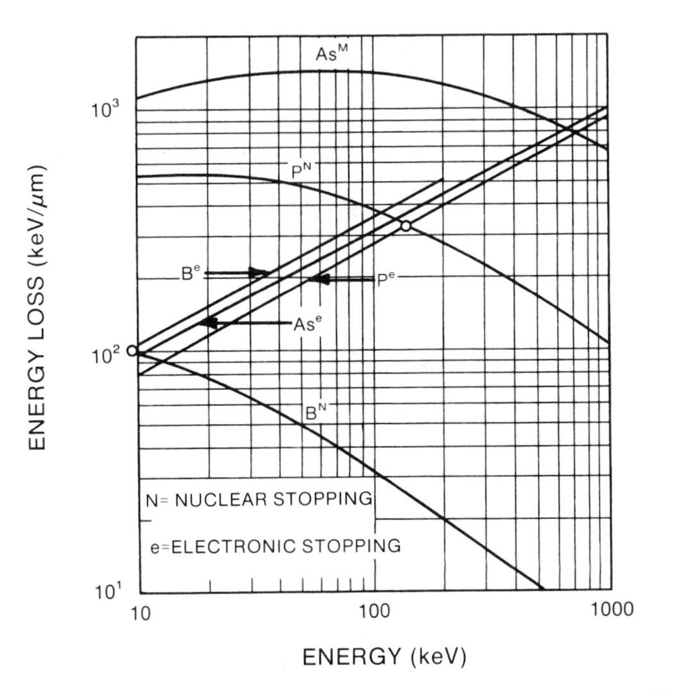

Fig. 5.4. Calculated values of dE/dx for As, P, and B at various energies. The nuclear (N) and electronic (e) components are shown. Note the points (o) at which nuclear and electronic stopping are equal. From Smith [3].

The projected range R_p and the projected straggle ΔR_p are similarly approximated and are given by [8]

$$R_p = R(1 + \frac{M_2}{3M_1})^{-1} \tag{5.8}$$

and

$$\Delta R_p = \frac{2R_p\sqrt{M_1 M_2}}{3(M_1 + M_2)}. \tag{5.9}$$

Figures 5.5 and 5.6 show the calculated R_p and ΔR_p and ΔR_L values for various ions as a function of the ion energy for implantation in silicon [3]. Figures 5.7 and 5.8 show similar curves for various dopants in GaAs [9]. Note that ΔR_p and ΔR_L are very high for lighter dopants and that they tend to be independent of ion energy greater than 100 keV for lighter ions.

5.2.2 Implanted Concentration Profiles

When a monoenergetic beam of ions is implanted into the substrate, there is depth dependence of the concentration. The depth distribution is best

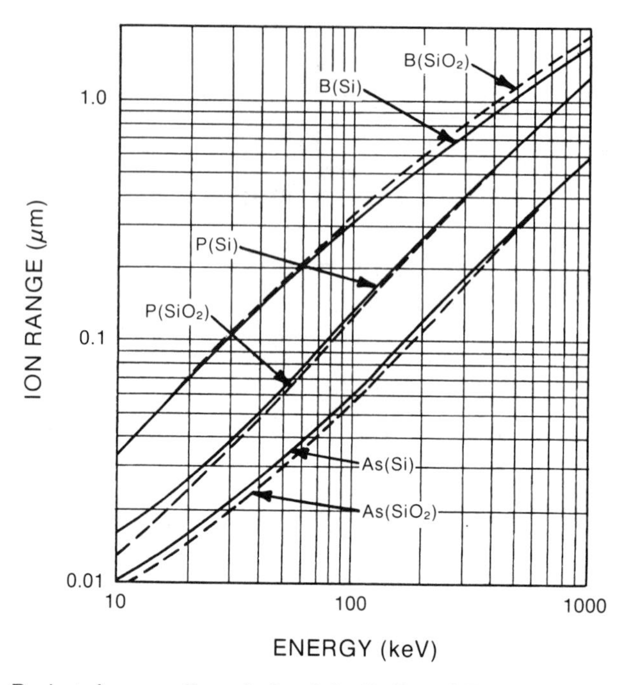

Fig. 5.5. Projected range, R_p, calculated for B, P, and B at various energies. The results pertain to amorphous silicon targets and thermal SiO_2 (2.27 g/cm^3). From Smith [3].

approximated by the Gaussian relationship

$$n(x) \;=\; N_{\max}\exp\left[-(x-R_{\mathrm{p}})^2/2\Delta R_{\mathrm{p}}^2\right], \tag{5.10}$$

where $n(x)$ is concentration of ions as a function of depth x measured in a direction perpendicular to the substrate surface and $N_{\max}$ is the maximum of the concentration at $x = R_{\mathrm{p}}$.

If ϕ is the total dose or flux of ions implanted into the substrate then

$$\phi = \int_{-\infty}^{\infty} n(x)\, dx \tag{5.11}$$

and

$$N_{\max} = \frac{\phi}{\sqrt{2\pi}\Delta R_{\mathrm{p}}} = \frac{0.4\phi}{\Delta R_{\mathrm{p}}}. \tag{5.12}$$

The total dose or flux of ions, ϕ, is determined experimentally by using the relationship

$$\phi = \frac{Q}{sqA}\text{ions/cm}^2 \tag{5.13}$$

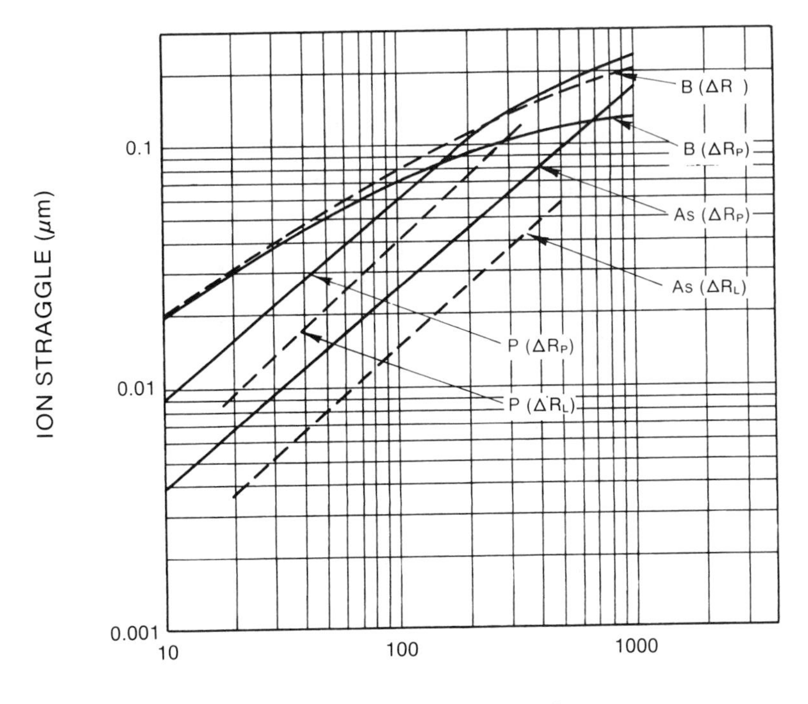

Fig. 5.6. Calculated ion straggle ΔR_{P} (vertical) and ΔR_{L} (transverse) for As, P, and B ions in silicon. From Smith [3].

where sq is the total charge on an ion (e.g., s is 1 for singly ionized ions and 2 for doubly ionized ions) and q is the electronic charge. A is the ion implanted surface area, and Q is the total integrated charge deposited in the substrate. Q is determined by measuring beam current I, and is given by

$$Q = \int_0^t I\,dt = It \tag{5.14}$$

where t is the time for which the implantation was carried out.

Problem 5.1: *Calculate the ion current density needed to implant 10^{16} per cm^2 of singly ionized ions in 1000s and 1s. If a 150 mm diameter wafer is to be implanted with ion flux of 10^{16} singly charged ions per cm^2, estimate the time needed for completing the implant on entire wafer surface.*

Solution: For the singly charged species

$$\phi = \frac{Q}{qA} = \frac{It}{Aq}$$

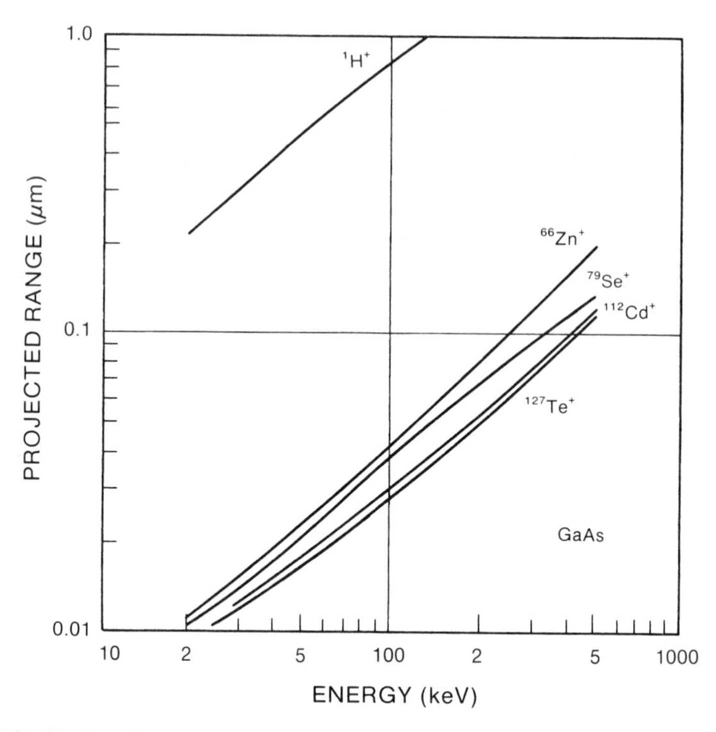

Fig. 5.7. Projected range of H, Zn, Se, Cd, and Te in GaAs. From Furukawa et al. [9].

$$I/A = \phi\ q/t \text{ amp per cm}^2,$$

when

$$t = 1000\ s,\ I/A = 1.6\ \times\ 10^{-6} \text{ amp per cm}^2;\text{ and}$$
$$t = 1\ s,\ I/A = 1.6\ \times\ 10^{-3} \text{ amp per cm}^2.$$

For a 150 mm diameter wafer, the surface area is 176.7 cm^2. Thus total flux required will be 1.767×10^{18} ions. At an ion flux of 10^{16} ions per cm^2 in 1000 s the time required will be 1.767×10^5 s or 49.2 h. On the other hand, at an ion flux of 10^{16} ions per cm^2 in 1 s, the time required will only be 176.7 s or ~3 min.

From the above example it is clear that to process large wafers at a required dose of 10^{16} per cm^2 per s, high-current (1-100 mamps per cm^2) implanters will be needed. The need for such high-current implanters becomes a must for materials modification using ion implantation because of the more than order of magnitude higher flux requirements. For ionic species with $s > 1$, higher beam currents will be necessary.

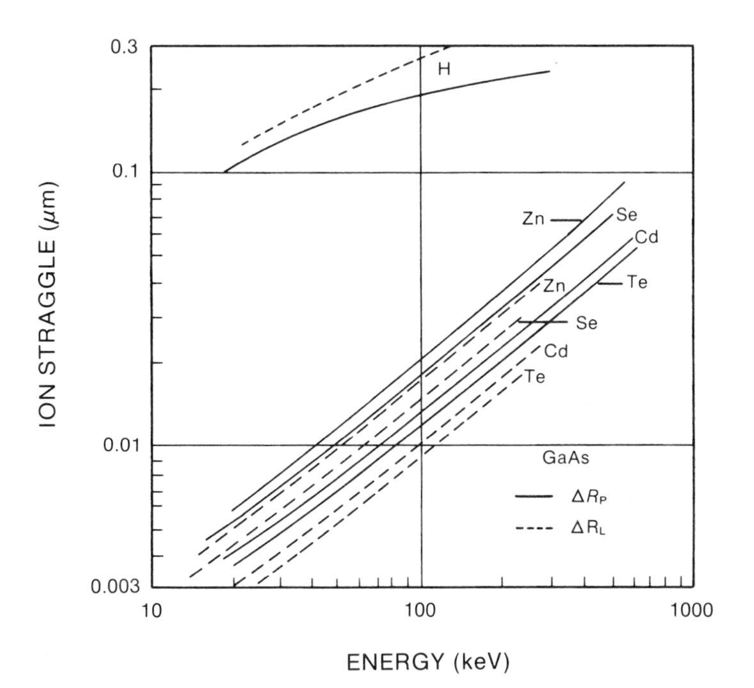

Fig. 5.8. Projected straggle and lateral straggle in GaAs. From Furukawa et al. [9].

Figure 5.9 show a linear plot of the Gaussian distribution [10], $n(x)/N_{\max}$ is plotted as a function of the distance x. R_p and ΔR_p are also shown. Note, by definition of ΔR_p, $n(x)/N_{\max} = 0.6065$ when $(x - R_p) = \pm\Delta R_p$. The ion distribution tail to the left of the origin represents fraction of the ions reflected by the surface. This amount is less than 5% of the total flux ϕ when $R_p/\sqrt{2}\Delta R_p \geq 1.5$. One can calculate the total number of implanted atoms $N(x_1)$ deposited at a depth greater than x_1, as given by the expression

$$N(x_1) = \int_{x_1}^{\infty} n(x)\, dx = \frac{\phi}{2}\,\mathrm{erfc}\left(\frac{x_1 R_p}{\sqrt{2}\Delta R_p}\right), \tag{5.15}$$

where $\mathrm{erfc}(\eta) = 1 - \mathrm{erf}(\eta)$ and

$$\mathrm{erf}(\eta) = \frac{2}{\sqrt{\pi}} \int_0^{\eta} e^{-r}\, dr. \tag{5.16}$$

Finally, it can be also shown that the implanted atom concentrations $n(x)$ at $x = R_p \pm 2\Delta R_p$ and $x = R_p \pm 3\Delta R_p$ are approximately 0.1 $N_{\max}$ and 0.01 $N_{\max}$, respectively. Thus practically all the implanted atoms are within the depth boundaries defined by $x = R_p \pm 3\Delta R_p$.

Equations (5.10) through (5.15) present an ideal implant-atom concentration distribution. In real life these equations give only a first-order approximation to the true concentration profile. Actual profiles show asymmetrical

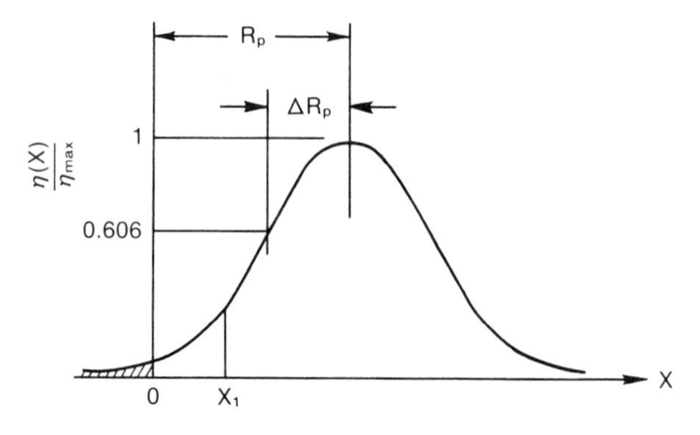

Fig. 5.9. Gaussian distribution.

distribution associated with the backscattering of implanting ions by the substrate atoms. Lighter ions will experience backscattering larger than predicted by eq. (5.10). On the other hand, heavier ions will experience less backscattering. Thus curves of the type shown in fig. 5.9 will have asymmetry (or skewness) on the surface side of the maximum for lighter ions and on the deep side of the maximum for the heavier ions.

If the skewness is not too large, the experimental asymmetric distributions could be fitted by adopting a so-called three-moment approach; whereas for the best fit, a four-moment approach has been used [11-13]. In the three-moment approach two half-Gaussian profiles with the projected straggles ΔR_{p1} and ΔR_{p2} are joined at a model range R_m such that at $x > R_m$ the projected straggle is ΔR_{p1} and at $x < R_m$ the projected straggle is ΔR_{p2}. In this case then

$$n(x) = \frac{2}{\sqrt{2\pi}(\Delta R_{p1} + \Delta R_{p2})} \exp\left[\frac{-(x - R_m)^2}{2\Delta R_{p1}^2}\right] \text{ for } x \geq R_m \quad (5.17)$$

$$n(x) = \frac{2}{\sqrt{2\pi}(\Delta R_{p1} + \Delta R_{p2})} \exp\left[\frac{-(x - R_m)^2}{2\Delta R_{p2}^2}\right] \text{ for } x < R_m. \quad (5.18)$$

R_m is obtained from the values of ΔR_{p1}, ΔR_p, ΔR_{p2}, R_p, and the third moment ratio skewness [14]. R_m is given as

$$R_m = R_p - 0.8(\Delta R_{p2} - \Delta R_{p1}) \text{ or} \quad (5.19a)$$

$$R_m = R_p - 0.8(\Delta R_{p1} - \Delta R_{p2}), \quad (5.19b)$$

respectively, for positive or negative skewness.

In the four-moment approach using projected range R_p, projected straggle ΔR_p, skewness γ, and kurtosis (β), the Pearson distribution

$$\frac{df(x)}{dx} = \frac{(x - a)f(x)}{b_0 + b_1 x + b_2 x^2} \quad (5.20)$$

with $\mathrm{x} = x - R_p$, and

$$\int_{-\infty}^{\infty} f(x)\, dx = 1 \tag{5.21}$$

is employed. β now describes the distribution at the tail end of the profile. a, b_0, b_1, and b_2 are constants related to various moments. The moments are given as

$$R_\mathrm{p} = \int_{-\infty}^{\infty} x f(x)\, dx, \tag{5.22a}$$

$$\Delta R_\mathrm{p}^2 = \int_{-\infty}^{\infty} (x - R_\mathrm{p})^2 f(x)\, dx, \tag{5.22b}$$

$$\gamma = \frac{1}{\Delta R_\mathrm{p}^3} \int_{-\infty}^{\infty} (x - R_\mathrm{p})^3 f(x)\, dx, \tag{5.22c}$$

and

$$\beta = \frac{1}{\Delta R_\mathrm{p}^4} \int_{-\infty}^{\infty} (x - R_\mathrm{p})^4 f(x)\, dx. \tag{5.22d}$$

Four constants can now be given as

$$a = b_1 = -\frac{\gamma(\Delta R_\mathrm{p}^2)(\beta + 3)}{A}, \tag{5.23a}$$

$$b_0 = \frac{(\Delta R_\mathrm{p}^4)(3\gamma^2 - 4\beta)}{A}, \tag{5.23b}$$

$$b_2 = \frac{(6 + 3\gamma^2 - 2\beta)}{A}. \tag{5.23c}$$

and

$$A = 10\beta - 12\gamma^2 - 18. \tag{5.23d}$$

A particular type of Pearson distribution is determined by the parameter

$$p = \frac{b_1^2}{4b_0 \cdot b_2}. \tag{5.24}$$

The Pearson IV solution [15] of the Pearson distribution is applicable when $0 < p < 1$. The solution is

$$\ell n \frac{n(x)}{N_0} = \frac{1}{2b_2}\, \ell n[b_2 \mathrm{x}^2 + b_1 \mathrm{x} + b_0] - \frac{b_1/b_2 + 2a}{(4b_2 b_0 - b_1^2)^{1/2}}$$
$$\arctan\left[\frac{2b_2 \mathrm{x} + b_1}{(4b_2 b_0 - b_1^2]^{1/2}}\right] \tag{5.25}$$

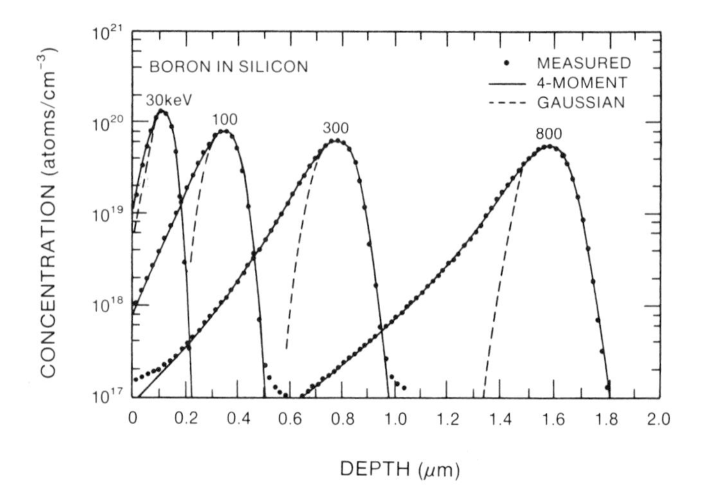

Fig. 5.10. Boron implantation atom distributions, with measured data points, and four-moment (Pearson-IV) and symmetric Gaussian curves. The boron was implanted into amorphous silicon without annealing. From Hofker [13].

where

$$\phi = N_0 \int_0^\infty f(x)\, dx. \tag{5.26}$$

The parameter p, for most applications of ion implantation, lies in the range $0 < p < 1$, and eq. (5.25) gives an accurate distribution for the implanted species.

Figure 5.10 shows the measured (points), four-moment (full line), and Gaussian (broken line) distributions of boron implanted in silicon. Excellent agreement between experimental and four-moment profiles is seen [13]. The Gaussian profile, as discussed earlier, failed to give a true profile near the surface, and the difference between the Gaussian and true profiles increases with increasing boron ion energies.

5.2.3 Effect of Crystallinity on the Range and the Depth Distribution

In the above, we calculated the range and the depth distribution assuming uniform density of atoms in all directions of the target material. In general, we are concerned with crystalline materials such as silicon and GaAs semiconductors. In crystals, the atomic density along various directions varies. Also, if oriented properly, the incident ion direction may match the direction of so-called open channels between atoms. In such a case, the ions can travel very large distances before being scattered significantly. Figure 5.11 shows an artist's conception of the ion travel in such a channel in the crystal. This is called *channeling*. Figure 5.12 shows channeling tunnels along <111>,

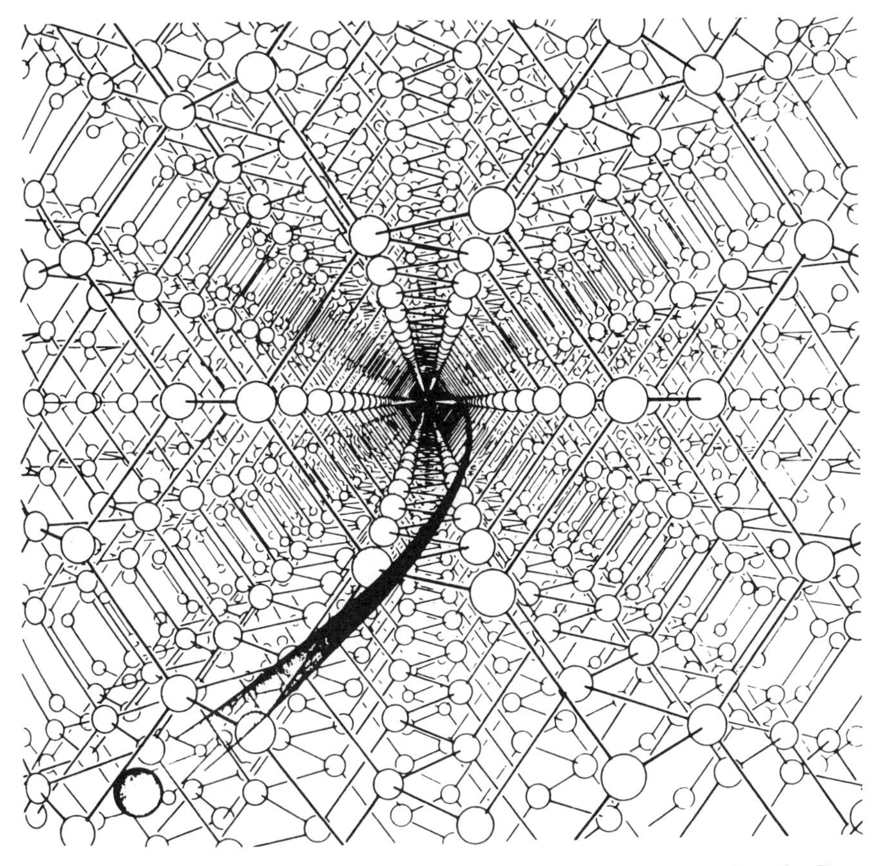

Fig. 5.11. Artist's conception of the channeling process on a microscopic scale. From *Channeling in Crystals* by W. Brandt. Copyright © 1968 by Scientific American, Inc. All rights reserved.

<100>, and <110> directions. These photographs were taken of a diamond crystal model by rotating the model in the appropriate directions. The crystal appears to the incident ion as this crystal model appears to the eye. On the other hand, if the incident ion-beam direction is random with respect to the crystal surface (as shown in fig. 5.13), the ion does not see any channels and the effect of channeling is significantly reduced.

When the incident ion beam direction is aligned with a channeling direction (such as <110> in fig. 5.12c), it can either travel a long distance, like a bullet through a tunnel, or get scattered by one of the atoms on the surface. When the ion enters the crystal at a very small angle to channeling directions, and as it approaches the row of atoms, it is repelled away from these atoms and thus steered back into the channel. Consequently, the rate of energy loss is considerably reduced, leading to a significantly larger penetration

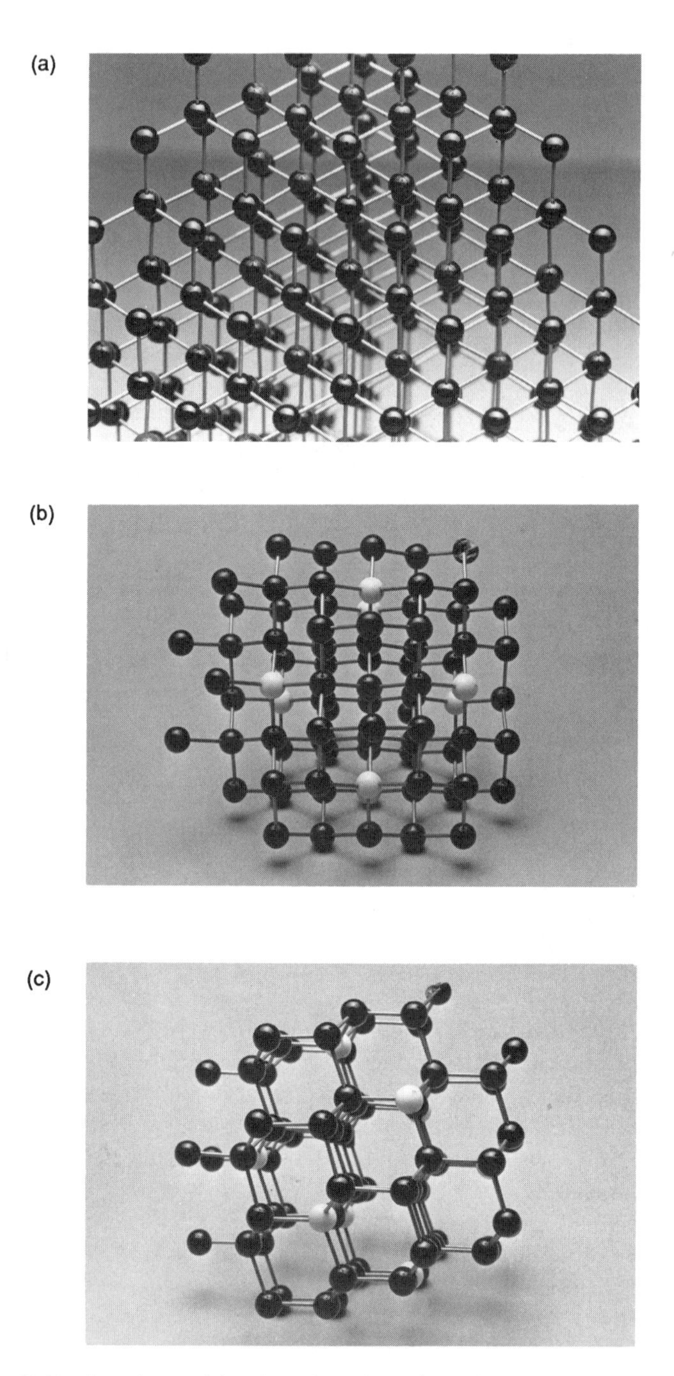

Fig. 5.12. Atomic models oriented to show channeling directions in diamond cubic lattice (a) < 111 >, (b) < 100 >, and (c) < 110 >.

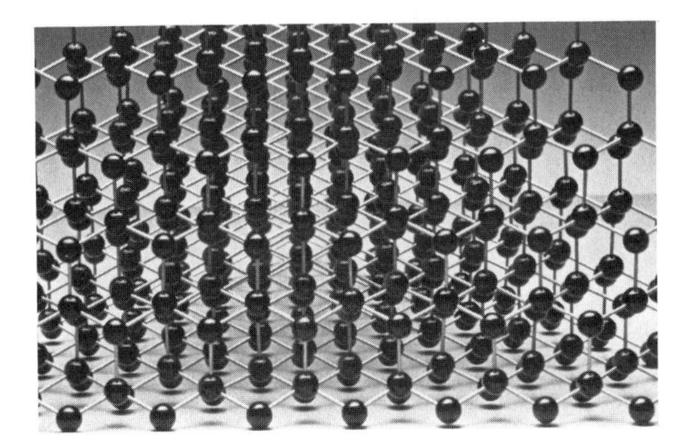

Fig. 5.13. Atomic model of diamond cubic lattice showing randomly oriented face.

in the crystal. The energy-loss mechanism is no longer associated with violent nuclear collisions. It is associated with gentle and glancing collisions with the row of atoms that form the wall of the channeling tunnel.

Ion channeling in crystals leads to a few advantages. It reduces the damage associated with violent collisions. It also can be used as a tool to reveal the crystalline structure [16], an area of great interest but beyond the scope of this chapter. Additionally, channeled ions can form deeper junctions. However, channeling makes it difficult to obtain reproducible range profiles. Even in those cases where ion implantations are carried out in a direction several degrees off the channeling direction, a few ions, following a few scattering events, end up (inside the crystal) in a channeling direction (see fig. 5.14). ψ_{crit}, defined as the critical angle (between the incident ion beam and the crystal's major channeling direction) within which channeling occurs, has been calculated for relatively heavy ions and high energies [16]. It is given as

$$\psi_{\text{crit}} = \left(\frac{2Z_1Z_2q^2}{Ed}\right)^{1/2}, \tag{5.27}$$

where d is the atomic spacing along the aligned row. For standard dopant ions of energies less than a few hundred keV, ψ_{crit} lies between 3° and 5°.

5.2.4 Lateral Spread of Implanted Ions

Ions entering the target are scattered in all directions. There are some that are scattered laterally, i.e., in a direction perpendicular to the incident beam direction, causing lateral straggling (ΔR_{L}). This lateral straggling leads to implantation away from the mask edges created to define or restrict the implantation in a desired region. Figure 5.15a shows typically defined coordinates for ion implantation SiO_2 mask on silicon. SiO_2 films

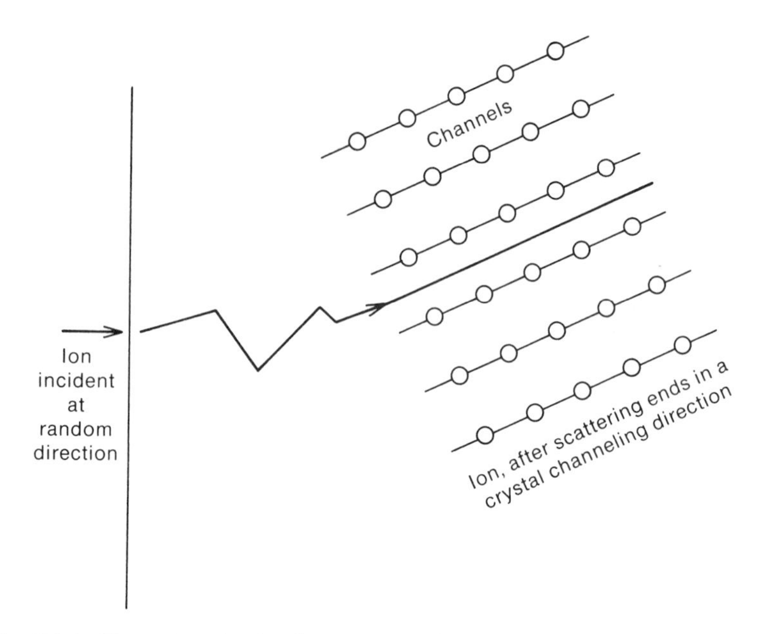

Fig. 5.14. Illustration showing how a projectile initially in a nonchanneling direction can end up in a channel after a few scattering events.

considerably thicker than the range of the ion implantation are grown or deposited on silicon and patterned to create a desired structure. Whole-area implantation is carried out. However, implantation in silicon occurs only in the regions where oxide has been etched off. Because of lateral straggling, the implantation does occur even under the oxide near the edge, as shown in fig. 5.15a. Finally, fig. 5.15b shows the contours of equal ion concentration in the silicon.

For the implantation situation described in fig. 5.15c, the concentration profile is given as

$$n(x,y) = \frac{n(x)}{2}\left[\operatorname{erfc}\left(\frac{y-a}{\sqrt{2}\Delta R_{\mathrm{L}}}\right) - \operatorname{erfc}\left(\frac{y+a}{\sqrt{2}\Delta R_{\mathrm{L}}}\right)\right] \tag{5.28}$$

for x, y and z coordinates shown in fig. 5.15b. Here 2a is the opening in the mask along y direction and $n(x)$ gives the concentration profile far away from the edge of the mask. The concentration profile in fig. 5.15b is derived for the case $a >> R_{\mathrm{L}}$.

5.2.5 Tables of Projected Range and Straggling

For the processing of semiconductors, it is important that the projected range and straggling for various dopants in the semiconductor and in the insulating

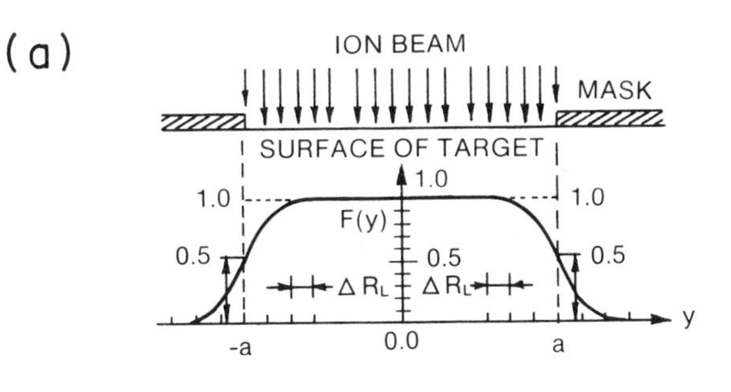

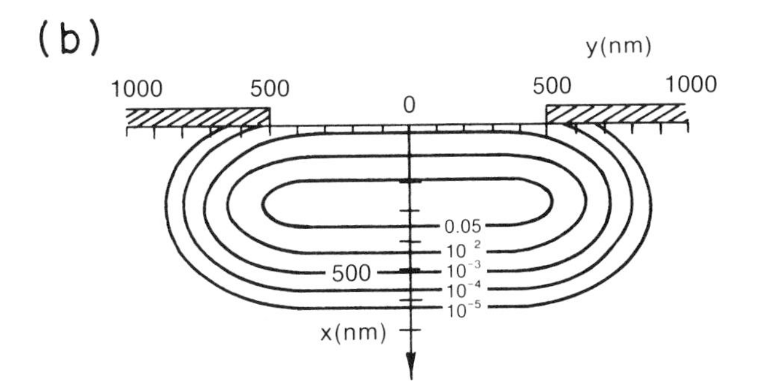

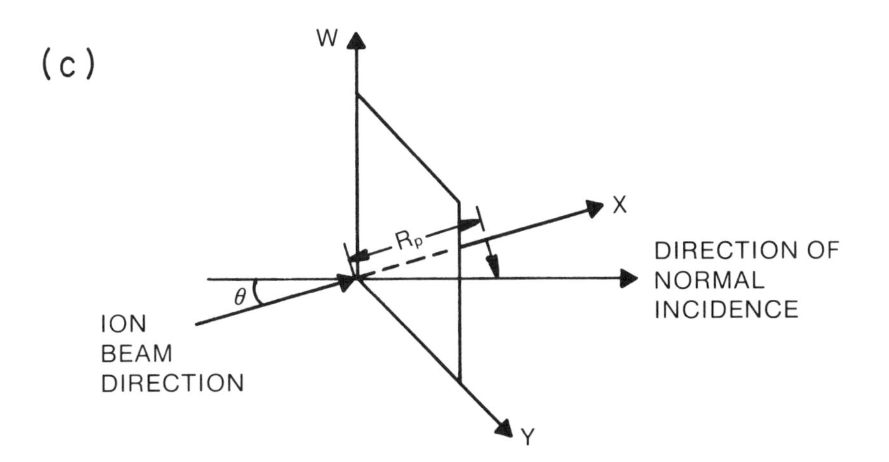

Fig. 5.15. Illustration of lateral profiles. (a) Ion concentration along the lateral direction (y) for a gate mask with $a >> \Delta R_{\rm L}$ and infinite extension in the w-direction. (b) Contours of equal-ion concentrations for 70-keV B^+ ($R_{\rm p}$ = 2710Å, $R_{\rm p}$ = 824Å, and $\Delta R_{\rm L}$ = 1006Å) incident into silicon through a 1-μm slit. From Furukawa, Matsumura, Ishiwara [9].

TABLE 5.1: Projected range statistics for ion implantation into amorphous Si

	Antimony		Arsenic		Boron		Phosphorus	
Energy (keV)	Projected Range (μm)	Projected Standard Deviation (μm)	Projected Range (μm)	Projected Standard Deviation (μm)	Projected Range (μm)	Projected Standard Deviation (μm)	Projected Range (μm)	Projected Standard Deviation (μm)
10	0.0088	0.0026	0.0097	0.0036	0.0333	0.0171	0.0139	0.0069
20	0.0141	0.0043	0.0159	0.0059	0.0662	0.0283	0.0253	0.0119
30	0.0187	0.0058	0.0215	0.0080	0.0987	0.0371	0.0368	0.0166
40	0.0230	0.0071	0.0269	0.0099	0.1302	0.0443	0.0486	0.0212
50	0.0271	0.0084	0.0322	0.0118	0.1608	0.0504	0.0607	0.0256
60	0.0310	0.0096	0.0374	0.0136	0.1903	0.0556	0.0730	0.0298
70	0.0347	0.0107	0.0426	0.0154	0.2188	0.0601	0.0855	0.0340
80	0.0385	0.0118	0.0478	0.0172	0.2465	0.0641	0.0981	0.0380
90	0.0421	0.0130	0.0530	0.0189	0.2733	0.0677	0.1109	0.0418
100	0.0457	0.0140	0.0582	0.0207	0.2994	0.0710	0.1238	0.0456
110	0.0493	0.0151	0.0634	0.0224	0.3248	0.0739	0.1367	0.0492
120	0.0529	0.0162	0.0686	0.0241	0.3496	0.0766	0.1497	0.0528
130	0.0564	0.0172	0.0739	0.0258	0.3737	0.0790	0.1627	0.0562
140	0.0599	0.0183	0.0791	0.0275	0.3974	0.0813	0.1757	0.0595
150	0.0634	0.0193	0.0845	0.0292	0.4205	0.0834	0.1888	0.0628
160	0.0669	0.0203	0.0898	0.0308	0.4432	0.0854	0.2019	0.0659
170	0.0704	0.0213	0.0952	0.0325	0.4654	0.0872	0.2149	0.0689
180	0.0739	0.0224	0.1005	0.0341	0.4872	0.0890	0.2279	0.0719
190	0.0773	0.0234	0.1060	0.0358	0.5086	0.0906	0.2409	0.0747
200	0.0808	0.0244	0.1114	0.0374	0.5297	0.0921	0.2539	0.0775
220	0.0878	0.0264	0.1223	0.0407	0.5708	0.0950	0.2798	0.0829
240	0.0947	0.0283	0.1334	0.0439	0.6108	0.0975	0.3054	0.0880
260	0.1017	0.0303	0.1445	0.0470	0.6496	0.0999	0.3309	0.0928
280	0.1086	0.0322	0.1558	0.0502	0.6875	0.1020	0.3562	0.0974
300	0.1156	0.0342	0.1671	0.0533	0.7245	0.1040	0.3812	0.1017

films that are used to mask the implant (see Chapter 3) be known. Table 5.1 lists the calculated values of the R_p and ΔR_p for various dopants in amorphous silicon [10]. R_p and ΔR_p values for the same dopants in SiO_2 and Si_3N_4 are given in table 3.3. of Chapter 3 and in table 4.8 of Chapter 4.

By using these tables it is easy to estimate the ion implantation depth for a given energy or vice versa. For example, to obtain an implantation depth of approximately 0.1 μm in silicon, ion energies of 250, 180, 30, and 80 keV will be necessary for antimony, arsenic, boron, and phosphorus ions, respectively.

5.3 Ion Damage, Amorphization, and Anneals

As ion travels through the solid, it looses energy essentially by two mechanisms, nuclear collisions and electronic interactions. The nuclear collisions cause a) scattering of the incoming ion, leading to changes in its trajectory, and b) displacement of the target atoms from their equilibrium positions. Depending upon the ion energy, the displaced target nuclei may acquire enough energy to produce their own trajectories prior to coming to a stop. Figure 5.2a showed the trajectory of the ion, the side branches representing trajectories of the displaced target nuclei, and the shaded areas representing the damage or disorder associated with this process. Each incoming ion produces its own "tree of disorder." When ion density increases, these disordered regions overlap, eventually leading to a noncrystalline region with no long range crystallographic order, as in the original target material. Such regions are called *amorphous regions* and the process leading to such regions is called *amorphization.*

As-implanted target will be highly disordered, even amorphous. Most of the as-implanted ions occupy electrically inactive (nonsubstitutional) sites in the semiconductor lattice. Also, the disorder renders large number of semiconductor atoms electrically inactive and introduces defects that effectively scatter charge carriers, leading to increased resistivity. Some of this disorder does anneal out because of the rise in temperature associated with the absorbed energy by the target. Most of it, however, must be annealed out at higher temperatures. Annealing at higher temperatures to restore the order and make the dopant species (implanted) electrically active (on substitutional sites) is called *electrical activation.*

The nature and the extent of the disorder/damage in the target depend on the type of ion species, energy, angle of incidence, dose or flux, temperature, and type of the target. The disorder generally consists of interstitials, vacancies, multiple vacancy aggregates, impurity-vacancy complexes, and dislocations, which generally form at higher substrate temperatures. In general, the larger the amount of energy transferred to the lattice per unit length of ion travel, the more the damage. If ΔE is the energy transferred to a target atom during the collision with the ion and E_d is the minimum energy required to displace the target atom (also called displacement energy, which is related to crystal binding energy for atom), the following results can be predicted:

1. $\Delta E < E_d$ Target atom is excited but there is no displacement
2. $\Delta E \gtrsim E_d$ Simple displacement occurs with isolated defect pair (vacancy on substitutional site and the displaced atom on interstitial site). Displaced atom does not have energy to produce cascading effect.
3. $\Delta E \gtrsim 2E_d$ Displacement followed by some secondary lattice disorder produced by energetic target atom.
4. $\Delta E >> E_d$ Results in large disorder and defect clusters associated with secondary displacements and cascading effects.

TABLE 5.2: Estimate ΔE values for implanted ions in silicon

Ion	ΔE(eV/0.3 nm) at ion energy (keV) of:			
	10	50	100	150
Boron	30	14.4	9.6	3.3
Phosphorous	159	147	117	54
Arsenic	360	450	420	285

Figure 5.4, which gives the energy loss per unit distance per ion, can be used to estimate ΔE. Table 5.2 lists ΔE per 0.3 nm of travel in silicon. The distance of 0.3 nm is chosen to represent an average interatomic distance in silicon. Also note these numbers are obtained by assuming nuclear scattering to be solely responsible for the damage. These ΔE values can be compared to the displacement energy E_{d} that is estimated to be $\sim$14 eV by using equation $E_{\mathrm{d}} = 4E_{\mathrm{b}}$ where E_{b} is the bond energy [18]. Assuming negligible effects of channeling, thermal diffusion, and saturation, it becomes obvious that high-energy ($>$ 50 keV) boron ions will not produce any significant damage. On the other hand, heavier ions, phosphorous and arsenic, will produce extensive damage in this energy range. It must be realized that as ions move deeper they lose energy, and, therefore, the degree of damage increases with depth. Thus, although high energy ions may not produce any damage near the surface, eventually they will cause damage prior to stopping in the target.

There are three quantities that define the degree of damage:

1. The total number of displaced atoms per ion, $N(E)$, which is given as [19]

$$N(E) = 0.4\frac{E}{E_{\mathrm{d}}}, \tag{5.29}$$

   where E is the ion energy. This equation is derived on the assumption that channeling and energy partition due to electronic collisions are neglected and that each atom is displaced only by breaking of four bonds.

2. A quantity $H_{\mathrm{D}}(E, x)$ is defined to represent the depth distribution of energy deposited into atomic processes (or damage) in the target for an ion of initial energy E. Thus H_{D} is also identified as the damage density. The damage-distribution profile in the target has been calculated by evaluating the integral [20],

$$H_{\mathrm{D}}(E, x) \;=\; \int_0^E \; S(E, E', x)\Sigma(E')\left[\frac{d\overline{R}}{dE'}\right] dE', \tag{5.30}$$

where

$$\sum(E') = N \int q(P)\, d\sigma(E', P).$$

The function $S(E, E', x)$ gives the relative depth distribution of the damage energy when the ion energy is reduced from incident energy, E, to E'. $\Sigma(E')$ gives the instantaneous rate of the energy deposition into the damage. dE' is the energy lost during the distance dR traveled by the ion of energy E'. $d\sigma(E', P)$ is the differential atomic cross section for the transfer of kinetic energy P to the recoiling target atom by an ion of energy E'. $q(P)$ is the portion P that is finally deposited into atomic processes (damage).

H_D has been calculated assuming negligible contributions from channeling, saturation effects, and thermal diffusion, and that all ions travel parallel to x axis. Figures 5.16 and 5.17 show the normalized damage density $(H_D R_P / E_T)$ as a function of the normalized distance x/R_p. E_T is the total energy deposited into damage and is given by the area under the depth distribution curves. Figure 5.18 gives plots of E_T as a function of the incident ion energy for various species implanted into silicon [20]. It is clear from the figures that ion damage is near surface for low-energy boron ions (fig. 5.16a) and deeper inside the substrate for high-energy ions. For heavier ions like As (fig. 5.17a) and Sb (fig. 5.17b), the location of damage is more or less independent of the incident energy.

3. The critical energy density, defined as [21]

$$E_c = fNE_d \tag{5.31}$$

represents the amount of energy per unit volume that must be deposited in the target to make it amorphous. It represents the energy necessary to replace N number of atoms. N for silicon is 4.88×10^{22} per cm^3. The factor f for silicon ranges from 0.1-0.5. Thus one can approximate the ion dose (ϕ_c) and energy (E_c) necessary to create an amorphous layer. If all the energy is absorbed in the materials, then it can be shown that

$$\phi_c \approx fNE_d \frac{R_p}{\Delta E_0} \tag{5.32}$$

where R_p is the range of the ions, ΔE_0 is the energy deposited per ion, and ϕ_c is given in ions per cm^2. Note that E_c will be a function of the substrate temperature, as discussed later.

The damage distribution relative to the ion distribution in the target depends on the relative mass ration M_1/M_2. For lighter atoms like boron $(M_1 < M_2)$ the peak in the defect and ion distributions occur at almost the same depth. For heavy ions like arsenic or antimony $(M_1 > M_2)$ the peak in

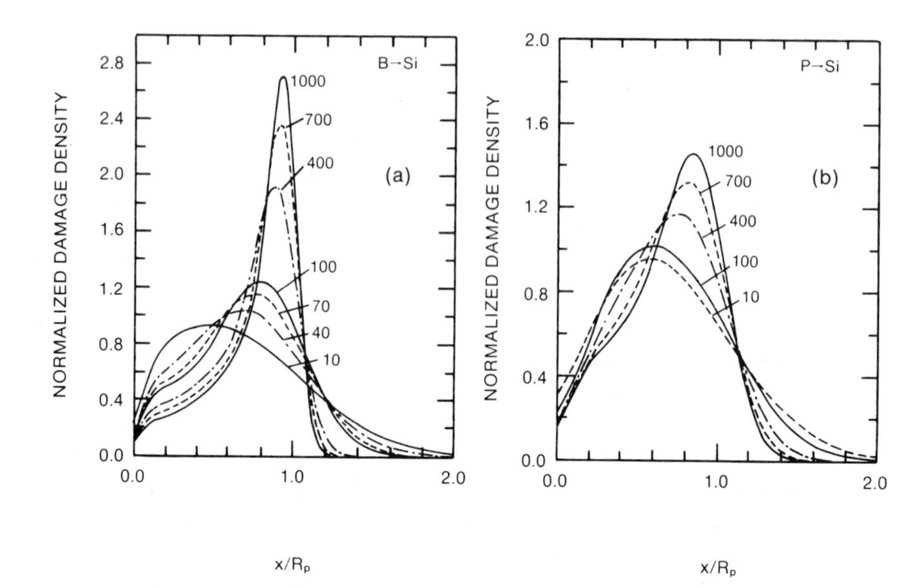

Fig. 5.16. (a) Normalized damage density, as a function of normalized depth, x/R_p, for boron ions incident on silicon. Energy indicated for each curve in keV. (b) Normalized damage density, as a function of normalized depth, x/R_p, for phosphorus ions incident on silicon. Energy indicated for each curve in keV. From Brice [20].

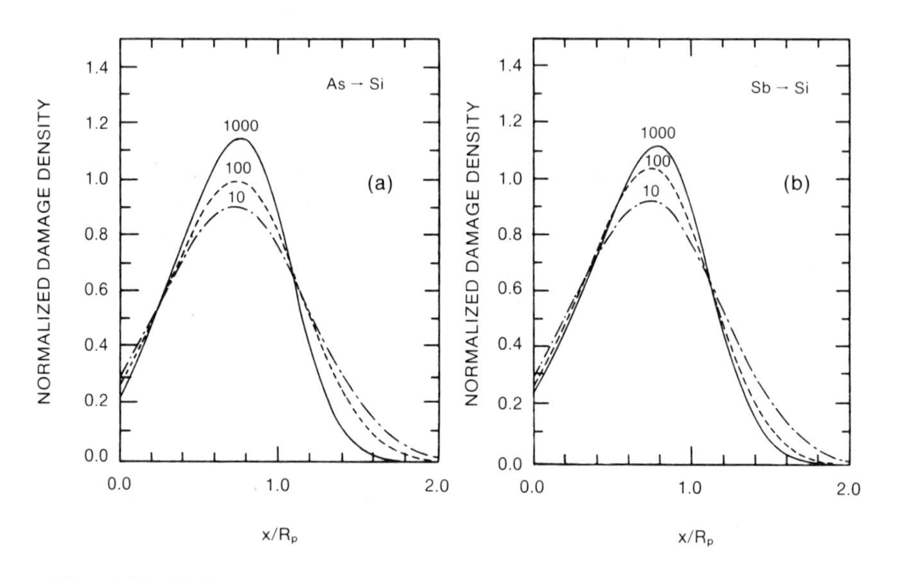

Fig. 5.17. (a) Normalized damage density, as a function of normalized depth, x/R_p, for arsenic ions incident on silicon. Energy indicated for each curve in keV. (b) Normalized damage density, as a function of normalized depth, x/R_p, for antimony ions incident on silicon. Energy indicated for each curve in keV. From Brice [20].

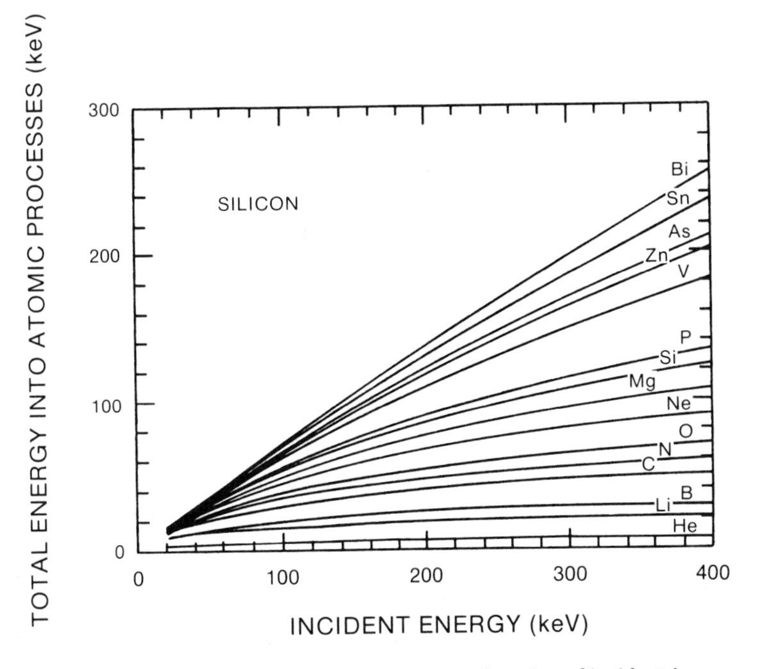

Fig. 5.18. Total energy deposited into damage as a function of incident ion energy for a variety of ions incident on silicon. From Brice [20].

the defect distribution lags the peak in the ion distribution. This is because the incident particle travels practically straight and the energy loss per unit distance is large.

Figure 5.19 schematically shows various stages of disorder/damage in the target as the ion flux increases. Note that the maximum damage, shown as intensity of points or as cross-hatched amorphous regions, precedes the ion range. Figure 5.20 shows actual disorder in an unannealed silicon crystal implanted with 1×10^{16} argon ions per cm^2 at an ion energy of 200 keV. The top, middle, and lower regions are the heavily microtwinned, amorphous, and lightly damaged regions, respectively. The lower region contains loops and tangles of dislocations and point defects. This result corresponds to the schematically shown region (c) in fig. 5.19.

It is possible to implant molecular ions instead of atomic species. One example is BF_2^+. BF_2^+ implantation has been instituted to reduce the range of boron ions in silicon. BF_2^+ is very heavy compared to boron (49 AMU versus 11 AMU). A 50-keV boron implant will have a projected range of ~160 nm, whereas 50-keV BF_2^+ dissociates as it enters the target. The distribution of energy results in a boron in energy of ~ 11 keV and the lower range. Flourine atoms still possess higher energy and are implanted deeper. This behavior leads to two distinct damage regions in silicon as shown in fig. 5.21.

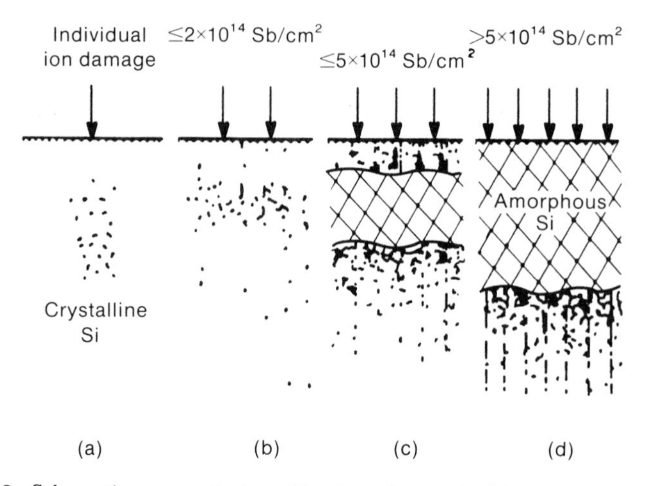

Fig. 5.19. Schematic representation of implant damage build-up as a function of dose for 250°C Sb implantation of Si. From Short et al. [23].

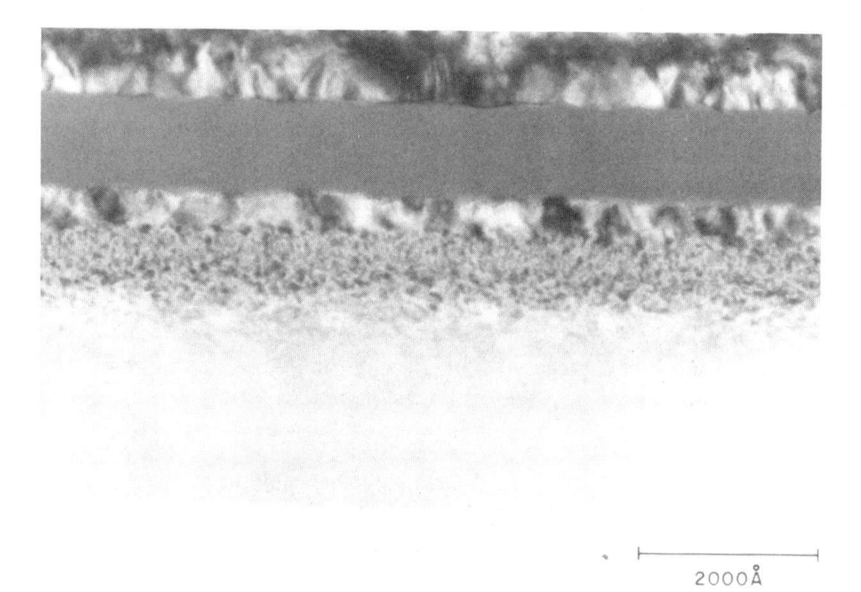

Fig. 5.20. Damage caused by 200-keV argon implantation into silicon. An amorphous layer 1000Å thick separates the upper damaged layer from the lower damaged layer. Courtesy of R.B. Marcus, Bell Comm. Res., NJ.

Substrate temperature has considerable influence on the degree of damage. The higher the substrate temperature, the lower the damage, as shown in fig. 5.22, which shows the disorder per Sb^+ ion as a function of implant temperature [22]. Figure 5.23 shows the plots of the critical dose, ϕ_c, necessary to make an amorphous layer as a function of the inverse of the

Fig. 5.21. Damage left after BF_2 implantation and annealing. Two distinct bands of defects are present: a narrow band 670 Å below the surface D_1 and a broad band that begins 1200Å below the surface D_2. Courtesy of R.B. Marcus, Bell Comm. Res., NJ.

temperature [24]. It is clear for these plots that at higher temperatures considerably greater ϕ_c is necessary to produce the amorphous layer. Also, above a certain substrate temperature (say, about 50°C for boron and 400° for Sb), the continuous amorphous layer can not be produced.

This effect of temperature on the damage can be understood by considering the non-equilibrium situation of the damaged layer. The point defects are frozen-in after production at low temperatures. At higher temperatures thermal diffusion permits the motion of these defects, leading to their annhilation.

5.3.1 Anneals

It is easy to carry out ion implantation at ambient temperatures that leads to damage in crystals. It is found that the damage is removed by post-implant

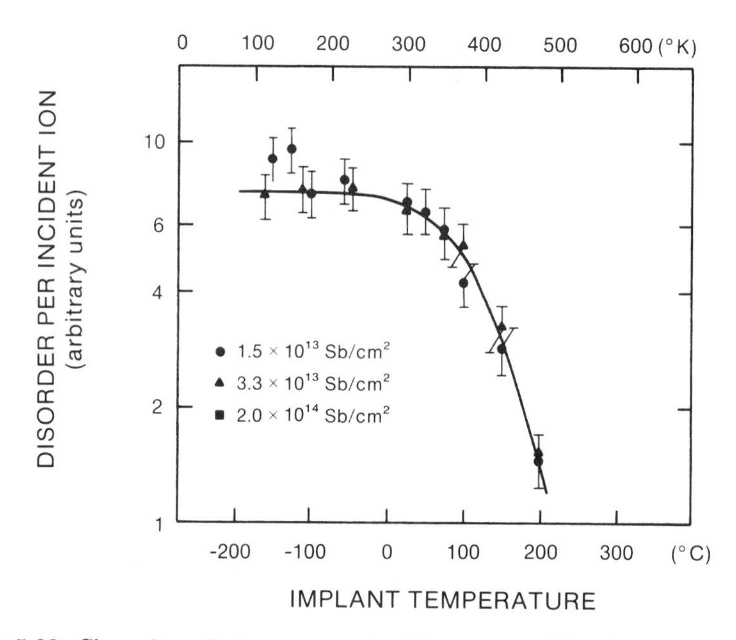

Fig. 5.22. Channeling-effect measurements of the amount of disorder per incident 40-keV antimony ion versus silicon-substrate temperature during implantation. Error bars represent the uncertainty in dose measurement. From Pucrayx et al. [22].

thermal annealing at temperatures as high as 600-900°C. For restoring electrical activity of the dopants, the thermal anneals are carried out at temperatures of about 900°C or higher. Later anneals may cause redistribution of the dopants depending on the duration of such heat treatments. Thus deeper dopant concentration profiles are obtained by annealing for a longer time or at higher temperatures or both. In such cases the ion-implanted material acts as a source of diffusing species (like a predeposition), and suitable diffusion equations can be employed to predict the concentration profiles and diffusion or junction depths (see Chapter 4).

Anneals can be carried out in a furnace or a rapid thermal annealer, or by a laser pulse or an electron beam (see Chapter 11). Choosing one over the other depends on the desired result. For damage anneal without affecting the dopant (as-implanted) profile, one could generally use a low-temperature (<700°C) and long-term furnace anneal or a high-temperature short-time rapid thermal, laser, or e-beam anneal. The electrical activation of dopants, however, is not complete at the low temperatures of furnace anneals. To ensure completely restored electrical activity without appreciable diffusion, therefore, furnace anneal followed by fast anneal by one of the above techniques will be required. In the following section we examine techniques that measure the disorder or crystal perfection and electrical activity

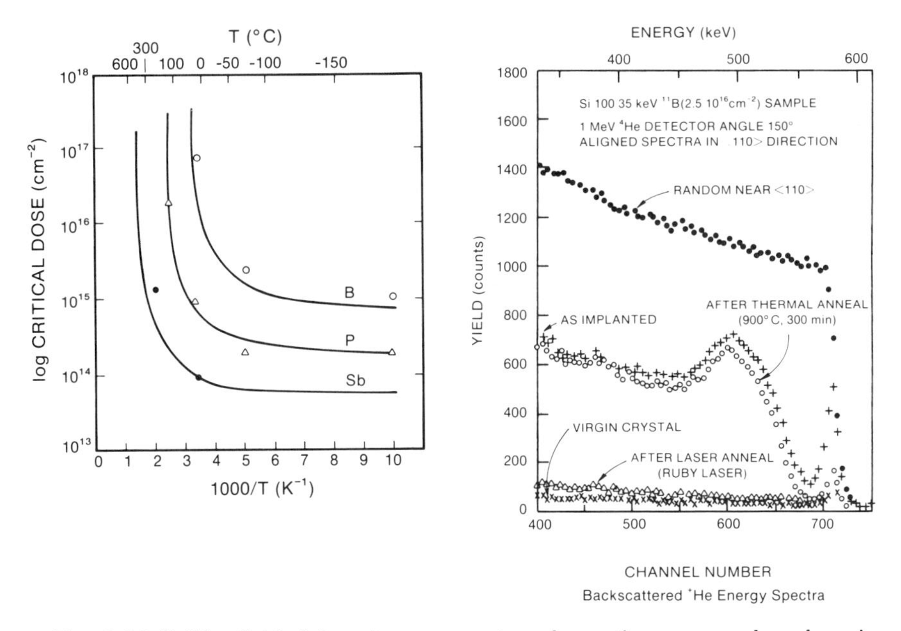

Fig. 5.23 (left). Critical dose ϕ_c necessary to make continuous amorphous layer in silicon as a function of the inverse of the substrate temperature. From Morehead, Crowder [24].

Fig. 5.24 (right). (110)-aligned backscattering spectra for implanted crystals in as-implanted, laser-annealed, thermally annealed conditions and a random spectrum from the virgin crystal. From Young et al. [25].

and show how thermal anneals restore both the crystal perfection and the electrical activity. In the subsequent section we discuss thermally induced diffusion of ion implanted impurities.

5.3.1.1 Damage Anneal and Electrical Activation

The damage anneal can best be studied by channeling Rutherford backscattering (RBS), although several other techniques have been employed [7]. RBS is briefly described in Section 4.3.7. Ion channeling was described in Section 5.2.3. Channeling of an incoming ion, such as a helium ion used in RBS technique, will occur if the crystal is properly oriented with resepct to the incident ion direction. These ions can remain in channels for long distances. If, however, the crystal is not perfect and has a displaced atom in the channel, the channeled ion will be scattered back into nonchanneling trajectories. This process is called *dechanneling.* The larger the number of displaced atoms, the more the dechanneling. All types of defects, interstitials, dislocations, stacking faults, twins, etc. cause dechanneling. Thus the

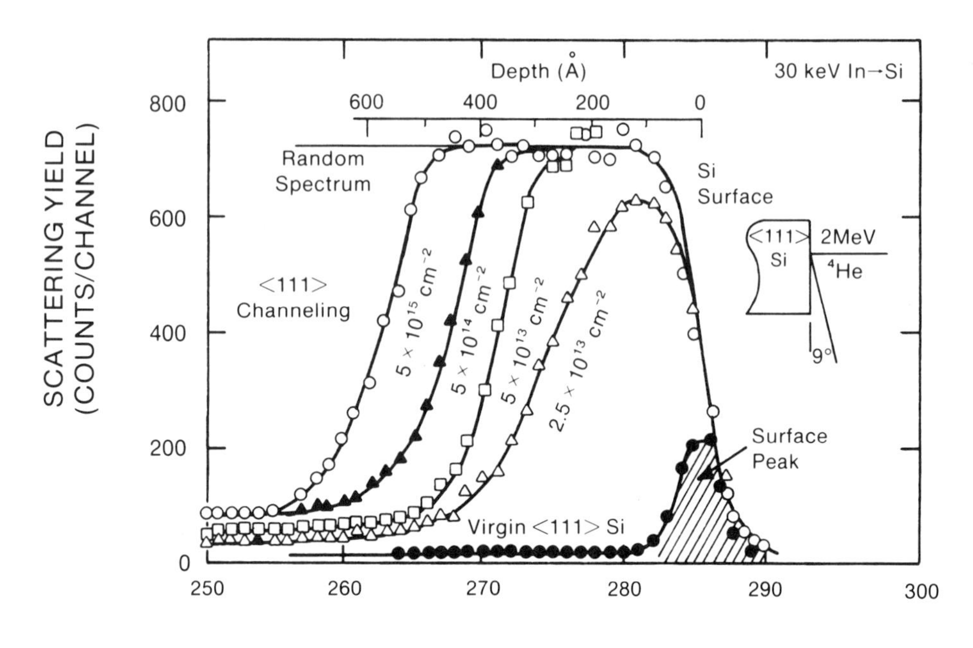

Fig. 5.25. Backscattering random and (111) channeling spectra for In implanted into Si at various fluences. From Thompson et al. [26].

ratio of dechanneling to channeling becomes a measure of the perfection of the crystal or the disorder in the crystal.

In practice, a channeling and a random RBS spectra are obtained for a perfect (unimplanted) crystal. These are then compared with the channeling RBS spectrum of the ion-implanted crystal. Figure 5.24 shows the [110]-aligned channeling RBS spectra after 35 keV, 2.5×10^{16} boron per cm^2 ion implantation and compares it with a) channeling spectra after the furnace and laser anneals and that for the virgin crystal and b) random spectrum of the same crystal. The disorder caused by ion implantation is clearly demonstrated. In this case, however, the amorphization is far from being complete. Also it is demonstrated that laser annealing was successfully annihilating the damage, whereas the furnace anneal was not.

Figure 5.25 compares a series of [111] channeling spectra of the virgin crystal and crystal ion implanted with increasing doses of 30 keV indium with the random spectrum [26]. At higher ion doses ($> 2.5 \times 10^{13}$ per cm^2 amorphous layer forms on the surface. The thickness of the amorphous layer increases with increased dose.

A comparison of figs. 5.24 and 5.25 point out the role of heavier ions in effectively producing the damage and hence the amorphous layers. Indium,

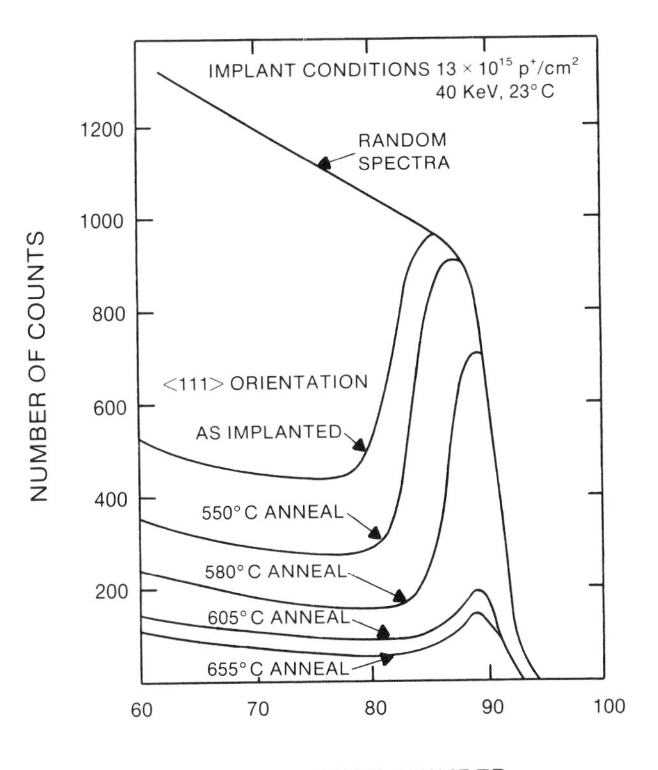

Fig. 5.26. Effect of annealing on the aligned backscattering spectra from a silicon sample implanted at room temperature with 1.3×10^{15} phosphorus ions/cm^2. All random spectra coincide within the statistical counting errors. The analyzing beam was 1.0-MeV helium ions. From Mayer et al. [27].

considerably heavier than boron, produced completely amorphized layers at a flux as low as 3-4 $\times$ 10^{13} per cm^2, whereas boron could not produce a completely amorphized surface layer, even at a flux of 2.5×10^{16} per cm^2.

Figure 5.26 shows the disorder and the effect of furnace anneal for a sample implanted with 1.3×10^{15} phosphorus per cm^2 at 40 keV [27]. In this case furnace anneal (in contrast to the case of boron in fig. 5.24) has been very effective in annealing out the damage. Annealing out the boron ion damage may require a temperature even greater than 900°C. This is because of the formation of dislocation structure [28, 29].

Thermal annealing of the ion-implanted substrates leads to the restoration of the crystallinity and enhanced electrical activity that is related to carrier concentration and the mobility. Thus the electrical conductivity and Hall effect measurements are used to determine the carrier concentration, which in turn becomes a measure of damage anneal. Sheet resistance measure-

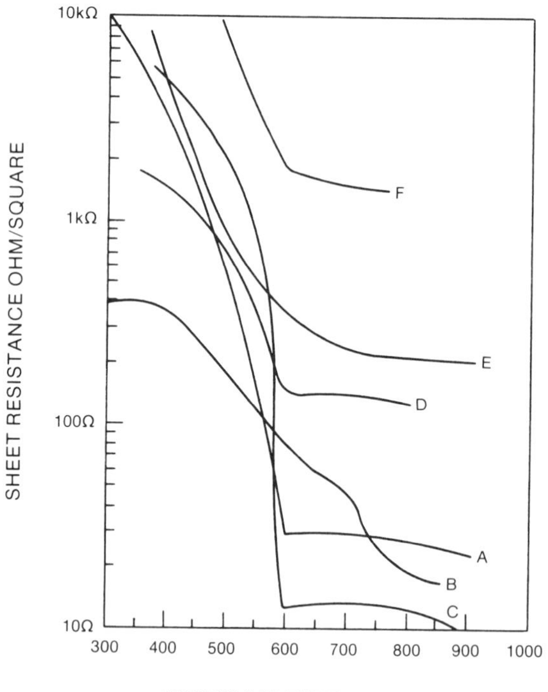

Fig. 5.27. Variation of sheet resistivity with annealing temperature for different phosphorus implants into silicon. (a) 100 keV, 6×10^{15} dose; (b) 100 keV, 6×10^{15} dose (450°C implant); (c) 300 keV, 2×10^{16} dose; (d) 20 keV plus 40 keV, 6×10^{14} dose; (e) 180 keV, 6×10^{14} dose; (f) 20 keV plus 40 keV, 1.1×10^{13} dose. Some implants were done at different substrate temperatures. From Dearnaley et al. [29].

ments, described in Chapter 4, are the easiest to carry out. Figure 5.27 shows a plot of the sheet resistance as a function of annealing temperature for various silicon samples implanted with phosphorus [29]. Note that the nearly complete electrical activation of phosphorus-implanted samples occur at anneal temperatures of 600°C or higher. The higher the implant dose, the lower is the final sheet resistance (indicating a higher concentration of donor atoms).

A similar conclusion is obtained by measuring the Hall effect and relating the measured Hall coefficient to the total effective carrier concentration, N_s. The effective carrier concentration is an averaged value over localized doping densities and carrier mobilities. Figure 5.28 shows the effective carrier concentration, normalized by dividing with the implanted flux, as a function of the annealing temperature [24]. A value of one on the ordinate indicates that all of the implanted species are electrically activated. Phosphorus ion doses that produced amorphous layers are shown as solid lines. Amorphized

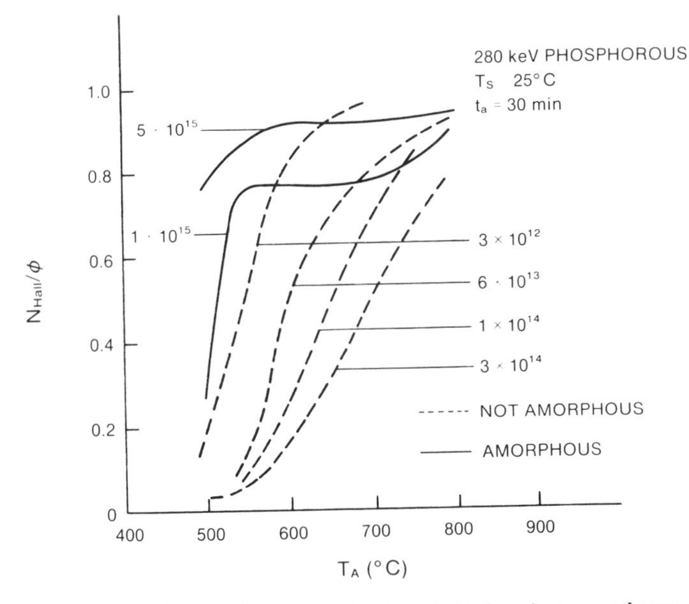

Fig. 5.28. The ratio of free-carrier content to dose plotted against anneal temperature (T_A) for various phosphorus doses. The solid curves represent amorphous layers that anneal by solid phase epitaxy. The dashed curves represent implantation where the damage is not amorphous. After Crowder and Morehead, Jr. [24].

layers anneal out at lower temperatures than non-amorphous layers. The effect is associated with the phenomenon of the solid-phase epitaxial (SPE) regrowth. SPE is an overgrowth of a crystalline film on a single-crystal substrate by means of solid-state reactions. Discontinuously amorphized layers (the nonamorphous case) have pockets of damaged regions and regrowth is complex. Individual damaged regions regrow and subsequently join each other, leading to mismatch at the interfaces. Such interfaces do not exist when a continuously uniform amorphized layer crystallizes, leading to more complete electrical activation at lower temperatures. Complete activation in all cases occurs at temperatures above 900°C.

At any temperature, the damage anneal and electrical activation are also a function of the time of anneal, as shown in fig. 5.29 [28]. For lighter implants, the damage density is small and requires lower time to anneal out. Higher damages require more annealing time at the same temperature.

SPE regrowth, following ion implantation, strongly depends on temperature. It also depends on the orientation of the surface on which regrowth occurs. Figure 5.30 shows a plot of the regrowth rate (determined using channeling RBS studies [30]) as a function of the inverse of temperature and the silicon-wafer surface orientation. Regrowth rates are almost two orders of magnitude higher on $< 100 >$ substrates. Thus, the higher the atomic density of planes, the slower is the rate of regrowth. However, the

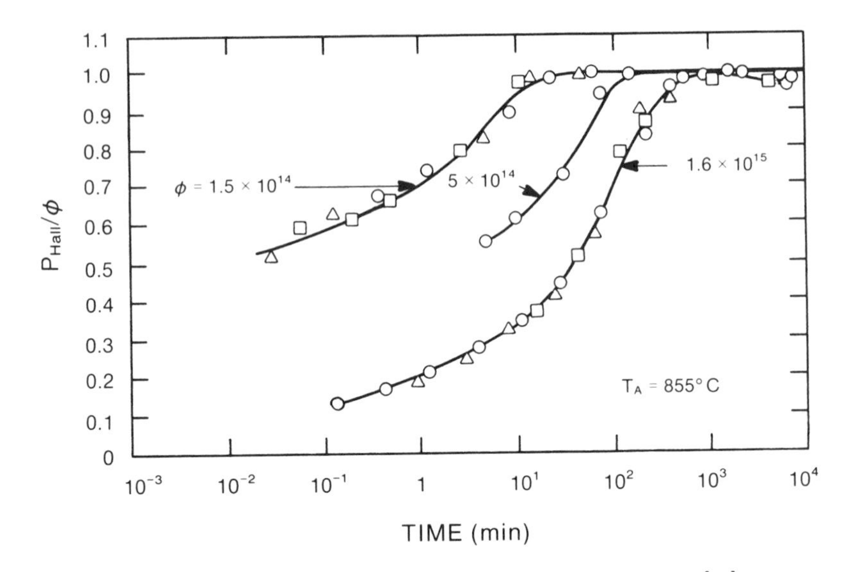

Fig. 5.29. Isothermal annealing of boron. From Seidel and MacRae [28].

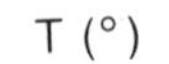

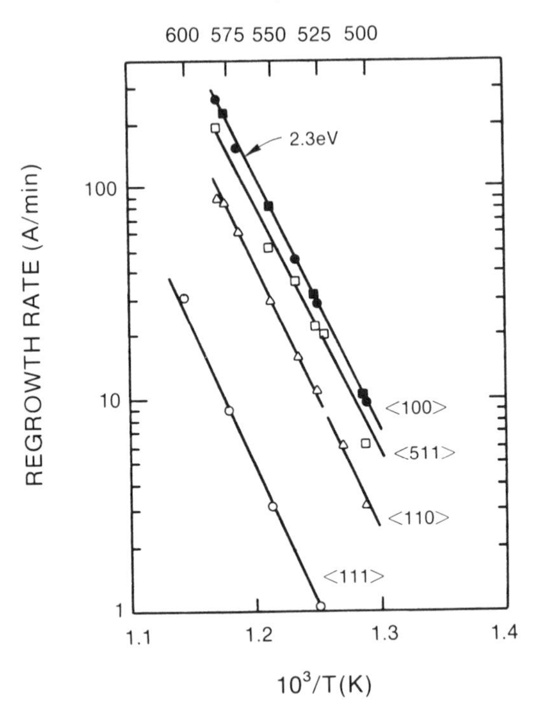

Fig. 5.30. The solid-phase epitaxial regrowth rate of amorphous silicon as a function of temperature for various crystal orientations. From Csepregi, Mayer, Sigmon [30].

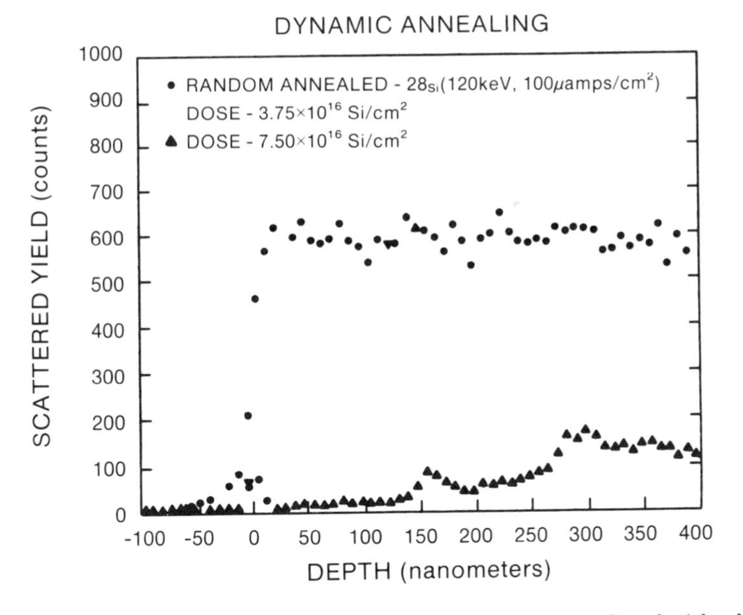

Fig. 5.31. Backscattering spectra from a Si(100) single crystal irradiated with a high-flux ^{28}Si ion beam. From Holland, Narayan [32].

temperature dependence is independent of the orientation. These results are similar to those for the growth of the oxidation-induced stacking faults (OISF) in silicon (Chapter 3). The effect is not clearly understood.

Ion implantation effectively deposits energy into the solid, causing a temperature rise [31]. If the temperature rise is sufficiently high, as will be the case for high-dose implants, self annealing of the ion damage (called dynamic annealing) will occur. Dynamic annealing will remove damage only partially, as shown in fig. 5.31 [32]. For low-dose (0.375×10^{16} cm^{-2}) silicon implants in silicon, amorphous layer forms. For high-doses (0.75×10^{16} cm^{-2}) implants, only random amorphous regions are formed and there is no evidence of a complete amorphous layer. Thus at higher doses dynamic annealing has occurred, eliminating most of the damage caused at lower doses. This process can be used to anneal out dopant implant damage and to activate them by use of high-dose neutral-species (such as Si, Ge, or even inert gas) implants. However, such implants will leave a rather extensive band of dislocation tangles and loops at the end of the range of the annealing ion beam [32].

5.3.1.2 Diffusion

Ion implantation is generally used to introduce dopants (or other impurities) in semiconductors (or other materials). Since the range of the implanted species is limited by the ion energy, deeper implants are generally not possible, except for very light ions. To produce deeper junctions, then, the diffusion anneals (see Chapter 4) are carried out. Thus ion implantation becomes a means of depositing a source of the impurities at a desired location near the surface of the solid. The post implantation diffusion profile is given by [33]

$$n(x,t) = \phi[2\pi(\Delta R_{\mathrm{p}}^2 + 2Dt)]^{-1/2} \exp[\frac{-(x - R_{\mathrm{p}})^2}{2(\Delta R_{\mathrm{p}}^2 + 2Dt)}], \tag{5.33}$$

where D is the diffusion coefficient at the temperature of diffusion anneal. The above equation has been obtained by assuming a Gaussian distribution of the initial ion-implanted deposit and a Gaussian solution of Fick's second law (Chapter 4). Since only a limited amount of material is available for diffusion, this solution [Eq. (5.33)] represents a reasonably accurate concentration profile. Note that contribution from the defects, due to ion damage, and loss from the surface are neglected.

Problem 5.2: *A silicon wafer is implanted with 10^{16} per cm^2 of boron and annealed at $1000°C$ for 1 hour. Calculate the concentration of boron at a distances of 0.5 μm and 1 μm from the surface.*

Solution: For 30-keV boron (From table 5.1) one finds $R_{\mathrm{p}} = 0.0987$ μm and $\Delta R_{\mathrm{p}} = 0.0371$ μm.

From Chapter 4, table 4.5, boron diffusivity (for V^+ vacancies) is given as $D = 0.76$ exp -3.46 eV/kT cm$^2/s$.

$$\text{At } 1000°\text{C}, D = 1.61 \times 10^{-14} \text{ cm}^2/\text{s},$$

using eq. (5.33), then by substitution of $\phi, R_{\mathrm{p}}, \Delta R_{\mathrm{p}}, D_1,\ t_1,$ and x one obtains

$$\Delta R_{\mathrm{p}}^2 + 2Dt = 1.30 \times 10^{-10}$$

and

$$(x - R_{\mathrm{p}})^2 = 1.61 \times 10^{-9} \text{ at } 0.5\ \mu\text{m}.$$

$$\begin{aligned} n(0.5\ \mu\text{m}, 3600s) &= 10^{16}\left[2\pi \times 1.30 \times 10^{-10}\right]^{-1/2} \exp\left(\frac{-1.61 \times 10^{-9}}{2 \times 1.3 \times 10^{-10}}\right) \\ &= 10^{16} \times 3.5 \times 10^4 \times 2.04 \times 10^{-3} \\ &= 7.2 \times 10^{17}\text{cm}^{-3}. \end{aligned}$$

For $x = 1$ μm, $(x - R_{\mathrm{p}})^2 = 8.12 \times 10^{-9}$, therefore, $n(1\ \mu\text{m}, 3600\ s) = 9.4 \times 10^6\text{cm}^{-3}$.

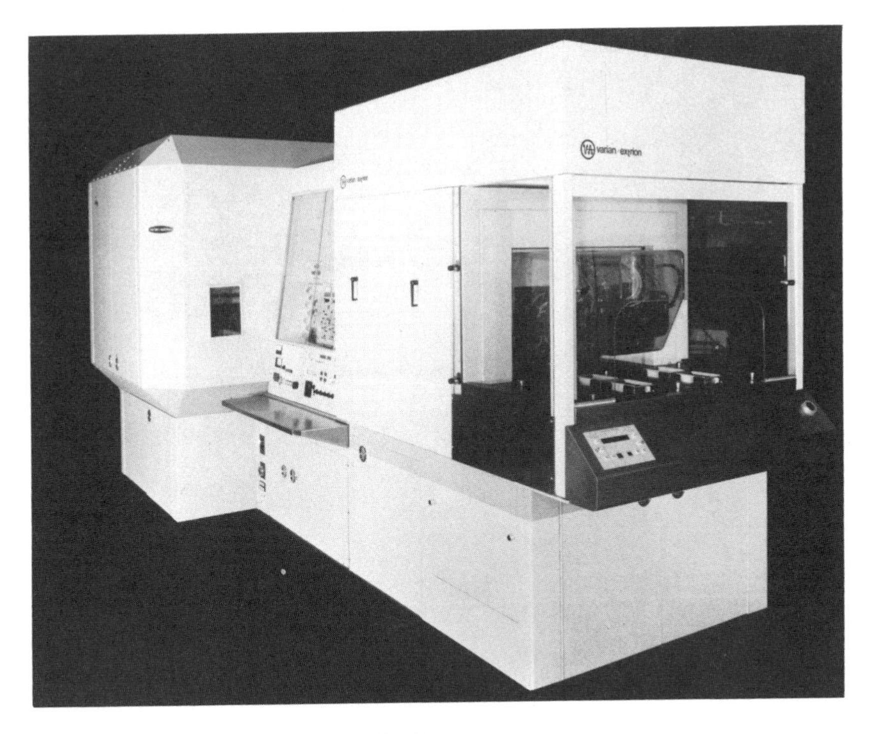

Fig. 5.32. A Varian/Extrion ion implanter.

Thus there is a very sharp fall in the concentration of the diffusing boron between 0.5 μm and 1 μm. Now at $x = R_p$, the $n(R_p,\ 3600s)$ is $3.5 \times 10^{20}\text{cm}^{-3}$.

5.4 Technique and Equipment

Figure 5.32 shows a Varian/extrion ion implanter. Most ion-implantation equipment consist of the following major components, as shown in fig. 5.33 [34]:

1. **The source of implanting ions**. This consists of a source of the gaseous material such as BF_3 for boron, PH_3 for phorphorous, AsH_3 for arsenic, and SiH_4 for silicon and an ionizing source to ionize the gas. The source produces an ion beam with very small energy spread enabling high mass resolution. While the gas is fed from a gas tank, the rest of the assembly is in a vacuum at pressures in the range of a pascal.
2. **An extracting and ion analyzing mechanism, generally consisting of electromagnets**. Ions are extracted from the source by the use of small accelerating voltages and then channeled into the analyzer magnets. A spatial separation of ions, due to the differences in the mass and charge, occur. Use of well-defined apertures screen the undesirable

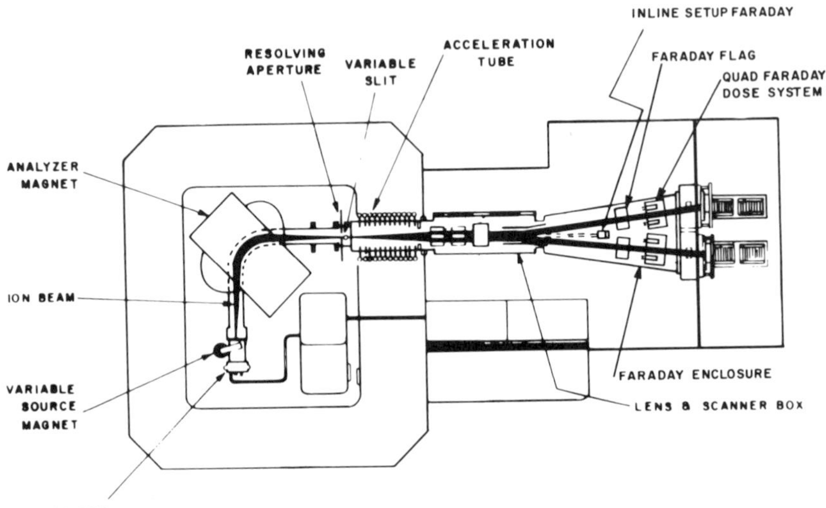

Fig. 5.33. Schematics of the Extrion implanter showing major components.

ions and only the selected ions are injected into the main accelerating column.

3. **An accelerating column**. Selected ions get accelerated by the use of the directed electric fields. In this part of the system the ion beam is also focused, shaped, and made ready for use with an ion energy as high as 200 keV.
4. **A scanning system**, usually consisting of sawtooth voltages applied in the x and y directions to deflect ions so that a uniform implantation can be achieved.
5. **An end station including area defining aperture, target transportation, positioning, and holding mechanisms, and a properly designed current integrator**. To obtain an accurate ion-current measurement, a negative electron suppressor is placed behind the aperture mask and also in front of the target to suppress secondary electrons. Similarly, a positively charged Faraday cup is used to suppress positive secondary ions that may leave the target.

Ion implanters are very sophisticated machines and different equipment builders build them differently, as can be learned from the equipment manuals. The basic features of the implanters of one kind (those with similar applications, e.g., ion current and accelerating voltages) are generally similar to those described above.

TABLE 5.3: A comparison of various types of implanters

Implanter type	Maxium ion current (mA)	Maximum energies (keV)	Use
Low-energy implanters (LEI)	~ 0.1-2	30	Shallow-predisposition type implants
Low-current implanters (LCI)	< 0.1	100-200	Experimental equipment where throughput is not important
Focused ion beam (FIB) implanters	< 0.001	100	Direct-writing implants (doping, materials modification, etc.)
Medium-current implanters (MCI)	2-10	200	Diffusion implants
High-current implanters (HCI)	100	100-200	Buried layer formations (see 5.5) Materials modification (see 5.6)
High-energy implanters (HEI)	< 0.05	Several MeV	Deep doping at low temperatures

Recently, to increase the throughput of the large-diameter wafers through an implanter, and for those applications where extremely high dosages ($> 10^{17}\text{cm}^{-2}$) are required, high-current implanters (HCI) have been developed. For deeper implants, especially those of heavy ions, high-energy implanters (HEI) have been developed. Table 5.3 compares all types of implanters. Of these FIB, HCI, and HEI are recent developments. FIB (focused ion beam), which is discussed in the last chapter of this book, promises new technologies such as maskless diffusions, metallization, and ion milling. HCI will be used to increase the throughput of large diameter wafers such as those used presently (150 mm diameter) and those projected to be used in the future (200 mm or larger in diameter). They are also needed to form buried layer formation such as those of SiO_2, Si_3N_4, or a silicide. HEI use is limited to deep implants such as those for forming buried collectors in a bipolar device, retrograde well, and buried layers to control latch up in CMOS devices (see Chapter 1). HEI can also be used to produce deeper damage in the crystal, the damage that can be used to getter defects and impurities to improve the devices formed on near-surface bulk. Near-surface material is practically undamaged during high energy implantation (see Section 5.3). A good description of appliations of ion implantation in semiconductors is given in references 7 and 35.

5.4.1 Limitations

There are several limitations to using ion implanters: dose accuracy, ion damage in wafers, wafer charging and heating, impurity contamination, and ion damage of the masking material. Other problems are associated with high voltages, x-rays produced in the columns, and gases employed in the ion sources.

The inaccuracy in the ion dose results from errors in the current integration and from neutral atom implantation. Neutral atoms are produced due to charge transfer during collisions with residual gas atoms in their path and by trapping thermal electrons produced in the beam. Generally such inaccuracies lead to a few percent error in the ion dose.

Charging of the substrate, inadequately flooded with an electron gun, could also lead to inaccuracies in the ion dose. Charging of the substrate can, however, create more severe problems by leading to electric field build-up that can destroy the electric circuits. Insulating surfaces are more prone to such damages.

Most severe problems are, however, associated with impurity contamination during ion implantation. Most serious of these are the metal contaminants produced from the ion sputtering of the metallic fixtures. In a recent study it has been clearly demonstrated that heavy-metal contaminants are implanted in silicon and that they can be practically eliminated by use of a thin ($\sim 10 - 20$ nm) surface oxide layer [36]. Practically all of these impurities have a very small range and are trapped in this oxide layer, which can be subsequently removed. Other contaminants such as sodium, carbon, or those from prior implants can be eliminated by adopting cleaner operating procedures and use of an oil-free pumping system.

Photoresists are generally used as ion implantation masks to prevent ion implantation in selected regions. Ions, penetrating such resists, cause sufficient damage to these materials. Generally the damage is severe enough to change the character of the resists, and it becomes difficult to remove the resists following a high-dose, high-energy implantation. In extreme cases, resist evaporation, blistering, flowing, and/or cracking can result. The recently employed use of multilevel resist system minimizes such problems [37].

Problems of high voltages, x-rays produced in the columns, and gases employed are all safety related. Although equipment designers and manufacturers use all available knowledge to make the use of machines safe, the safe practices, adopted by equipment users, are invaluable. Only well-trained and experienced persons should use such machines. During the change of gas cylinders and maintenance, care should be exercised not to be exposed to gases such as arsine, phosphine, diborane, and silane. Those and many similar hydrides of group III, IV, V, and VI elements are extremely dangerous and can be fatal if inhaled. Thus the system carrying such gases should be completely leak free and should be opened to the atmosphere only after all toxic gases are flushed out.

5.5 Ion-Beam Mixing and Materials Modification

5.5.1 Ion-Beam Mixing

When a two (or more) component system, such as a metal film on a silicon substrate, is irradiated with high energy ions, such as those from an ion implanter, mixing results, leading to compositional and structural changes. The process has been called *ion-beam induced mixing* or simply *ion-beam mixing*. Ion-beam mixing offers the advantage of modifying materials with significantly lower ion doses compared to those required for material modification by direct implantation. For example, a platinum silicide film can be formed by irradiating a 400 to 500 Å thick platinum film on a silicon substrate with a low dose ($\sim 10^{15}\text{cm}^{-2}$) of inert gas or silicon ions [38,39]. If, on the other hand, a platinum implant into silicon or silicon implant into platinum film is carried out to form the similarly thick film of platinum silicide, without any ion mixing, a flux of about two orders of magnitude higher will be required.

Each ion causes mixing primarily over its range or the dimensions of collision cascade. A *collision cascade* is defined as the region of disturbance caused by the collisions of the incoming ion or by the energetically displaced atoms until all have come to rest (see fig. 5.2a). Overlap of the individual cascade regions leads to uniform ion-mixed layers in a manner similar to the production uniform amorphized layers discussed in Section 5.3.

Ion-beam mixing is a result of a) the direct collisions and b) the enhanced mobility (of the target atoms) due to the defect generation associated with collisions. Sample temperature plays an important role in determining which of the above two mechanisms contribute significantly. There appears to be a critical temperature, T_c, below which the ion mixing results are insensitive to temperature. Below T_c, the mixing is primarily a result of the ion-solid interactions (direct collisions), which can be examined using collision and associated energy-loss concepts. The temperature is too low to cause thermally induced motion of atoms and defects. Above T_c thermally induced atomic motions cause rearrangements that are temperature and material dependent. Thus above T_c the ion mixing is temperature dependent.

The amount of ion-beam mixing is directly related to the energy loss per unit distance of ion travel. The more the energy loss is, the more is the ion mixing. Thus, the higher the dose and the mass of the ions, the more effective is the mixing. Figure 5.34 shows the thickness of the Pt_2Si formed as a function of the square root of the ion dose for various ions used in the experiment [38]. In the experiment a 45-nm thick platinum film on the silicon substrate is exposed to various ions, resulting in Pt_2Si formation. The ordinate represents the thickness of the silicide in terms of silicon atoms in the silicide film. It is clear that the higher doses leads to more silicide formation and that for a given dose the Pt_2Si thickness increases with increasing mass of the mixing ion. The thickness ratio (3 : 2.2 : 1) is similar to the mass ratio (3.3 : 2.1 : 1) of the Xe, Kr, and Ar ions.

Ion energy also plays an important role in ion mixing, a role that is similar to that discussed for the production of the ion damage. Low-energy ions produce mixing near the surface whereas high-energy ions cause the mixing into the interior of the substrate. In addition, for a high-energy ion there is more energy to dissipate, resulting in a large amount of mixing.

Ion mixing can generally be induced in most systems. Experience has, however, shown [40] that immiscible systems, such as Au-W or Cu-W, tend not to mix, whereas systems with equilibrium phases, such as silicides, mix well. In the latter cases, metastable phases, such as Pt_2Si_3 which decomposes into PtSi and Si at $\sim 500°C$, have been formed by ion mixing [40].

5.5.2 Materials Modification

Ion implantation is increasingly used for controlled modification of the surface-sensitive properties. Table 5.4 lists the material properties influenced by the surface composition [41]. Surface compositional changes can be induced by a preferential sputtering from an alloy surface, ion mixing and/or metastable alloy formation, ion mixing leading to a surface alloy formation,

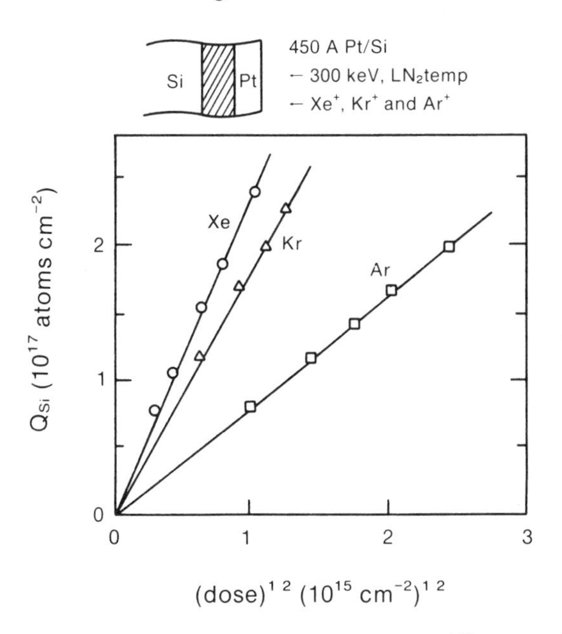

Fig. 5.34. Average thickness of Pt_2Si as a function of dose$^{1/2}$ for 470-Å-thick Pt film on silicon substrates implanted with different ions (Ar, Kr, and Xe). From Tsaur, Liau, Mayer [38].

TABLE 5.4: Materials properties influenced by ion-implantation induced changes of surface composition[a]

Adhesion	Fatigue
Bonding	Friction
Catalysis	Harding
Corrosion resistance	Lubrication
Decorative finish	Reflectance
Electrochemistry	Wear

[a] From ref. 41

and defect generation and related enhanced mobility of atoms near the surface. Discussion of each of these areas is beyond the scope of this text and the interested reader is referred to reference [42].

The advantages and disadvantages of materials modification by ion implantation are compared in table 5.5. Some specific applications are listed in table 5.6. Of these applications, the formation of buried oxide and nitride in silicon is considered in the following section.

TABLE 5.5: Advantages and disadvantages of ion implantation as a surface modification technique[a]

Advantages	Disadvantages
1. Solid solubility limit can be exceeded	1. Line-of-sight process
2. Alloy preparation independent of diffusion	2. Shallow penetration
3. Allows fast screening of the effects of changes in alloy composition	3. Relatively expensive equipment and processing costs
4. No sacrifice of bulk properties	
5. Low temperature process	
6. No significant dimensional changes	
7. No adhesion problems since there is no sharp interface	
8. Controllable depth concentrations	
9. Clean vacuum process	
10. Highly controllable and reproducible	

[a] From ref. 42

5.6 Dielectric Film Formation

Ion implantation has been used to enhance the oxidation of metal surfaces [43]. Ion implantation into surface layers causes high defect concentration, which in turn enhances the oxidant-diffusion and oxidation-reaction kinetics.

More recently, considerable interest has developed in forming buried dielectric (oxide or nitride) layers in silicon so that the silicon-on-insulator (SOI) structures can be realized. Ideally, a perfectly crystalline silicon layer on SiO_2 or Si_3N_4 is desired. This silicon layer can be used as an active semiconductor to fabricate devices, offering several advantages, including improvements in device performance and circuit design.

The concept of the buried oxide layer formation by oxygen ion implantation is schematically shown in fig. 5.35 [44]. Figure 5.35a shows the Gaussian concentration profile of the implanted oxygen ions of a given energy. The R_p and ΔR_p for a 150 keV oxygen ion is 371 nm and 98 nm, respectively. Figure 5.35b shows the damage profile, which is closer to the surface of the silicon. The ion energy, dose, and the substrate temperature are controlled

TABLE 5.6: Specific applications of ion implantation for materials modification

Application	Material	Treatment	Result
Forming Tools[a]	12 Cr 2C steel	$4 \times 10^{17} N/cm^2$	Much reduced adhesive wear
Bearing alloy[b]	AlSI 52100 steel	High-dose Ti and C implants	Very significant reduction of friction
Ion mixing[b]	Pt on Si	Ar, Kr, Xe $10^{14} - 10^{16}/cm^2$	Silicide formation
Fatigue[a]	AISI 1018 steel	$2 \times 10^{17} N_2/cm^2$	Improved fatigue strength
Buried oxide[c] layer formation	Silicon	6-13$\times$ 10^{17} O_2+/cm^2	Silicon on insulator
Buried Si_3N_4[d]	Silicon	1-10 $\times$ $10^{17} N/cm^2$	Silicon on insulator

[a] Ref. 42
[b] Ref. 40
[c] Ref. 44
[d] Ref. 45

to avoid amorphization of the damaged surface silicon layer. Higher temperatures allow enough atomic mobility to keep this condition maintained. By increasing the oxygen dose to the desired value for SiO_2 formation, a stoichiometric composition can be achieved. Further increase in the dose results in thicker oxide layer. Eventually the oxide layer width becomes greater than the implant profile width ($R_p \pm R_p$) and a structure of the type shown in fig. 5.35c results. To ensure the cyrstallization of the top silicon layers and annhilation of ion damage and to ensure SiO_2 formation, a high-temperature anneal (usually at a temperature greater than 1100°C) is then carried out.

A similar concept can be used to form buried Si_3N_4 layers using nitrogen implantation [45]. In either case a high-current implanter will be required to carry out the implantation processes in reasonable times. A flux of 4.5×10^{17} 0^+ (or 2.25×10^{17} 0_2^+) ions per cm^2 or 5.3×10^{17} N^+ (or 2.65×10^{17} N_2^+) ions per cm^2 will be required to form a 100-nm thick oxide or nitride layer, respectively. The ion energies are chosen to suit the desired depth of the dielectric film formation. For example, the R_p and ΔR_p values for a 180-keV implant in silicon are approximately 0.22 and 0.7 μm for 0_2^+ ions and 0.47 and 0.11 μm for N^+ ions, respectively. Thus a fully formed oxide or nitride layer will be 0.15 or 0.36 μm below the silicon surface and 0.36 μm or 0.69 μm thick, respectively.

Figure 5.36 shows a TEM cross section of a buried oxide layer formed by

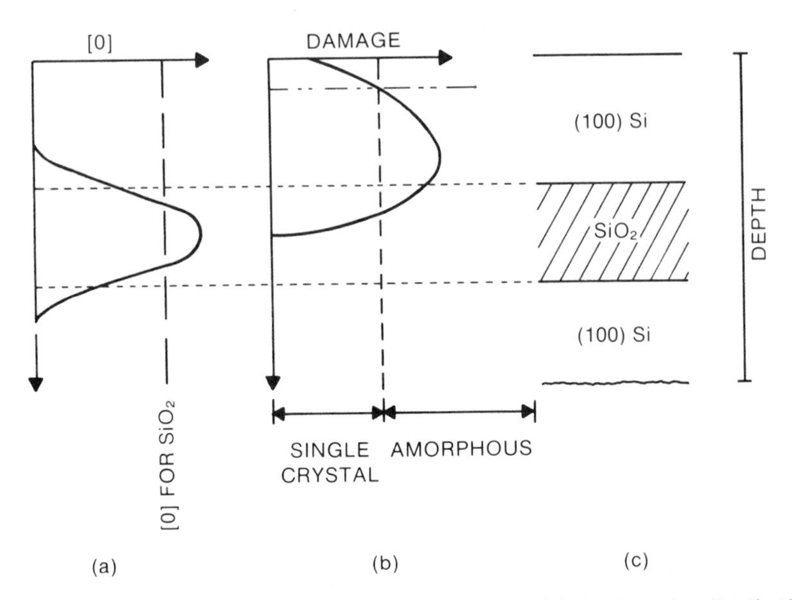

Fig. 5.35. The basics of buried oxide SOI formation: (a) the impurity distribution profile after implantation; (b) the damage distribution profile; (c) a schematic of the final structure. From Pinizzotto [44].

Fig. 5.36. A cross-sectional transmission electron micrograph of buried oxide SOI. The original sample surface is marked with arrows. Note the abrupt silicon/oxide interfaces. (a) the epitaxial layer; (b) the top single crystal Si layer; (c) a damaged polysilicon layer; (d) the buried oxide; (e) a damaged substrate layer; and (f) the substrate. From Pinizzotto [46].

ion implantation. A few things are clearly apparent [44]. The top silicon layer is defective although of good crystalline quality. The oxide layer is amorphous and the Si-SiO_2 boundaries are sharp. The silicon adjacent to the SiO_2 layer is damaged.

The concept of forming buried layers can be extended to form a variety of materials such as silicides and carbides, provided a high current ion source is available.

5.7 Summary

In this chapter the phenomenon of introducing ions into semiconductors and other materials has been reviewed. We examined:

- The interaction of energetic ions into a solid and the role of energy, mass, charge, and proximity of the ion with respect to those of the atomic or molecular species in the solid.
- Range and concentration profiles of the implanted species and the effect of substrate crystallinity on range and concentation profiles.
- Lattice damage and amorphization caused by ions and their relationship to ion energy and crystal binding. Annealing of ion damage and electrical activation.
- Equipment for ion implantation and limitations of the process.
- Ion-beam mixing and materials modification, including dielectric film formation.

References for Chapter 5

1. "Ion Implantation for Materials Processing," F.A. Schmidt, ed., (Noyes Data Corp., Park-Ridge, NJ (1983).
2. J.K. Hirvonen, C.R. Clayton, in "Surface Modification and Alloying by Laser, Ion, and Electron Beams," J.M. Poate, G. Foti, D.C. Jacobson, Eds., Plenum Press, NY (1983).
3. B. Smith, "Ion Implantation Range Data for Silicon and Germanium Device Technologies," Research Studies, Forest Grove, Oregon (1977).
4. P. Sigmund, J.B. Sanders, "Proc. Intern. Conf. Appl. Ion Beams Semiconductor Techn." P. Glotin, Ed., Editions Ophrys, Grenoble, p. 215 (1967).
5. J. Lindhard, M. Scharff, H.E. Schiott, Kgl. Danske Videnskab Selskab. Mat. Fys. Medd. 33, (14) (1963).
6. K.O. Nielsen, in "Electromagnetically Enriched Isotopes and Mass Spectroscopy," M.L. Smith, Ed., Academic Press, NY, p. 68 (1956).
7. J.W. Mayer, L. Erickson, J.A. Davies, "Ion Implantation in Semiconductors," Academic Press, NY, p. 25 (1970).
8. H.E. Schiott, Can. J. Phys. 46, 449 (1968).
9. S. Furukawa, H. Matsumura, H. Ishiwara, Jpn. J. Appl. Phys. 11, 134 (1972).
10. W.E. Beadle, J.C.C. Tsai, R.D. Plummer, in "Quick Reference Manual for Silicon Integrated Circuit Technology," Wiley, NY, p. 7-3 (1985).

11. J.F. Gibbons, S. Mylroie, Appl. Phys. Lett. 11, 568 (1973).
12. K.B. Winterbon, "Ion Implanation Range and Energy Deposition Distribution, Vol. 2, Low Incident Ion Energies,"IFI/Plenum, London (1975).
13. W.K. Hofker, Philips Research Reports, Supplement No. 8 (1975).
14. J.F. Gibbons, W.S. Johnson, S.W. Mylroie, "Projected Range Statistics,"2nd ed. Dowden, Hutchinson, and Ross, Stroudsburg, PA (1975).
15. W.P. Eldrton, N.L. Johnson, "Systems of Frequency Curves,"Cambridge Press, London (1914).
16. L.C. Feldman, J.W. Mayer, S.T. Picraux, "Materials Analysis by Ion Channeling,"Academic Press, NY (1982).
17. H. Okabayashi, D. Shinoda, J. Appl. Phys. 44, 4220 (1973).
18. R. Bauerlein, in "Radiation Damage in Solids,"D.S. Billington, ed., Academic Press, NY, p. 358 (1962).
19. J.W. Mayer, L. Erickson, J.A. Davies, "Ion Implantation in Semiconductors,"Academic Press, NY, p. 72 (1970).
20. D.K. Brice, J. Appl. Phys. 46, 3885 (1975).
21. H.J. Stein, F.L. Vook, D.K. Brice, J.A. Borders, S.T. Picreaux, in "First International Conf. on Ion Implantation, Thousand Oaks,"F. Eisen, L. Chadderton, eds., Gordon and Breach, NY, p. 17 (1971).
22. S.T. Picraux, J.E. Westmoreland, J.W. Mayer, R.R. Hart, O.J. Marsh, Appl. Phys. Lett. 14, 7 (1969).
23. K.T. Short, D.J. Chivers, R.G. Elliman, J. Liu, A.P. Pogany, H. Wagenfeld, J.S. Williams, Mat. Res. Soc. Symp. Proc. 27, 247 (1984).
24. F.F. Morehead, B.L. Crowder, in Proc. First International Conf. on Ion Implantation, Thousand Oaks, F. Eisen, L. Chadderton, eds., Gordon and Breach, NY (1971). Also see Appl. Phys. Lett. 14, 313 (1969).
25. R.T. Young, C.W. White, G.J. Clark, J. Narayan, W.H. Christie, M. Murakami, P.W. King, S.D. Kramer, Appl. Phys. Lett. 32, 139 (1978).
26. D.A. Thompson, A. Golanski, H.K. Haugen, D.V. Stevanovic, G. Cater, C.E. Christodoulides, Radiat. Effects 52, 69 (1980).
27. J.W. Mayer, L. Ericksson, S.T. Picraux, J.A. Davies, Can. J. Phys. 46, 663 (1968).
28. T.E. Seidel, A.U. McRae, in "First International Conf. on Ion Implantation, Thousand Oaks,"F. Eisen, L. Chadderton, eds., Gordon and Breach, NY, (1971).
29. G. Dearnaley, J.H. Freeman, R.S. Nelson, J. Stephen, in "Ion Implantation Defects in Crystalline Solids,"Vol. 8, North-Holland, Amsterdam (1973).
30. L. Csepregi, J.W. Mayer, T.W. Sigmon, Appl. Phys. Lett. 29, 92 (1976).
31. E.P. Donovan, F. Spaepen, D. Turnbull, J.M. Poate, D.C. Jacobson, Mat. Res. Soc. Symp. Proc. 27, 211 (1984).
32. O.W. Holland, J. Narayan, Mat. Res. Soc. Symp. Proc. 27, 235 (1984).
33. T.E. Seidel, A.U. McRae, Trans. Met. Soc. AIME, 245, 491 (1969).
34. Courtesy of Varian Associates Inc.

35. J.L. Stone, J.C. Plunkett, in "Impurity Doping Processes in Silicon," F.F.Y. Wang, ed., North-Holland, NY, p. 54 (1981).
36. Ion Contamination - J.M. Andrews, private communication.
37. Trilevel resist system, see Chapter 9.
38. B.Y. Tsaur, Z.L. Liau, J.W. Mayer, Appl. Phys. Lett. 34, 168 (1979).
39. G.E. Chapman, S.S. Lau, S. Matteson, J.W. Mayer, J. Appl. Phys. 50, 6321 (1979).
40. J.W. Mayer, S.S. Lau, in "Surface Modification and Alloying by Laser, Ion, and Electron Beams," J.M. Poate, G. Foti, D.C. Jacobson, eds., Plenum, NY, p. 241 (1983).
41. G. Dearnaley, Materials in Engineering Applications 1, 28 (1978).
42. J.K. Hirvonen, C.R. Clayton, in "Surface Modification and Alloying by Laser, Ion, and Electron Beams," J.M. Poate, G. Foti, D.C. Jacobson, eds., Plenum, NY, p. 323 (1983).
43. G. Dearnaley, P.D. Goode, Nucl. Instrum. Methods 189, 117 (1981).
44. R.F. Pinizzotto, Mat. Res. Soc. Symp. Proc. 27, 265 (1984).
45. P. Bourguet, J.M. Dupart, E. LeTiran, P. Auvray, A. Guivarc'h, M. Salvi, G. Pelous, P. Henoc, J. Appl. Phys. 51, 6169 (1980).

Problem Set for Chapter 5

5.1 Show that during an elastic collision of an ion of mass M_1 with the substrate atom of mass M_2, the maximum energy transferred is

$$E_{\mathrm{T}}(\max) = 4E\frac{M_1 M_2}{(M_1 + M_2)^2},$$

where E is the energy of incident ion prior to collision.

5.2 Calculate the range of 100-keV ions of hydrogen, aluminum, and tin in silicon. Explain the differences in the value of the ranges.

5.3 Plot the concentration versus depth profile for 10-and 100-keV boron in silicon. The flux of boron implant is 1×10^{16} per cm^2 per second.

5.4 Assuming a Gaussian profile for ion implanted P^+, show that the junction depth is

$$x_{\mathrm{j}} = R_{\mathrm{p}} + \Delta R_{\mathrm{p}} \left(2\ell n \left[\frac{\phi}{\sqrt{2\pi} \cdot \Delta R_{\mathrm{p}} N_{\mathrm{B}}} \right] \right)^{1/2}$$

where N_B is the background dopant concentration in silicon. For P^+ implants carried out in two different machines, the ion currents were measured to be 1.6×10^{-6} and 4.8×10^{-6} A per cm^2. Implants were carried out for 1000 seconds. Compare the junction depths if $R_p = 2 \times 10^{-5}$cm, $\Delta R_p = 0.4 \times 10^{-5}$cm, and $N_B = 1 \times 10^{15}$ per cm^3.

5.5 The range and straggling of a boron implant in silicon (substrate doping 1×10^{14} per cm^3 of phosphorus) are 1000Å and 200Å, respectively. The flux was 1×10^{15} B^+ per cm^2. After implant the sample is annealed at 950°C for a time of 4 hours to yield a junction depth of $x\mu$m. If D at this temperature is 6×10^{-15} cm^2/sec calculate x.

5.6 A high-current ion implanter is used to form a 75-nm thick buried oxide layer in silicon. The top surface of the oxide is buried 62.5 nm below the surface of silicon.

a. If the wafer surface area is 150 cm^2 and is to be implanted in 100 s, calculate the ion current necessary to achieve this implant. (Hint: Assume $2\Delta R_p = 74$ nm and that the oxygen atom is singly ionized).
b. Draw appropriate concentration and damage profiles of the as-implanted material.
c. Suggest an annealing process that will convert implanted oxygen to SiO_2.
d. Calculate the stress generated in silicon assuming that no stress relief occurs during the processing and Young's modulus of silicon is $\simeq 1 \times 10^{12}$ dyn/cm^2. $\rho_{Si} = 2.33$ g/cm^3 and $\rho_{SiO_2} = 2.27$ g/cm^3.

5.7 Calculate the flux of Pt ions and implantation time necessary to form a 1 μm × 1 μm × 100 nm silicide layer on silicon. The maximum ion current for the implanter is 10 nA, with an ion beam diameter of 100 nm. (Density of PtSi is 12.4g/cm^3.)

5.8 Assuming that the stopping power of a compound is a weighted average of the individual elements it is made of, calculate the minimum thickness of a hypothetical organic material $C_{50}O_{20}Si_{19}$ required to screen the underlying substrate from a 80-keV arsenic implant. Density of this hypothetical material is 4 g/cm^3.

5.9 Thin oxide masking layers are used on the surface of the semiconductor to randomize the ion implants and to reduce channeling. If a boron-implanted depth of only 70 nm in silicon is required and the energies available are 10, 30, 80, and 120 keV, calculate the thickness of the SiO_2 layers that must be used to obtain desired depth.

5.10 Explain why high-energy implants are used to produce gettering implants. If a junction is 300 nm deep, suggest a gettering implant and necessary treatments, if any, to produce a built-in impurity gettering layer somewhere below the junction. Explain your selection of the technique to achieve the desired end result.

5.11 Submicron VLSI and ULSI circuits may require junction depths of less than 100 nm. Comment on the applicability of ion implantation to produce p on n and n on p junctions.

5.12 Assume the depth of an implant is large enough that surface concentration effects can be neglected in analyzing redistribution effects. Prove that the distribution of impurities is still Gaussian after activation or drive. What is the new ΔR_{p}? Hint: express the diffusion equation in its integral form using an impulse response approach.

$$\frac{dN(x)}{dt} = D\frac{d^2N(x)}{dx^2} \longrightarrow N(x,t) = \frac{1}{2\sqrt{\pi}}\int_{-\infty}^{\infty}\frac{N(x',0)e^{-(x-x')^2/4Dt}dx'}{\sqrt{Dt}}$$

where $N(x',0)$ is the initial implant distribution.

6 Metallization

6.1 Introduction

Metallization in integrated circuits is required a) to provide a means of supplying power; b) to provide communication between devices and between devices and the outside world; c) to control the device characteristics such as the Schottky barrier height, contact resistance, metal-oxide-semiconductor properties, back contact, and other not-so-specific uses that also control device properties such as shields, guard rings, field plates, and provide redundancy in memories; and d) to provide high stability and protection from the environment, such as diffusion barriers and corrosion resistant coatings.

Figure 6.1 shows a cross section, a simplistic schematic, and a designer's symbolic representation of a typical MOSFET (metal-oxide-semiconductor field effect transistor). The central region is called the *gate*. In this region the substrate silicon is isolated from the metal electrode by an insulating layer, which is, generally, a thin thermally grown SiO_2 layer (see Chapter 3). The metal electrode, generally a polysilicon (see Chapter 7) layer, is called the *gate electrode* and it controls the on-off properties working of the MOSFET. The work function of the metal plays an important role in determining the voltage required to activate the device. The two regions adjacent to the gate region are called the *source* and the *drain*. The metal contacting these regions has generally been aluminum. The metal work function again plays an important role in determining the characteristics of current flow between these regions.

In applications involving compound semiconductors such as GaAs and InP, there is no good material that can be used as the gate oxide. Therefore a different type of device structure, called a metal-semiconductor field effect transistor (MESFET), is made in which the gate metal is directly deposited on the semiconductor. The metal forms a Schottky barrier with the semiconductor and thus forms the gate. The Schottky barrier height is related to the work function of the metal. Thus even in MESFETs, the work function of the metal plays an important role.

Metallization usage in bipolar devices is very similar to usage in MOSFETs. In bipolar devices the central region is the semiconductor (base), which directly contacts the metal. The two neighboring regions are called the *emitter* and the *collector*, also directly in contact with the metal. Thus a bipolar device, in cross section, looks very similar to a MOSFET with no

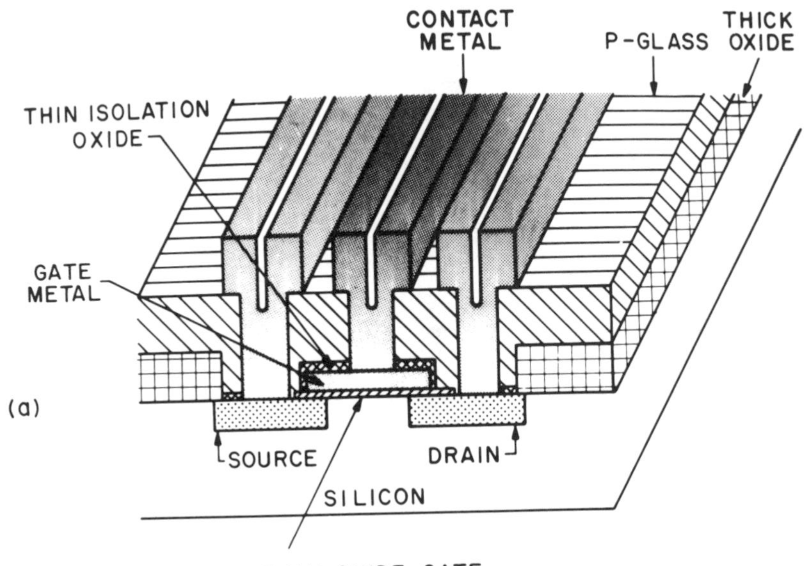

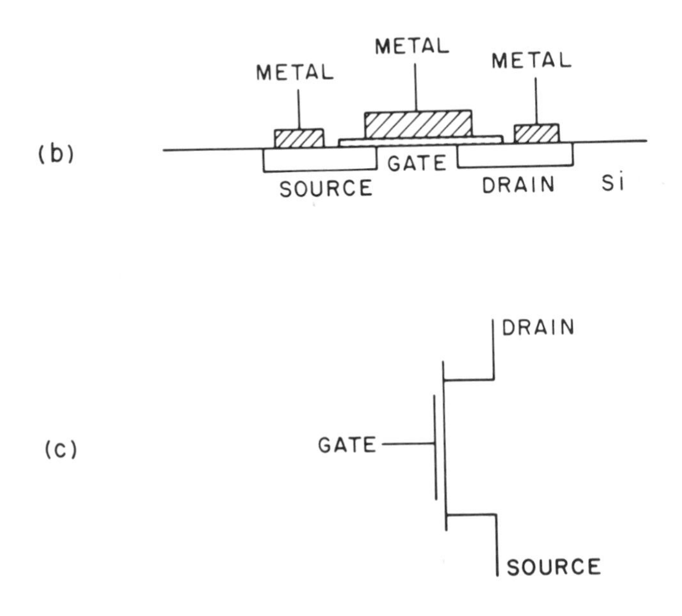

Fig. 6.1. (a) A schematic cross section of a typical MOSFET, (b) a simple schematic representation of the same, and (c) a designer's symbolic representation.

gate oxide. The metal work function plays an important role in all three regions in determining the current flow characteristics in a manner similar to the role it plays in the source and drain regions of the MOSFET.

Primary metallization applications can be divided in two groups: a) gate and interconnection and b) contact. The interconnection metallization is generally the same as the gate metallization. All metallizations involving direct contact with the semiconductor surface are called *contact metallizations.* As mentioned earlier, polysilicon film is generally the metal used for both gates and interconnections in MOS devices [1]. In constructing MESFETs on compound semiconductors, various metals have been tried. Metal-rich silicides such as W_5Si_3 have been recently reported to produce the best MESFETs [2]. Aluminum is used as the contact metal for devices and also as a second-level connection to the outside world [3]. Several other metallization schemes have been proposed to produce ohmic contacts to semiconductors [4]. In several cases multiple-layer structures involving a diffusion barrier [5], have been recommended. PtSi has been used as a Schottky barrier contact and also simply as an ohmic contact for deep junctions [6]. Titanium/palladium/gold or titanium/platinum/gold beam lead technology was successful in providing high-reliability connection to the outside world [7-9].

Continuing advances in the fields of the very-large-scale integration (VLSI), of ultra-large-scale integration (ULSI), and of the continued development of smaller and smaller devices aroused concern about existing metallization schemes for gates, interconnections, and ohmic contacts and about the reliability of aluminum and its alloys as the current barrier. However, the applicability of any metallization scheme to integrated circuits (IC) depends on several requirements (table 6.1). Most important of these requirements is the stability of the metallization throughout the IC fabrication procedure and during the actual usage of the finished product. Thus, both the mechanical (physical) and electrical characteristics of the metallization and the devices it covers must be preserved. In addition, metallization schemes should be easy and economical to implement, i.e., the metal should be easy to deposit and pattern. A considerable amount of research and development, therefore, goes into looking for a metallization scheme. Such efforts have led to a large volume of work involving thin metallic films including film deposition, film characterization, interactions between one film and another film, and interactions between films and substrates.

In the following section the role of the metal resistance and work function in determining the device properties will be examined. In subsequent sections various other aspects of metallization such as choices of metal, characterization, deposition, patterning, problems, and new applications will be discussed.

TABLE 6.1: Desired properties of metallization for integrated circuits

1.	Low resistivity
2.	Easy to form
3.	Easy to etch for pattern generation
4.	Should be stable in oxidizing ambients, and oxidizable
5.	Mechanical stability, good adherence, and low stress
6.	Surface smoothness
7.	Stability throughout processing, including high-temperature sinter, dry or wet oxidation, gettering, phosphorus glass (or any other material) passivation, and metallization
8.	No reation with final metal, aluminum
9.	Should not contaminate devices, wafers, or working apparatus
10.	Good device characteristics and lifetimes
11.	For window contacts—low contact resistance, minimal junction penetration, and low electromigration

6.1.1 The Metallization Resistance

Resistance of the metal lines is the most important factor in determining the utility of a given metal. Resistance of these lines is generally measured in terms of a quantity R_s, called sheet resistance (see Chapter 4, and below). R_s is determined only by the thickness t of the film and its resistivity ρ and is given by

$$R_s = \frac{\rho}{t} \tag{6.1}$$

in units of ohms per square. In eq. (6.1), ρ and t are given in Ω-cm and cm, respectively. This definition of sheet resistance means that each square of the metal film of a given thickness will have the same R_s. Thus a metal line can be divided quickly into a total number of squares that multiplied by R_s gives the total resistance of the line in ohms. Since R_s depends on metal resistivity, anything that will influence resistivity will affect R_s.

Resistivity of a thin metallic film is a function of several parameters—temperature, purity, film thickness, crystallinity, defect structure, and any applied field. Matthiessen's rule states that the total resistivity of a sample is sum of all the individual contributions made by the above factors. Traditionally resistivity is then written as

$$\rho = \rho_{\text{temp}} + \sum_i \rho_{\text{i}}. \tag{6.2}$$

ρ_{temp} is the temperature-dependent term and ρ_{i} is the resistivity contribution due to i^{th} factor. One can write

$$\sum \rho_i = \rho_{\text{ residual, so that}}$$

$$\rho = \rho_{\text{temp}} + \rho_{\text{ residual}}. \tag{6.3}$$

The *temperature dependence* of the resistivity arises due to interactions between the charge carriers and the lattice vibration modes. Thus in metals, the resistivity, and hence the resistance, of a given sample increases with temperature. The temperature coefficient of the resistance (TCR) is generally defined as

$$\text{TCR} = \frac{R_1 - R_2}{R_{\text{T}}(T_1 - T_2)}, \tag{6.4}$$

where $T_1 > T > T_2$ and R_1, R_2, and R_{T} are the resistance at temperatures T_1, T_2, and T (T is normally taken as 20°C), respectively. For continuous metal films the TCR is always positive. In discontinuous metal films and in continuous films with large excess of defects (such as vacancies, interstitials, and dislocations), however, annealing at higher temperatures reduces the resistance, leading to negative TCR values. In discontinuous films, the electron transport between the islands of the film material is governed by a mechanism that is itself determined by the island size and the separation between the islands [10]. For continuous film with defects in concentration larger than predicted by thermodynamic equilibrium relationships, annealing at higher temperature results in reduction of the defect concentrations and therefore of the contribution to the resistance by such defects.

Impurities change resistivity of a film. This effect is associated with several factors. a) The impurity atom has a different valence than the host metal atom. b) The impurity atom has a significantly different size than the host atom, causing strain in the host metal lattice. The resulting strain will cause electron scattering. c) Even for the same valence, the screening of the impurity ions by the electron cloud is different, thus affecting the electron distribution and scattering around the impurity. d) The impurity may form a compound with the host metal. Such compound formation will lead to significant changes in the resistivity. In most cases, the resistivity increases; the majority of the increase is due to a) and c). For a dilute solid solution the resistivity is related to composition by Matthiessen's rule with only one $\rho_{\text{i}}(\text{c})$ term in ρ_{residual}. $\rho_{\text{i}}(\text{c})$ is the temperature-insensitive, concentration-dependent residual resistivity due to the impurity atoms. For dilute alloys, where impurity atoms can be considered to be placed far apart from each other, this ρ_{i} term can be approximated by [11]

$$\rho_{\text{i}}(\text{c}) = n_{\text{i}}(\Delta z \cdot q)^2 \tag{6.5}$$

where n_{i} is the electron density per unit volume and $\Delta\ zq$ is the difference between the ionic charges of the solvent and solute atoms.

For dilute alloys with concentrations less than a few atomic percent, one can determine experimentally a relationship between $\rho_i(c)$ and the impurity concentration c. The relationship is generally of the form

$$\rho_i(c) = Kc, \tag{6.6}$$

where K is the constant determined from experiments. Thus for dilute solutions the total resistivity is a sum of the pure metal resistivity and ρ_i estimated by eqs. (6.5) or (6.6).

For very thin films, the free surfaces contribute significantly to the resistivity of the film. This is especially true when the film thickness t is comparable to the mean free path of the electrons. The mean free path (mfp), l_0, is given by the free-electron model as

$$l_0 = \frac{(3\pi^2)^{1/3}\hbar}{\rho_0 q^2 N_p^{2/3}} \tag{6.7}$$

where ρ_0 is the resistivity of pure crystalline material. N_p is the charge carrier density, q is the electronic charge, and $\hbar$ is Planck's constant divided by 2π. It has been shown [10] that the resistivity of a film ρ_F and the resistivity of the same pure crystalline material ρ_0 are related by the following expressions:

$$\frac{\rho_F}{\rho_0} = 1 + \frac{3}{8\gamma} \text{ for } (\gamma >> 1) \tag{6.8}$$

and

$$\frac{\rho_F}{\rho_0} = \frac{4}{3\gamma \ell n(1/\gamma)} \text{ for } (\gamma << 1), \tag{6.9}$$

where $\gamma = t/\ell_0$. In deriving these simplified equations it was assumed that the surfaces caused diffused scattering of electrons and that the relaxation process for surface scattering is the same as that for the bulk. If, however, a fraction p of the electrons are scattered elastically, then eqs. (6.8) and (6.9) are modified to read

$$\frac{\rho_F}{\rho_0} = 1 + \frac{3}{8\gamma}(1-p) \text{ for } (\gamma > 1) \tag{6.8a}$$

and

$$\frac{\rho_F}{\rho_0} = \frac{4}{3\gamma \ell n(1/\gamma)} \frac{1}{(1+2p)} \text{ for } (\gamma << 1;\ p < 1). \tag{6.9a}$$

Thus when γ is large, i.e., the thickness of the film is significantly larger than the mfp, ρ_F approaches ρ_0. When γ is small $\rho_F > \rho_0$.

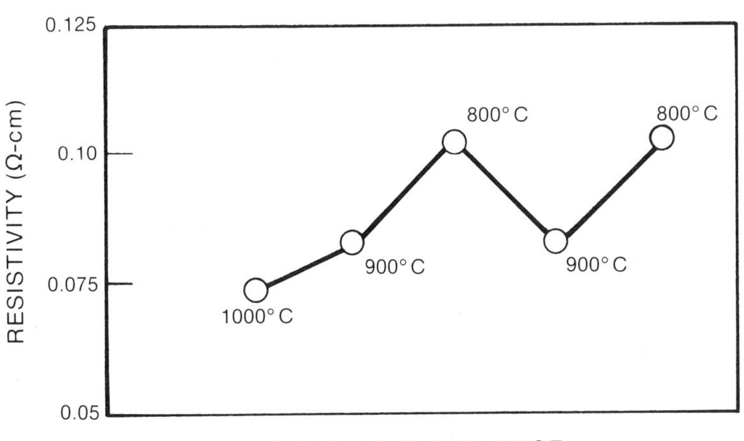

Fig. 6.2. Resistivity of arsenic-implanted polysilicon as a function of the thermal cycling temperature. From Mandurah et al. [13].

In polycrystalline films, grain boundaries increase the resistivity. Mayadas and Shatzke [12] derived the following expression for the resistivity ρ_g of a polycrystalline film with a large number of grain boundaries.

$$\rho_g = \rho_0 \left\{ 1 - \frac{3}{2}\alpha + 3\alpha^2 - 3\alpha^3 \, \ell n \left(1 + \frac{\alpha}{2}\right) \right\}^{-1}, \tag{6.10}$$

where α is defined as

$$\alpha = \frac{\ell_0}{d} \frac{R}{1-R}. \tag{6.11}$$

In eq. (6.11), ℓ_0 is mfp within a grain, R is the reflection coefficient at the grain boundaries, and d is the average spacing between the reflecting planes or the grain boundary width. Knowing the values of d and R, one can estimate the resistivity of a polycrystalline film. R for aluminum and copper were given as 0.17 and 0.24, respectively [12].

Grain boundaries, in addition, play a role that also contributes to resistivity. Impurities present in grain boundaries and segregation at these grain boundaries contribute significantly to the resistivity of polycrystalline silicon. Doped polysilicon films are used as gate material. At very high concentrations (above the solid solubility limit) dopants segregate in grain boundaries and therefore do not participate in the conduction process. Dopants can move in and out of the grains and grain boundaries reversibly when annealed at different temperatures. This dopant movement will affect the resistivity, as shown in fig. 6.2 [13]. After a lower temperature anneal, resistivity is higher because of smaller solubility in the grains and (rejection of more dopants into the grain boundaries). After high-temperature anneals, dopants move back in the grains and lower the resistivity.

For very thin films a phenomenon called *field effect* associated with the electrostatic charges induced by an electric field normal to the sample surface has been observed [14]. The magnitude of the effect is small and difficult to measure, especially for metals where electron densities are high.

For thin film metallization, where resistance is very important, one must therefore be aware of the variations in purity, grain size, and thickness of the film. Temperature of operation of devices will also affect the resistance since resistance is proportional to temperature. Usually during device operation, the heat dissipation increases the device temperature causing increased resistance in the metal lines. Table 6.2 lists various metals and alloys and their properties. For most of these materials a 30°C rise in temperature will increase the resistance by about 10%.

Why is the resistance so important? In addition to controlling the current through the interconnection lines, resistance plays another important role. The characteristics of MOS (metal oxide semiconductor) devices depend on several parameters of which the RC time constant is the most important. The higher the RC value, the slower is the operating speed of the device. The R and C are, respectively, the effective total resistance and capacitance of the device at the gate and interconnection level. The relationship of the RC factor to device scaling is rather complicated, especially because of the dependence of the total capacitance on the feature size. To a simple approximation

$$R = R_S \frac{L}{W} \tag{6.12}$$

and

$$C = LW\ \epsilon_{ox}/t_{ox} \tag{6.13}$$

so that

$$RC = R_S L^2\ \epsilon_{ox}/t_{ox}. \tag{6.14}$$

Here L and W are the length and width of the line, and ϵ_{ox} and t_{ox} are the oxide permittivity and the oxide thickness to the silicon ground plane, respectively. R_S is the ohm per square (sheet) resistance given by eq. (6.1). Thus scaling in vertical direction, i.e., reduction in t and t_{ox}, will increase RC. For a given t and t_{ox}, however, RC depends only on L and is independent of the width W. Thus reducing line width does not change RC. Indirectly this is a disadvantage since the equivalent RC times can offset the speed advantages of device miniaturization.

As the line widths and line separations get smaller interactive effects occur. Therefore the above conclusion is not valid for very small linewidths because of the effect of fringing fields. The effect is negligible at larger linewidths

TABLE 6.2: Properties of the metals of interest[a]

Metal	Resistivity[a] ($\mu\Omega$-cm)	Temperature coefficient of resistance (per °C)	ϕ_m (V)
Ag	1.59	0.0041	4.73
Al	2.65	0.00429	4.08
Au	2.35	0.004	4.82
Co	6.24	0.00604	4.40
Cr	12.9	0.003	4.60
Cu	1.67	0.0068	–
Fe	9.71	0.00651	4.04
Hf	35.1	0.0038	3.53
In	5.3	0.003925	5.3
Mn(α)	185.0	–	3.83
Mo	5.2	–	4.20
Nb	12.5	–	4.01
Ni	6.84	0.0069	5.02
Os	9.5	0.0042	4.55
Pd	10.8	0.00377	4.98
Pt	10.6	0.003927	5.34
Re	19.3	0.00395	5.1
Rh	4.51	0.0042	4.8
Ta	12.45	0.00383	4.19
Ti	42.0	–	~ 4
W	5.65	–	4.52 (001) face
Zr	40.0	0.0044	4.21

[a] From Handbook of Chemistry and Physics, 55th edition, CRC Cleveland (1974-75).

but contributes significantly to total capacitance at widths lower than 5 μm. Figure 6.3 shows the access time as a function of the feature size for a 4K bit static memory with a 20-$\Omega/\square$ polysilicon gate and 1-$\Omega/\square$ refractory gate materials. It is apparent that the access time increases for devices with feature size less than 2 μm. These calculations reflect solely the effects of high interconnection resistance. More recently, Sinha et al. [16] calculated the total capacitance as the sum of intrinsic parallel-plate-type capacitance as given by eq. (6.13); C_A; metal over field oxide capacitance or bottom capacitance, C_B, and the side-wall capacitance, C_S, as a function of the

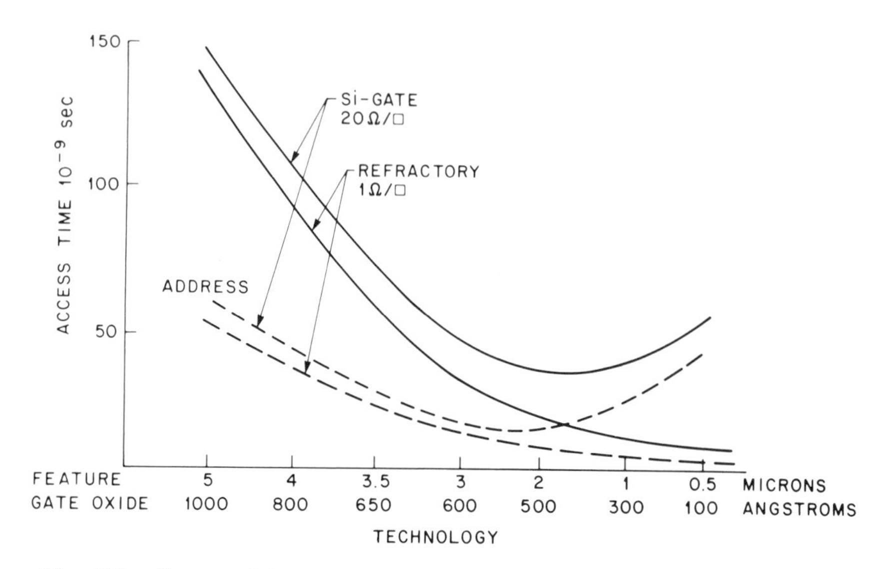

Fig. 6.3. Impact of interconnect resistance on the performance of 4K bit static memory. From Ghate and Fuller [15], reprinted by permission of the publisher, The Electrochemical Society, Inc. This figure was originally presented at the 1981 Spring meeting of the Electrochemical Society, Inc. held in Minneapolis, MN.

linewidth. It is shown that the side-wall capacitance dominates the total capacitance, at line widths of less than 3 μm. Figure 6.4 shows various capacitances, resistance, and the RC time constant as a function of the design rule (or the linewidth) [16]. Note that C now represents total capacitance and C_A is the parallel-plate-type intrinsic capacitance given by eq. (6.13). The calculations were made for one micron of silicon nitride covering one micron of aluminum on one micron of silicon dioxide. As is apparent, RC, which would have been independent of the design rule but for the contribution from C_B and C_S, increases with decreasing line widths. As expected at linewidths greater than 5 μm, RC is nearly independent of the linewidth. The RC time constant was also shown to be affected by the dielectric constant of the passivation layer (or the interlevel insulator in two-level Al interconnects), the field oxide thickness, and the metal thickness for feature size of less than 2 μm.

The above considerations show that for optimum results, the lowest resistivity metal should be used and that it should be pure, single crystal, and of a thickness greater than the electron mean free path.

6.1.2 Metal Work Function

Work function ($q\phi_m$) is defined as the energy required to move an electron from the Fermi level of the metal to vacuum. ϕ_m, which is given in

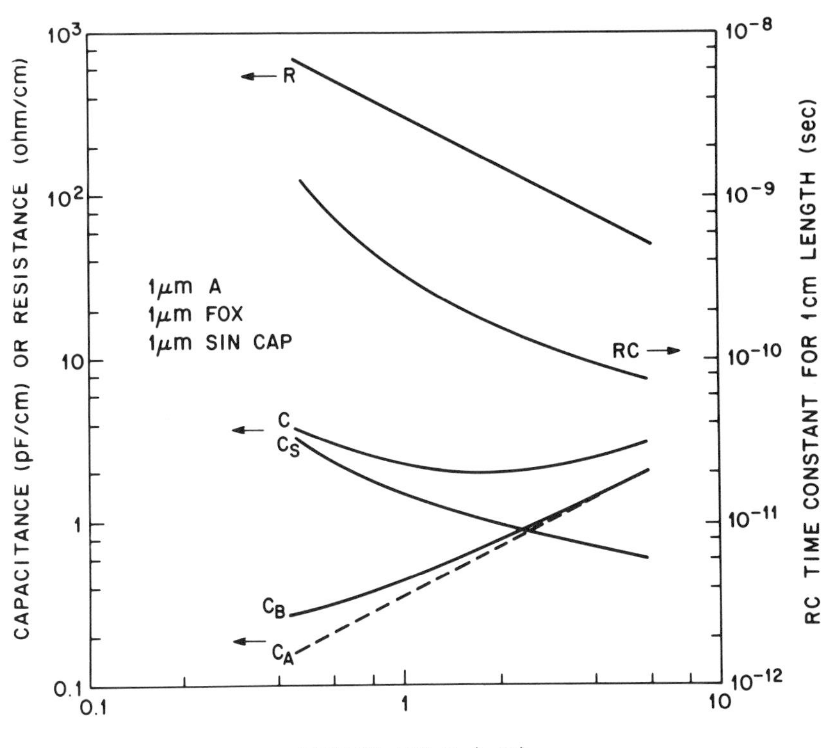

Fig. 6.4. Effect of aluminum interconnect width and spacing (design rule) on the line capacitance C, resistance R, and RC time constant for 1-cm long runner. C_A = parallel plate capacitance; C_B = bottom capacitance; C_S = sidewall capacitance; C, total capacitance. From Sinha et al. [16].

volts, ranges from 2 to 6 volts for various metals. The measured value of ϕ_m depends strongly on the cleanliness of the surface. The role of ϕ_m in controlling device function is described below for a MOS capacitor and a semiconductor-metal contact.

For MOS capacitors, the metallization controls the so-called flat-band voltage V_{FB}. V_{FB} is the voltage required to counterbalance the work function difference between the metal and the semiconductor so that a flat-band condition (see Chapter 3) is maintained in the semiconductor. In the absence of any charge in the oxide or at the oxide-semiconductor interface

$$V_{FB} = \phi_m - \phi_S = \phi_{mS}, \tag{6.15}$$

where ϕ_m and ϕ_S are the work functions of the metallization (at the gate) and the semiconductor, respectively. Generally, there are various types of charges in the oxide affecting the flat-band voltage. For details reference 17 should be examined.

The flat-band voltage V_{FB} contributes to the threshold voltage V_{T}. V_{T} is the voltage required at the gate metal (measured with respect to source voltage) to achieve a conduction path between the source and drain regions. It determines the gate-to-source voltage that will switch the MOSFET to the on condition and allow the current to flow. Thus it is important that the gate metallization be chosen very judiciously to achieve the desired MOS behavior and circuit speed.

A good contact, usually formed by depositing a metal on the semiconductor, does not perturb the device characteristics and is stable both electrically and mechanically. It has a resistance, called *contact resistance*, which is negligible compared to the device resistance. The specific contact resistance R_{C} (Ω-cm^2) is defined as [18]

$$R_{\mathrm{C}} = \left(\frac{dV}{dJ}\right)_{v\,=\,0}, \tag{6.16}$$

which can be obtained from current density (J) and voltage (V) characteristics.

R_{C} is related to ϕ_{B}, the so-called Schottky barrier height [18] of the metal, and doping density N_{D} in the semiconductor. For lower doping density R_{C} is shown to be [18]

$$R_{\mathrm{C}} = \frac{k}{qA^{*}T}\exp\left(\frac{q\phi_{\mathrm{B}}}{\mathrm{k}T}\right). \tag{6.17}$$

For higher doping densities ($N_{\mathrm{D}} \gtrsim 10^{19}$ cm^{-3}); R_{C} is dominated by the electron tunneling across the barrier and is given by

$$R_{\mathrm{C}} \approx \exp\left[\frac{\alpha(\epsilon_{\mathrm{S}}m^{*})^{1/2}}{\hbar}\left(\frac{\phi_{\mathrm{B}}}{\sqrt{N_{\mathrm{D}}}}\right)\right], \tag{6.18}$$

where k is the Boltzmann constant, A^{*} is the Richardson constant, T is the temperature in degrees Kelvin, ϵ_{S} is the permittivity of the semiconductor, $\hbar$ is Planck's constant divided by 2π, and m^{*} is the effective electron mass.

Figure 6.5 shows the relationship between R_{C}, ϕ_{B}, and N_{D} for contacts on a n-type silicon substrate. Both experimental points and theoretical curves are shown for metals of different barrier heights [18]. As is evident, R_{C} decreases very rapidly with increasing doping density N_{D} and decreasing barrier height ϕ_{B}. Also note from eq. (6.17) that for lower doping densities, the higher the temperature the lower the contact resistance, whereas at higher doping densities the barrier height dependence only dominates.

6.1.3 Schottky Barrier Height

The occurrence of the so-called Schottky barrier between the metal and semiconductor arises due to the requirement that the Fermi levels in the two

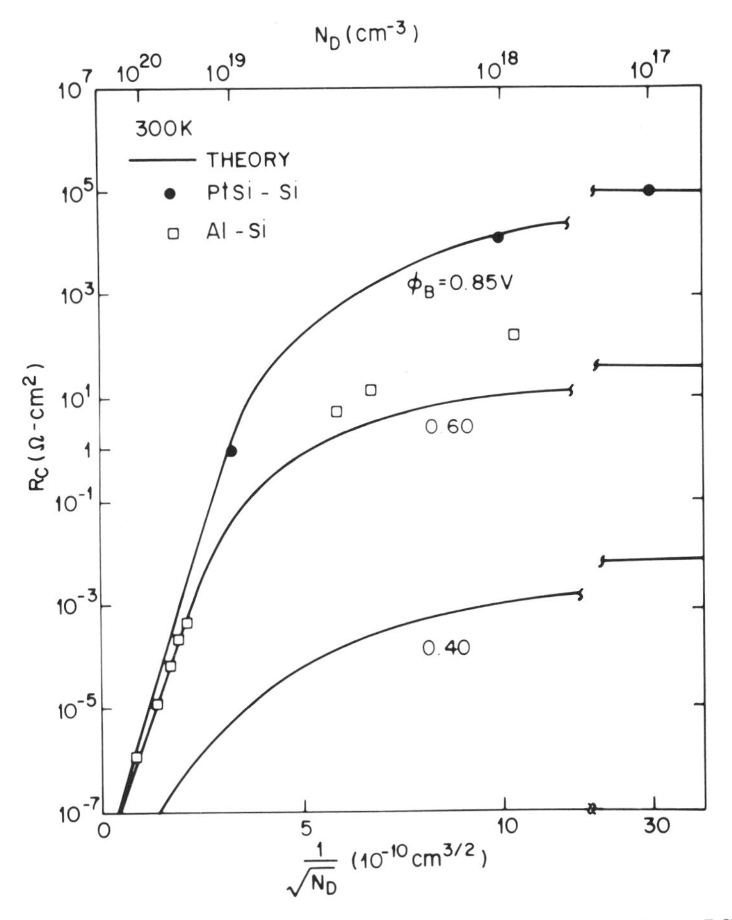

Fig. 6.5. Theoretical and experimental values of specific contact resistance RC as a function of the dopant concentration N_D and barrier height ϕ_β. From Sze [18].

materials match up. Such a barrier to charge transfer between the metal and the semiconductor is the result of the unequal work functions of these materials. Ideally we can calculate the difference and predict the behavior of the metal on a semiconductor. For example, for an n-type semiconductor, if the work function ϕ_m of the metal is greater than the work function ϕ_S of the semiconductor, the contact between the two is rectifying. On the other hand, if ϕ_m is less than ϕ_S the contact is ohmic. The calculation of the Schottky barrier height ϕ_B is, however, considerably complicated and must include the effects of the semiconductor surface states, the image force lowering, and the effects of interfacial layers. For a detailed discussion of these calculations the books by Sze [18] and Rhoderick [19] are useful. A brief summary of the role of the interfacial oxide layer is given here. Semiconductor surfaces, unless cleaved or cleaned in UHV conditions prior to the deposition of metal, have

TABLE 6.3: Effect of phosphorus implant on the Schottky barrier height for Ti on p-silicon[a]

Implant			Barrier Height (eV)		
Energy (keV)	Dose (cm^{-2})	Depth (Å)	Experimental $C-V$	$I-V$	Calculated
0	0	0	0.59	0.60	0.60
20	5.20×10^{10}	300	0.60	0.61	0.60
20	8.2×10^{10}	700	0.67	0.64	0.65
40	5.04×10^{10}				
20	3.3×10^{11}	780	0.77	0.78	0.84
40	2.02×10^{11}				
20	1.26×10^{12}	400	0.96	0.96	0.93

[a] From Li et al. [22]. Copyright © 1980, IEEE.

a native oxide 5 to 20 Å thick. This leads to three effects [19]: a) a potential drop across the oxide, leading to a lowering of the zero bias ϕ_B, b) since electrons now have to tunnel through this oxide, the current for a given bias is lowered; and c) ϕ_B becomes a function of applied bias. In addition, if the oxide leads to a fixed surface state charge at the oxide-semiconductor interface, this will affect the charge distribution in the semiconductor and hence ϕ_B. A positive fixed charge (in the case of oxide on n-semiconductor) will cause a reduction in ϕ_B [20].

One can control ϕ_B by controlling the doping in the surface layers of semiconductors. Shannon [21] has shown that the implantation of donors into an n-type material, followed by the usual annealing to remove the damage and activate the dopants, can be used to reduce the barrier height. Table 6.3 lists experimental and calculated values of ϕ_B of titanium on p-silicon for various phosphorus implant conditions [22]. A remarkable increase in the ϕ_B values is clearly seen.

For lightly doped semiconductors, the metal-semiconductor contact may not be ohmic. This is true when ϕ_B is high. In such cases, the contact is rectifying, allowing current flow only in the direction of applied voltage that decreases the barrier. Thus when a metal contact to n-semiconductor is made positive with respect to semiconductor current flows, the situation is termed the *forward bias condition*. On the other hand, when the same contact is made negative with respect to the semiconductor, no current flows. In the reverse bias case, the barrier to electron flow increases by the applied bias restricting the current flow.

The barrier height ϕ_B is determined by measuring the current density J (current per unit area) in terms of the applied voltage V across a Schottky

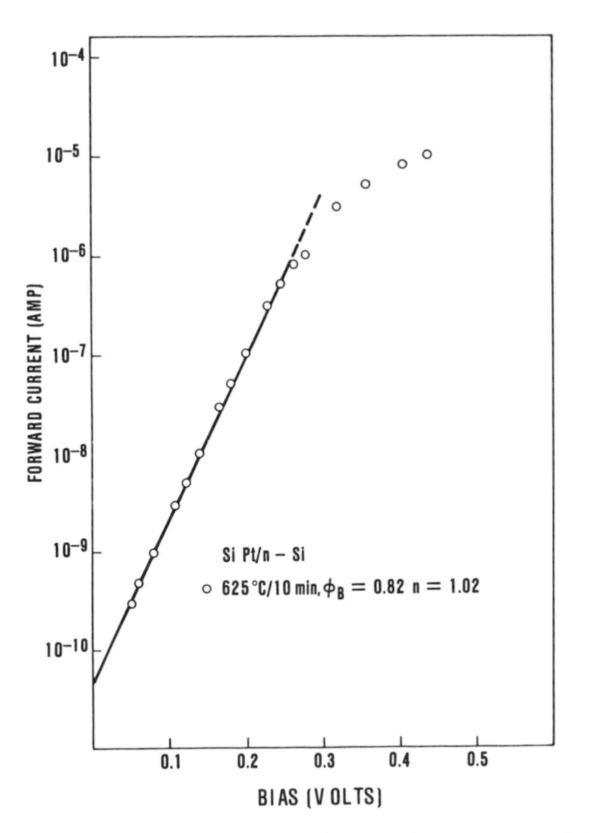

Fig. 6.6. A typical current-voltage ($I - V$) forward bias characteristic of a Pt on n-GaAs Schottky barrier diode.

barrier. J is related to V by the equation [19]

$$J = A^*T^2 \left[\exp\left(-\frac{q\phi_B}{kT}\right)\right] \left[\exp\left(\frac{qV}{nkT}\right) - 1\right]. \quad (6.19)$$

In this equation, n is the diode ideality factor, which is unity for the ideal Schottky barrier diode. The plots of J versus V (in the forward bias condition) allow the determination of both ϕ_B and n. Figure 6.6 shows a typical plot for platinum on n-GaAs. The intercept at $V = 0$ yields the saturation current that gives ϕ_B. The slope of the linear region yields the value of n. Other methods to determine ϕ_B include $C - V$ measurements, the temperature dependence of $I - V$, and photoelectric measurements. ϕ_B values for various metals and silicides are listed in reference 18.

As shown, the resistance and the work function of the metal play an important role in determining the metal's applicability to integrated circuits and devices. However, as pointed out earlier, there are other factors that may dictate a given metal's usefulness. These factors, related to stability and processability of the material, are discussed in the next section.

There are several other applications of metallization in integrated circuits, the most important one being the top-level metal that connects devices to the outside world. In general, this metal film is deposited and patterned during the last stages of processing, whereas the gate interconnection and contact metallization are formed during the earlier stages. The top-level metal is simply a current carrier and is therefore very thick. This metal film also acts as a corrosion-resistant coating for protecting active devices. To reduce interconnection resistance and save area on a chip, multilevel metallization may replace single-level metal.

6.2 Metallization Choices

Table 6.1 lists the desired properties of the metallization for integrated circuit applications. Most metals do not satisfy these requirements, as shown in table 6.4. For example, the lowest resistivity metal, silver, is unstable on most surfaces. It oxidizes, agglomerates, and does not bond well with SiO_2. Gold does not bond with SiO_2 and forms an eutectic with silicon that melts at 370°C. Gold is also undesirable because it is a fast diffuser in silicon and it has donor and acceptor levels in silicon causing undesirable diode leakage. Aluminum, in spite of its low melting point, is the only metal that satisfies most requirements. In order to circumvent the problems presented by its low melting point, it is applied towards the end of processing.

For present and future integrated circuits, different metallization schemes for different applications may be desirable. The changing world of silicon integrated-circuit fabrication has led to the fast shrinkage of device dimensions—both vertically and laterally—requiring the processing temperatures to be 900°C or lower. In addition, wet chemical etching techniques except for those used in self-aligned selective etching, are a thing of the past. Thus a metallization scheme must employ an etching process able to achieve desired selectivity and anisotropy in etching. These demanding changes have opened doors to a reexamination of alloys and metals that were not considered fit for application in ICs.

6.2.1 Metals or Alloys

Table 6.5 lists the metallization applications together with the possible metallization choices. Polysilicon is the gate metallization for MOS devices. Polysilicon films when doped with phophorus, arsenic, or boron become conducting with resistivity as low as 500 $\mu\Omega$-cm. These films can be oxidized easily to form passivating SiO_2 layers on their surface and are easy to pattern (chemically and now by dry etching).

TABLE 6.4: Properties that make metals unsuitable for VLSI application

Undesirable property	Metal
Low eutectic temperature (< 800°C)	Au, Pd, Al, Mg
Medium eutectic temperature (800-1100°C)	Ni, Pt, Ag, Cu
High diffusivity in silicon	All
High oxidation rate, poor oxidation stability	Refractory metals; rare earths; Mg, Fe, Cu, Ag
Low melting point	Al, Mg
Interaction with substrate or polysilicon at temperature less than 450°[a]	Pt, Pd, Rh, V(?) Mo(?), Cr(?)
Interaction with substrate polysilicon at temperatures up to 1000°C	All
Interaction with SiO_2	Hf, Zr, Ti, Ta, Nb V, Mg[b], Al[b]
Poor chemical stability, especially in HF-containing solutions	Refractory metals, Fe, Co, Ni, Cu, Mg, Al
Poor etchability	Pt, Pd, Ni, Co, Au
Electromigration problems	Al
Contact spiking due to interdiffusion	Al

[a] Typical last high temperature in device fabrication.

[b] Interact with SiO_2 to form metal oxide that (self) limits the further interaction.

They make self-aligned gate and source doping easy. All these advantages of polysilicon made it the best choice as the gate metal. Only recently have the polysilicon/refractory metal silicide bilayers replaced polysilicon so that lower resistance can be achieved at the gate and interconnection level. Bilayer application preserved the use of polysilicon as the metal in contact with the gate oxide. Well-established device characteristics and processes were unaltered. Since polysilicon and silicide layers form resistors in par-

TABLE 6.5: Possible metallization choices for integrated circuits

Application	Choices
Gate and interconnection and contacts	Polysilicon, silicides nitrides, carbides, borides refractory metals, aluminum and a combination of two or more of above
Diffusion barrier	Nitrides, carbides, borides, Ti-W alloy silicides
Top level	Aluminum
Selectively formed metallization on silicon only	Some silicides, tungsten, aluminum

allel, the lowest resistance resistor (silicide) dominated the total resistance. Refractory silicides formed directly on polysilicon provided the highest process compatibility. Thus disilicides of molybdenum [23], tantalum [24], and tungsten [25] were developed and found their way into the production of microprocessors and random access memories. More recently, $TiSi_2$ [26], [27] and $CoSi_2$ [28] have been suggested to replace $MoSi_2$, $TaSi_2$, and WSi_2. Refractory metals W [29], Mo [30], and Al [31] are once again being evaluated as gate metal.

For contacts, Al has been the preferred metal because of a) the ease of processing, b) its ability to reduce native SiO_2 that is present on the silicon surface exposed to atmosphere, and c) its low resistivity. For VLSI and ULSI applications, several other factors that were ignored and effects that were tolerated become important. These are the preservaton of shallow junctions, step coverage, electromigration at increased current densities, and high contact resistance. Each of these is discussed below.

Aluminum-silicon interactions and problems caused by such interactions are discussed in Section 6.6.3. These interactions, which led to the penetration of Al in silicon and silicon dissolution into Al, were tolerated because the junction depths were considerably larger than the metal penetration depths. As junction depths of 2500 A or less become desirable, Al penetration leading to spiking shorts or to contact depths that are a significant fraction of the junction depth can result in reliability problems.

The decreasing area of the contacts also causes step-coverage problems. Windows in the passivating layer are deep, and most metals are deposited using physical vapor deposition techniques (evaporation and sputtering) that do not produce the good step coverage required for these deep holes (see Section 6.4.2).

The increased current density in the reduced-size conductors and the increased chip temperature promote electromigration-induced problems in the presently employed aluminum lines and contacts. Thus, the contact metallization must also be resistant to electromigration (see Section 6.6.4).

Finally, a contact, although stable both with respect to electromigration resistance and junction spiking, could still be poor due to high contact resistance that could result from the presence of an interfacial layer between the silicon and the contact metallization. Thus, elimination of this interfacial layer is essential for good contacts.

Contact failure can thus be a result of junction spiking, incomplete step coverage, electromigration-induced open circuits, and high contact resistance. Numerous investigations, aimed at providing understanding of these phenomena, have been carried out. Simultaneously several possible solutions to the contact problems have been considered. These include use of: a) dilute Si-Al alloy; b) polysilicon layer between source, drain, or gate and top-level aluminum; c) tungsten that is deposited by CVD methods in such a manner that the metal is deposited only on silicon and not on oxide; and d) a diffusion barrier between silicon and aluminum, using a silicide, nitride, carbide, or a combination thereof. The use of self-aligned silicide, such as PtSi [4] employed for almost two decades, guarantees extremely good metallurgical contact between silicon and silicide. Silicides have been recommended in processes where shallow junctions and contacts are formed at the same time [32].

Even use of a silicide as contact metallization will require a diffusion barrier to protect the silicide from interaction with aluminum to be used as the top metal. Aluminum interacts with most silicides in the temperature range of 200 – 500°C. Nicolet [33], citing the high chemical and thermodynamical stability of transition metal nitrides, carbides, and borides, recommends that use of these compounds as diffusion barriers between silicide (or silicon) and aluminum. Among other suggested barriers, Ti-W alloy [34] has been used in various metallization schemes.

The most important requirement of an effective metallization scheme in SIC is that the metal must adhere to the silicon in the windows (or polysilicon or polysilicon/silicide on gate) and to the oxide that defines the window. In this respect, metals, e.g., Al, Ti, Ta, etc., that form oxides with a heat of formation higher than that of SiO_2 are the best. This is the reason why Ti is the most commonly used adhesion promoter. Titatinum reduces SiO_2 at relatively low temperatures and forms a strong bond with it and thus acts as the desired "glue" layer. On the other hand, metals like Au, Pt, W, and Mo that form oxides with a heat of formation lower than that of SiO_2, do not reduce SiO_2 and therefore have poor adhesion. It is for this reason that W and Mo, although recommended for MOS gate metal application more than two decades ago, have not found this type of application.

TABLE 6.6: Properties of interest of various metallizations

Metal or alloy	ρ^{a} ($\mu\Omega$-cm)	$T_m{}^{b}$ (°C)	α^{c} (ppm/°C)	Reaction with Si at (°C)	Stable on Si up to (°C)
Al	2.7-3.0	660	23	~250	~250
Mo	6-15	2620	5	400-700	~400
W	6-15	3410	4.5	600-700	~600
$MoSi_2$	40-100	1980	8.25	–	>1000
$TaSi_2$	38-50	~2200	8.8-10.7	–	≥1000
$TiSi_2$	13-16	1540	12.5	–	≥950
WSi_2	30-70	2165	6.25, 7.9	–	≥1000
$CoSi_2$	10-18	1326	10.14	–	≤950
$NiSi_2$	~50	993	12.06	–	≤850
PtSi	28-35	1229	–	–	≤750
Pt_2Si	30-35	1398	–	–	≤700
HfN	30-100	~3000		450-500	450
ZrN	20-100	2980		450-500	450
TiN	40-150	2950		450-500	450
TaN	~200	3087		450-500	450
NbN	~50	2300		450-500	450
TiC	~100	3257		450-500	450
TaC	~100	3985		–	–
TiB_2	6-10			>600	>600

[a] ρ = resistivity, typical thin film value.
[b] T_m = melting point.
[c] α = Linear thermal expansion coefficient.

6.2.2 Properties

Properties of various metallizations of interest are compared in table 6.6. Resistivity values given in the second column are typical values for the films deposited in the thickness range of 100 Å to 10,000 Å. Lower resistivity values belong to purer, thicker, and large-grained (or single crystalline) films. This is because impurities, grain boundaries, and surfaces all scatter electrons and thus increase resistivity (see Section 6.1.1). For silicides, nitrides, carbides, and borides, the variation in composition will also affect the electrical resistivity. Deviation from stoichiometry of the compound leads to defects such as vacancies or interstitials that behave like impurities and increase resistivity. Thus, for metallization applications, the resistivity of a given film may slightly vary from one deposition to another unless all deposition parameters are controlled to yield identical films every time. Finally, the resistivity of the deposited film may change when subjected to a higher temperature anneal. Such anneals may lead to impurity contamination, interaction with the local environment, grain growth, or simply annealing of the defects. Most as-deposited films will exhibit an initial reduction in defects and increase in grain size during low-temperature anneals, leading to lower resistivities. This occurs prior to any chemical or metallurgical interaction. Interaction may cause resistivity to decrease in cases where the resulting product has lower resistivity and to increase in cases where the resulting product has higher resistivity. Titanium's resistivity increases when it absorbs oxygen or reacts with it to form an oxide. Also, if during high temperature anneal the metal dissolves small amounts of foreign material such as silicon, phosphorus, arsenic, or boron, all possible ingredients of the substrate silicon, the resistivity will increase (see Section 6.1.6). On the other hand, when titanium reacts with silicon to form $TiSi_2$, the resistivity is lowered. For other metals, however, the resistivity increases due to a silicide formation.

Grain growth, annealing of defects, and interactions in the solid state are all diffusion controlled (see Chapter 4). Solid-state diffusion is appreciable only at temperatures that are larger than one-third the melting point of the solid in which diffusion is occurring. The melting points of various materials used for metallization are given in the third column of table 6.6. Reaction temperatures with silicon and the highest temperature at which these materials are considered stable on silicon are given in the fourth and fifth columns, respectively. Note that the reaction temperatures are very low compared to the melting points of the metal-silicon intermetallics, pure Mo, or W. This can be understood if one considers the melting point of the silicon, 1410°C, and the fact that the temperatures cited in this table refer to thin polycrystalline metal films on silicon. Diffusion, leading to reaction, in polycrystalline films can occur, even at low temperatures, via grain boundaries and dislocations.

Thin films on any substrate are in a condition of stress determined by two factors. The first, called the *intrinsic* stress is related to the lattice mismatch between the substrate and the film, the film structure and purity, and the defects in films. Some of these stresses can be changed by annealing. The second factor that causes stress is the difference between thermal expansion coefficients of the film and the substrate. The larger the difference in the thermal coefficients and the larger the change in temperature, the greater the stress, silicides which in general have large thermal coefficients compared to Si, SiO_2, or polysilicon substrates and are generally formed at high temperatures, are formed with large stress conditions. The higher the temperature

TABLE 6.7: Processing capabilities of various metallizations

Metal or Alloy	Deposition Method[a]	Patterning [b]	Oxidation for films on Si	Oxidation for films on SiO_2	Stability of films On SiO_2	Stability of films With Al as top metal
Al	E,S	DE	Yes	Yes	Good	—
Mo	E,S,CVD	DE	No	No	Poor	500
W	E,S,CVD	DE	No	No	Poor	500
$MoSi_2$	CD(E,S,CVD)	DE	Yes	No	OK[d]	500
$TaSi_2$	CD(E,S,CVD)	DE	Yes	No[c]	OK[d]	500
$TiSi_2$	CD(E,S,CVD),R	DE,SA	Yes	No[c]	OK[d]	500
WSi_2	CD(E,S,CVD)	DE	Yes	No	OK[d]	500
$CoSi_2$	(E or S)+R	SA	Yes	No	OK[d]	400
PtSi	(E or S)+R	SA	Yes	No	OK[d]	250
TiN	S,R.S.	DE	No	No	Good	450

[a] E = evaporation, S = sputtering, CVD = checmical vapor deposition, CD = codeposition, R = reaction with Si, R.S. = reactive sputtering
[b] DE = Dry etching, SA = self-aligned
[c] Silicide oxidizes to form solid oxides (on surface) that may prevent further oxidation.
[d] In inert ambients.

at which the silicide is formed, the larger will be the stress in the resulting film. Because the stress in the film must be balanced by an opposite stress induced in the substrate, the stress in the film will not only determine the mechanical stability of the film, but it may affect the electrical properties of the substrate and the devices formed on such substrates. Thus knowledge of the film stresses is essential in the proper utilization of a particular film in integrated circuits.

In addition to the above mentioned desirable properties, the chosen metallization scheme must be one that can be implemented in a processing sequence. Table 6.7 compares the processing capabilities of various metallizations. Section 6.4, 6.5, and 6.6 discuss these in detail and Section 6.2.3 discusses the stability on semiconductors and insulators.

6.2.3 Stability on Semiconductors and Insulators

The last column in table 6.6 gives the highest temperature applicability of various metallization materials. These temperatures were found through experience in using these materials on silicon. At higher temperatures, either the materials react with silicon or they lose their character and properties due to decomposition, agglomeration, or mechanical failure. For example, a Pd_2Si film will agglomerate at temperatures above 700°C, a tungsten film will react with silicon at about 600°C, and all nitrides will react with silicon above ~500°C, leading to the decomposition of the metal-nitrogen bonds. Thus the use of these metallizations on silicon (or polysilicon) above the cited temperatures is not recommended.

The stability of various popular metallizations on SiO_2 is summarized in table 6.7. Only Al and TiN are stable on SiO_2 because both Al and Ti reduce SiO_2 to form interfacial metal-oxide bonds that promote adhesion and stability. Al and TiN are unstable at high temperatures and in oxidizing ambients, respectively. This is due to the low melting point of Al and the very high oxygen affinity of Ti. WSi_2, $MoSi_2$, $CoSi_2$, and PtSi adhere well to SiO_2 surfaces, but all have poor stability when exposed to oxidizing ambients at high temperatures. These silicides readily decompose, in such ambients, to form SiO_2 and metal. On continued oxidation, they form SiO_2 and metal oxide. $TiSi_2$ and $TaSi_2$ on Si, on the other hand, are oxidized to form silicon and metal oxide, and, on continued oxidation, to form SiO_2 and metal oxide. Silicides on silicon are stable in all ambients up to temperatures given in table 6.6.

The last column in table 6.7 gives the temperature to which various metallization/aluminum composites can be taken to without noticeable change in the properties. Aluminum is a natural choice for top-level metal. Thus it is used on top of all metallization schemes. After the deposition of an Al film, the structure is generally subjected to an anneal at 400-450°C to promote adhesion and to eliminate electrical damage associated with the metal deposition process (see Section 6.4). As can be seen, except for $CoSi_2$ and PtSi, all other selected metallization materials will survive such anneals. $CoSi_2$ and PtSi react with aluminum at 400 and 250°C, respectively. Thus when using aluminum on top of these silicides, a diffusion barrier of a nitride, carbide, or a refractory metal will be necessary to stabilize the structure at 400-450°C. We shall discuss this again in Section 6.6.

6.3 Characterization of Thin Metallic Films

Thickness, purity, grain size and grain-size distribution, stoichiometry of the alloy, stress, adhesion, surface morphology, thermal expansion coefficient, resistivity, and stability are important parameters in determining the performance of a deposited film. Most of these depend on the film deposition method and on subsequent processing treatments. Thus it is imperative to learn about these properties at every step of the processing so that optimum deposition and processing conditions can be defined. There are several methods that can be employed to obtain this information. Optical, transmission, and scanning electron microscopy (OM, TEM, or SEM) provide information on surface morphology, grain size and distribution, and thickness. Auger electron spectroscopy (AES), Rutherford backscattering spectroscopy (RBS), secondary-ion mass spectroscopy (SIMS), and neutron activation analysis (NAA) can be used to give information on composition and purity. All of these four techniques will also provide information as a function of depth. Electron or x-ray diffraction techniques will provide information on the compound or alloy formation, the grain size and distribution, and the stress in the film; careful analysis can also lead to compositional information about the solid solutions. Stress can also be measured by use of the substrate curvature measurements before and after film deposition. Resistivity is generally measured by the use of four-point probe sheet resistance measurements (see Chapter 4). Adhesion, which is seldomly quantified, is most commonly checked by the so-called scotch-tape test. The thickness of the film can be measured by creating a step and measuring the step height using surface profilometer. This tool is also useful in a quantitative evaluation of the surface roughness. Thickness alternatively can be measured with TEM or SEM cross sections, resistance measurements, and in certain cases by use of beta backscattering and ellipsometry. The concept of ellipsometry has been applied in an automated thickness measuring tool called *nanometrics*.

Nondestructive techniques are the best in providing information on a sample. Thus sheet resistance measurements using a four-point probe, curvature measurements using lasers, large area microscopes (OM or SEM), nanometrics, surface profilometers, RBS, and x-ray diffraction have become indispensible tools. For testing adhesion, the scotch-tape peel test could be placed in this category as well.

Sheet-resistance measurements are the simplest to carry out. The wafer is placed under a four-point probe (see Chapter 4), and the sheet resistance is measured directly from the voltmeter that has been set to read this number. One must, however, realize that the probes can penetrate the film. Thus soft and spring-loaded probes are most frequently used. Alternatively, a contactless probe method can be used to calculate the total conductivity of the substrate and the film followed by simple calculation to extract the information about the film [35]. For films on a substrate, the measured sheet resistance R_s is a function of the contributions from all the films and the substrate, i.e.,

$$\frac{1}{R_s} = \frac{1}{R_{s1}} + \frac{1}{R_{s2}} + \frac{1}{R_{s3}} + - - -.$$

This relationship assumes that all the films and the substrate form resistors in parallel and that their thicknesses are uniform across the measurement area. If one of the films is significantly more conducting than the others, the measured R_s will reflect the R_s value of this highly conducting film. For example, consider an aluminum film on SiO_2 film on silicon. Since SiO_2 is an insulator, only the Al film will contribute to R_s measurement. Consider an Al film on $TaSi_2$ on polysilicon. Unless the Al film is very thin, it will dominate R_s measurements since the resistivity of Al is considerably lower compared to those of $TaSi_2$ and doped polysilicon (2.7, 50, and 500 $\mu\Omega$-cm, respectively).

Sheet resistance measurements can also be used to determine the thickness t of a film of known resistivity by use of the simple relationship, $t = \rho/R_s$. Such measurements have also been used to follow metallurgical interactions between metal and silicon or between one metal and another [34, 36]. Limitations of such applications should, however, be carefully reviewed.

Stress in the film can be measured by studying the changes in the radius of curvature of the substrate resulting from film deposition. Stress in the film will deform the substrate. Strain, and thereby stress, can be found by measuring this deformation. This requires the derivation of a relationship between film stress and the measurable deformation. Figure 6.7 illustrates the beam-bending theory applied to derive such a relation. The film exerts a force that is considered here to act at the midpoint of the substrate, at $D/2$, where D is the thickness of the substrate. The plate is assumed to be in equilibrium, which requires that the sum of the forces, ΣF, and the sum of the moments, ΣM, both equal to 0. The force of the film per unit width, F_f, is

$$F_f = \sigma_f t, \tag{6.20}$$

where σ_f = film stress and t = film thickness. Moment from the film is then

$$M_f = \int_0^{D/2} \sigma_f \, t dy = \sigma_f \frac{tD}{2}, \tag{6.21}$$

where y = the distance from the neutral plane.

Assuming that the wafer is uniformly thick, elastic, homogeneous, and thin compared to its radius, the expression for the substrate stress, σ_s,

$$\sigma_s = \frac{E}{1-\nu}\varepsilon_s, \tag{6.22}$$

is valid. Where E and ν are the Young's modulus and the Poisson's ratio of the substrate, respectively, and ε_s is the strain in the substrate. Substrate strain is

$$\varepsilon_s = \frac{y}{R}, \tag{6.23}$$

where R is radius of curvature of the wafer. Moment from the bulk of the crystal is

$$\begin{aligned} M_s &= \int_{-D/2}^{D/2} \frac{E}{R(1-\nu)} y^2 \, dy \\ &= \frac{E}{3R(1-\nu)} \frac{D^3}{4}. \end{aligned} \tag{6.24}$$

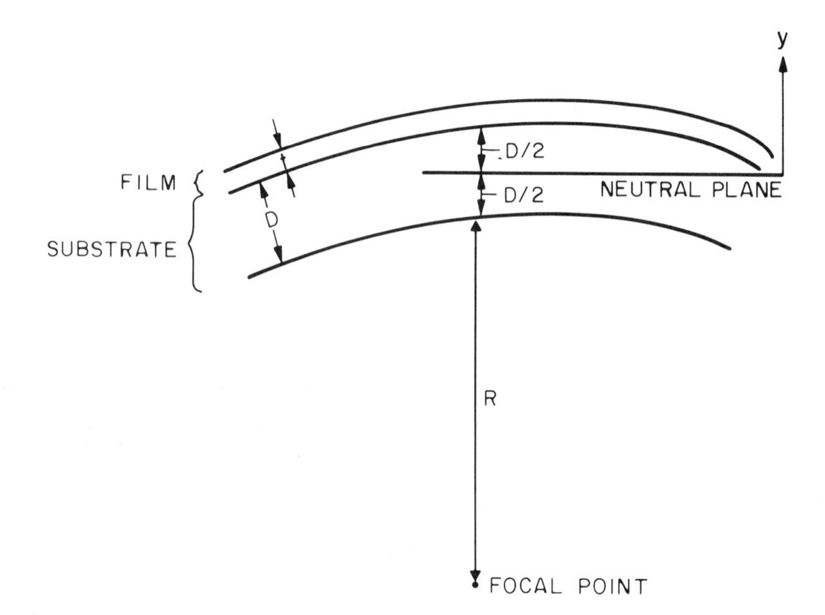

Fig. 6.7. A schematic relationship between wafer thickness (D), film thickness (t), and the radius of curvature (R) of the composite.

Equating (6.21) and (6.24) yields

$$\sigma_f = \frac{ED^2}{6(1 - \nu)Rt} \tag{6.25}$$

E and ν are known; D and t can be easily measured. Now the film stress can be found by measuring the radius of curvature R.

Equation (6.25) is only valid when $t \ll D$. This allows the small force required to bend the thin film to be neglected. If the film were thicker, the deformation would depend on the elastic constants of both the substrate and the film.

This equation was originally derived by Stony in 1909 [37] without the Poisson term, $(1\text{-}\nu)^{-1}$. This term was added by Finnegan and Hoffman [38] to take into account biaxial stresses and the fact that the substrate is a thin plate rather than a beam.

Experimentally it is easy to use two parallel laser beams aimed at the wafer. The application of some simple geometry can be used to find R by measuring the distance between the two reflected beams. The configuration is shown in fig. 6.8. From similar triangles:

$$R = 2L \frac{\Delta X_w}{\Delta d} \tag{6.26}$$

where $\Delta X_w = \frac{1}{2}$ beam separation, $\Delta d =$ deviation of one beam from parallel, and $L =$ wafer-to-screen distance. Note that although the calculation of Δd

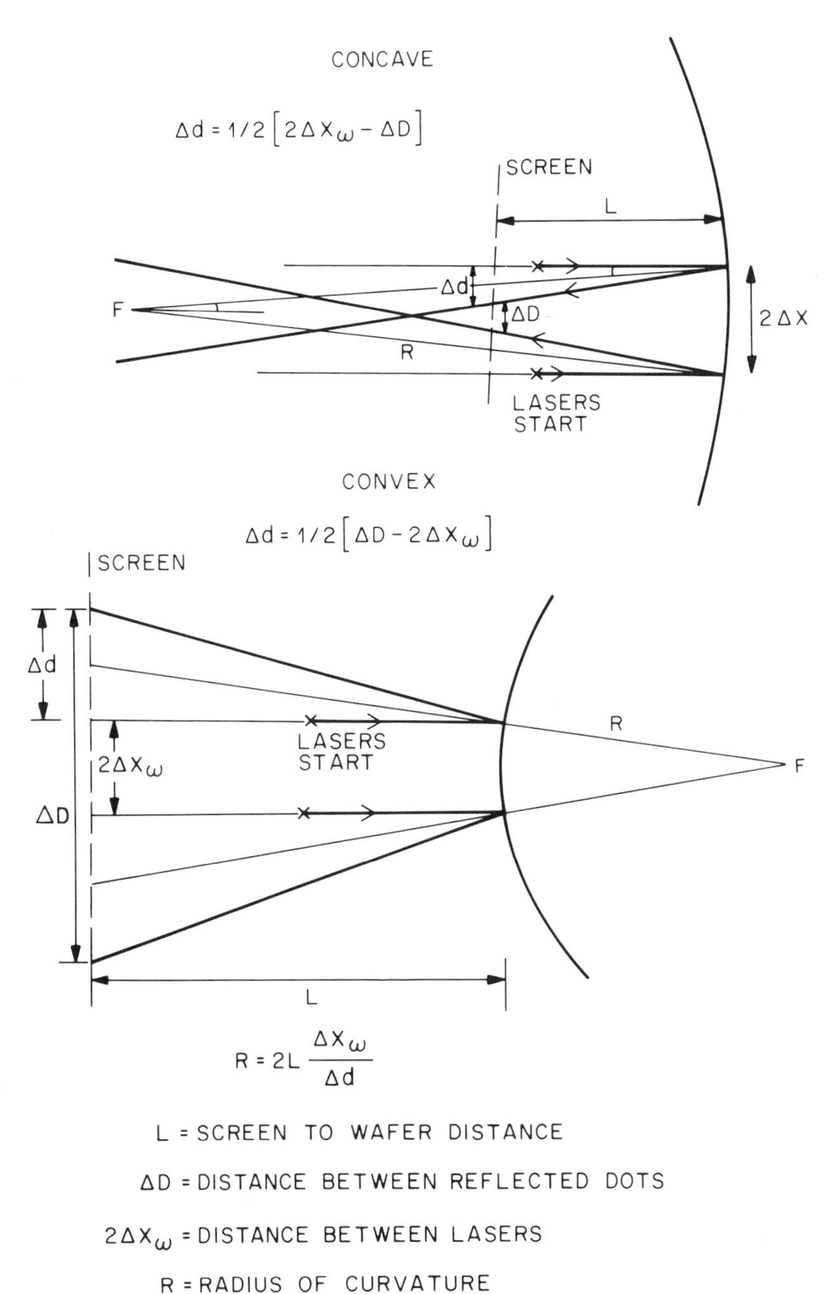

Fig. 6.8. Various factors that are measured in an actual set-up to find the radius of curvature. The top diagram is for concave surfaces, whereas the bottom one is for convex surfaces. Parameters are also defined.

is different depending on whether the wafer is concave or convex, the same relation between Δd and R exists in both cases.

Accuracy [39] of these measurements is generally very good, and results are found to be reproducible within a given level of accuracy. Generally stresses of the order of $(1 \pm 0.5) \times 10^9$ dyn per cm^2 can be measured. Sensitivity and accuracy can be improved by increasing the sample-to-recording-screen distance and by eliminating vibrations and film thickness variability across the wafer.

The method described above is simple and inexpensive. However, it does require lasers and a large substrate-to-screen distance. To eliminate this situation, a single laser beam unit, which is compact and works on electronic detection of the reflected laser beam, was devised [40]. In one of the most commonly used designs, an optically levered laser beam measures the stress-induced change in the radius of curvature. This nondestructive method requires no special sample preparation and minimal wafer handling. The wafer is placed on a stage driven by a constant-speed motor. A laser beam hits the wafer and is reflected back into a detector, which responds by moving. Electronic circuitry records the detector position (y axis) as a function of the stage position (x axis). A trace, representing the curvature of the wafer, is thus obtained. Similar traces before and after film deposition (or sintering) are obtained, and R is then calculated. Figure 6.9 shows typical traces obtained from a) the substrate, b) the substrate plus the as-deposited silicide film, and c) after sintering the silicide film. After the change in the radius of curvature has been found, a simple calculation can determine the stress.

One of the most accurate methods for determining stress is to determine changes in the lattice parameter and line broading using x-ray or electron diffraction techniques. The stress σ is given by

$$\sigma = \frac{E}{2\nu} \frac{a_0 - a}{a_0} \tag{6.27a}$$

or by

$$\sigma = \frac{E}{1-\nu} \frac{a - a_0}{a_0}, \tag{6.27b}$$

where E is Young's modulus, ν is Poisson's ratio of the film, and a_0 and a are the lattice constants of the unstrained and the strained material, respectively. Equation (6.27a) refers to the value of a perpendicular to the film plane, and eq. (6.27b) refers to the case where a is the lattice constant in the plane of the film. The technique, although accurate, is time consuming and requires a good understanding of the diffraction process and the equipment. This technique has therefore lost its appeal for routinely monitoring stress in films.

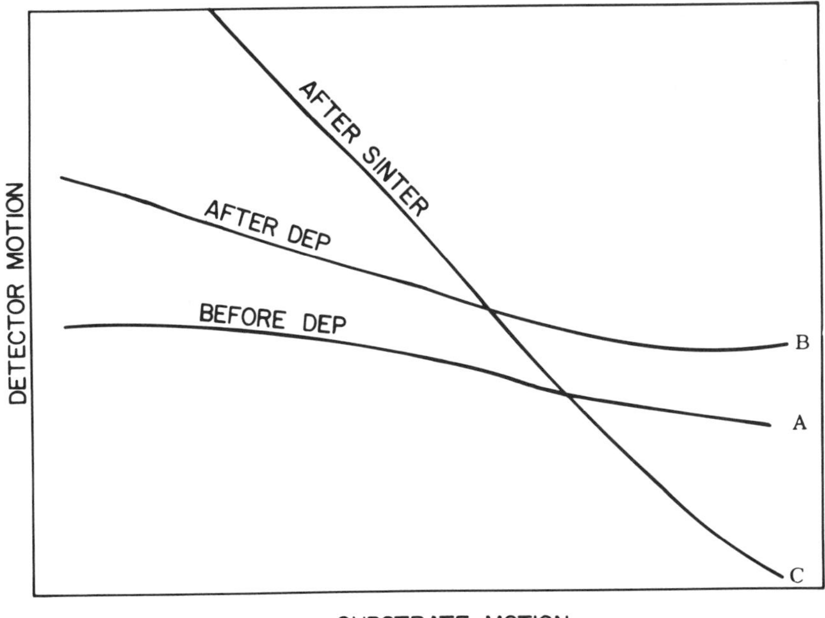

Fig. 6.9. Typical curvature measuring traces obtained by using an optically levered laser beam (a) before film deposition, (b) after film deposition, and (c) after sinter to form a silicide.

Adhesion of thin films to a substrate is difficult to measure. For most applications a qualitative (or semiquantitative) test using scotch tape has been found to be very effective and least time consuming [10]. In this test the tape is applied to the surface of the film and then peeled off. Adhesion to the substrate is considered good when the tape lifts off the film surface clean. On the other hand, if any or all parts of the film lifts off with the tape, the adhesion to the substrate is poor. The test can be made semiquantitative by the use of the variable pull forces during peeling.

X-ray diffraction, a nondestructive technique, provides a quick means of detecting intermetallic formations and making grain-size determinations. The latter information is obtained from the diffraction peak broadening, whereas intermetallic formation is followed simply by locating the positions of the diffraction peaks, obtaining corresponding interplanar spacings and matching them with appropriate known spacings listed in x-ray diffraction handbooks [41]. Information on random or preferred orientation can also be derived by examining relative intensitities of various peaks and matching them with the tabulated relative intensities of peaks of a powder sample.

6.4 Thin Film Deposition

There are two types of deposition processes that are useful. The first one, called *chemical vapor deposition* (CVD), has been presented in Chapter 7. CVD offers several advantages of which three are very important: a) excellent step coverage, b) large throughput, and c) low temperature processing. W and WSi_2 have been very successfully deposited using LPCVD. Mo metal, $MoSi_2$, $TiSi_2$, and $TaSi_2$ have also been deposited by LPCVD process, although the general applicability of such processes has not yet been established. Selective CVD depositions, in which the metallization is selectively deposited only on top of the silicon surface and not on exposed SiO_2 surfaces, are very promising techniques. Laser CVD of metals holds promise for both direct writing and area deposition of metals. Chapter 7 describes these processes in detail.

In this chapter, we shall consider only physical vapor deposition methods, namely, evaporation and sputtering. In both cases, the formation of a deposit on a substrate away from the source consists of three steps: a) converting the condensed phase (generally a solid) into a gaseous or vapor phase, b) transporting the gaseous phase from the source to the substrate, and c) condensing the gaseous phase on the substrate followed by the nucleation and growth of the film. In cases where a compound, such as a silicide, nitride, or a carbide, is deposited, one of the components is a gas and the deposition process is termed *reactive evaporation* or *sputtering.* In both types of deposition, the deposition material is transformed into a gas phase where interactions with the residual gases and sputtering gases become important in determining the property of the deposited film. More detailed information on deposition processes can be found in reference 42.

6.4.1 Deposition Methods

Deposition of films by thermal evaporation is the simplest. In the evaporation method, a film is deposited by the condensation of the vapor on the substrate, which is maintained at a lower temperature than that of the vapor. All metals vaporize when heated to sufficiently high temperatures. Several methods, such as resistive, induction (or radio frequency, RF), electron bombardment, or laser heating, can be used to attain these temperatures. For transition metals, especially the refractory metals, evaporation using an electron gun (e-gun) is very common. For low melting point metals like aluminum, any of the above heating methods can be utilized. Resistive heating is, however, limited in application because of the lower throughput associated with the smaller metal charge generally used in such systems. On the other hand, e-gun evaporations cause radiation damage and, therefore, require a high-temperature postdeposition heat treatment to anneal out the radiation damage. This method is advantageous because the evaporation takes place at considerably lower pressures than the sputtering pressures, and gas entrapment in the film is negligible or nonexistent. RF heating of the evaporating source in an appropriate crucible could prove to be the best compromise for providing both large throughput and a clean environment.

Evaporation methods can also be used to deposit an alloy or a mixture of two or more materials by the use of two or more independently controlled evaporation sources. Experiments are carried out to determine the individual component evaporation rates independently under different conditions. Given this information, the condition is chosen to deposit an alloy or a mixture in the desired composition. This method of coevaporation, which has been used to deposit refractory silicides, however, requires good calibration, which must be maintained during a run and from run to run. Coevaporation from a single source containing the alloy constituents is generally not possible, because fractionation occurs as the evaporation proceeds. Fractionation is the phenomenon in which one component evaporates preferentially, leaving the source rich in the other component or components and thus leading to films richer in the preferentially evaporated component.

In sputter deposition, the target material is bombarded by energetic ions. Some target atoms are knocked off the surface. These atoms are then condensed on the substrate to form a film. Sputtering processes, unlike evaporation, are very well controlled and are generally applicable to all materials—metals, alloys, semiconductors, and insulators. RF, direct current (DC), and DC magnetron sputtering can be used for metal deposition. Alloy film deposition by sputtering from an alloy target is possible because the composition of the film is locked to the composition of the target. The difference between the target and film compositions depends on the type of the equipment, the sputtering parameters, and the alloy constituent. By proper choice of the equipment and sputtering parameters, several complex alloys have been deposited with identical composition in the target and film. Alloys can also be deposited by sputtering metal in a reactive environment. Thus gases such as methane, ammonia or nitrogen, and diborane can be used in the sputtering chamber to deposit carbide, nitride, and boride, respectively.

Sputtering is carried out at relatively higher pressures (0.1 to 1 Pa range). Because gas ions are the bombarding species, the films usually include small amounts of the sputtering gas. These gases in the film are nonreactive except when the sputtering gas is contaminated with chemically active gases. However, these gases cause stress changes, which are a function of the sputtering conditions [42]. Sputtering is a physical process in which the deposited film is also exposed to ion bombardment. Such ion bombardments cause so-called sputtering damage, which leads to unwanted charges and internal electric fields that affect device properties. Such damages are generally corrected by annealing at relatively low temperatures (500°C).

6.4.2 *Deposition Equipment and Process*

The decision to select or build a given vacuum system depends on the specific deposition needs. A mechanical pump, usually called a *roughing pump*, brings the system from atmospheric pressure to about 10-0.1 Pa (7.5×10^{-2} to 7.5×10^{-4} torr). Such pressures may be sufficient for low-pressure chemical vapor depositions. An oil-diffusion pump, backed up by the roughing pump, can bring the pressure down to 10^{-5} Pa and, with the help of liquid nitrogen traps, to as low as 10^{-7} Pa. Liquid nitrogen traps are essential for minimizing the oil contamination that streams into the main working chamber. For most sputtering and evaporation systems a combination of a liquid nitrogen trap, an oil-diffusion pump, and a mechanical pump should be adequate. For faster pumping an oil-diffusion pump can be replaced with a) a turbo-molecular pump, which could bring the pressure down to 10^{-8} -10^{-9} Pa or b) a cryopump or sputter-ion pump, which are capable of bringing the pressure down to 10^{-9} Pa. The latter group of pumps are oil free and are recommended where oil contamination must be avoided, such as in an MBE system.

In choosing a vacuum pump system, the pumping speed of the pumps and the resistance offered by the plumbing between the main chamber and the pump must be carefully evaluated. The pumping speed is defined as the ratio Q_a/p in units of liters per second at 20°C. Q_a is the quantity of the gas that flows through the intake cross section of the pump in a unit time and is pressure dependent. p is the partial pressure of the gas at a point near the intake port of the pump. Most pumps operate at a constant speed in a large pressure range above and below which the speed drops very quickly. Thus during the early pumpdown period (from atmospheric pressure) the pumping speed is low. The pump downtime t in seconds can be calculated using the following equation [42]

$$t(s) = \frac{276V}{S_o + S} \ln \frac{p_o}{p} \tag{6.28}$$

where V the volume of the vacuum chamber, S_o and S are the pumping speed in litres per second at starting and final pressures p_o and p, respectively. Values for S_o and S can be obtained from the pump manufacturer.

Since pumping speed is determined at the intake cross section of the pump, the pump intake should be as close to the main vacuum chamber as possible. This is not possible because of the need to isolate the chamber from the pumping station. To optimize pumping, therefore, interconnecting tubes and valves should be as large as possible. In addition, the outgassing from the interconnections and the main chamber will increase the pump down-time and ultimate achievable pressure. Outgassing is temperature dependent and this effect is utilized in minimizing the effect of the outgassing. This is done by heating or baking the chamber to temperatures higher than operating temperatures while pumping is continued. After an ultimate pressure at high temperature is achieved, the baking is discontinued, leading to a significant improvement in the final ultimate pressure. More on outgassing is presented later.

Besides the pumping system, the need for pressure gauges and controls, residual gas analyzers, temperature sensors, the ability to clean the surface of the wafers by backsputtering, contamination control, and gas manifolds and that of automation should be carefully evaluated. These are discussed in detail in reference 42.

Physical vapor deposition of thin metallic films is carried out in a vacuum chamber [42]. Typical equipment is shown schematically in fig. 6.10. Generally the equipment can be divided in four sections: a) the main vacuum chamber with the evaporation source (or sources) or sputtering target (or targets), substrates on the substrate holders, and a shutter mechanism to isolate substrates from the sources or targets; b) a vacuum system, which may consist of a liquid nitrogen trap nearest to the main valve isolating the chamber from the pumping system, an oil diffusion, turbomolecular, closed-cycle cryogenic pump, or an ion pump and a roughing pump (mechanical pump); c) a gas-handling system for sputtering and reactive depositions; and finally d) all the necessary power supplies and electronic switching, routing, and control systems. In better but costlier systems an additional chamber is placed between the main chamber and vacuum system, each isolated by gate valves. This is called a *loading chamber.* The substrates are loaded in or unloaded from this chamber, leaving the main chamber— the deposition chamber—in high vacuum. This arrangement eliminates the exposure of the deposition chamber to airborne contamination during each loading and unloading cycle and minimizes oil-vapor contamination from the roughing pump. Oil-vapor contamination is severe during the early part of evacuation from atmospheric pressure. In addition, the presence of an additional space in the loading chamber allows for certain vacuum processing such as annealing without disturbing the main chamber.

For high-quality film deposition, one must minimize the carbon (from oil and other organic matters), nitrogen (from air), sodium (generally human related), and sputtering gas contamination. Use of oil-free vacuum systems and/or liquid-nitrogen coldtraps will minimize carbon contamination. Both carbon and moisture contamination are reduced significantly by not permitting their adsorption onto the chamber walls. This is generally achieved by maintaining the chamber walls at a temperature above 100°C during pumpdown to high vacuum and at a low temperature ($<25°$C) during actual deposition process. During pumpdown, heating of chamber walls leads to desorption of trapped gases that are pumped out. During deposition, cold walls and fresh deposits forms on the wall trap impurities by adsorption, minimizing the impurity's presence in the deposited film. Moisture and

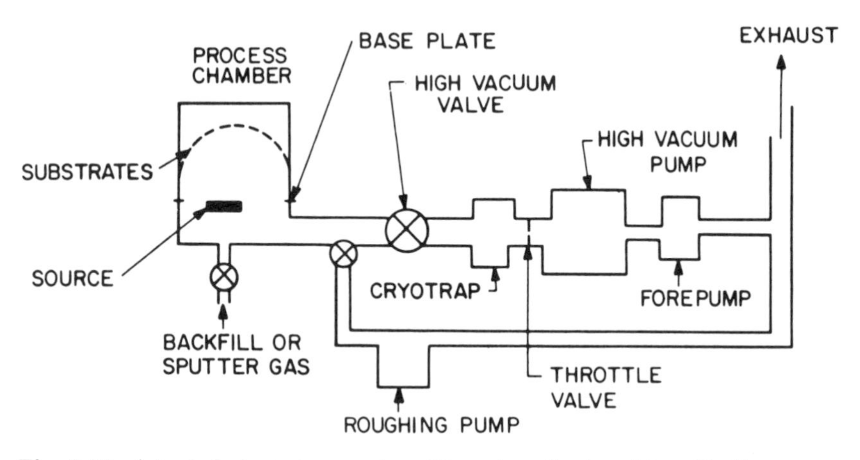

Fig. 6.10. A typical physical vapor deposition set-up for depositing thin films.

oxygen contamination are also minimized by using the purest gases available and sometimes by the use of a titanium sublimation pump. Nitrogen contamination is eliminated by the use of nitrogen-free gases to purge the system. Sodium contamination generally arises from human touch or sodium-contaminated cleaning solutions. A good acid clean of the chamber and of the substrates eliminates this problem. Sputter-gas contamination can be reduced by optimization of the sputtering condition.

To deposit good-quality films reproducibly, one must characterize the evaporation or the sputtering equipment (for the given film depostion) very carefully. Deposition powers and rates, pressure, temperature, substrate biasing conditions, and substrate-to-target spacing all must be characterized by depositing films under various conditions and measuring the film characteristics. Each experimental equipment has its own characteristics. This leads to considerable variation in film properties deposited under seemingly identical conditions.

Figures 6.11 and 6.12 show, for example, the effect of deposition variables on the properties of Pt films deposited on silicon substrates [43]. Figure 6.11 shows the resistivity of the film as a function of a) argon sputtering pressure, b) forward power (RF) on the target, c) bias voltage on the substrate, and d) substrate temperature. Figure 6.12 shows the stress in the film as a function of the first three of these factors. Thus, after a careful evaluation of the film deposition and properties, the optimum conditions can be adopted to deposit the best films. Similar calibration and optimization experiments are carried out for pure metal films or alloy films.

For the deposition of silicide films, cosputtering from two or more elemental targets or one single silicide target is carried out. For cosputtering using two or more targets, a sputtering chamber that can accomodate independently powered sputtering targets is needed. The deposition rate from

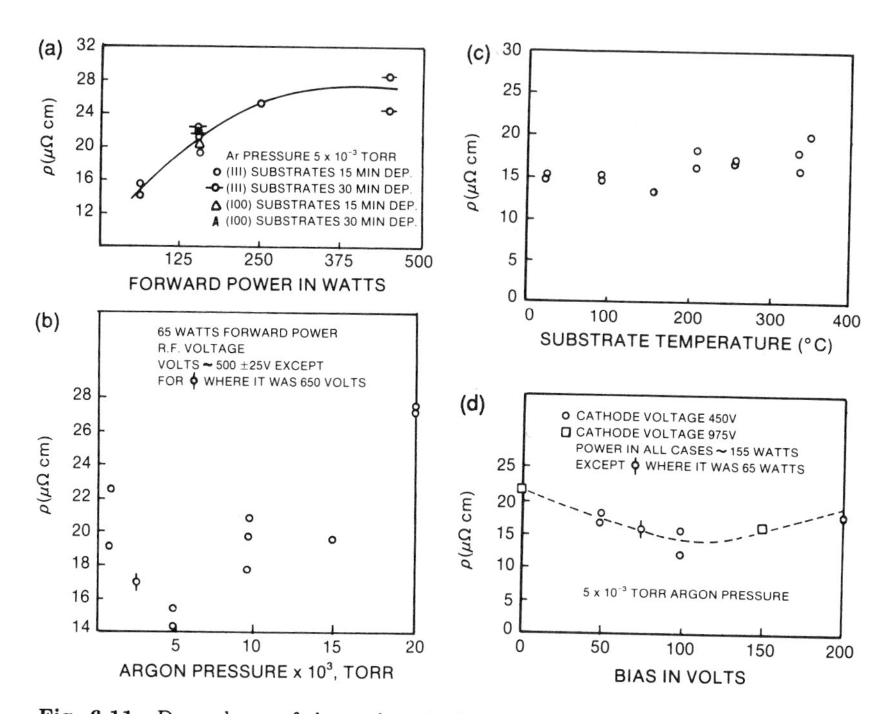

Fig. 6.11. Dependence of the as-deposited resistivity of the RF sputtered platinum films as a function of (a) forward power on target, (b) sputtering argon pressure, (c) substrate temperature, and (d) substrate bias.

the individual metal and silicon targets for a given sputtering condition is determined. These rates are then used to determine the desired Si : M atomic ratio in the deposit [44]. Independent calibration of the Si : M ratio in the film is, however, necessary because mutual scattering of the species and their mean free paths are affected by deposition parameters leading to a variation from the conditions determined from independent elemental calibrations. In addition, codeposited films will require the optimization of experimental conditions, as shown for pure elemental film in figs. 6.11 and 6.12.

A control of conductive film thickness is essential, because a film thinner than desired can lead to excess current density and device failure during operation. Conversely, excessive thickness can lead to difficulties in etching. The use of thickness monitors is common in evaporation deposition and in magnetron sputter deposition, where planetary systems support the substrates [45]. In some magnetron deposition processes, the film is deposited without monitoring during the deposition but is checked after the deposition.

The most common thickness monitor is a resonator plate made from a quartz crystal. The plate is oriented relative to the major crystal axies so that its resonance frequency is relatively insensitive to small temperature

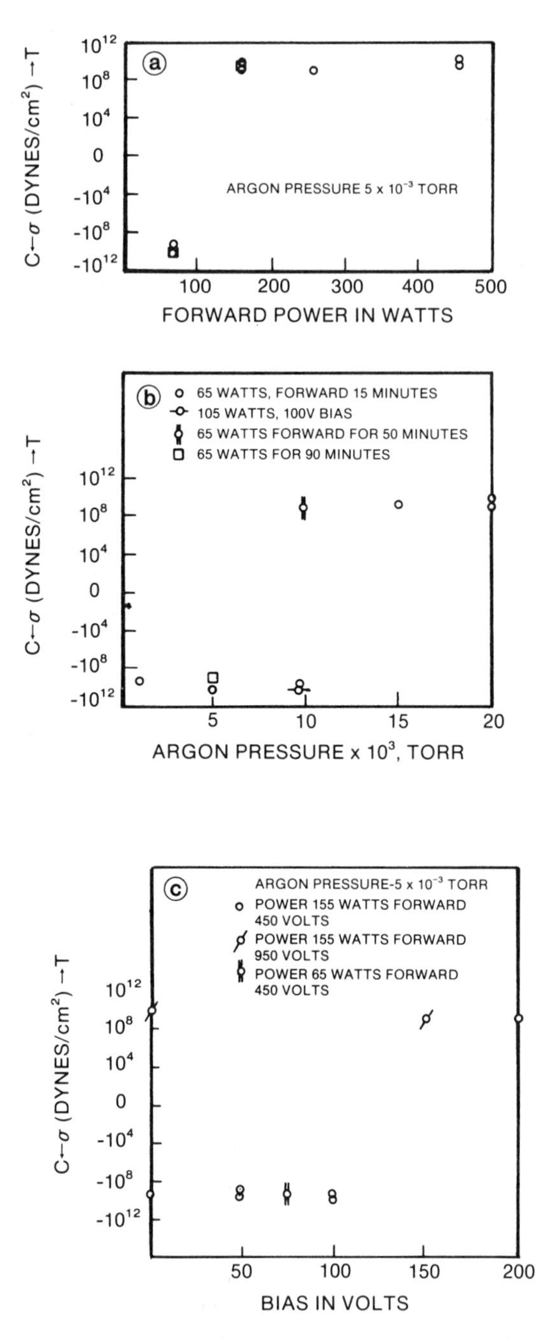

Fig. 6.12. The as-deposited stress of the RF sputtered platinum films as a function of (a) forward power on target, (b) sputtering argon pressure, and (c) substrate bias. C and T, respectively, represents compressive and tensile stress conditions.

changes. The acoustic impedance and the additional mass of any film deposited on the resonator cause a frequency change that can be measured accurately. After calibrating the monitor in the deposition system, it may be used to control the deposition rate as well as the final thickness of the deposited film. The resonator crystal has a finite useful life and must be replaced; however, no recalibration is necessary if the deposition system has not been modified. The resonator has a finite useful life because $\Delta f \alpha \Delta m$ holds true for $\Delta f / f_0 \leq 0.05$, where Δf is the resonator frequency change, Δm is the additional deposited mass, and f_0 is the initial resonator frequency.

We can measure the thickness of unmonitered films in at least two different ways. The simplest way is to use a microbalance and weigh the substrate before and after the film deposition. The film is assumed to have bulk density ρ_{D}, so that the increase in mass Δ_{M} is related to the film thickness t by

$$\text{Volume} = \frac{\Delta_{\mathrm{M}}}{\rho_{\mathrm{D}}} = At \tag{6.29}$$

and

$$t = \frac{\Delta_{\mathrm{M}}}{\rho_{\mathrm{D}} A} \tag{6.30}$$

where A is the area of the film. The other direct method will be to measure thickness by use of cross-sectional SEM or TEM.

6.4.3 Fundamentals of Physical Vapor Depositions

Although evaporation and sputtering processes are physically very different, certain behavior of the species in the gas phase are governed by the same principles. One is the phenomenon of the atomic or molecular scattering and randomization during their travel from the source to the substrate. Scattering occurs due to collisions with atoms or molecules of all kinds—vapor species and the residual gas molecules in the chamber. Thus scattering is related to the density or pressure of atoms or molecules in the gas phase and is defined in terms of a quantity called the *mean free path* (mfp). The mfp is defined as the average distance of travel between subsequent collisions. From the kinetic theory of gases mfp (λ) is calculated to be given as

$$\lambda = \frac{\mathrm{k}T}{p\pi\sigma^2\sqrt{2}}, \tag{6.31}$$

where k, T, p, and σ are Boltzmann's constant, the temperature in degrees Kelvin, pressure, and molecular diameter, respectively. Thus mean free path is directly proportional to the temperature of the gas and inversely proportional to the pressure and the square of the molecular diameter. At room temperature and for a typical diameter of 3 Å, one obtains

$$\lambda = \frac{1.455}{p(Pa)} \text{ cm.} \tag{6.32}$$

For a typical evaporation chamber with a pressure of 10^{-4} Pa, λ equals $1.455 \times 10^{+4}$ cm, whereas for a typical sputtering pressure of 0.5 Pa, λ equals 2.9 cm. Thus evaporated species can travel very long distances before they collide with anything. Therefore, many molecules get from the source to the substrate unscattered. On the other hand, during sputtering the mfp is small, and atoms or molecules can be scattered several times prior to deposition on the substrate. The scattering probability can then be defined as the fraction n/n_0 of the molecules that are scattered in a distance d during their travel through the gas. n/n_0 is given as

$$\frac{n}{n_0} = 1 - \exp(-d/\lambda). \tag{6.33}$$

Here n_0 is the total number of molecules and n is the number that suffered collision. Then, for the above example, if the source-to-substrate distance is 50 cm (a typical distance in an equipment with a planetary substrate holder), during evaporation in 10^{-4} Pa pressure, only about 0.3% molecules will undergo collision, whereas during sputtering at 0.5 Pa pressure practically all molecules will undergo collision. This means that during evaporation molecular motion is more or less nonrandomized and there is line-of-sight deposition. During sputtering, however, there is considerable randomization of travel direction (unless a bias is applied to provide directionality to charged species), leading to better uniformity of deposition on stepped surfaces.

6.4.3.1 The Cosine Law of Deposition

The evaporation rate R from a clean surface is related to the equilibrium vapor pressure p_e of the evaporating species by the Langmuir expression

$$R = 4.43 \times 10^{-4}\left(\frac{M}{T}\right)^{1/2} p_e \ g/\text{cm}^2\text{s}, \tag{6.34}$$

where M is the molecular weight of the evaporating species in grams and T is the temperature in degrees Kelvin. The directionality and spatial distribution of the evaporated species in the evaporating chamber can be calculated following Holland's discussion of evaporation from various types of sources [46]. For evaporation from a small-area source and the deposition

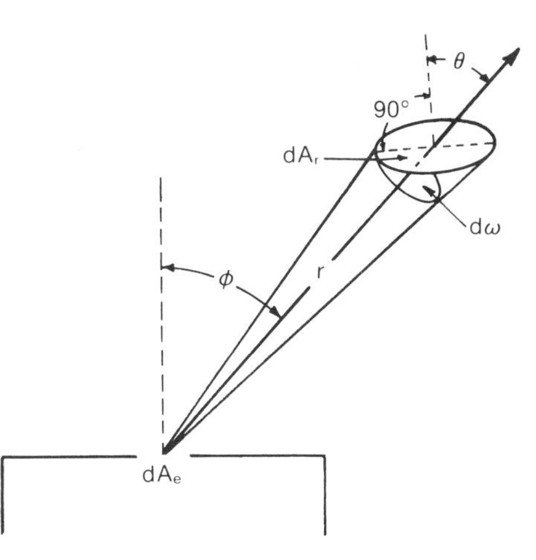

Fig. 6.13. Angular relationship between the substrate and the source. Areas of the source and the substrate are *d*Ae and *d*Ar, respectively. *dW* is the solid angle of the cone as shown.

on a plane receiver, the geometrical relationships are shown in fig. 6.13. In such a case, the mass deposited per unit area is given as

$$R_D = \frac{M_e}{\pi\, r^2} \cos \phi .\ \cos \theta, \tag{6.35}$$

where M_e is the total mass of the evaporated material, and r, ϕ, and θ, are shown in fig. 6.13.

Equation (6.35), known as the *cosine law of deposition*, clearly shows that emission from finite area evaporation sources is not spherically symmetrical, as would occur in the case of a point source. Maximum deposition occurs in directions normal to the emitting surface where cos ϕ is maximum, i.e., ϕ is zero. In practice, small-area evaporation sources and, therefore, eq. (6.35) are realistic. However, if the deposition surface and the evaporation source are both mounted on a spherical surface of radius r_0, it can be shown that $\cos\phi = \cos\theta = r/2r_0$ so that the mass deposited per unit area is now given by

$$R_D = \frac{M_e}{4\pi\, r_0^2}. \tag{6.36}$$

Thus the amount of deposit should be the same at all points on the spherical surface. This is the reason that most modern deposition systems use planetary wafer holders.

6.4.3.2 Impurity Trapping During Deposition

There are several sources of impurities, the deposition source and the gaseous environment being the most important ones. Each source releases impurities in the gas phase. The impurities in the deposition source and those in the environment bombard the surface of the growing film and become trapped during continued deposition. The fraction f_i of the species i trapped in the film is given by [42]:

$$f_i = \frac{\alpha_i N_i}{(\alpha_i N_i + R)}, \tag{6.37}$$

where N_i is the number of atoms of the species i bombarding per unit area per unit time, α_i is the effective sticking coeffient of species i during deposition, and R is the deposition rate of the film.

In the case of sputtering, where a bias may be applied to the substrate, eq. (6.37) is modified [42].

$$f_i = \frac{\alpha_i N_i - (j/q)(\mathrm{As} - \beta)}{\alpha_i N_i - (j/q)(\mathrm{As} - \beta) + R}, \tag{6.38}$$

where

$$A = \frac{\alpha_i N_i + \beta j q^{-1}}{\alpha_i N_i + j(s + \beta) q^{-1}} \tag{6.39}$$

and β is the fraction of the bias current due to the impurity ions, j is the bias current density, q is the electronic charge, and s is the sputtering yield for the impurities.

It is clear from eqs. (6.37) and (6.38) that to reduce contamination the deposition rate should be increased to its maximum and α_i and N_i should be reduced. Application of bias during sputtering will increase or decrease the impurity contamination depending on the contribution of impurities to the bias current. For generally applied negative bias, positively ionized impurities will be trapped, whereas negatively ionized impurities will be repelled.

6.4.3.3 Nucleation and Growth

Other important processes must be considered in order to achieve a complete understanding of film growth. Film growth can be considered as condensation, nucleation, and growth, the events that occur in sequence. Molecules from the gas phase interact with the substrate. When the kinetic energy is small, the impinging molecule is absorbed on the surface with a certain probability of getting desorbed. Adsorption is favored when the binding energy between the adsorbed molecule and surface is large, as occurs when some type of chemical reaction happens. If the binding energy is small, the nucleation of the thin film may never occur, depending on the temperature and vibrational energy on the surface. Nucleation of the film will occur when adsorbed atoms diffuse on the surface and form aggregates. Growth of the film will occur when the nucleus size exceeds a certain critical size that is determined by the condensate-species interfacial free energy and the free energy per unit volume of the growing phase. The formation of aggregates and critical nuclei during condensation from the vapor phase has been theoretically treated in several ways. The simplest concept revolves around the growth of aggregates leading to a change in the free energy of the system. During growth two opposing forces are at work. One of these is due to the surface free energy, which increases as the surface area of the aggregate increases. The other is due to the free energy difference between the condensate (the solid) and the vapor species. This second factor decreases the free energy, and the free energy decrease is directly related to the aggregate volume. Additionally, the larger the supersaturation in the vapor phase, the larger the decrease due to the volume term. Thus the total free energy during an aggregate formation is given by

$$\Delta G = \Delta G_{\mathrm{s}} + \Delta G_{\mathrm{v}}. \tag{6.40}$$

ΔG_{s}, the positive surface contribution, for a spherical aggregate of radius r is given as

$$\Delta G_{\mathrm{s}} = 4\pi r^2 \sigma, \tag{6.41}$$

where σ is the condensate-vapor interfacial energy per unit area. ΔG_{v}, the total volume free energy, could be written for a spherical aggregate as

$$\Delta G_{\mathrm{v}} = \frac{4}{3}\pi r^3 \, G_{\mathrm{v}}, \tag{6.42}$$

where G_{v} is the free energy per unit volume. G_{v} is a negative term. Thus total ΔG is then given as

$$\Delta G = 4\pi r^2 \sigma + \frac{4}{3}\pi \, r^3 \, G_{\mathrm{v}}. \tag{6.43}$$

It is apparent from this equation that during the formation of aggregates (or nuclei) the surface term dominates when r is small and the volume term dominates at a larger value of r. The maximum in ΔG will occur at a value r_{c}, called critical radius, given by

$$r_{\mathrm{c}} = -\frac{2\sigma}{G_{\mathrm{v}}}. \tag{6.44}$$

Thus the growth of nuclei from a zero radius to a radius of r_{c} is not favored thermodynamically. However, once a nucleus of radius r_{c} is formed, further growth leads to a decrease in free energy and hence further growth is favored. More on the nucleation process is discussed in the Chapter 7.

For a circular disc of radius r and height h, one can similarly calculate a critical nuclei radius (of the given height or thickness) as

$$r_{\mathrm{c}} = \frac{-h\sigma}{2\sigma + hG_{\mathrm{v}}}. \tag{6.45}$$

This critical nuclei equation can be used for the growth in lateral direction.

The above simple thermodynamic theory explains the nucleation and growth behavior, except for the fact that thermodynamic concepts are applied to a few atom size aggregates. For other theories, where individual atoms and bonding between such atoms (to form nuclei) are considered, reference 42 should be examined.

6.4.4 Optimization of the Deposition Process

Deposition processes for metallization need to be tailored for each specific device processing scheme. For example, for gate metallization, a polysilicon, polysilicon/silicide sandwich, silicide alone, or refractory metal alone may be appropriate. Similarly, for contacts, aluminum, aluminum on a silicide, aluminum on a silicide with a diffusion barrier in between, or aluminum on tungsten could be used. Top metal, generally, is aluminum. Individual or multiple deposition processes must be optimized to yield the best product. Engineering considerations in such applications are related to the mechanical stability of the films, the uniformity of the thickness and the properties in a run and from run to run, coverage of the stepped surfaces, the equipment cost, and the cost per chip. Each of these considerations vary from metal to metal and from substrate to substrate and therefore must be carefully evaluated for each metallization scheme. Sputtering from a single target or multiple targets offers flexibility and versatility and is thus easy to adopt. Evaporation, on the other hand, provides purer films. The use of rotating and planetary substrate holders enhance the uniformity and step coverage. In certain cases, in-situ cleaning of the substrates (by backsputtering) and heating helps adhesion and improves properties of the deposit and the devices.

Backsputter etching of the substrate is frequently employed to remove the native oxide surface layers and other surface contaminants. This is generally carried out by reversing the sputtering process. The substrates and the shutter, which isolates the target from the substrates, are biased in such a way that a controlled sputtering of the substrate surface occurs, leading to removal of undesirable surface layers and impurities.

Backsputter etching is not a panacea for solving all surface contamination problems. In fact, if carried out without proper characterization and calibration, it creates more problems than it is supposed to take care of.

For example, an organic contaminant on the surface, instead of being sputter etched, could polymerize under inert-gas ion bombardment. Energetic inert-gas ions could be imbedded in the substrate. Cross contamination from different materials on surfaces such as SiO_2, Si_3N_4, Si, etc. on a patterned wafer, from the substrate table, and from the shutters may result in unwanted residues on surface. Redeposition of backsputtered material could be an important factor if the backsputtered species are not pumped away effectively, and contamination of the metal target could lead to impure deposits on subsequent film deposition following the sputter etching. In addition, removal of large amounts of materials, generally carried out to ensure surface cleanliness, may lead to a very rough surface [47].

A successful backsputter etching procedure is carefully established for each equipment and varies from machine to machine. To minimize contamination the oil-free chambers, a substrate table made of substrate or film material (generally done by sputter coating the bare table), clean gases, and very low bias (preferably below a hundred volts) on the substrate should be employed. The preferred way of cleaning a substrate surface is to carry out sputter etching in an ultra-high vacuum chamber followed by heating to temperatures above 800°C. Annealing at these temperatures allows desorption of sputter gas and recrystallizes the damaged surface. In most conventional sputtering or evaporation equipment, however, in-situ cleaning methods cannot be adopted. Only simple backsputter etching can be of use in such cases.

6.5 Patterning

VLSI and ULSI applications require very demanding controls on the circuit, and, therefore, on the metallization dimensions. This necessitates the use of anisotropic etching techniques, as discussed in Chapter 10. Practically all wet-chemical etchings are isotropic and therefore not suitable for fine-line patterning. In this section we shall briefly look at dry etching processes used to etch various metallization materials and also give some chemical recipes for etching. The latter are useful in cases where the large-area etching becomes necessary—such as removal of metal for reworking on the wafers.

6.5.1 Dry and Chemical Etching

Reactive ion etching (RIE) processes are carried out in a manner similar to backsputtering. Accelerated ions and neutrals hit the substrate and cause sputtering of the material from the surface, effectively etching the material on the surface. The difference lies in the nature of the gases employed. For backsputtering or sputter etching or ion milling, inert gases are used. For RIE, reactive gases are used, hence the name *reactive ion etching.* Such

TABLE 6.8: Gases for reactive ion etching of metallizations

Metal	Gases
Polysilicon	Fluorine or chlorine-containing gases (CF_4, SF_6, Cl_2, CCl_3F, etc.) with/without oxygen.
Al and Al containing small amounts of Si, Cu, Ti	CCl_4, $CCl_4 + Cl_2 + BCl_3$, $BCl_3 + Cl_2$, $CCl_4 + BCl_3$
Tungsten	Fluorinated gases
Refractory silicides	Fluorinated plus chlorinated gas with or without oxygen
Thin layers of TiN, TiC	Al - etch

gases play a dual role and both sputter etch and react with the material. The reaction product is volatile so that it is pumped out of the system. Because the etching medium is a flux of species directed towards the substrate, anisotropy in etching, i.e., preferential etching in the direction normal to the surface compared to that tangential to the surface, is achieved. Table 6.8 lists popular gases used to reactive ion etch various metals.

Aluminum and aluminum doped with small amounts of additives like Cu, Si, Ti, etc. have been the most difficult materials to dry etch. The difficulty is related to the presence of native aluminum oxide on such materials and the oxidizability of fresh aluminum surfaces in the presence of moisture and oxygen. Fluorine also reacts with aluminum to form stable fluoride on surfaces. Chlorine and chlorine-containing gases are found to etch aluminum. However, the reaction product $AlCl_3$ absorbs water readily. To minimize these problems BCl_3 is added to the reactive gas mixture due to its ability to scavenge water vapor and oxygen. The etch rate of Al in BCl_3 alone is unfortunately very low ($\sim$ few hundred angstroms/min) and therefore BCl_3 is never used alone. The most popular gas mixtures for Al etching are CCl_4/BCl_3, Cl_2/BCl_3, and $CCl_4/Cl_2/BCl_3$.

Fluorine containing gases like CF_4/O_2, SF_6, and NF_3, and chlorine-based gas mixtures like Cl_2, CF_4/Cl_2, CCl_4, and SF_6/Cl_2 have been used for etching refractory metals and their silicides. The etching of single-layer metals (or silicides) is less complicated compared to the etching of multilayer (silicide on polysilicon) structure. Thus the etching of a silicide-polysilicon bilayer sandwich is not trivial. This task becomes even more demanding when the substrate is a thin ($<$250 Å) layer of the gate oxide that must be preserved. Thus the etching of silicide-polysilicon bilayers on very thin oxides requires extremely good control of the process, a good etch selectivity between silicide and polysilicon, and a high degree of selectivity between

TABLE 6.9: Chemical etches for metals

Metal	Chemical etch and conditions
Polysilicon	HF + an oxidizing acid, usually HNO_3 or CrO_3. Various compositions to achieve desired etch behavior. Acetic acid is added to reduce the etch rate and enhance the uniformity across the surface. Also 23.4% KOH (by weight), 13.3% n-propanol, and 63.3% water will remove polysilicon without attacking oxide.
Al and Al containing small amounts of Si, Cu, Ti	$Ch_3COOH : HNO_3 : H_3PO_4 : H_2O$ in 5:5:85:5 mixture at 40-45°C. Does not attack SiO_2 or silicon; will attack Si_3N_4
W	0.25 M KH_2PO_4/0.24 M KOH/0.1M $K_3FE(CN)_6$. Will attack polysilicon or silicon. $HNO_3 : HF$ in 1 : 1 or diluted solution. (Will attack Si, polysilicon, SiO_2, Si_3N_4.)
Mo	$CH_3COOH : HNO_3 : H_3PO_4 : H_2O$ in 5 : 1 : 85 : 5 mixture followed by $HCl : H_2O$ in 1 : 3 mixture
Ti	H_2O_2 : EDTA in 1 : 2 mixture at 65°C, H_2O_2 - HCl, or H_2O_2 - H_2SO_4

polysilicon and the oxide. The requirement of etching vertical walls without undercutting the polysilicon and attacking the gate oxide is extremely severe because of the overetching required to clear both the silicide-polysilicon from the sloped surfaces and also the deposits on the oxide walls. This is because the vertical height of the deposits on such surfaces is much higher than that on plane surfaces.

Table 6.9 lists the popular chemical etches that can be used to etch metals. All silicide etches contain HF and HNO_3 acid in varying ratios. They are not listed here because such etches attack everything else on the substrate and are to be used cautiously for specific experimental purposes only.

6.5.2 Self-Aligned Silicides

The initial thrust of the silicide use in the silicon integrated circuits (SIC) concentrated on using PtSi as the contact metallization [4]. This application was concerned with the reliability of contacts. The formation of the PtSi (and of Pd_2Si) by a self-aligned process yielded two important advantanges: a) the process did not require any additional lithography and etching and alignment was predetermined and b) it resulted in a very clean silicon-silicide interface and thus highly reproducible contacts. The process was easily adaptable in fabrication lines having a sputtering equipment withbacksputter etch capability. Annealing was easy because both platinum and palladium were rather insensitive to small amounts of oxygen. In fact, small amounts of oxygen were added during silicide formation in order to form a thin SiO_2 layer on the silicide surface. The SiO_2 layer improved the selectivity of the chemical etch for the metal over the silicide. PtSi was adopted more readily because of better reproducibility. Contact to the outside world was made using a Ti/Pt or Ti/Pd diffusion barrier between the silicide and gold or aluminum as the outermost layer.

Presently titanium disilicide and a few lower resistivity group VIII metal silicides, namely, those of cobalt and nickel, are being evaluated. The resistivities of these silicides are included in table 6.6. $CoSi_2$ offers the lowest resistivity. A comparison of these self-aligned silicides and their processes is made in reference 48.

The conventional self-aligned silicide process for the contacts is described here. The metal film is deposited over the patterned oxide and silicon in the window. In-situ cleaning, such as by backsputtering in the sputtering chamber, is essential for reproducibility. After metal deposition, the silicide is formed by annealing in the desired ambient at the desired temperature. Finally, unreacted metal over SiO_2 is removed by a selective wet etch that etches metal without attacking silicide or SiO_2.

In order to use the same approach at the gate level, a few changes are necessary. In fig. 6.14, a refined sequence for forming the silicide at gate, source, and drain simultaneously is described. In this sequence polysilicon is deposited on the gate oxide and doped. Then the oxidation mask Si_3N_4 is deposited by the LPCVD process. Now the polysilicon-Si_3N_4 sandwich is defined to form the gate and interconnection pattern. The source and drain are formed by ion implantation, and the photoresist is removed. Following this, oxidation is carried out to form the oxide sidewalls on polysilicon. This heat treatment activates and diffuses the dopants. The oxide is now removed anisotropically from the source and drain region leaving oxide sidewall on polysilicon. The remaining nitride is removed by the selective chemical etch, leaving exposed polysilicon and source and drain surfaces. Metal is then deposited following an in-situ surface cleaning. A silicide is formed by annealing and the unreacted metal is removed by selective chemical etch, leaving silicide at the gate, source, and drain regions and on the interconnection lines. Note that the use of Si_3N_4 on top of the polysilicon film serves two important roles. First, it serves as the oxidation mask for the polysilicon film, allowing the oxidation to occur only on the sidewalls of polysilicon and in source and drain regions. Secondly, it allows the formation of the silicide with excellent step coverage that matches polysilicon step coverage.

A successful application of the self-aligned silicide technology will, however, occur only after a clear understanding of the metal-silicon reaction

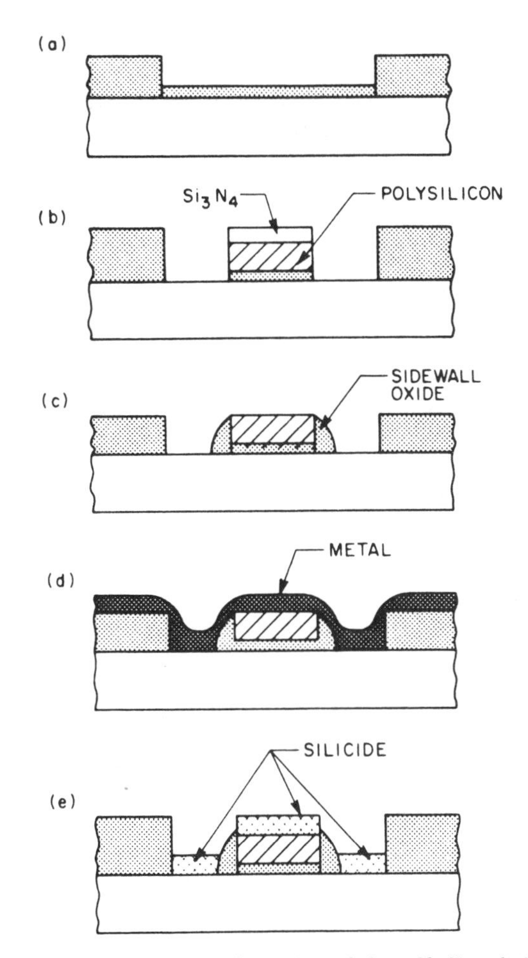

Fig. 6.14. A refined sequence of the formation of the self-aligned silicide simultaneously on gate, source, and drain of a MOSFET structure (see text for details).

kinetics and the role of various factors (such as interfaces, dopants, and impurities) on the reaction kinetics has been realized. On the other hand, silicide formation by a reaction of metal with the doped substrate leads to dopant redistribution between the growing silicide phase and the substrate, possibly affecting the electrical characteristics of the metal-semiconductor contact. The factors that will influence the dopant redistribution between a silicide film and polysilicon or silicon substrate are: a) the diffusivity of the metal, silicon, and dopant in the metal or silicide relative to the silicon; b) the solid solubility of dopant in the various phases; c) the segregation coefficient at the surface and at the silicide/silicon (or polysilicon) interface; d) the surfaces and interfaces; and e) the evaporation and reactive losses to the heat-treating environments. Each of these factors are determined by various parameters that are both related to experimental conditions and to material characteristics, which in turn are not mutually exclusive. Thus implementation of a well-understood silicide-silicon contact system formed by the self-aligned method will require considerable research and development effort.

6.6 Metallization Problems

In this section, various metallization problems associated with a) deposition, b) processing, c) metallurgical and chemical interactions, d) electromigration, and e) device performance are considered. Each is examined separately. Solutions, in some cases, are related to careful experimentation and control. In other cases there are no simple solutions and research continues.

6.6.1 Deposition

Impurities in the film, adhesion, stress, cracks, grain size, incorrect stoichiometry in alloys, poor step coverage, and thickness nonuniformity are all problems associated with deposition. With the exception of the step-coverage problems, all other problems can be solved by tailoring the physical vapor-deposition process. Impurities in the films can be minimized by using pure evaporation and sputtering sources, high vacuum or pure gases, and clean surfaces. Use of an oil-free vacuum system eliminates contamination resulting from hydrocarbons. Surface treatments that reduce sticking coefficients may be formulated to minimize impurity trapping in the film. Adhesion of the film is promoted by a) inducing strong atom-atom bonding (between the film and the substrate) within the interfacial region, b) reducing the stresses to lowest levels, c) the absence of easy deformation or fracture mode such as those caused by surface unevenness and particulates, and d) the absence of long-term degradation modes such as those that may happen due to the exposure to moisture and air. Surface cleanliness and localized deposit-surface interaction will improve localized atom-atom bonding. Deposition parameters are mainly responsible for the stresses, although stresses caused by thermal expansion coefficient differences and intrinsic factors could play an important role.

A thin film deposited on a substrate is in a state of stress that can be tensile (i.e., the film wants to contract) or compressive (i.e., the film wants to expand parallel to the surface). Such stresses arise from several factors and are generally grouped into two components, the thermal stress σ_{t} and the intrinsic stress σ_{i}, where

$$\sigma = \sigma_{\mathrm{t}} + \sigma_{\mathrm{i}}. \tag{6.46}$$

σ_t results from a difference between the coefficient of thermal expansion of the film, α_f, and that of the substrate, α_s. Qualitatively, one material expands with temperature at a different rate from the other. Quantitatively, σ_t, is given by

$$\sigma_t = \frac{E_f}{1 - \nu_f} \int_{T1}^{T2} (\alpha_s - \alpha_f)\, dT, \tag{6.47}$$

where E_f and ν_f are the Young's modulus and the Poisson's ratio of the film, respectively, and T is the temperature.

Equation (6.47) is strictly true only if the coefficients of thermal expansion do not change with temperature (i.e., $d\alpha/dT = o$), which is not the case for silicon. α_{Si} goes from 2.6×10^{-6} °C^{-1} and 25°C to $\sim 3.3 \times 10^{-6}$ °C^{-1} at 900°C.

Intrinsic stress, also referred to as growth stress, is less quantitatively understood. There are a variety of sources of intrinsic stresses in polycrystalline films. Among these are volume changes, grain growth, microstructure, impurities, defects, and film composition. These stresses are low when the temperature is greater than $T_m/3(T_m$ = melting temperature)). At this point, atom mobility is high and atoms readily diffuse along grain boundary paths. Holding the film above this temperature during an anneal will minimize growth stresses.

In sputtered films, the entrapment of the sputtering gases (e.g., argon) causes the development of intrinsic stresses. Thus the sputtering pressure-stress and sputter-rate-stress relationships must be carefully established for each sputtering system and sputtering conditions found that yield the lowest stress condition in the deposits.

Cracks in the deposits are the result of poor adhesion, the presence of particulates or unevenness, and/or stresses in the film. Control of these parameters will eliminate cracks in practically all cases. Cracks may, however, occur following a heat treatment of the films, usually due to a volume change or a chemical interaction.

Grain-size variation in the deposits is of little concern in most cases. However, in cases where the film is subjected to postdeposition anneal or electromigration, the grain-size variation and grain growth may cause failure, as discussed below for aluminum films. Larger grains minimize grain boundary area, and therefore diffusion and electromigration is reduced. Larger grains provide more stability. However, large grain size may lead to surface roughness affecting photolithography and mask alignments.

When depositing alloy films, stoichiometry of the alloy in the film should be controlled. As discussed earlier in this chapter, the use of an alloy sputtering target provides the best control for a given sputtering condition. Varying sputtering rates and chamber pressure has a small effect on the composition of the deposit. However, variability in the composition of the alloy target with depth could lead to compositional variations in the films deposited in different runs.

Step coverage is a serious problem for physical vapor-deposited films. Poor step coverage results because of a) the directionality of the deposition from the evaporating or sputtering sources, b) low mobility of the deposited atoms, molecules, or ions, and c) the enhanced topography (deeper steps) resulting from processing prior to metallization.

The step-coverage problem has been approached in several ways. Raising the temperature of the substrate during film deposition ($\sim 300°C$) creates greater surface mobility of the deposited material, thus reducing the severity of cracks that exist in the corner regions. Furthermore, the orientation of the substrate relative to the source can be optimized [49, 50]. Optimization is especially important since shadowing occurs in the deposition process when using source such as an e beam or an inductively heated melt. Computer simulation techniqes have been useful in finding apropriate ways to modify the supporting planetary system.

Since most planetary systems do not use rotation of the individual substrate about its own axis, orientation within the planet is significant in reducing step-coverage problems [49]. Step edges that are parallel to the planet radius are coated symmetrically. Steps with edges placed perpendicular to the planet radius tend to be coated asymmetrically and also tend to exhibit more cracks (fig. 6.15) [49, 45].

If small contact windows are to be coated, the course of action may be different than outlined above. For VLSI, a plane surface may be approximated by depositing the interlevel dielectric by bias-sputter deposition or by using planarization [51]. Planarization is a low-temperature process that reduces surface features (see section 6.7.1). A thick resist layer is applied to the dielectric, and a plasma-etch process is used that attacks the dielectric and the resist at equal rates. To accommodate this process a thicker than normal (usually by a factor of two) intermediate dielectric layer is needed. The extensive heat treatment that normally would be used to make the dielectric flow, thus reducing the severity of the step contours, cannot be tolerated in VLSI where implanted dopants are not permitted to diffuse extensively. Contact-window step coverage remains a problem, even on planar surfaces, because extensive taper etching of the window edge consumes excessive area.

The use of sources that have larger areas than point sources, such as magnetrons, relieves many of the step-coverage problems. If the substrates are relatively distant (20 to 30 cm) from the source, such as planetary-mounted wafers, the directionality of the sputtered metal vapor becomes more random. Randomness occurs because at pressures of abnout 0.5 Pa the mean free path of the Ar atoms is of the order of 1 cm. Thus the vapor incident on the planetary-mounted substrate during magnetron sputtering is more random in direction than evaporated vapor but the vapor is "colder" because it transfers energy to the Ar gas. The vapor's lower energy leads to less

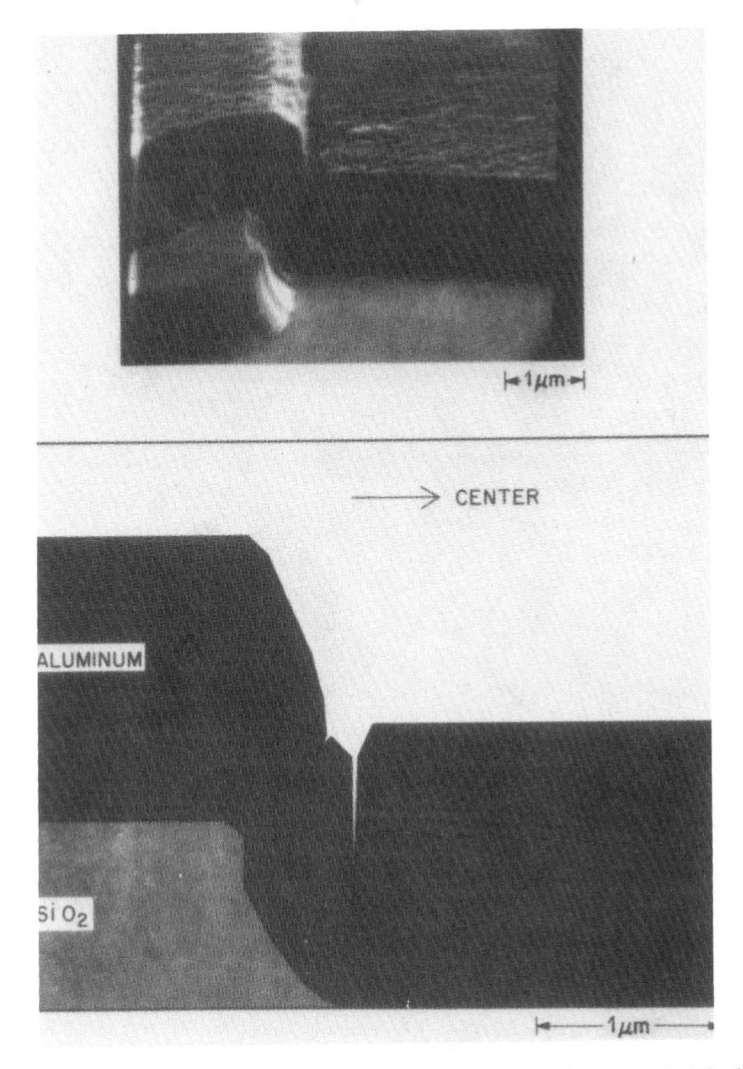

Fig. 6.15. (a) The photograph on top shows what is actually obtained. The bottom sketch shows the prediction by a computer model. The step in SiO_2 is perpendicular to the planet center.

movement of the deposited species on the substrate surface. Decreased movement can limit grain growth and the development of ordered (fiber texture) structures. The substrates can be relatively close and stationary, or they can move slowly before a large-area magnetron. Proximity to the source permits high deposition rates with material that has undergone an order of magnitude less travel through the Ar. Significantly more heating of the substrate can be achieved, resulting in improved step coverage. Sidewall to flat-surface film thickness ratios ranging from 1 : 2 to 1 : 1 have been

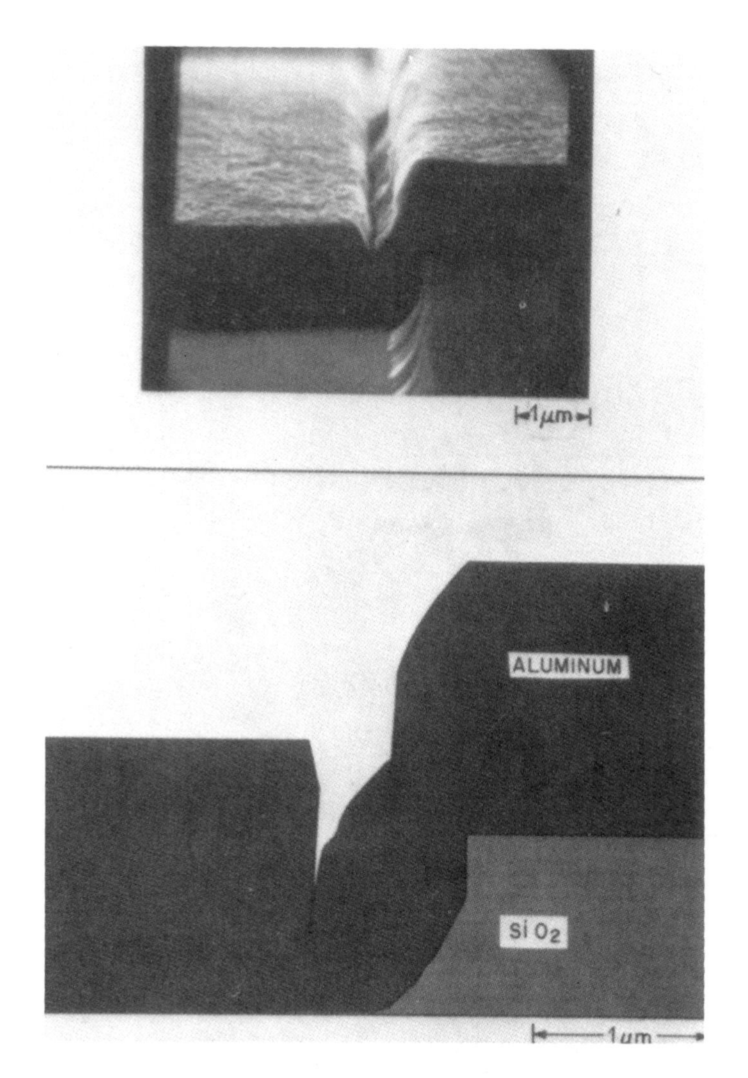

Fig. 6.15. (b) Actual (top) and computer model prediction (bottom) are shown for the step parallel to the planet radius.

obtained on steps. In windows this ratio is dependent on the aspect ratio (depth/width).

6.6.2 Processing

Metallization is applied at different stages of SIC processing. Gate metal is deposited and patterned during early stages. Contact metal is formed towards the end. If contact metal is different from the top-level metal, the

TABLE 6.10: Post-metal processing steps

Gate and Interconnection Metal	Contact Metal	Top Metal
1. Patterning by use of RIE	1. P-glass deposition flow or planarization	1. Metal patterning
2. Silicide anneal	2. Diffusion barrier	2. Si_3N_4 deposition if necessary
3. Sidewall oxide formation by thermal oxidation or CVD followed by selective aniostropic etch	3. Aluminum (top metal) deposition	3. Si_3N_4 patterning
4. Ion implantation for source drain		4. 450° C/H_2 anneal for 30-60 minutes
5. Source and drain diffusion		
6. Contact metallization		

latter is generally applied after the former. Thus a gate metal is subjected to considerable processing, whereas the contact metal experiences very little processing. Post-metal processing steps are listed in table 6.10.

It is evident from table 6.10 that gate metallization is subjected to several high-temperature steps and at least one ion-implantation step. It is also subjected to oxidizing environments. All these lead to stability problems. Polysilicon alone has been used as a metal at this level with practically all problems solved. In recent years the polysilicon/silicide sandwich has replaced polysilicon in VLSI devices. The stability of this sandwich during processing and during device operation is related to the tight control required of the silicide deposition parameters. Each silicide has its limitation [52]. Impurities, especially carbon and oxygen, play varying roles in different silicides. Silicides of Ti and Ta tolerate small amounts of oxygen without noticeable effects on mechanical stability. Other silicides behave differently. Silicides of tungsten and molybdenum require special precautions, especially during oxidation. This is due to the inability of these metals to reduce native interfacial oxide layers on polysilicon. The stability of the refractory silicides can be enhanced by the use of silicon-rich silicides. Excess silicon provides bonding at the interface and improved oxidizability. Silicides of titanium and cobalt, recently recommended for gate applications because of their self-aligned features and low resistivities, need special attention because of their oxygen affinity and low (<900°C) temperature stability, respectively [48].

Contact metallization stability requirements are less severe. Shallow junctions (100-2000 Å deep) require contacts that will not penetrate and ruin such contacts. In addition, dopant redistribution during silicide formation must be controlled or compensated for by reliable junction/contact formation.

Aluminum is the most popular top-level metal. Since its melting point is low, the 350-450°C post-metal anneal leads to various metallugical and chemical interactions, as discussed in Section 6.6.3. Besides these, Al is a very stable material.

6.6.3 Metallurgical and Chemical Interactions

An introduction to the metallization problems associated with the stability of various metallization materials was given in Section 6.2.3. Metallization can be completely destroyed due to reactions induced by thermally activated processes with the substrates or the layers on top. Pure silicide films on polysilicon or silicon are fairly stable with respect to their metallurgical interactions with the substrate. Most silicides undergo very limited chemical attack from the various solutions used in IC fabrication. $TiSi_2$, however, dissolves at a very fast rate in HF-containing solutions. Thus the exposure of $TiSi_2$ to such solutions should be avoided.

Aluminum on silicon or silicide, however, leads to metallurgical interactions leading to serious instabilities. Annealing of aluminum (typically at 450°C) on silicon causes the dissolution of silicon by diffusion into the metal and leads to pit formation (fig. 6.16) [53-54]. The dissolution proces is highly nonuniform, leading to isolated, crystallographic etch pits, bounded by <111> surfaces [55]. Vaidya [53] has shown that the pit growth, which could lead to contact and junction failure, is a thermally activated process with 0.8 eV activation energy. The median pit size is directly related to the square root of the anneal time (fig. 6.17) and to $W^{-2/5}$, where W is the window area (fig. 6.18). Note that aluminum anneal at 450°C is carried out after aluminum deposition and pattering into long interconnect lines. Thus for all practical purposes, silicon from the substrate in the window opening, has to satisfy the solubility requirements of infinitely large amounts of aluminum, diffusion permitting. Diffusion in aluminum at 450°C is high and, therefore, not limiting. With the increasing window size the Si to Al contact area increases allowing more pits to form and, therefore, reducing the pit size since the amount of silicon dissolved in Al is independent of window size and depends only on Al film thickness, the length of the runners, and the anneal temperature and time.

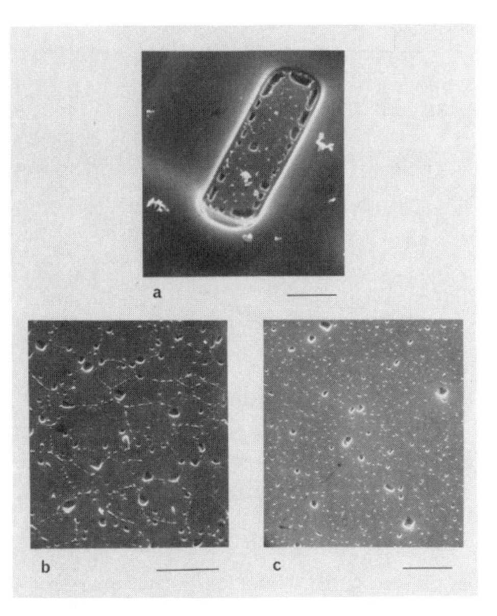

Fig. 6.16. Aluminum/silicon interpenetration structures observed after a 450°C/30 min/H_2 anneal and aluminum etch. (a) Typical contact window that had been covered with In-source aluminum. Recrystallized silicon particles can be seen dispersed randomly on the wafer surface. (b, c) Pit structure and silicon-precipitate distribution associated with e gun and In-source aluminum, respectively. The micrographs correspond to the central region of a large contact pad. The marker in each case represents 5 μm. From Vaidya [53].

Contact failures in Al-Si systems also occur due to the precipitation of dissolved silicon (from aluminum) on cooling. The silicon precipitates (fig. 6.19) [56] may cause an undesirable increase in the contact resistance, especially for Al-n-Si contacts.

To prevent junction shorts caused by the preferential dissolution of silicon into aluminum, silicon is added to aluminum during the deposition of the metal film. The amount of silicon required in the aluminum is determined by the maximum process temperature and the solid solubility that can be obtained from the Al-Si phase diagram. Normally, slightly more than 1 wt.% silicon is added. If the contacts are all to p-silicon, this method of solving the junction spiking problem is acceptable. Once again, there is a problem if n-silicon is to be contacted. Because of excess silicon present in aluminum, on cooling, some precipitation of silicon occurs in the contact window. This leads to a non-ohmic contact to n-silicon, because the recrystallized silicon precipitate contains Al, which is a p-type dopant. Figure 6.20 shows the resistance changes in various width resistors as a function of the anneal time at 200°C. Thus even at 200°C, Si-Al interaction occurs, leading to changes in the properties of the metal.

To solve this problem of Al-Si interaction, a diffusion barrier between Si and Al is recommended. Of the barriers suggested in Section 6.2.1, self-

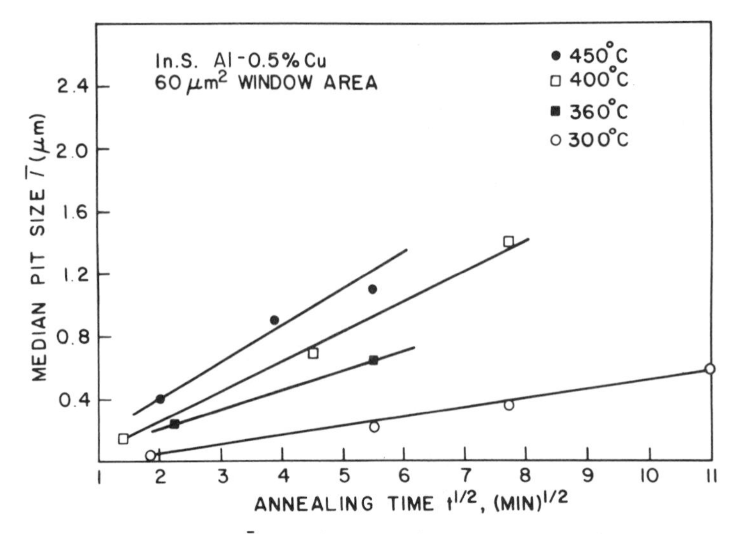

Fig. 6.17. Median pit size ($\bar{l}$) as a function of the square root of the annealing time at 300, 360, 400, and 450°C. The window size was 60 μm^2. From Vaidya [53].

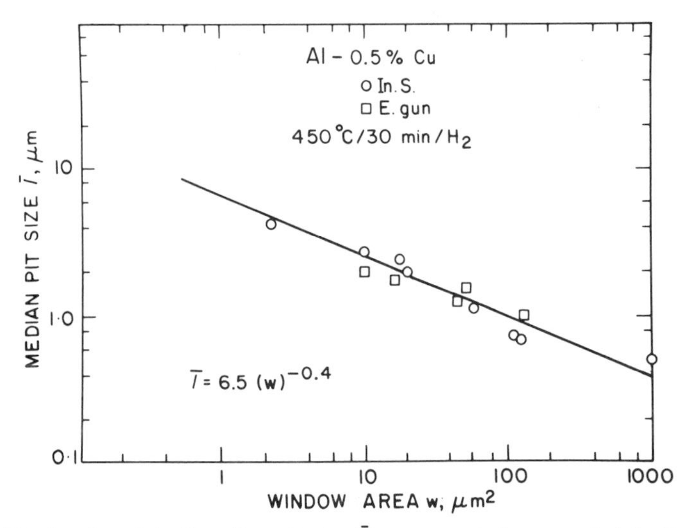

Fig. 6.18. Log-log plot of median pit size ($\bar{l}$) versus. window area (w) for In-source and e-gun aluminum afer 450°C/30 min/H_2 anneal. From Vaidya [53].

aligned silicides of Pt, Pd, Co, Ni, and Ti have been considered most often. They have been used both as Schottky and Ohmic contacts to silicon. The formation of silicides by metallurgical interaction between pure metal film and silicon leads to the most reliable and reproducible Schottky barriers. The properties of such silicide contacts are reliable and reproducible because silicide formation by metal-silicon interaction frees the silicide-silicon interface of surface imperfections and contamination. Similarly, the formation of

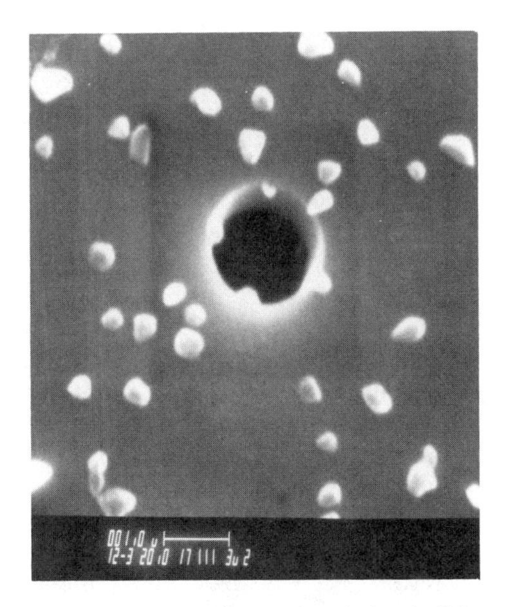

Fig. 6.19. Silicon precipitates in and around a window in SiO_2 on silicon; aluminum has been etched off. From Vaidya [56].

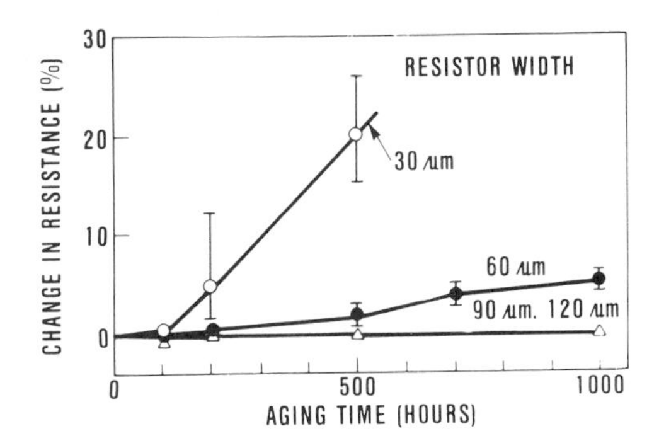

Fig. 6.20. Change in the resistance of the diffused resistors with silicon doped aluminum metallization during a 125 mA and 200°C ambient stress test. From Mori, Kanamori, and Ueki [54].

the silicide-silicon contact atomically cleans the interface, thus avoiding the variability in contact properties that may otherwise occur when the surface is contaminated or imperfect.

The formation of self-aligned silicides, however, consumes silicon from the shallow junction regions. The possibility of using deposited silicides directly into contact windows offers the advantages of preserving shallow junctions, which may be penetrated by a conventional silicide formed by reacting metals with the silicon. For deposited silicides one must develop a patterning process. Several ways in which the use of silicides have been suggested for the formation of shallow contacts are: a) the use of the snowplowing effect during silicide formaton, b) the diffusion of dopant from the deposited silicide, c) the process of solid-phase epitaxy, d) the use of metal-rich codeposited silicide, and e) the use of two-metal deposits. Of these ways the first three form the shallow junction the same time silicide is formed. The last technique forms the contact silicide (such as Pt or Pd silicide) together with the diffusion barrier (such as W metal) on top. All these techniques have advantages and disadvantages that need further investigation.

Aluminum interacts with silicides in a manner similar to the interaction behavior of aluminum and silicon. The reaction leads to the decomposition of the silicide, the dissolution of silicon in aluminum, and the formation of binary aluminum-metal or ternary aluminum-metal-silicon intermetallic compounds depending on the heat treatments involved. The result is the Al penetration of the silicide and a penetration into the underlying silicon substrate. Various silicides react with aluminum at different temperatures. Refractory silicides of Mo, Ta, Ti, and W react with Al at temperatures $> 450°C$. Pd_2Si and PtSi react at about 250°C. $CoSi_2$ reacts at about 400°C. The use of the last three silicides will therefore require a diffusion barrier between them and Al. Diffusion barriers like TiN, Ti-W, and TiC are found to be stable at 450°C and no metallurgical or electrical degradation of the silicide-barrier-aluminum contact metallization has been reported at 450°C as long as these barrier films are pinhole free. Generally a thickness in the range of 500-1000 Å is appropriate.

6.6.4 Electromigration

Electromigration can lead to considerable material transport in metals. It occurs due to high electric currents in the conductor, resulting in enhanced anisotropic atom migration. The enhanced and directional mobility of atoms are caused by a) the direct influence of the electric field and b) the collision of electrons with atoms, which leads to momentum transfer. In thin-film conductors that carry sufficient current density during device operations, the mode of material transport can occur at much lower temperatures (compared to bulk metals) because of the presence of grain boundaries, dislocations, and point defects that aid the material transport. Thus in devices using aluminum, failures associated with discontinuities in the conductor caused by electromigration have been observed [54-63]. Figure 6.21 shows electromigration failures in several Al-Si and Al-polysilicon structures. Al is accumulated in the direction of electron flow, leading to a discontinuity in the current carryling lines. This type of device failure seriously affects the reliability of the silicon integrated circuits.

The drift velocity v of migrating ions is given by

$$v = j\rho \frac{ez^*}{\mathrm{k}T} D_o \exp(-\frac{Q}{\mathrm{k}T}). \tag{6.48}$$

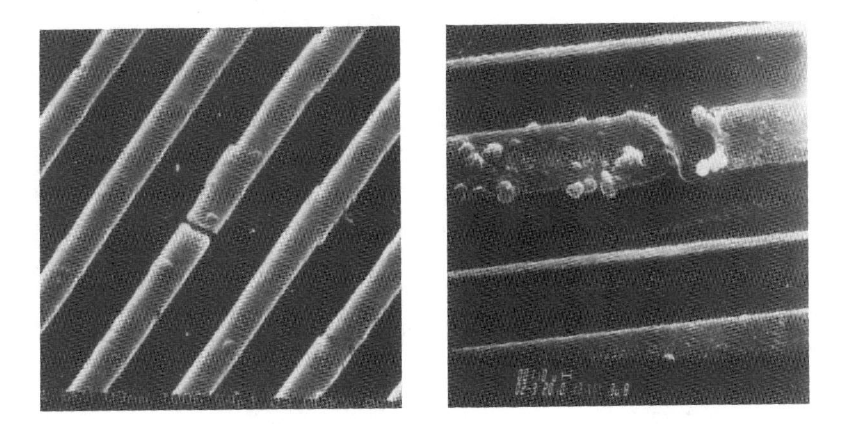

Fig. 6.21. SEM micrographs of electromigration failure in aluminum runners (a) s-gun magnetron deposited Al-0.5% Cu alloy and (b) In-source evaporated Al-0.5% Cu alloy. From Vaidya, Fraser, and Sinha [58].

This equation, derived theoretically from Einstein's drift velocity equation [61], has been modified by Black [62] for use in determining the relationship between the median time to failure (MTF) and j and Q. MTF is the time, for a given testing condition, at which 50% of the testing sites fail. The general expression for the MTF is

$$\mathrm{MTF}\alpha j^{-n} \exp(\frac{Q}{\mathrm{k}T}). \tag{6.49}$$

In eqs. (6.48) and (6.49) j is the current density, n an exponent between 1 and 3, ρ the film resistivity, eZ^* the effective ion charge [58], k the Boltzmann's constant, T the absolute temperature, D_o the pre-exponential diffusivity, and Q the diffusion activation energy.

Papers published in the last 20 years clearly point to electromigration-induced failure as the most important mode of failure in the aluminum lines [59]. Most studies have measured MTF, for a given current, as a function of the temperature and the linewidth of the conductor. The results are found to be strongly influenced by the method of deposition of the metal, the temperature of the substrate during deposition, the presence of the alloying element, if any, the substrate type, the film thickness, and the length of the conductor.

The activation energy Q measured for electromigration in aluminum and its alloys ranges from 0.4 to 0.8 eV. This can be compared with the activation energies of 1.4 eV for bulk diffusion, 0.28 eV for surface diffusion, 0.4-0.5 eV for grain-boundary diffusion, 0.62 eV for the grain boundary plus bulk diffusion, and a value $0.62 < Q < 1.4$ for defect-assisted bulk diffusion in Al [63]. Thus, the measured activation energy lies in the range of the pure grain boundary diffusion and the grain boundary (and defect) plus bulk diffusion. This is very significant since the relative contributions of the grain boundary and bulk diffusions depend on the grain size and the grain structure of the film. This explains the observed spread in the Q values obtained for a variety of films prepared by different methods. Within experimental error and allowing for grain size variations, the activation energy Q is found to be independent of the film deposition condition and the alloying element (or elements) [59, 63].

Vaidya et al. [58] found, from a detailed investigation of the electromigration, that the MTF increases with the increasing grain size S, the degree of 111-preferred orientation, and the decreasing spread in the grain size distribution σ. Thus MTF is proportional to an empirical microstructural quantity [58] η, given by

$$\eta = \frac{S}{\sigma^2} \log\,(I_{111} I_{200})^3, \tag{6.50}$$

where I represents the intensity of corresponding x-ray diffraction peak. This is the first time a rationale has been presented that takes into account various microstructural factors that influence electromigration resistance.

Figure 6.22 shows the MTF as a function of the linewidth [57]. In general, the MTF decreases with the linewidth. However, at linewidths of less than 2 μm the MTF increases, thus reversing the trend observed in wider lines. This effect has been rationalized on the basis of the grain structure and size distribution. TEM studies [57] revealed that in narrow lines, the grain structure takes on a "bamboo" appearance with grain boundaries generally running perpendicular to the direction of the current flow. In contrast, the wider, less stable lines of the same material contained much more heterogeneous grain structure, with several triple points across the width (fig. 6.23). Thus in narrower lines the Al diffusion in the direction of current flow, i.e., along the length of lines, was considerably reduced, enhancing the MTF. Only grain boundaries that were not perpendicular to the direction of current flow contributed to atom flux and hence to reduction of the MTF.

The addition of copper to Al or Al-Si alloys considerably increases the electromigration resistance that is also found to strongly depend on the methods of the deposition of the metal film in addition to the linewidth (see fig. 6.22). Apparent superiority of the Al-Cu alloys [64] and e-gun evaporated metals or alloys is associated with the resulting preferred (111) texture and improved grain size distribution such as the bamboo structure [57]. At present, one can speculate that in addition to changing the texture of the aluminum films, the addition of copper increases the electromigration resistance by enhancing the activation energy of the self-diffusion of aluminum. Other elements that have shown a similar effect are Ni [65], Cr [66], Mg [67], and O [68].

Agarwala et al. [69] have shown that MTF first decreases with the length of the aluminum runner and then becomes independent of the length for all

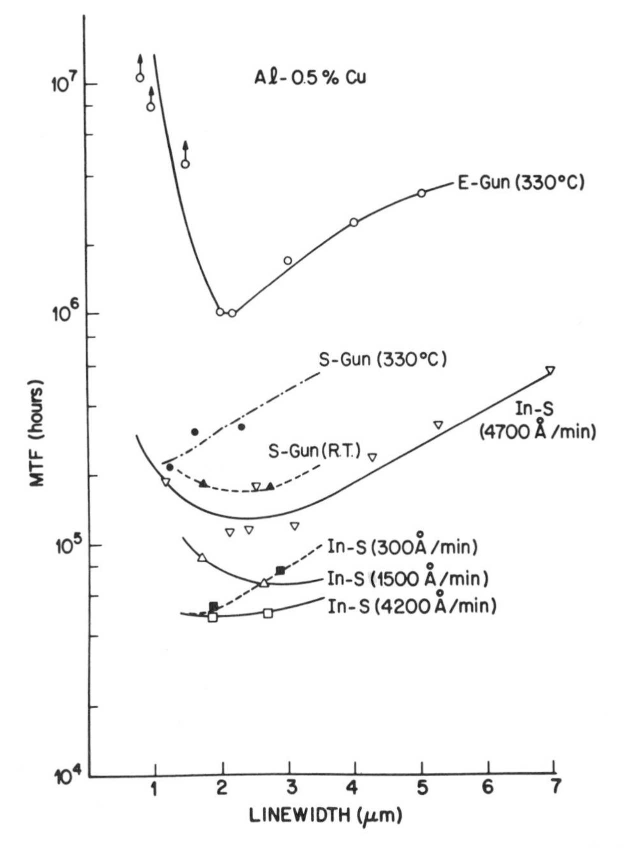

Fig. 6.22. Extrapolated median-time-to-failure (MTF) as a function of a linewidth for e-gun, s-gun, and In-source (In-s) aluminum films at 80°C and 1 x 10^5A cm^{-2}. The film deposition conditions are shown on the plot. The in-source films were evaporated onto heated (300°C) substrates at different rates. RT = room temperature. From Vaidya, Sheng, and Sinha [57].

practical purposes. Blech [70, 71] reported that a current density threshold, which is found to be inversely proportional to the aluminum length, exists, below which the aluminum mass transport due to electromigration stops. Thus, for a given current through the runner, the electromigration-induced mass transport is higher in the small runners. It was suggested that the stress gradients in the film are the prime cause for this reverse mass transport and current density threshold [70, 71].

Clemens [72] suggested the use of doped polysilicon between aluminum and silicon. The use of polysilicon as a sacrificial layer between the source, drain, or gate and the top-level aluminum offers two very important advantages. First, polysilicon is deposited by LPCVD techniques that produce excellent step coverage, thus eliminating the possibility of contact break at the window edges. Second, it provides enough silicon to satisfy the solubility require-

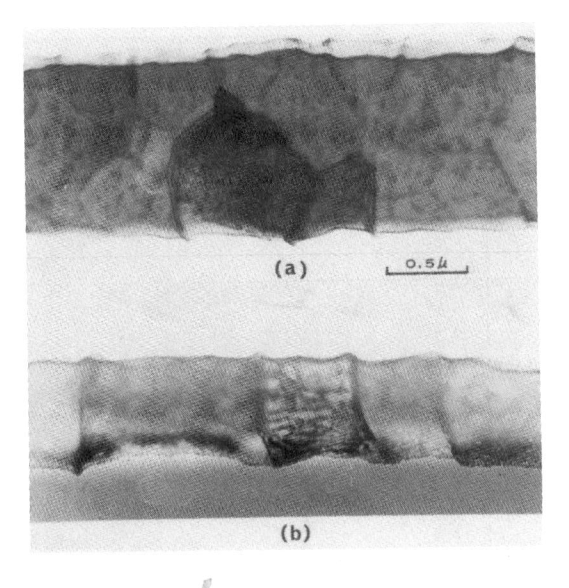

Fig. 6.23. Transmission electron micrographs of the grain structure of e-gun aluminum-0.5% copper in (a) 2.2-μm, and (b) 1.0-μm wide lines. From Vaidya, Sheng, and Sinha [57].

ments of aluminum. Doping of polysilicon can be achieved by autodoping from the source and drain areas, in-situ doping during deposition, or by ion implantation. This process, called the *poly-plug process*, is used in fabrication of NMOS devices [72]. For CMOS devices requiring both n- and p-type polysilicon, extra masking and lithographic steps are necessary to implant n and p dopants in the desired windows.

Junction leakage and electromigration studies of the aluminum over n^+-polysilicon composite have been carried out [70-76]. It is found that while such a composite can provide excellent step coverage and prevent thermally induced Al-Si interpenetration, the devices are still susceptible to failure by junction leakage at positively biased windows resulting from the reduced electromigration resistance of such composite. Figure 6.24 shows this reduced lifetime for e gun or in-source aluminum over polysilicon compared to that for e-gun or in-source aluminum alone. The reduction in lifetime of aluminum over polysilicon is due to a process involving a localized diffusion of silicon (from polysilicon) into aluminum, followed by the transport of silicon across the metal semiconductor interface. The lifetimes exhibit a strong dependence on current, varying as I^{-n} with $n = 10$, indicative of steep temperature gradients in the vicinity of the contacts [68]. These studies [73-76] indicated that the thermal-activation energy associated with this failure mode is 0.9 eV, corresponding to the diffusion of silicon in the aluminum alloy that contains silicon in excess of 0.4% [77].

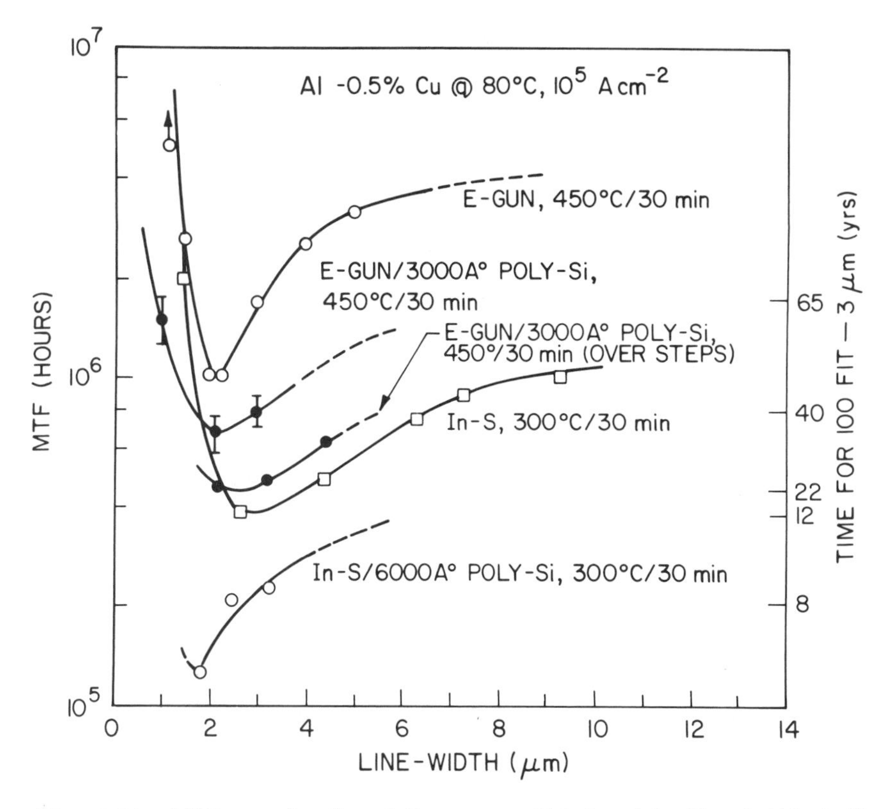

Fig. 6.24. MTF as a function of the runner width for plain Al and Al/poly Si composites at 80°C and 1 x 10^5 A cm^{-2}. The respective heat treatments are shown on the plot. The time taken to react a 100-FIT (failure in time) failure rate for the 3-μm lines of each structure is listed on the right-hand side. From Vaidya and Sinha [73].

6.7 Role of New Metallizations

New metallizations will play increasingly important roles in electronic devices and integrated circuits. Metallizations will not only yield current carrier but will also play an active role in determining device properties, as in the case of the gate electrode or Schottky barrier diode.

6.7.1 Multilevel Structures

To minimize the interconnection resistance and save the valuable area on a chip, multilevel metallization schemes have been proposed. In this concept, as demonstrated in fig. 6.25 where a three-level metallization scheme is shown, metal interconnects run in the third dimension as well. Theoretically several levels of metals can be employed. Dielectric films are used as the

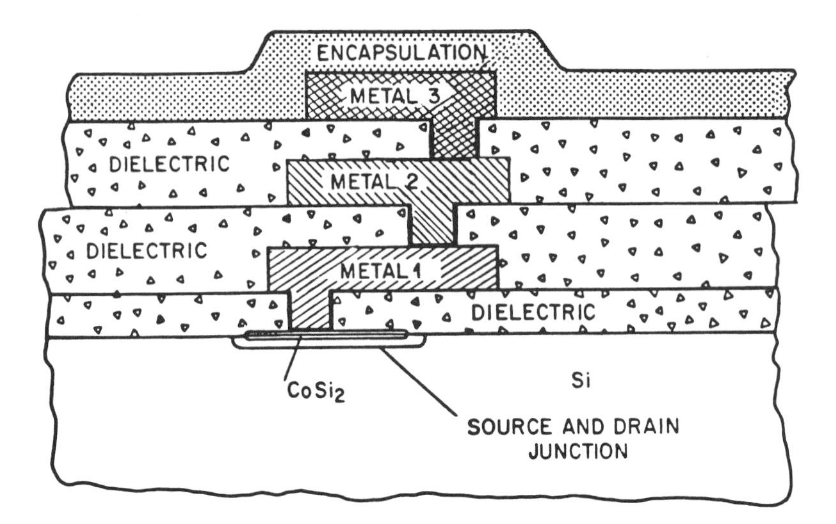

Fig. 6.25. A schematic drawing of the multilevel metallization structure.

insulator between various metal layers. The interconnection is made by opening holes in the dielectric layer.

There are several problems associated with making a multilevel metallization scheme work. These are all related to deposition, etching, and metal characteristics. Aluminum is the favored metal. Its use limits the maximum temperature for dielectric film deposition to 400-450°C. The most serious processing problems are associated with step coverage and via filling such that each level is flat and parallel to the original substrate surface. It is very difficult to produce such deposits—metal or dielectric. CVD methods, known for their ideal step coverage, have not yet been developed for Al. Reported CVD Al deposits [78] are found to contain large amounts of impurities (carbon and oxygen) that increase the resistivity. Also they are very rough deposits, making patterning very difficult. Similarly, although low-temperature CVD depositions of SiO_2 leads to reasonably good coverage, these depositions have been found to be inadequate due to their poor dielectric strength. Several other dielectric like polyimides and spin-on-glasses have been tried with success only in limited cases.

Among the new developments that may lead to realization of structures of the type shown in fig. 6.25 are deposition by bias sputtering and planarization processes. In bias sputtering the substrate is biased so that back sputtering is facilitated on the surface. Thus sputter deposition and back-sputter etching occur simultaneously. Back sputtering is asymmetric and is favored at sharp corners and edges in such a manner that when optimized it results in excellent step coverage of the deposit ending up as a flat surface. Bias-sputter deposition rates are, however, low. Also ion damage occurs during these depositions. More work is necessary to improve this process.

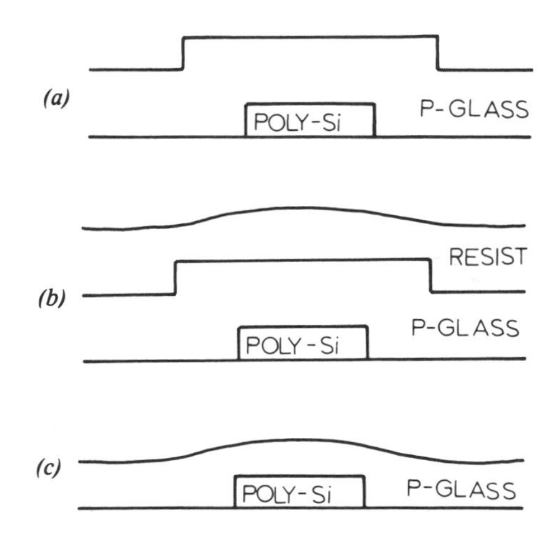

Fig. 6.26. Planarization process is shown: (a) as-deposited film, (b) thick planarizing resist layer, and (c) after planarization etch. From Adams [51].

In the planarization process (see Chapter 9), the surface is planarized parallel to the substrate surface. After thick CVD deposits that fill and cover around vias have been made, a thick photoresist is deposited. Photoresist is usually so thick that it covers all the steps well and the top surface is flat. Now a plasma or reactive ion etch process is chosen to etch the resist and the CVD deposit at the same rate. The surface is etched until all the resist and some glass is etched off, leaving the surface flat. The process steps are shown schematically in fig. 6.26. The role of the sacrificial photoresist layer is to make the top surface flat prior to etching. The flatness can easily be achieved with the spin-on applications of the viscos photoresist. During the subsequent isotropic etching of the surface, the flatness is retained because the underlying CVD deposit etches at the rate equal to the sacrificial photoresist layer. The planarization process is becoming popular in VLSI fabrication.

Finally, a mention of selective metal deposition, a technology yet to be proven, must be made. In this type of deposition [79-81], tungsten metal is selectively deposited on silicon only, and no deposit occurs on the SiO_2 layers. Such a method of deposition, if proven reliable and reproducible, will probably prove to be the best for multilevel structures as well as for contact metallization. It could provide excellent contact to silicon and the deposit could fill up vias. In addition, the deposit of W will not require a diffusion barrier when contacted with the top Al layer. Finally, selective tungsten technology may be useful in the formation of gate and interconnection metallizations on polysilicon. In fact, selective tungsten could be deposited simultaneously on polysilicon gate and interconnection, source, and drain, in a manner similar to that described for the self-aligned silicide technology. The advantage over silicide technology lies in the minimal consumption (100-200 Å) of the silicon in the source and drain regions of the technique. However, several problems remain to be solved at this writing. These include: a) reproducibility of the selectivity of the deposition; b) occurrence of microstructural defects at the W/Si interface, such defects are associated with the deposition process; c) preferential erosion of n^+ silicon compared to that of p^+ silicon; and d) sporadically observed contact instability [80].

6.7.2 Epitaxial Metals, Three Dimensional Devices and Heterostructures

Recent developments in forming epitaxial silicides [82-84] and aluminum [85, 86] open a door to new hope for contact metallization and for the possibility of forming three-dimensional devices and heterostructures. Epitaxial schemes will provide grain-boundary-free and possibly dislocation-free films. Such films, therefore, practically eliminate the metallugical interaction problems associated with grain boundary (and dislocation) diffusion. Thus metallizations will be stable at higher temperatures and have longer electromigration lifetimes. Epitaxial growth of metallization on silicon also suggests the possible epitaxial growth of silicon on top of the metal. Thus heteroepitaxial structures of silicon-metal-silicon can be created. This scheme points to the possibility of making three-dimensional circuits and new devices. Metal-base transistors [18] and permeable-base transistors [18] are two examples of new devices. Silicon-$CoSi_2$-silicon heteroepitaxial structures, and even metal-base transistors, have recently been fabricated[84].

Epitaxy of aluminum on silicon has recently been demonstrated [85, 86]. In this case it was demonstrated that the Al-Si interaction did not occur, at least up to 450°C, which is a remarkable improvement on the stability of Al-Si contacts. Since the diffusion activation energy in single-crystal aluminum films is nearly 1.4 eV, electromigration will be significantly reduced in such epitaxial lines on silicon as well.

6.7.3 Diffusion Barriers

In the absence of epitaxial metal films, diffusion barriers will be required between silicon or silicide and aluminum for devices that will be subjected to post-aluminum anneals at 400°C or higher temperature. As discussed earlier, transition metal borides, carbides, and nitrides have been suggested for this application with reasonably good success reported. A new approach employs a codeposited two-metal mixture. A low-temperature reaction between a substrate silicon with a refractory-metal-silicon alloy mixture always leads to the formation of the noble-metal silicide on silicon and to the accumulation of the refractory metal (or the silicide formed at sufficiently high temperatures) at the top. The amount of silicon consumed can be controlled by varying the composition of the alloy and limiting the reaction temperature so that only noble-metal silicides are formed. The advantage lies in the observed layered phase separation [87], so that the refractory metal top layer can act as the diffusion barrier between the silicide and subsequently deposited aluminum. Such an approach, however, will require the development of a pattern generation process for the two-metal deposit.

An extension of the two-metal approach was made by the use of metal-rich titanium carbide films [88]. Titanium-rich carbide films that can be reactively sputtered react with silicon, leading to a phase separation with $TiSi_2$ near the silicon substrate plus an outer carbide layer. Such a heat-treated structure with an Al top layer maintains its integrity even after 550°/30-minute heat treatment. Thus, a heat treatment of titanium-rich carbide ($Ti_{3.1}C$) film on silicon at 650-750°C results in a good $TiSi_2$ contact and a built-in diffusion barrier to Al penetration.

Recently titanium boride films [89] of excellent quality have been deposited. They were found to provide excellent stability to the silicon-TiB_2-aluminum contact system. One can envision using TiB_2 on silicon and heating it to temperatures high enough to initiate a reaction between the boride and silicon at the interface, leading to release of boron to dope silicon and forming silicide at this interface. Thus in one anneal a shallow (p-n) junction, a silicide contact, and a boride diffusion barrier could result.

6.7.4 Redundant Metal Links

Many large MOS memory chips, in order to improve yield, are designed with redundant metal (fusible) links. These spare metal rows and columns of bits, which can be exchanged for faulty ones, are part of the circuit design. They can be opened by laser or electrical means, allowing for the disconnection of faulty links. After the faulty links are disconnected, their previous memory in the memory array is transferred to the spare row or column by opening additional fusible links in the memory-decoding circuitry. Thus use of metallization in the form of these fusible links allows for considerably enhanced yields and a lower cost of the large memory chips.

6.8 Summary and Future Trends

Metallization for VLSI and ULSI require considerable understanding of the application, metallization choices, properties, stability, and selection of optimum deposition processes. Future applications demand lower resistance at all levels and sophisticated deposition and patterning techniques. The use of metallization in creating three dimensional device structures and super-fast, so-called ballistic electron devices is just being considered. In this chapter the above was presented, together with an understanding of the limitations of various possible metallization schemes.

It is apparent that presently used or conceived metallization schemes are inadequate for futuristic device concepts. Selective tungsten or aluminum deposition, epitaxial silicides or aluminum, self-aligned silicides, liquid nitrogen (or higher) temperature super-conductors, and even copper show promise and could pave the way for better and faster devices and circuits. Multilevel metal networks will ease the interconnection resistance problems and reduce the chip area significantly.

References for Chapter 6

1. For earlier works on polycrystalline silicon see papers in Semiconductor Silicon 1969, 1973, and 1977 (The Electrochemical Society, Princeton, NJ, 1969, 1973, 1977).
2. T. Chung, J. Electrochem. Soc. 109, 229 (1962); also see references listed in ref. 3.
3. A.G. Milnes, D.L. Feucht, "Heterojunctions and Metal-Semiconductor Junctions," Academic Press, New York, pp. 293-305 (1972).
4. M.P. Lepselter, J.M. Andrews, in "Ohmic Contacts to Semiconductors," B. Schwartz ed., The Electrochemical Society, Princeton, NJ, p. 159 (1969).
5. M-A. Nicolet, Thin Solid Films 52, 415 (1978).
6. M.P. Lepselter, Bell Syst. Tech. J. XLV, 233 (1966).
7. S.P. Murarka, H.J. Levinstein, I. Blech, T.T. Sheng, M.H. Read, J. Electrochem. Soc. 125, 156 (1966).
8. S.P. Murarka, I.A. Blech, H.J. Levinstein, J. Appl. Phys. 47, 5175 (1976).
9. S.P. Murarka, "Silicides for VLSI Applications," Academic Press, New York 1983).
10. L.I. Maissel, in "Handbook of Thin Film Technology," L.I. Maissel, R. Glang, eds., McGraw-Hill, NY, Chapter 13 (1970).
11. N.F. Mott, Proc. Cambridge Philos, Soc. 32, 281 (1938).
12. A.F. Mayadas, M. Schatzkes, Phys. Rev. B1, 1382 (1970)
13. M.M. Mandurah, K.S. Saraswat, T.I. Kamins, Appl. Phys. Lett. 36, 683, (1980).
14. K.L. Chopra, "Thin Film Phenomena," McGraw-Hill, NY, p. 376 (1969).
15. P.B. Ghate, C.R. Fuller, in "Semiconductor Silicon-1981," H. Huff, R. Kriegler, Y. Takeishi, eds., Electrochemical Soc., Princeton, NJ, p. 680 (1981).
16. A.K. Sinha, J.A. Cooper, Jr., J.H. Levinstein, Electron Dev. Lett. EDL-3, 90 (1982).

17. E.F. Nicollian, J.R. Brews "Mos Physics and Technology," Wiley, NY (1982).
18. S.M. Sze, "Physics of Semiconductor Devices," 2nd ed., Wiley, NY (1981).
19. E.H. Rhoderic, "Metal-Semiconductor Contacts," Oxford Univ. Press, Oxford (1978).
20. M. Peckerar, J. Appl. Phys. 45, 4652 (1974).
21. J.M. Shannon, Appl. Phys. Lett. 25, 75 (1974).
22. S.S. Li, J.S. Kim, K.L. Wang, IEEE Trans. Electron Dev. ED-27, 1310 (1980).
23. T. Mochizuki, K. Shibata, T. Inoue, K. Ohuchi, Jpn. J. Appl. Phys. Suppl. 17-1, 37 (1978).
24. S.P. Murarka, D.B. Fraser, J. Appl. Phys. 51, 1593 (1980).
25. A.L. Crowder, S. Zirinsky, IEEE J. Solid State Circuits SC-14, 291 (1979).
26. R. Hanken, J. Vac. Sci. Technol B3, 1657 (1985).
27. S.P. Murarka, D.B. Frazer, A.K. Sinha, H.J. Levinstein, EEE J. Solid State Circuits SC-15, 474 (1980).
28. H.J. Levinstein, S.P. Murarka, A.K. Sinha, U.S. Patent #4,378,628, dated April 5, 1983.
29. D.M. Brown, W.E. Engler, N. Garfinkel, P.V. Gray, J. Electrochem. Soc. 115, 874 (1968).
30. P.L. Shah, IEEE Trans. Electron Devices, Ed-26, 631 (1979).
31. S.P. Murarka, unpublished, paper given at "Refractory Metals and Silicides for VLSI IV," San Juan Baufista, CA.
32. S.P. Murarka "Silicides for VLSI Applications," Academic Press, NY (1983).
33. M.A.Nicolet, Thin Solid Films, 52, 415 (1978).
34. P.B. Ghate, J.C. Blair, C.R. Fuller, G.E. McGuire, Thin Solid Films, 53, 117 (1979).
35. J.D. Crowley, T.A. Rabson, Rev. Sci. Instrum. 47, 712 (1976).
36. S.P. Murarka, I.A. Blech, J.H. Levinstein, J. Appl. Phys. 47, 5175 (1976).
37. G.G. Stony, Proc. Roy. Soc. (London), A82, 172 (1909).
38. R.W. Hoffman, in "Measurement Techniques for thin films," B. Schwarz, N. Schwartz, eds., Electrochem. Soc. Inc., NY, p. 312 (1967).
39. V. Teal, S.P. Murarka, J. Appl. Phys. 61, 5038 (1987).
40. T.E. Smith, unpublished work.
41. Standard X-ray Diffaction Listings by Joint Committee for Powder Diffraction Standards.
42. "The Handbook of Thin Film Technology," L.I. Maissel, R. Glang, eds., McGraw-Hill, NY (1970).
43. S.P. Murarka, Thin Solid Films 23, 323 (1974).
44. S.P. Murarka, Thin Solid Films 140, 35 (1986).

45. D.B. Fraser, in "VLSI Technology," S.M.Sze, ed., McGraw-Hill, NY, Chapter 9, (1983).
46. L. Holland, "Vacuum Deposition of Thin Films," Wiley, NY (1956).
47. J.L. Vossen, J.J. Cuomo, in "Thin Film Processes," J.L. Vossen, W. Kern, eds., Academic Press, NY, Chapter II (1978).
48. S.P. Murarka, J. Vac. Sci. Technol. B4, 1325 (1986).
49. I.A. Blech, D.B. Fraser, S.E. Haszko, J. Vac. Sci. Technol. 15, 13 (1978); Errata J. Vac. Sci. Technol. 15, 1856 (1978).
50. W. Fichtner, in "VLSI Technology," 2nd edition, S.M. Sze, ed., McGraw-Hill, NY (1988).
51. A.C. Adams, Solid State Technol. 24, 178 (1981).
52. S.P. Murarka, J. Vac. Sci. Technol. 17, 775 (1980).
53. S. Vaidya, J. Electron. Mater. 10, 337 (1981).
54. A.M. Mori, S. Kanamori, T. Ueki, IEEE Trans. Components, Hybrids, Manfacturing Technicol. CHMT-6, 159 (1983).
55. J. Black, RADC-TR-77-410, Final Tech. Report, Motorola Semiconductor Group, (1977).
56. S. Vaidya, unpublished.
57. S. Vaidya, T.T. Sheng, A.K. Sinha, Appl. Phys. Lett. 36, 464 (1980).
58. S. Vaidya, D.B. Fraser, A.K. Sinha, "Proceedings of the 18th Annual Reliability Physics Symposium," IEEE, New York, p. 165 (1980).
59. P.B. Ghate, Solid Technol. 26, 113 (1983); see other references of electromigration listed in this paper.
60. D.Pramanik, A.N. Saxena, Solid StateTechnol. 26, 131 (1983).
61. D.A. Blackburn, in "Electro-and Thermo-transport in Metals and Alloys," R.E. Hummel, H.B. Huntington, eds., AIME, New York, p. 20 (1977).
62. J.R. Black, "Proceedings of the 3rd International Congress on Microelectronics, Munich, Nov. 1968," p. 141 (1968).
63. H.U. Schrieber, B. Grabe, Solid State Electron, 24, 1135 (1981).
64. F.M. d'Heurle, Metal. Trans. 2, 683 (1971).
65. F.M. d'Heurle, A. Gangulee, in "Nature and Behavior of Grain Boundaries," H. Hu, ed., Plenum Press, New York, p. 339 (1972).
66. F.M. d'Heurle, A. Gangulee, C.F. Aliotta, V. Ranieri, J. Appl. Phys. 46, 4845 (1975).
67. F.M. d'Heurle, A. Gangulee, C.F. Aliotta, V. Ranieri, J. Electron. Mater. 4, 497 (1975).
68. H.J. Blatt, Appl. Phys. Lett. 19, 30 (1971).
69. B.N. Agarwala, M.J. Attardo, A.P. Ingraham, J. Appl. Phys. 41, 3954 (1970).
70. I.A. Blech, C. Herring, Appl. Phys. Lett. 29, 131 (1976).
71. I.A. Blech, J. Appl. Phys. 47, 1203 (1976).
72. J.T. Clemens, U.S. Patent #4 291 322 (1981).

73. S. Vaidya, A.K. Sinha, "Proceedings of the 20th Annual Reliability Physics Symposium," IEEE, New York, p. 50 (1982).
74. H.M. Naguib, L.H. Hobbs, J. Electrochem. Soc. 125, 169 (1978).
75. J.R. Lloyd, M.R. Polcari, G.A. MacKenzie, Appl. Phys. Lett. 36, 428 (1980).
76. Y.Fukuda, S. Kohda, Appl. Phys. Lett. 42, 68 (1983).
77. G.J. van Gurp, J. Appl. Phys. 44, 2040 (1973).
78. R.A. Levy, P.K. Gallagher, R. Contolini, F. Schrey, J. Electrochem. Soc. 132, 457 (1985).
79. R.S. Blewer, V.A. Wells, "Proceedings 1st IEEE VLSI Multilevel Interconnection Conference," New Orleans, LA, June 1984), IEEE Cat. # 84CH1999-2, p. 153 (1984).
80. M.L. Green, R.A. Levy, J. Electrochem. Soc. 132, 1243 (1985).
81. R.H. Wilson, "The Use of Selective Tungsten Deposition. Achieving the Goals of the 1/4 um CMOS Technology" presented at the SRC Topical Research Conference, "1/4 CMOS Technology," held at Cornell University, Ithaca, Dec. 5-6, 1985.
82. R.T. Tung, J.M. Givson, J.M. Poate, Phys. Rev. Lett. 50, 429 (1983).
83. M. Liehr, P.E. Schmidt, F.K. LeGoues, P.S. Ho, J. Vac. Sci. Technol. 4, 855 (1986).
84. J.C. Hensel, R.T. Tung, J.M. Poate, F.C. Unterwald, Appl. Phys. Lett. 44, 913 (1984).
85. A.S. Ignatier, V.G. Mokerov, A.G. Petrova, A.V. Rybin, N.M. Manzha, Sov. Tech. Phys. Lett. 8, 174 (1982).
86. T-M. Lu, Physics Department, RPI, Troy, NY, private communication.
87. K.N-Tu, J. Vac. Sci. Technol. 19, 766 (1981).
88. M. Eizenberg, S.P. Murarka, J. Appl. Phys. 54, 3190 (1983).
89. J.R. Shappiro, J.J. Ginnegan, R.A. Lux, J. Vac. Sci. Technol. 134, 1409 (1986).

Problem Set for Chapter 6

6.1 Show that the scaling down to smaller linewidths does not affect the RC time constant except for the effects of fringing fields.

6.2 List the advantages and disadvantages of using thicker metal layers.

6.3 Pure aluminum, cobalt disilicide, tantalum disilicide, or doped polysilicon is to be used as an interconnection metallization on 1 μ thick SiO_2. Respective polycrystalline film resistivities are 2.8, 15, 55, or 1000 $\mu\Omega$-cm. If the length of the metal line is 1 cm, calculate the thickness of each metal film that need be deposited to yield a RC time constant of 3.5×10^{-9}s ((ϵ_{ox} = 3.5 x 10^{-13} F/cm).

Suppose you have a task to determine if intermetallics $TiAl_3$ and $CoSi_2$ can be used as the contact material on silicon and GaAs, respectively. What are the factors that you must carefully evaluate before recommending or rejecting these combinations?

6.4 Assuming there is no charge in the oxide and Q_f (the fixed oxide charge) density at the oxide (1000 Å thick)-silicon interface is 2×10^{11} q per cm^2: calculate the ϕ_m that will be necessary to produce a condition V_{FB} = o. ϕ_s for silicon is 4.8 V.

6.5 High-temperature stability of metal films on lightly doped silicon can be investigated by measuring the resistivity changes of the metal film as a function of the annealing temperature. Design the experiment and state the assumptions you need to make for this study.

6.6 Metal interconnection lines of varying widths are formed. Prior to patterning, the sheet resistance was measured to be 2.5 Ω/□. Calculate the total resistance of 0.5-, 1-, 2-, and 5-μm wide and 1-cm long lines. What other information will you need to guess the type of metal used?

6.7 Calculate the mean free paths of Al, Si, Co, and Pt atoms sputtered in a pressure of 0.5 and 1 Pa at 300°K. Estimate the number of collisions an aluminum atom will encounter in a path length of a) 2.5 cm and b) 40 cm. Which of the two deposits will have larger grains?

6.8 Schottky barrier height is generally determined by measuring $I - V$ characteristics of a Schottky diode. Invariably, the error in defining the contact area leads to erroneous results. Derive an expression for the error in the Schottky barrier height as a function of the error in the area.

6.9 Calculate the theoretical contact resistance of PtSi and $CoSi_2$ contacts on n-type silicon with a) low doping of 10^{16}cm^{-3} and b) high doping 10^{20}cm^{-3}.

6.10 Calculate the volume change associated with the formation of PtSi, $CoSi_2$, and $IrSi_3$ when a metal film reacts with the substrate silicon according to the reaction

$$\mathrm{M} + x\mathrm{Si} = \mathrm{MSi}_x$$

6.11 A metal film ($\alpha_v = 27 \times 10^{-6}$ per degree K) is deposited on silicon ($\alpha_v = 9 \times 10^{-6}$ per degree K) and heated to 1000°C. Calculate the stress in the substrate at 25°C. (Given that α_v is the volume expansion coefficient and $E = 1 \times 10^{12}$ dyn/cm^2.) If a metallurgical reaction occurs between the metallic film and the substrate, how will the room-temperature stress be affected?

6.12 Aluminum metal is deposited over a window (to silicon) in SiO_2. Assume that aluminum is in direct contact with silicon at n sites per unit window area (i.e., there are n pinholes per unit area of the native oxide on silicon). Also assume that the amount of aluminum covering silicon surfaces (i.e., the window opening) is very small compared to that on oxide surfaces. During heat treatment at 450°C, silicon is dissolved in aluminum and heat-treatment time is long enough to satisfy the solid solubility requirements of silicon dissolution in aluminum. Assuming that silicon that diffused out through the pinhole and dissolved in aluminum came from a cube of side l under the hole, show that l is proportional to $w^{1/3}$, $h^{1/3}$, $L^{1/3}$, $n^{-1/3}$, and $A^{-1/3}$ where w, h, and L are the width, height, and length of the aluminum film and A is window are in the oxide. How do the above relationships compare with the observed experimental results?

6.13 Activation energy Q of the electromigration in aluminum has been reported to be 0.3, 0.4, 0.6, 0.8, 1.1, and 1.4 eV by different authors. At room temperature, plot the MTF versus Q curve for three current densities through the conductor 10^3, 10^5, 10^7 A/cm^2. Compare the current density and lifetime of these with a typical tungsten-filament 100-watt bulb. (Use a constant of proportionality $= 2.78 \times 10^{-6}$ h.cm^4/A^2, and $n = 2$).

6.14 In an experiment, several 1 μm thick aluminum runners of 0.5-, 1-, 2-, 4- and 6-μm widths are tested for electromigration failure. If the diffusion activation energies in these runners are 1.3, 1.0, 0.7, 0.6, and 0.5 eV, respectively, plot MTF as a function of width. A fixed current of 5 mA is forced through conductors at 25°C. (Use factors given in problem 6.13.)

6.15 Do problem 6.13 at 100°C and -100°C. Explain the significance of these results.

6.16 A circuit design requires a maximum permissible current density of 5 $\times$ 10^5 A/cm^2 through a conductor 1-mm long, 1-μm wide, and nominally 0.5-μm thick. Assume that 10% of the conductor length passes oversteps and is 50% of the nominal metal-film thickness. What maximum voltage may be used across the conductor if the sheet resistance is $5.6 \times 10^{-2} \Omega/\square$? (Neglecting that the thinner cross sections at steps can lead to reliability problems.)

7 Chemical Vapor Deposition Nucleation and Growth of Thin Solid Films

7.1 Introduction

Chemical vapor deposition (CVD) is the growth of thin films using gaseous compounds containing the necessary elements to form the films. The grown films can be amorphous, polycrystalline, or crystalline. Metals, semiconductors, and insulators have been grown with CVD techniques. The goals of any CVD process can be summarized as:

- To provide films that are uniform in thickness, chemical, electrical, and mechanical properties.
- To provide films that are pinhole free.
- To provide films that adhere to the host substrate.
- To provide films with controlled surface roughness.

The chemical, electrical, and mechanical properties sought after depend on the intended film applications. Most CVD films must be patterned. Because of this, uniformity of etch rate is an important chemical property. Achieving good etch uniformity eliminates the need to overetch the work piece in order to assure that the process is completed everywhere. Over etching leads to a loss of etched-feature size control. Furthermore, the films are, in many cases, relied on to provide diffusion barriers. CVD films are frequently used as dielectrics in the active and passive parts of an integrated circuit. The electrical conductivity of these films must be controlled as it can influence device performance. Crystalline semiconductors grown on semiconductor substrates form the medium in which the active devices are fabricated. In addition to conductivity, mobility must also be controlled. Mechanically, films tend to stress underlying substrates. The magnitude of this stress is important in determining adhesion and wafer flatness. Techniques for analyzing and controlling these properties are discussed below.

Any CVD process involves the following operations. First, reacting gas is bled into the reactor. The gas moves towards its thermal equilibrium temperature and composition through gas-phase reactions. Equilibrated (or near-equilibrated) species are transported to the reaction surface. The surface chemical reactions then commence and the film is formed. Each of these operations is discussed, in turn, below. This is followed by a section on film properties and characterization and a section on applications.

7.2 CVD Apparatus and Techniques

CVD processes create a thin film on the surface of the work piece through the pyrolytic decomposition and subsequent surface reaction of molecules introduced as a gas. In pyrolysis, heat drives the breakdown of the feed gases. The CVD apparatus must accomplish the following.

- Allow the introduction of desired gases at desired rates and prevent their escape during the reaction process.
- Prevent the introduction of undesired gases.
- Supply heat to the reaction system while maintaining constant pressure and temperature.

Control of gas flows, temperature, and base pressure must be emphasized to get uniformity of film thickness and properties. In this section, some typical CVD reactor designs and processes are presented.

There are three basic reactor configurations available today [1]. These are referred to as horizontal, vertical, and cylindrical in fig. 7.1. The designations *horizontal* and *vertical* refer to the way the velocity vector of the gas stream points in the vicinity of the wafer. In the horizontal case, the mean flow is parallel to the wafer surface. In the vertical case, it is perpendicular. The *cylindrical* designation refers to the susceptor shape. Heating is accomplished in one of three ways. Early CVD systems were RF heated. The placement of the RF coils are shown in fig. 7.1. In the case of horizontal reactors, the whole reaction tube can be placed in a resistance heated furnace. Lamp heating is currently employed in the case of the cylindrical reactor. As the lamp energy is deposited on the surface of the wafer, the wafer is not heated through its depth. Rather, a thin surface volume is heated. This can reduce wafer warpage and minimize dopant redistribution effects. Both crystalline and noncrystalline films can be grown in any of these apparatus. Systems used for crystal growth are generally operated hotter (about 1100°C) than those used for growth of polycrystalline or amorphous materials.

Current variants on CVD processes involve the use of lasers. Such work, pioneered by Lincoln Laboratories, uses an intense, focused, laser beam to initiate the CVD surface reaction. This is illustrated in fig. 7.2 [2]. The process can proceed in a number of ways. It can be initiated through local heating effects. Alternatively, the surface bonding configuration can

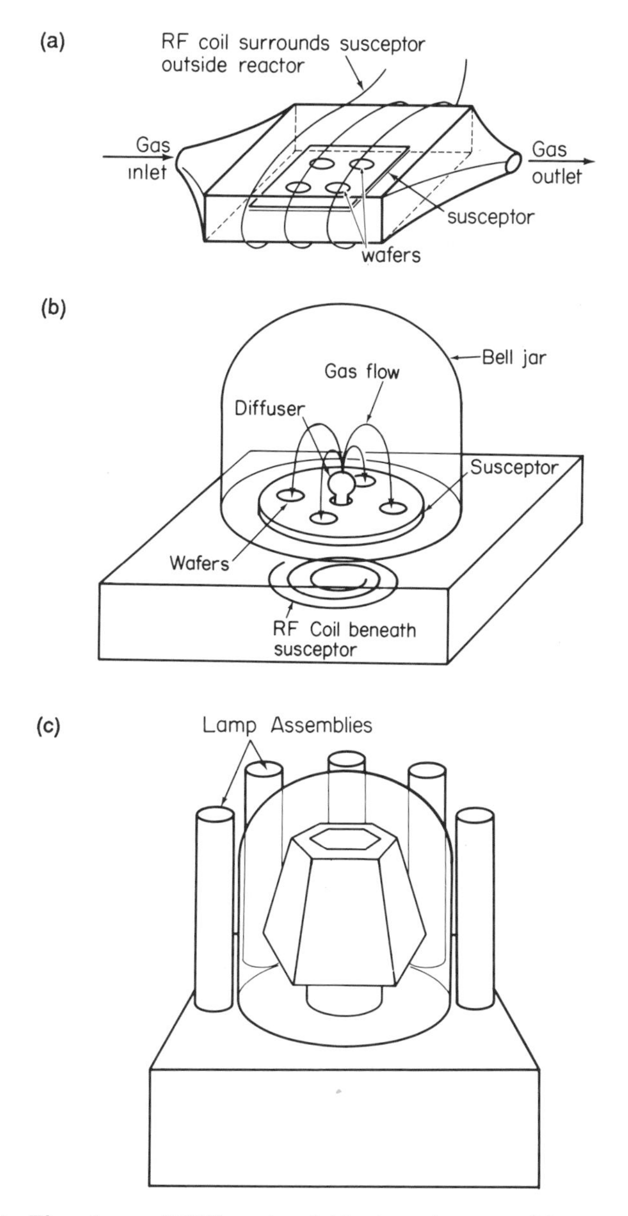

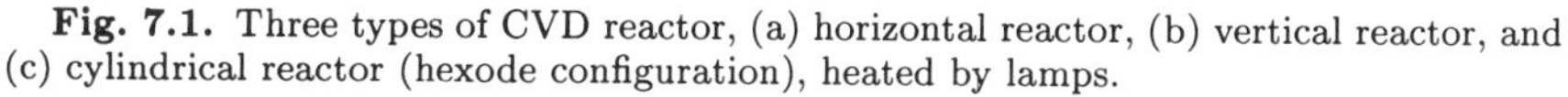

Fig. 7.1. Three types of CVD reactor, (a) horizontal reactor, (b) vertical reactor, and (c) cylindrical reactor (hexode configuration), heated by lamps.

be changed and made more reactive by the light-matter interaction. In any event, a "direct-write" deposition of a number of materials is possible using this technique. This and other beam deposition techniques are discussed in Chapter 11.

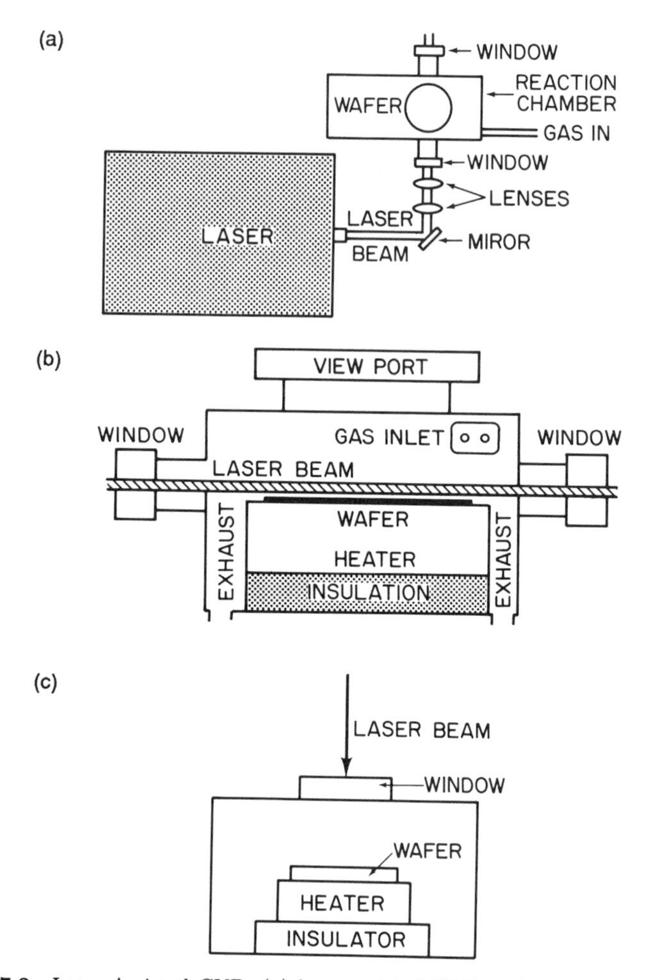

Fig. 7.2. Laser Assisted CVD, (a) laser-assisted CVD and associated beam-steering optics, (b) configuration in which beam is not incident on wafer but induces reactions above the wafer, and (c) beam normally incident for direct writing.

Many variants on these fundamental designs can be found. The operating pressure of the system can be lowered to the torr level creating a low-pressure (LP) CVD system. The accompanying reduction in gas viscosity and increased gas diffusivity both help to provide a more uniform deposition. This is elaborated below. The reacting gases can be broken down to create a plasma using an RF field. The breakdown of the gas in the plasma lowers the chemical reaction activation barrier. This allows deposition to occur at a lower temperature. Plasma deposition temperatures of oxides and nitrides lower than 300 °C have been reported [3] (as opposed to "normal" deposition temperatures higher than 450 °C). Low-temperature processing is particularly attractive in today's VLSI technology because it creates less diffusion

TABLE 7.1: Typical overall reactions for chemical vapor deposition of materials of interest in microelectronics processing

1.	$SiH_4 \longrightarrow Si + 2H_2$
2.	$SiH_2Cl_2 \longrightarrow Si + 2HCI$
3.	$3SiH_4 + 4NH_3 \longrightarrow Si_3N_4 + 12H_2$
4.	$3SiH_2Cl_2 + 4\ NH_3 \longrightarrow Si_3N_4 + 6H_2 + 6\ HCl$
5.	$SiH_4 + O_2 \longrightarrow SiO_2 + 2\ H_2$

spreading, less chance of shallow junctions disappearing, and minimal wafer warpage.

Oxides, nitrides, semiconductors, and metals have been deposited using CVD techniques. Oxides, nitrides, and semiconductors are deposited from inorganic materials, such as silane, chlorinated silanes, oxygen, and ammonia. The fabrication of many heterostructure and quantum-well devices relies on the deposition of metals from metallorganic precursors (this is called MOCVD). The major CVD reactions currently employed in the industry are shown in table 7.1.

Continuous films that tend to conform to underlying topography are the most common CVD products. Recently, selective CVD processes have been developed. For example, tungsten metal is directly deposited on silicon. This occurs because tungsten hexafluoride and silicon react to form free tungsten and a volatile silicon fluoride. The free tungsten deposits as a solid film on the surface. The gas-phase flourine cannot replace the oxygen in silicon dioxide, and the reaction cannot take place over gate or field oxide regions. Similarly, it is possible to initiate the process over some metals, like aluminum. More will be said on this in the advanced applications and techniques section of this chapter.

7.3 Gas Phase Boundary Layer Transport and CVD

Consider the simplest form of CVD currently available: atmospheric-pressure CVD done in a horizontal reactor. Gas molecules containing the deposited material are transported to the surface of the work piece. These molecules decompose, yielding the desired film. For example, silane will decompose on the reacting surface to form solid silicon for silicon epitaxy or for polysilicon film growth, or the decomposition products of one species of molecule combine with the decomposition species of another to yield the desired substance. Silane and ammonia must both decompose to provide the silicon and nitrogen necessary to make silicon nitride. Obviously, there are a number of steps

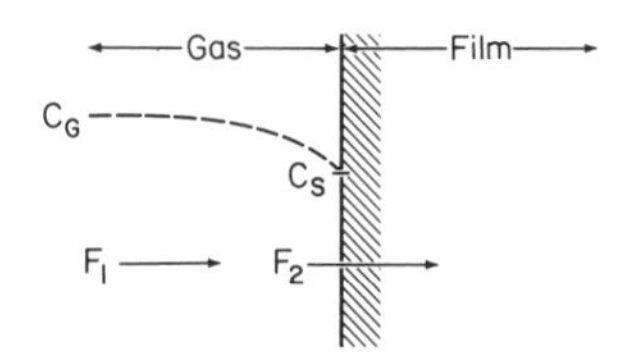

Fig. 7.3. Reacting gas concentration in front of reaction surface.

that must be taken to accomplish the deposition. First, active molecules must be transported to the reacting surface. Second, these molecules must decompose on the surface. Third, the decomposition byproducts must then react chemically with the existing surface or with each other. In this section, the process of gas-phase transport to the growth surface is described.

First, we analyze the process through which fresh reacting molecules are brought to the growth surface [4]. The surface reaction depletes the gas immediately above it of reacting molecules. A plot of reacting molecule concentration as a function of distance from the surface is shown as fig. 7.3. The concentration gradient that appears is, in fact, the determining factor which supplies the surface with new reacting species. The concentration gradient gives rise to a net diffusion current in the direction of the surface. The molecule current flux is given as

$$F_1 = D_g \left(\frac{dc}{dy} \right), \tag{7.1}$$

where D_g is the gas-phase diffusion coefficient for the reacting molecule and dc/dy is the concentration gradient at the growth surface.

The problem is simplified if we assume that the change in concentration going from the bulk of the gas to the interface is linear. Then we can write

$$F_1 = \frac{D_g(c_g - c_s)}{\delta} = h_g(c_g - c_s). \tag{7.2}$$

Here, c_g is the concentration of reacting species in the feed gas; c_s is the species concentration at the growth interface; h_g is a constant (called the gas-phase mass transfer coefficient); and δ is the characteristic length over which the gradient occurs. The region of extent δ over which the gradient occurs is called the stagnant layer. More will be said on δ later.

We can further define a *reaction flux* that occurs at the growth interface. This is the rate, per unit area, at which reacting molecules are destroyed at this interface through the surface reaction process. To first approximation, we say that this rate is simply proportional to the surface concentration of active species

$$F_2 = k_s c_s, \tag{7.3}$$

where k_s is termed the surface-reaction rate constant. In equilibrium, the two fluxes are equal, and we can solve (7.2) and (7.3) for the surface concentration

$$c_s = \left(\frac{h_g k_s}{h_g + k_s}\right) c_g. \tag{7.4}$$

The film growth rate is directly proportional to the surface-reaction flux (which is equal to the gas-phase transport flux, F_1). If we consider the chemical vapor deposition of silicon, the proportionality constant is the reciprocal of the number, N_{si}, of silicon atoms in a unit volume of film. N_{si} is 5×10^{22} per cc. Thus, we have the overall growth rate, V, as

$$V = \frac{1}{N_{si}}\left(\frac{k_s h_g}{h_g + k_s}\right) c_g. \tag{7.5}$$

Two different operating regimes can be defined for the CVD process, based on eq. (7.5). First, assume $h_g >> k_s$. This implies a very rapid transport of reacting species to the reaction surface. The rate-limiting step in film growth is, thus, the surface reaction rate. In the surface-reaction rate-limited regime, we have

$$V = \frac{k_s c_g}{N_{si}}. \tag{7.6}$$

Next, assume $h_g << k_s$. Then we have

$$V = \frac{h_g c_g}{N_{si}}, \tag{7.7}$$

and we say that we are in the gas-phase transport-limited regime. Either of these conditions can prevail in a CVD reactor, depending on gas flow rates and gas pressures. To see how this occurs, some thought must be given to the factors that determine transport through the stagnant layer.

The simplest way to obtain reasonable estimates of δ is through what is known as the *stagnant-layer model.* Begin by considering the flow of a fluid or a gas over a solid surface. It is well known from fluid dynamics that the frictional force of the solid on the moving medium causes the flow velocity to be zero at the fluid (or gas) interface. The flow reaches its maximum at some distance away from the surface in the moving medium. The region over which the velocity differs from the maximum is called the *stagnant layer.* The distance away from the surface at which 99% of the maximum velocity is achieved is usually taken as δ. The reason for this is not immediately obvious. The main justification is as follows. The bulk of the gas, moving with maximum velocity, is constantly supplied with fresh reactants from the feed source. Thus, *we make the approximation* that the reactant concentration is at its maximum up to the edge of the stagnant layer. This concentration will decline monotonically until the reaction surface is reached. At this surface,

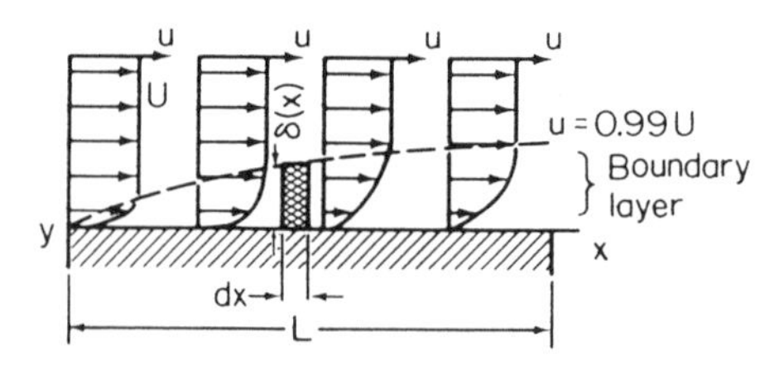

Fig. 7.4. Thickness of transition layer as a function of position along the reacting surface (after ref. [4]).

reactant molecules are continuously decomposed and the minimum reactant concentration is reached. A graphic representation of the concentration and gas velocity field is shown in fig. 7.4.

The stagnant-layer model also assumes laminar flow of the gas; that is to say, the velocity vectors representing the mean gas velocity at any point in the stream all point in the same direction and are all parallel to the reacting surface. The notion that the gas flows in laminar sheets of constant velocity is an unfortunate misconception created by the terminology. The frictive force supplied by the surface acts to slow the gas down all along the surface. Thus, the disturbance in the flow will grow as the gas travels down along the susceptor; that is, the stagnant layer thickness increases, as shown in fig. 7.4. This can be a serious problem in CVD reactor design. As a result of the variable $\delta(x)$, the supply flux, F, is a function of position in the reactor tube. This leads to a nonuniform deposition rate. Analysis of the factors that determine δ must be used to minimize this variation.

We begin the analysis with an expression for the frictional force per unit area created by the surface/gas interaction

$$F_{\text{frict}} = \mu \left(\frac{du}{dy} \right), \tag{7.8}$$

where μ is the gas viscosity and (du/dy) is the gas velocity gradient along a direction normal to the surface. Here, u is the magnitude of the gas velocity. This force will bring about a deceleration of the gas. How much deceleration is achieved is determined using Newton's second law, $F = ma$. To apply Newton's law in this instance requires some further approximation. First, break the stagnant layer up into infinitesmal "blocks," whose height is equal to the stagnant layer thickness, whose width is dx (an infinitesmal distance element along the susceptor surface), and whose depth is unity. The problem simplifies into the determination of how fast the frictive force slows down one of these fictional blocks as it moves down the susceptor. Care must be taken to realize that the height of the block grows (and the mass of the block as well) as the block slows down.

The mass of the block is given by

$$m = \rho\delta(x)\ dx, \tag{7.9}$$

where ρ is the density of the gas. The deceleration is the change in velocity with respect to time. We are more concerned with the change in velocity over distance. Newton's law can be rewritten to accomodate this demand

$$\frac{du}{dt} = \left(\frac{du}{dx}\right)\left(\frac{dx}{dt}\right) = \left(\frac{du}{dx}\right)u. \tag{7.10}$$

By substituting eqs. (7.9) and (7.10) into Newton's law and by equating the Newton's law of force to the total frictive force on the block ($F_{\text{frict}}\ dx$), the following differential equation is obtained:

$$\mu\left(\frac{du}{dy}\right) = \rho\delta(x)\ u\ \left(\frac{du}{dx}\right). \tag{7.11}$$

While the equation is first order, it represents a two dimensional boundary value problem. We follow Grove's approach [4] and linearize the problem to obtain a solution. This allows us to replace differentials by differences and the following is obtained:

$$\mu\left[\frac{U}{\delta(x)}\right] = \rho\delta(x)U(\frac{U}{x}), \tag{7.12}$$

where U is the maximum velocity of the gas. This can be solved to obtain the position-dependent stagnant-layer thickness

$$\delta(x) = \sqrt{\frac{\mu x}{\rho U}}. \tag{7.13}$$

The average layer thickness is gotten by integrating over x and dividing by the susceptor length, L

$$<\delta>= \left(\frac{2}{3}\right)L\sqrt{\frac{\mu}{\rho UL}}. \tag{7.14}$$

This can be rewritten in somewhat simpler form using one of the fundamental dimensionless parameters of fluid dynamics, the Reynolds number, R_e

$$<\delta> = \left(\frac{2}{3}\right)L/\sqrt{\mathrm{R_e}}, \tag{7.15}$$

where

$$\mathrm{R_e} = \frac{\rho UL}{\mu}. \tag{7.16}$$

Inspection of eqs. (7.15) and (7.16) indicate that as the maximum stream velocity, U, gets bigger, the average stagnant layer thickness gets smaller.

This, in turn, makes h_g bigger. Thus, by elevating the flow rate, it is possible to change the deposition process from the gas-transport rate-limited case to the surface-reaction rate-limited case. This minimizes the impact of stagnant-layer thickness variation on thickness uniformity. It seems apparent that running at the highest stream velocity possible is best. This is true within certain limits. When the Reynolds number is about 2000, turbulence sets in; that is, there is a tendency for the gas flow to break into eddies, or vortices, and to become chaotic. The laminar flow approximation used above breaks down. Furthermore, since these vortices are "nucleated" at randomly appearing rough spots on the susceptor, uniformity of film characteristics also suffers.

There are other ways to achieve maximization of the gas-phase mass-transport coefficient. One popular way, now in widespread use, is to use low-pressure reactors. By reducing the pressure, the gas diffusivity increases, raising h_g. The gas viscoscity declines somewhat, allowing higher flow rates.

To summarize, in this section, we have studied the impact of gas-phase—mass-transport and surface-reaction rates in defining the CVD film-deposition process. Two regimes of operation are possible for the CVD reactor. In the gas-phase mass-transport-limited regime, the speed with which reacting gas can be moved from the feed to the reacting surface defines the growth rate. In the surface-reaction rate-limited regime, the speed of molecular decomposition and surface chemical reaction defines the growth rate. Since the stagnant-layer thickness varies, it is best to work in the surface-reaction rate-limited regime. This is achieved by increasing flow rates or by reducing pressure.

7.4 Gas-Phase Chemical Equilibria in CVD

In the preceeding section, transport kinetics and surface reaction rates were used to determine the growth rate of a chemical-vapor-deposited thin film. In addition to these factors, there may be limitations due to the thermodynamics of gas-phase equilibrium chemistry. This occurs in two ways. First, the solid-deposited surface attempts to achieve thermodynamic equilibrium with the gases above it. Consider the deposition of silicon as an example. Thermodynamics demands that there should be a certain amount of silicon tied up in its various gaseous compounds over the reacting surface. Silicon-containing molecules will decompose on the wafer surface, increasing the volume of the deposited layer. If the amount of silicon in the feed gas exceeds the equilibrium requirement, deposition can take place. If the feed gas contains less silicon than equilibrium requires, solid silicon will return to the gas phase. In addition, certain gas compounds can be formed that will "etch" the growing silicon film. Specifically, HCl will attack silicon, returning the solid deposit to the gas phase. Methods of analyzing these effects are given below.

Let us continue with our example of the CVD of silicon. The feed gas to be considered is SiH_2Cl_2 in an inert carrier (such as nitrogen). A large number of reactions can occur in this heated gas mixture. Spectroscopic analysis indicates the following five reactions occur at measurable rates:

$$SiCl_4 + H_2 \rightleftarrows SiHCl_3 + HCl \tag{7.17}$$
$$SiHCl_3 + H_2 \rightleftarrows SiH_2Cl_2 + HCl \tag{7.18}$$
$$SiH_2Cl_2 \rightleftarrows SiCl_2 + H_2 \tag{7.19}$$
$$SiHCl_3 \rightleftarrows SiCl_2 + HCl \tag{7.20}$$
$$SiCl_2 + H_2 \rightleftarrows Si_{[solid]} + 2HCl. \tag{7.21}$$

By analyzing these reactions, we hope to answer the questions:

1. How much silicon is in the equilibrium gas phase for a given amount of silicon in the feed?
2. How much "etchant" gas, such as HCl, is present?
3. What effect will the addition of other reaction gases, such as hydrogen, have on the overall reaction?

The analysis is begun by writing the equilibrium constants for each of the five reactions:

$$K_1 = [SiHCl_3][HCl]/[SiCl_4][H_2] \tag{7.22}$$
$$K_2 = [SiH_2Cl_2][HCl]/[SiHCl_3][H_2] \tag{7.23}$$
$$K_3 = [SiCl_2][H_2]/[SiH_2Cl_2] \tag{7.24}$$
$$K_4 = [SiCl_2][HCl]/[SiHCl_3] \tag{7.25}$$
$$K_5 = [HCl]^2/[SiCl_2][H_2]. \tag{7.26}$$

In the gas phase, the brackets refer to the partial pressures of the molecules cited. Note that the solid silicon activity does not appear in eq. (7.26). The solid activities are taken as 1.

Numerical values can be assigned to each of the equilibrium coefficients using thermodynamic tables. The JANAF Thermochemical Tables [5] available through the National Standards Reference Data Service of the National Bureau of Standards were used for this example. The equilibrium constant for a given reaction can be evaluated with the aid of the following formula:

$$K_i = \exp\left(-\frac{\Delta G}{RT}\right), \tag{7.27}$$

TABLE 7.2: Free energy of formation ΔG_f at 1100°C of various compounds from elemental constituents in their standard states

Compound	Free Energy of Formation ΔG_f (kcal/mol)
1. Trichlorosilane ($SiHCl_3$)	-83.78
2. Silicon Tetrachloride ($SiCl_4$)	-122.55
3. Dichlorosilane ($SiCl_2$)	-50.75
4. Silicon Dichloride ($SiCl_2$)	-49.22
5. Hydrochloric Acid (HCl)	-24.24
6. Hydrogen (H_2)	Reference state (0 Kcal/mol)
7. Silicon (Si)	Reference state (0 Kcal/mol)

where ΔG = free energy change on reaction, T = temperature (°K), R = universal gas constant [1.987 kcal/mole °K], and K_i = equilibrium constant for the ith species. The free-energy change on reaction is derived by subtracting the sum of the free energies of formation of the compounds on the right-hand side of the chemical equation from the sum of the free energies of formation of the compounds on the left-hand side. The relevant data (taken from the JANAF tables) are displayed in table 7.2.

The equilibrium constants (at 1100 °C) from the data in table 7.2 are:

$$K_1 = 5.39 \times 10^{-3}$$
$$K_2 = 4.22 \times 10^{-2}$$
$$K_3 = 5.76 \times 10^{-1}$$
$$K_4 = 2.45 \times 10^{-2}$$
$$K_5 = 7.66 \times 10^{-1}.$$

Inspection of these constants allows us to simplify the problem somewhat.

First, note that all of these constants are less than one. This means that higher partial pressures would be expected for molecular species appearing on the left-hand side of eqs. (7.17) to (7.21). If we feed dichlorosilane into the reactor there will be some tendency for the molecule to dissociate (in the gas phase) to become silicon dichloride and hydrogen through reaction (7.19). Formation of trichlorosilane or silicon tetrachloride requires hydrochloric acid to be present in the gas phase. None has been formed at this point. Silicon dichloride and hydrogen can combine to form solid silicon and hydrochloric acid through reaction (7.21).

Thus the sequence of events is as follows. A little dichlorosilane will dissociate to form a little silicon dichloride. This, in turn, forms an even smaller amount of HCl, which can enable the other reactions. These reactions will produce other species. The "rate-limiting step" for the formation of trichlorosilane and silicon tetrachloride is the formation of HCl through reactions (7.19) and (7.21). Thus, we consider dichlorosilane, silicon dichloride, HCl, and hydrogen as the dominant species present in the equilibrium mix. We evaluate the partial pressures of these species using reactions (7.19) and (7.21), ignoring the remaining equations. For a more precise solution, we can use these results as the starting point for a numerical analysis of the full equation set. The approximation allows us to see major trends without getting hung up in the nuances of numerical analysis.

The computation can be simplified by using the following notation for the equilibrium partial pressures: silicon dichloride = x_1, hydrogen = x_2, dichlorosilane = x_3, HCl = x_4. The feed partial pressures of dichlorosilane and nitrogen are x_f and y, respectively. We assume that the feed partial pressures are known. This leaves the four unknowns $x_1, \ldots, x_4$ to be determined. Equations (7.19) and (7.21) can be rewritten:

$$K_3 = \left(\frac{x_1 x_2}{x_3}\right), \tag{7.28}$$

and

$$K_5 = \frac{x_4^2}{x_1 x_2}. \tag{7.29}$$

Most CVD reactions take place at constant pressure. For this example, assume atmospheric pressure prevails in both the feed and in the equilibrium mixture. For the equilibrium mixture, this works out to a third equation,

$$x_1 + x_2 + x_3 + x_4 + z = 1, \tag{7.30}$$

where z is the partial pressure of nitrogen in the equilibrium mixture. While the number of moles of nitrogen is taken to be fixed and known throughout the reaction, y and z are not the same. The reasons for this will be shown shortly.

The fourth equation is derived by demanding conservation of atoms in the gas phase. The numbers of hydrogen and chlorine atoms in the reactor must equal the numbers of these species in the feed. We do not consider silicon, as these atoms are continuously lost to the gas through the deposition process.

At first glance it appears that the system is overdetermined. Conservation of hydrogen and chlorine lead to two more equations: five in total for a four-unknown system! In actuality, there is a fifth unknown variable. This variable is the reaction volume. To see how this comes about, view the system in the following light. Say that the reactions take place in a cylinder of some inert material that is sealed from the outside world by a light (massless?) stopper. This stopper is in intimate contact with the cylinder walls but is free to slide along the cylinder axis without friction.

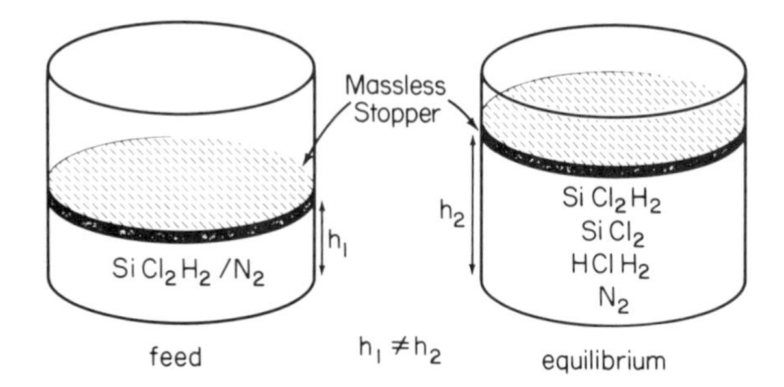

Fig. 7.5. Volume change on reaction.

This frictionless, massless cork seals the reaction but allows the gases in the cylinder to remain at atmospheric pressure. It is further assumed that controllers are available to keep the temperature constant.

Now imagine that the feed gases are bled into the cylinder (see fig. 7.5). The stopper moves up along the cylinder axis and stops at some level. Next, the reaction is initiated and equilibrium is reached. Since the total number of molecules present is different after the reaction, the stopper level is changed, but the temperature and cylinder pressure are the same as they were before the reaction took place. This is a pretty fair approximation of what happens inside an atmospheric-pressure CVD reactor.

All of these considerations are important in determining the atom conservation equations. We begin by expressing these equations verbally:

\# of moles of H in feed = # of moles of H in H_2 part of equilibrium mix
\+ # of moles of H in HCl part of equilibrium mix
\+ # of moles of H in SiH_2Cl_2 part of equilibrium mix.

We give this statement mathematical substance through the ideal gas law

$$n_x = \frac{p_x V}{RT}, \tag{7.31}$$

where the subscript x refers to the species, p_x is the partial pressure of the x species, V is the volume of the gases, and RT is as defined in eq. (7.27). Thus, we can write

$$2\left(\frac{x_\mathrm{f} V_\mathrm{f}}{RT}\right) = (2x_2 + 2x_3 + 1x_4)\frac{V_\mathrm{e}}{RT}. \tag{7.32}$$

Here, V_f refers to the volume of feed gas, and V_e refers to the equilibrium reaction volume. These volumes are not the same, for reasons given above. The factor of 2 on the left-hand side occurs because there are two hydrogen atoms in dichlorosilane. Similarly x_2 gets a 2 multiplier because there are 2 hydrogen atoms in H_2, etc.

Continuing along these lines, we find for the chlorine atom conservation

$$2\left(\frac{x_f V_f}{RT}\right) = (2x_1 \ + \ 2x_3 \ + \ 1x_4)\frac{V_e}{RT}. \tag{7.33}$$

Thus, the fifth unknown is the ratio of the feed to equilibrium volumes. This unknown is eliminated and the fourth equation is gotten by equating the right-hand sides of eqs. (7.32) and (7.33):

$$x_1 \ = \ x_2. \tag{7.34}$$

The four equations can be rearranged to form the following set:

$$x_1^2 + \left(2k_3 \ + \ k_3\sqrt{k_5}\right)x_1 \ + \ (z-1)k_3 = 0, \tag{7.35}$$

$$x_2 = x_1, \tag{7.36}$$

$$x_3 = \frac{x_1^2}{k_3}, \tag{7.37}$$

$$x_4 = \sqrt{k_5}x_1. \tag{7.38}$$

Equation (7.35) can be solved for x_1, and all the other unknowns can be recovered. The root of the quadratic [eq. (7.35)] is chosen so as to make all the partial pressures positive.

These solutions are displayed as a function of y for the 1100°C reaction temperature case. To make this plot, the equilibrium partial pressure of nitrogen, z, must be converted to the feed partial pressure, y. This is done by realizing that the number of moles of nitrogen are the same before and after reaction. Thus, the following is obtained:

$$y = \left(\frac{V_e}{V_f}\right)z. \tag{7.39}$$

Equations (7.32) or (7.33) can be used to solve for the feed volume to equilibrium volume ratio. Since $x_f + y \ = \ 1$ (assumption of atmospheric pressure before and after reaction), we immediately know x_f.

As stated above, one of the goals of this calculation is to see if solid silicon can be deposited. As stated above, for this to happen, the number of moles of silicon in the feed must be greater than the number of moles of silicon in the equilibrium gas mixture. If the opposite were true, solid silicon would be etched from the depositing surface and enter the gas phase. To see if this is the case, we evaluate the ratio of the mole number of silicon atoms in equilibrium to the mole number of silicon atoms in the feed gas. The number

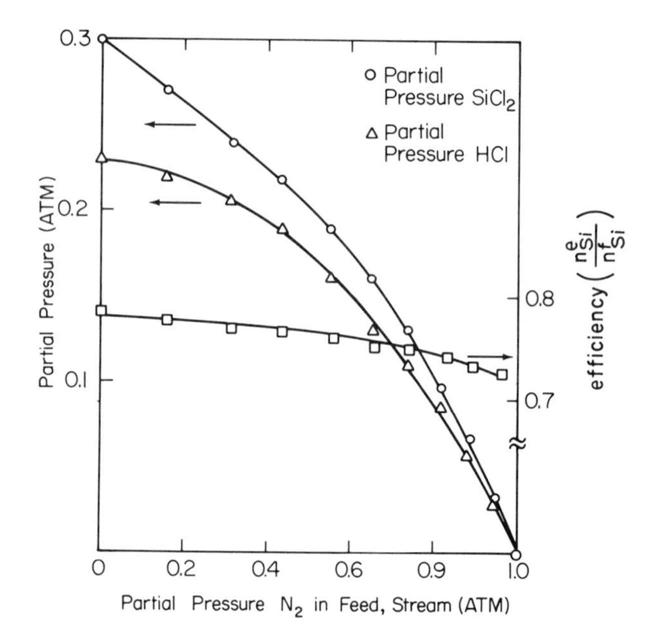

Fig. 7.6. Gas phase equilibrium partial pressures and equilibrium to feed mole ratios for CVD silicon at 1100°C from $SiCl_2H_2$.

of moles of silicon atoms in equilibrium per unit volume is just the number of moles of silicon dichloride plus the number of moles of dichlorosilane (per unit volume)

$$n_{\rm si}^{\rm e} = (x_1 + x_3)\frac{V_{\rm e}}{RT}. \tag{7.40}$$

The number of moles of silicon in the feed (accounting for volume changes) is

$$n_{\rm si}^{\rm f} = x_{\rm f}\left(\frac{V_{\rm f}}{RT}\right). \tag{7.41}$$

If $n_{\rm si}^{\rm e}/n_{\rm si}^{\rm f} < 1$, deposition will take place. From fig. 7.6, we see that *at the temperatures and pressures under consideration*, the ratio is always less than 0.8. Thus, deposition is always thermodynamically possible.

The fact that deposition is thermodynamically possible does not mean that it will always occur. The calculations described in this section specify the availability of reactants in the gas phase. The deposition occurs on the surface of the work piece. To calculate net deposition rate, an analysis based on the gas-phase mass transfer rate and surface reaction rate must be employed. Such an analysis was performed in the preceeding section. The techniques described in this section supply the $C_{\rm g}$ terms. An added dimension appears in the introduction of HCl. The arrival of silicon dichloride can be viewed as the driver for deposition. The arrival of HCl from the equilibrium gas phase can be viewed as the driver for *reverse deposition*, i.e., etching. Each of these reactions is characterized by a different gas-phase mass-transfer coefficient and surface-reaction rate. The net growth rate is just the sum of the growth rates of the two processes. The HCl "growth rate" is negative as it is a removal process.

Finally, we must address the question of what happens when the feed is enriched with some other species present in the equilibrium phase. For example, consider reaction (7.19) alone. According to eq. (7.24), addition of molecular hydrogen would move the reaction to the left (creating more dichlorosilane), but consideration of eq. (7.21) indicates that more solid silicon would be formed. What actually happens? The results can be calculated by realizing that in the feed gas mixture we now have:

$$x_{\mathrm{f}} + y + x_{\mathrm{H}_2} = 1, \tag{7.42}$$

where x_{H2} is the partial pressure of molecular hydrogen in the feed. Equation (6.32) is modified to read:

$$2\left[(x_{\mathrm{f}} + x_{\mathrm{H}_2})\frac{V_{\mathrm{f}}}{RT}\right] = (2x_2 \ + \ 2x_3 \ + \ 1x_4)\frac{V_{\mathrm{e}}}{RT}. \tag{7.43}$$

All other equations are unaffected.

While the mathematics of the simultaneous nonlinear equations is a bit more difficult, solutions are obtainable. The results are shown in fig. 7.7. Inspection of fig. 7.7 indicates the following. Addition of hydrogen to the feed decreases the equilibrium partial pressure of silicon dichloride. This occurs through the reversal of reaction (7.19). Along with this, the increase in the partial pressure of dichlorosilane occurs. This is also evident in the figure. Somewhat more surprising is the substantial increase in hydrochloric acid, which could lead to a lower overall growth rate (or even etching). This enhancement occurs through the activation of the forward part of reaction (7.21).

To summarize, in this section we have shown the importance of equilibrium gas chemistry in determining whether or not a given deposition reaction proceeds. We have also used this analysis to show how to calculate the effect of enriching the feed with gases in the mixture above the reacting surface. It must emphasized that this is an *equilibrium* analysis. When the feed-gas flow rate is very high, or if the reacting volume through which the gas flows is very small, this condition may not prevail. In this case the analysis can be performed by defining another variable: the ratio of the number of moles of product gas to the number of moles of feed gas. New constraining conditions on the thermodynamic variables must be sought to allow solution for the extra variable. For example, we can write a new equation demanding that there should be no net change in the heat content of the feed and reacting mixtures (adiabatic approximation).

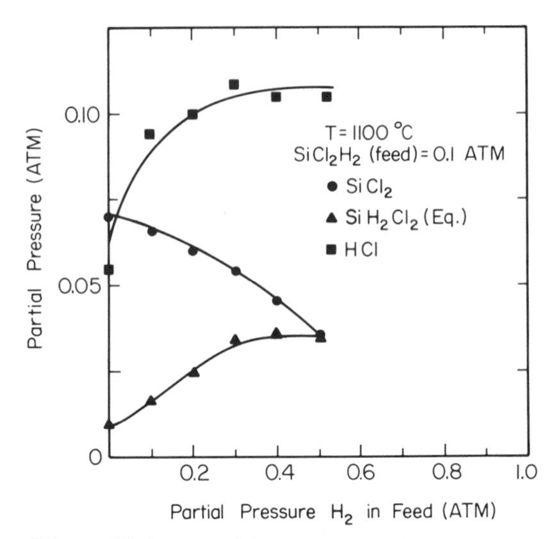

Fig. 7.7. Effect of H_2 equilibrium partial pressures on silicon epitaxy from Dichlorosilane.

The results cited in this section do not supplant the transport/surface reaction analysis done in the preceeding section. Rather, they augment it and provide a more in-depth look at the CVD process. The example chosen, dichlorosilane decomposition, is rather elementary. Both simpler and more complicated systems exist in practice. Silane (SiH_4) will decompose to form free silicon and two molecules of hydrogen. This process can be modeled with a single reaction equation. Silicon tetrachloride/hydrogen mixtures require reaction eqs. (7.17) to (7.21) for a complete analysis. Accounting for more reactions means solving more nonlinear equations in more unknowns. Techniques for performing such analyses through Newton-Raphson iteration are given in reference [6]. This technique is general and can be applied to the case of enrichment of the feed with hydrogen or HCl. A series of analyses on CVD deposition based on the considerations presented here have appeared (see refs. [7], [8], [9], and [10]). It is instructive for the student to review these papers. In addition, more thorough modeling of gas flow and the effects of gas flow on CVD have been presented by Rosner [11].

7.5 Nucleation and Growth of Amorphous, Polycrystalline and Crystalline Thin Films from Chemical Vapors

In this section, we deal with the final phase of CVD thin-film growth: the incorporation of the constituent atoms or molecules into the solid surface. We will take as a given that the requisite atoms or molecules have arrived at the growth interface and that they are in a suitable state for this incorporation. Consider, as an example, the chemical vapor deposition of silicon from silane.

The following process is envisioned [12]. Free silicon exists on the depositing surface through the breakdown of the incident silane molecules. These atoms are only weakly bound to this surface, largely through Van Der Waals forces. This is called *physisorption*, as opposed to chemisorption (in which there is charge transfer between the atom and the substrate). The enthalpy of desorption of physisorbed species is, generally, less than 0.1 eV. As such, physisorbed species can move about relatively freely. A number of things can happen to the weakly bound silicon atoms as they travel about. A mobile atom can encounter a strong chemical binding site, at which point the atom's migration ceases and the film has grown by one atom. Evaporation processes are also possible. A surface phonon can interact with a physisorbed atom and kick the atom back into the gas phase. Most CVD films are considered to be the result of nucleation and growth processes that include some features of the first two processes mentioned (single-atom binding and evaporation).

We encountered nucleation and growth phenomena in our study of starting materials. The formation of oxygen precipitates is an example of such a process. Let us review what we mean by nucleation and growth. Nuclei are small islands of the growing film. There is a distinction between stable and unstable nuclei. The free energy of unstable nuclei increases as we add atoms or molecules and decreases when we remove them. For stable nuclei, the opposite is true. In a true nucleation and growth process, there is a nucleus of critical size. Nuclei smaller than the critical size are unstable and will tend to disappear. Larger nuclei are viewed as stable for all time and can grow as they encounter more film atoms or molecules. The growth process is illustrated in fig. 7.8.

Since atoms or molecules attach themselves to the nuclei one at a time, the question arises as to how stable nuclei are ever formed. The answer lies in the realm of probability and statistics. Any time an atom or molecule encounters a nucleus, it will attach itself for a brief period of time before thermal agitation can shake it off. This occurs whether or not the nucleus has reached critical size. The nucleus is under constant bombardment by active species. Sometimes a nucleus can encounter many potential members in the time it takes to shake off a single one. This gives rise to a distribution of cluster sizes. The tail of this distribution can extend over the critical size limit. The factors affecting the shape of the distribution are the size of the critical nuclei and the dwell time of a single member on an unstable nucleus. If the dwell time is extremely short, it is less likely that the nucleus will ever contain the critical number. The smaller the size of the critical nucleus, the more likely it should be to create a stable cluster. The rate of impingement of incorporating species is also important in defining the number of stable clusters present. The impingement rate is determined by two factors. First, the arrival rate of active species at the surface from the gas must be specified.

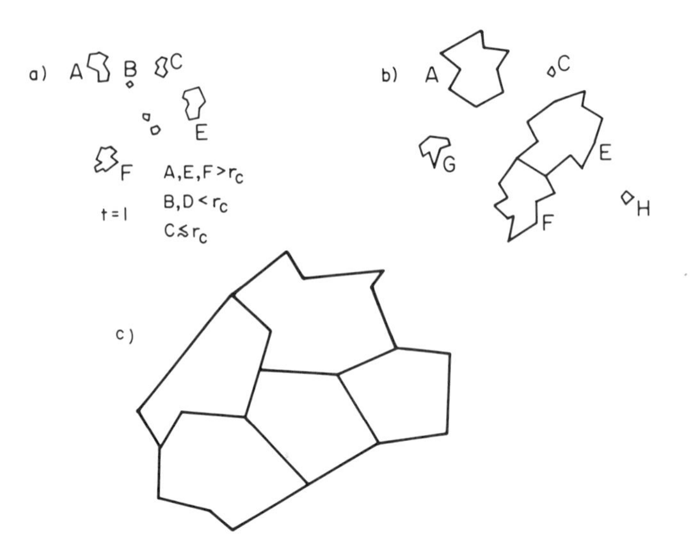

Fig. 7.8. Nucleation and growth of polycrystalline films. (a) Distribution of embryos at some time, t. (b) Embryo B is incorporated into A; C has shrunk; D has disappeared; coalescence has begun (E and F). New embryos appear (G, H). (c) Final phase; complete coalescence.

In addition, the surface mobility of the species is also important. If the atom or molecule is not really free to travel around the growth surface, it is less likely to encounter a growing nucleus. It will probably be desorbed.

The nucleation and growth process is "driven" by the change in free energy accompanying growth. For thin-film growth there are two parts to the free energy that must be considered: an interface part and a bulk part. The bulk free energy is largely determined by the energy of bond formation. It is negative. Take the case of CVD deposition of one material on a substrate of the same material. The surface term is the free energy of formation of the ambient/film interface plus an added component due to possible misalignment of the nucleus with the underlying substrate. In the case of a film of one material grown on another, there will be a bound length mismatch stress component to the surface free energy. The free energy of surface formation is a positive term. The bulk free energy lowers as the nucleus grows in proportion to the volume of the nucleus. Surface free energy raises in proportion to the surface area. Typical free energy behavior is shown in fig. 7.9. The free energy increases to some maximum. Then, the volume term begins to dominate, lowering the system free energy. The nucleus size at which the system free energy starts to lower is the critical size of a stable nucleus.

The nucleus formation process can be viewed as a chemical reaction [13]:

$$n\,A \rightleftarrows N_n, \tag{7.44}$$

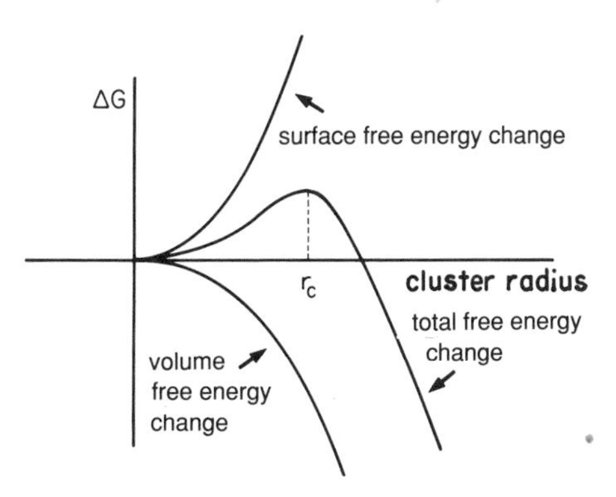

Fig. 7.9. Change in free energy on incorporating new atoms in a growing nucleus.

n atoms of A combine to form a nucleus of size n. From our knowledge of equilibrium constants, we might anticipate that the number of nuclei of size n could be written

$$N_n = C \exp\left(-\frac{\Delta G}{\mathrm{k}T}\right), \tag{7.45}$$

where ΔG is the free energy change of the nucleus on assembly from free atoms, k is the Boltzman constant, T is the temperature, and C is the proportionality constant. Typically, C is about equal to the number of atoms per unit area, N, present. Since surface area and nucleus volume are directly related to the number of atoms in the nucleus, a plot of ΔG as a function of n can be made. This, in turn, can be used to create a distribution function giving the number of nuclei of size n as a function of n. It is left as an exercise for the reader to do this calculation for some simple-geometry nuclei.

The appearance of eq. (7.45) raises some questions. As the free energy swings negative, N_n increases without bound. This would imply a tremendous number of large, stable nuclei. Such nuclei are not observed, implying that there is a limit on the applicability of the theory so far presented. In what is known as the Volmer-Weber (V-W) theory [14], the distribution is taken to be valid only for subcritical nuclei. The distribution curve is terminated abruptly (and unnaturally) when that size cluster is reached. Thus, the model holds only for the early phases of the growth process, when the number of stable nuclei are minimal. The V-W theory proceeds to calculate a nucleation rate, I, (the number of stable clusters formed per unit area per unit time) as

$$I = p_c A_c N \exp\left(-\frac{\Delta G(n_c)}{\mathrm{k}T}\right), \tag{7.46}$$

where p_c is the probability per unit time per unit of critical nucleus area of capturing an atom or molecule, and A_c is the area of the critical nucleus.

Note that ΔG is to be evaluated at the critical nucleus size, specified in terms of n_c. Here, n_c is the number of atoms in a critical nucleus.

An alternate theory has been put forth by Becker and Döring [14] that does not unnaturally terminate the nucleus size distribution. They form a number of coupled equations of the following form.

$$\begin{aligned} I_n &= p_n\ A_n\ N_n \ - \ d_{n+1}\ A_{n+1}\ N_{n+1} \\ I_{n+1} &= p_{n+1}\ A_{n+1}\ N_{n+1} \ - \ d_{n+2}\ A_{n+2}\ N_{n+2} \\ &\cdots \\ &\vdots \end{aligned} \tag{7.47}$$

The first term on the right-hand side of each equation represents the rate at which a cluster of some size captures an atom or molecule to become a cluster of some other size (n, in the case of the first equation). The second term represents the rate at which a cluster of some size ($n+1$ in the first equation) loses an atom to become a smaller cluster. The d factor represents the cluster desorption probability. Becker and Döring show that the nucleation rate for critical clusters is

$$I = \left(\frac{p_c A_c N}{N_c}\right)\left(\frac{\Delta G(n_c)}{3\pi \mathrm{k}T}\right)^{1/2} \exp\left(-\frac{\Delta G(N_c)}{\mathrm{k}T}\right). \tag{7.48}$$

This expression differs from that of the V-W theory by the exponential prefactor, providing a better model temperature dependence of the nucleation rate.

The above analysis presents a model of the kinetics of film growth. We must also be concerned with the basic structure of films that the nucleation and growth process provides. There are three basic atomic structures encountered in the solid state: amorphous, polycrystalline, and crystalline. In the amorphous state, atoms are displaced by some randomly occurring distance from their ideal crystal lattice sites. The atomic positions obviously cannot be totally random. Rather, the nearest neighbor and next nearest neighbor, etc., separation distances can be plotted as a series of peaked distributions. The separation at which each peak occurs is called a range parameter for the amorphous solid. This is shown in fig. 7.10. Amorphous solids are called glasses. Their atomic arrangement is referred to as a network, rather than a lattice. CVD-deposited SiO_2 and Si_2N_4 tend to be glasses. Both materials can still be viewed as tetrahedrally bonded solids. In amorphous SiO_2 each silicon atom is coordinated with four surrounding oxygen atoms. However, the bond angles vary from tetrahedron to tetrahedron and even within a tetrahedron.

The polycrystalline solid is composed of an aggregate of many nearly perfect crystallites. Each crystallite is called a grain. The grains butt up

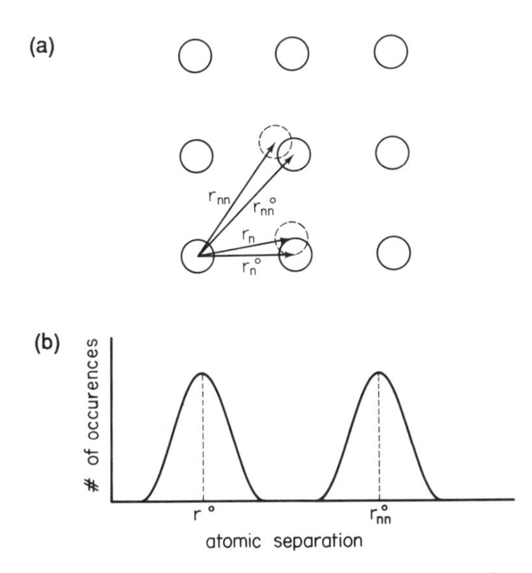

Fig. 7.10. Atomic separation in amorphous materials. (a) Nearest neighbor (r_n) and next nearest neighbor (r_{nn}) separation of atoms. There are 4 r_n^s and 4 r_{nn} for each atomic site. Note that r_n and r_{nn} may be greater or less than the r_n^o and r_{nn}^o. (b) In the amorphous solid there are "distributions" of nearest and next-nearest neighbor sites.

against each other at grain boundaries. Each tiny crystal has the same basic structure as every other. They differ only in their spatial orientations. The size of the grains depends on a number of factors. The primary factors are the rate of formation of stable clusters and the growth rate of these clusters. A large nucleation rate favors the formation of many small grains, but if the growth rate of these critical nuclei is very large, a single stable nucleus will engulf many small subcritical clusters. This increases the mean grain size. The cluster growth rate is determined by two other factors: the surface mobility of physisorbed species and the arrival rate of these species. A high surface mobility increases the probability that an atom or molecule will encounter a growing cluster. A high arrival rate ensures a large number of available constituents for incorporation into the cluster.

Both silicon and metals deposited on glassy substrates tend to form polycrystalline films. Silicon deposited on crystalline silicon at temperatures below 1000°C also tends to form polycrystal aggragates. At temperatures of about 1000°C or higher, the films can be viewed as crystal extensions of the substrate.

The process of thin-film crystal growth on a crystal substrate is called epitaxy [15] (from the Greek, meaning to "arrange upon"). Crystal growth of one material on a material of the same type is called *homoepitaxy*. Silicon-on-silicon growth is an example of homoepitaxy. Growth of material on a material of another type is called *heteroepitaxy*. Silicon-on-sapphire is an example of heteroepitaxy. Most epitaxial processing in silicon-based technology is done through chemical vapor deposition. Other techniques are possible. For example, one crystal can be placed in a melt and a crystal thin film can be grown on it. This is called *liquid-phase epitaxy* (LPE). Epitaxy can also be accomplished in evaporators. In addition, the use of molecular beams for epitaxy (MBE) is now being actively pursued. Both LPE and MBE are techniques of great importance in compound semiconductor work and will not be covered in great depth here, as our main focus is on silicon processing. However, much of what is said about the epitaxy process in general also holds for LPE and MBE.

It is interesting to ask what determines whether a given CVD thin film will be amorphous, polycrystalline, or crystalline. It is also interesting to note that, as of this writing, there is no good theory available to answer this question. We can, though, list some of the factors that must go into the creation of such a theory. Two factors figure prominently in any discussion of thin film structure: strain and surface mobility. The impact of these terms is discussed below.

One popular way to obtain an amorphous film is to reduce the substrate temperature. This does two things. First, it reduces the surface mobility of physisorbed species. The thermal-agitation driven motion of the substrate atoms is reduced. The dwell time of captured atoms is increased for the same reason. Thus, an atom can land in a metastable site (not a true minimum energy site) and rest there long enough to form a bond. Subsequent atoms attaching themselves to the growth surface are also displaced from their ideal lattice positions. There is minimal thermally driven reorganization of the metastable assembly. The result is a highly disordered growth—an amorphous film.

As the temperature is elevated, atoms are freer to migrate about the surface and to find more stable positions. Clusters tend to grow larger by attachment through surface migration. The cluster orientations do not have to coincide. In fact, such concidence is rare. There are a number of reasons for this. Think of the case of deposition on an amorphous solid. A number of cluster orientations, each with the same interface energy, is possible due to the random nature of the substrate atomic positions. Such a growth tends to be polycrystalline.

Even in the case of a crystalline substrate, there are many reasons why the growth tends to be polycrystalline. Dislocation formation figures strongly in the discussion of such structures. There is usually a lattice mismatch between the growth film and the substrate. This, of necessity, leads to a strained interface. How is this strain relieved? In the case of SiO_2, the bonds are flexible enough that the stress can be accomodated through a change in the bond angle. This leads to an amorphous growth. In the case of silicon on sapphire, though, this flexibility is not present. The strain is accomodated through the formation of dislocations. The dislocations can appear in two places (see fig. 7.11). They appear at the silicon/sapphire interface, where

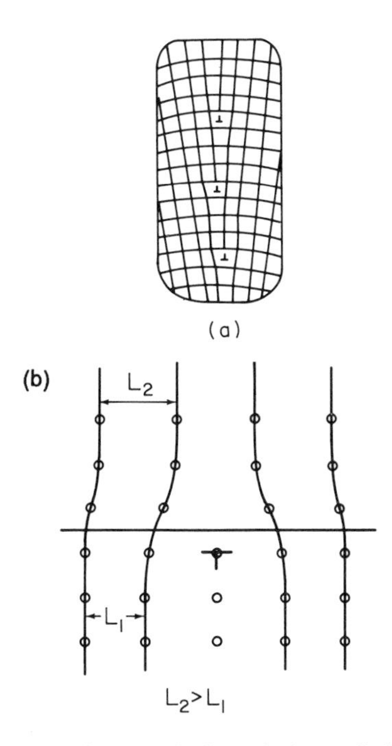

Fig. 7.11. Lattice mismatch at grain boundaries and through misfit dislocations. (a) Accommodation of rotational misfit at a grain boundary. (b) Accommodation of lattice mismatch through dislocation formation.

they are termed misfit dislocations; or, in the case of polycrystalline growth, they form low-angle grain boundaries. Both processes are shown in fig. 7.11.

Of course, both homo- and heteroepitaxy are possible. Above some critical temperature, the physisorbed species are mobile enough to wander the surface lattice and fall into their "best" minimum energy positions. In such a case, it is a matter of debate as to whether or not epitaxy is a nucleation and growth process. There is no "critical cluster" formed. An alternative way of looking at this is to say that the critical cluster size is one atom big. It is also possible that whole clusters can reorient themselves into lower energy configurations at high enough temperatures. The critical temperature for epitaxy is not simply material dependent. It is a function of the defect structure pre-existing on the host substrate, the surface cleanliness, and the deposition rate.

For epitaxy to proceed, it might be anticipated that the interface stress terms should be minimal. This consideration has led to what is known as the Royer rules for epitaxy. They state:

1. The epitaxial film lattice constants are the same or are multiples of those of the host substrate.

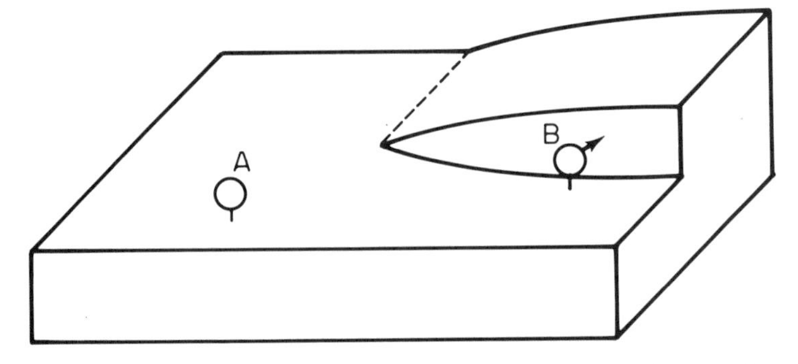

Fig. 7.12. Thin film growth through the screw dislocation mechanism. Atom B (at the screw ledge) as 2× the number of bonds as A.

2. For ionic crystals, the polarity of film ions should be the same as those of the substrate had the substrate extended to that level in the film.
3. Substrate and film should have the same type of chemical bonding.

These rules have been observed as violated so many times that they can be viewed as being mainly of historic interest. Some more reasonable modifications have appeared. In the Van Der Merwe theory [16], the affinity of the substrate host for the captured atom is large (much larger than the deposited atom-deposited atom interaction energy). There is very little atomic redistribution during growth, and the film grows at a monolayer at a time. Here, the lattice constants may differ by 4%. For a lattice constant difference of about 12% an array of misfit dislocations forms (as shown in fig. 7.11b).

One of the main reasons a comprehensive model of epitaxy has not emerged is the poorly understood role of substrate defects in the process. Defects frequently serve as nucleation sites of the critical clusters. The free energy of film formation is usually lower at these sites. The reasons for this are evident by inspection of fig. 7.12. Here we see substrate attachment of a film atom at a growth ledge caused by a screw dislocation. While the surface free-energy term may be the same, at the screw ledge two surface bonds are formed (rather than one). Surface bonding is usually exothermic, lowering the system free energy. The extra bond at the screw ledge provides a significant lowering of free energy. In addition, strained regions of surface can provide better lattice match in certain circumstances, further lowering the free energy of interface formation. Surface contaminants can block nucleation sites, impeding crystal growth.

Practical considerations associated with surface cleaniliness and machine maintenance frequently dominate in determining the quality of epifilms. Surface cleans, such as those described in Chapter 2, are usually mandatory. The walls of a heavily used CVD system will flake and cause epilayer defects.

Thus, the CVD reactor walls must frequently be scrubbed. In addition, at high growth rates, gas-phase nucleation can occur. This creates particle defects in the growing surfaces. Native oxide films must be removed completely to obtain a good crystal overgrowth. This is usually accomplished in an H_2 or HCl gas preclean done in the epireactor just prior to the deposition. This usually is a relatively high-temperature step (done at temperatures greater than or equal to $1150°C$). Short time periods (5-10 minutes) are common for this preclean to prevent outdiffusion of substrate dopant. Lowering the pressure to less than 25 torr enables a lowering of the native oxide stripping temperature to 950°C in H_2 without increasing the time of the cycle.

7.6 CVD Film Properties and Characterization

In this section, we consider the main CVD film properties that are of importance in IC processing. In addition, some common techniques used to characterize these properties are mentioned. The properties are grouped under the following headings: thickness and chemical, electrical, mechanical, and structural properties. Ellipsometric and transmission electron microscopic measurements are highlighted here. Specific results and applications for CVD films are cited.

7.6.1 Film Thickness Measurement

The most common technique used for thickness determination of optically transparent thin films (such as oxides, nitrides, and thin polysilicon films) is ellipsometry [17]. The technique is nondestructive and fairly localized, that is, the thickness measured is an average thickness over the diameter of an incident beam of polarized light (usually about a millimeter). The thickness uniformity from point to point across a wafer and from wafer to wafer can be determined by the ellipsometer. The mean thickness and standard deviations thus obtained can be used as process control variables. The ellipsometer also provides a measurement of the optical index of refraction. Some important chemical and structural information is included in this parameter, as discussed below.

Consider the basis of the ellipsometric measurement. An elliptically polarized beam of light is incident on a sample. The sample consists of a thin, partially transparent film on a reflecting, or partially reflecting, substrate. The beam undergoes a change in polarization state as it passes through the film and is reflected by the substrate. The ellipsometer measures this change in polarization state. First, we present the mathematical formalism used in the analysis of ellipsometric data. In succeeding paragraphs, we describe the apparatus used to determine the change in polarization state of the beam on reflection.

First, we must find a way to describe the beam's polarization. This is done by considering the electric-field vector, **E**, of the electromagnetic light wave. From elementary electromagnetic theory, it is known that there is no component of **E** in the direction of wave propagation. Typically, we specify **E** in terms of its projections parallel to and perpendicular to the film surface. These are designated as the p and s components, respectively. The instantaneous value of **E** projected along these axes is specified with a sine function. Each sine is characterized by an amplitude and a phase. Any state of polarization can be characterized by giving the amplitudes and phases of the p and s waves. If we view the electric-field vector along an axis pendendicular to the p and s directions, we find that this vector appears to rotate about this axis. The magnitude of **E** changes during the course of this rotation, and the tip of the vector appears to trace out an ellipse in the p-s plane. This is the origin of the term *elliptically polarized light.* The polarization state is specified by two quantities: the ratio of the semimajor to semiminor axes lengths and the angle made by the semimajor axis and either the p or s axes.

The change in polarization state of the beam on reflection is usually specified in terms of two quantities: ψ and Δ. The ψ term is related to the ratio of the reflected to incident p and s wave amplitudes

$$\psi = \arctan\left[\left(\frac{A_{\mathrm{p,r}}}{A_{\mathrm{s,r}}}\right) / \left(\frac{A_{\mathrm{p,i}}}{A_{\mathrm{s,i}}}\right)\right], \tag{7.49}$$

where the r and i subscripts refer to the reflected and incident waves. The Δ term is a function of the change in phase β of the p and s waves on reflection

$$\Delta = (\beta_{\mathrm{p,r}} - \beta_{\mathrm{s,r}}) - (\beta_{\mathrm{p,i}} - \beta_{\mathrm{s,i}}). \tag{7.50}$$

The intensity of the light is given as the square of the amplitude of the sine wave. From this fact and eqs. (7.49) and (7.50), the ratio of the reflection coefficients for the p and s waves can be expressed in terms of ψ and Δ

$$\frac{r_{\mathrm{p}}}{r_{\mathrm{s}}} = \tan\,\psi e^{i\Delta}, \tag{7.51}$$

where r refers to the reflection coefficient and i is $\sqrt{-1}$. The individual p and s reflection coefficients can be obtained from the Fresnel formulas. This yields the following relationship

$$\tan\,\psi e^{i\Delta} = \left(\frac{r_{1\mathrm{p}} + r_{2\mathrm{p}}e^{-2i\delta}}{1 + r_{1\mathrm{p}}r_{2\mathrm{p}}e^{-2i\delta}}\right)\left(\frac{1 + r_{1\mathrm{s}}r_{2\mathrm{s}}e^{-2i\delta}}{r_{1\mathrm{s}} + r_{2\mathrm{s}}e^{-2i\delta}}\right), \tag{7.52}$$

where the subscript 1 refers to the coefficient for reflection at the air/film interface and the subscript 2 refers to the coefficient for reflection at the film/substrate interface. The δ term is evaluated using the relationship

$$\delta = \left(\frac{360}{\lambda}\right) d(n_f^2 - \sin \phi_i)^{1/2} \text{ [degrees]}, \tag{7.53}$$

where λ = wavelength of the light incident, ϕ_i = angle of beam incident in the ambient (i.e., the angle between the incident beam and the interface normal), n_f = optical index of the measured film, and d = film thickness. Note that the film thickness is included in this expression. Usually, the film is so thin that the imaginary part of the index (the absorption term) is ignored and only the real part is considered.

The individual reflection coefficients can be evaluated using the Fresnel coefficients themselves

$$r_p = \frac{n_a \cos \phi_b - n_b \cos \phi_a}{n_a \cos \phi_b + n_b \cos\phi_a}, \tag{7.54}$$

and

$$r_s = \frac{n_a \cos \phi_a - n_b \cos \phi_b}{n_a \cos \phi_a + n_b \cos \phi_b}, \tag{7.55}$$

where n_a = refractive index of the ambient medium above the reflecting surface, n_b = refractive index of the medium below the reflecting surface, ϕ_a = angle of incidence, and ϕ_b = angle of refraction. The angle of incidence is related to the angle of refraction through Snell's law:

$$\frac{n_a}{n_b} = \frac{\sin \phi_b}{\sin \phi_a}. \tag{7.56}$$

The ellipsometer is used to measure ψ and Δ directly. In the usual case, the substrate optical constants are well characterized and the optical index of the ambient above the film is that of air. The imaginary part of the optical index is related to the absorption in the film. For thin films, this is taken to be zero. Separating the real and imaginary parts of (7.52) yields two equations that can be solved for the two unknowns: the index of refraction and the film thickness. The solution is usually done graphically. A family of plots of ψ and Δ for different δ's and film refractive indices is prepared. Such a plot is shown in fig. 7.13. Each tick mark on the curves shown in this figure indicates a different δ. Since the film optical index and the beam incidence angle are specified for each curve, δ can be converted to a thickness through eq. (7.53). To find n_f and d for a given ellipsometer measurement, the experimentally determined ψ and Δ point is indicated on the graph. The

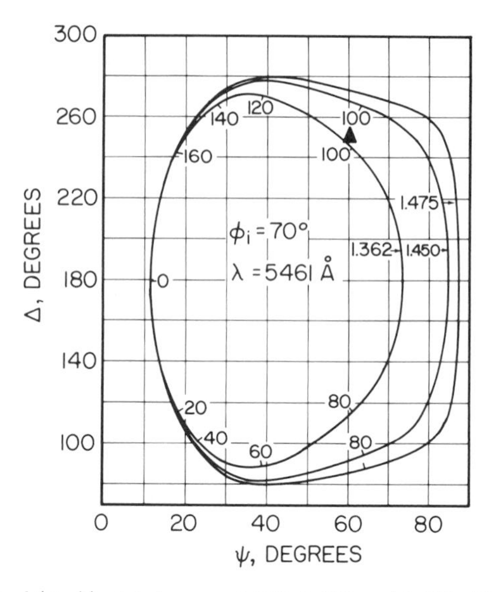

Fig. 7.13. A typical (ψ, Δ) plot for $n_f = 1.362$, 1.450 and 1.475. The individual curves for fixed n_f are collections of (χ, Δ) points for a variety of δ values. The triangular point is approximately at $\delta = 101°$ for an index of 1.362. This would thus correspond to a film thickness of 1674 Å for the λ and ϕ_i listed. A better estimate can be obtained by fitting a new (ψ, Δ) plot for a slightly higher n_f.

curve upon which this point lies tells us the film index. The position along the curve gives us δ (and, as a result, film thickness).

The ellipsometer is extremely sensitive to the presence of surface films. Films as thin as a monolayer can cause measurable polarization shift. There are problems in data analysis, though. For example, refer to fig. 7.13. There are points on the ψ - Δ plots in which a number of different index curves converge. It becomes difficult to separate out an unambiguous index curve for the determination of thickness. In addition, the technique used to fit a given ψ - Δ point to a given plot is somewhat arbitrary. When the curves are widely spaced, this is not a problem, though. With these considerations in mind it is frequently useful to inspect the (Δ, ψ) plots when interpreting ellipsometric data, even when commercial computer programs are used to reduce these data.

Frequently, layers of dielectric film are encountered in IC work. Nitride on oxide layers are typical examples. The Fresnel reflection equations can be extended to account for this. This leads to a modification of the right-hand side of eq. (7.52). The separation of this equation into real and imaginary parts leads to convenient determination of the real part of the optical index and thickness for a single layer only. Thus, the thickness and index of the second layer must be supplied to determine the unknown parameters of a second film using one ellipsometric reading. It is also possible to vary the wavelength of the incident light or the incident angle of the beam to get

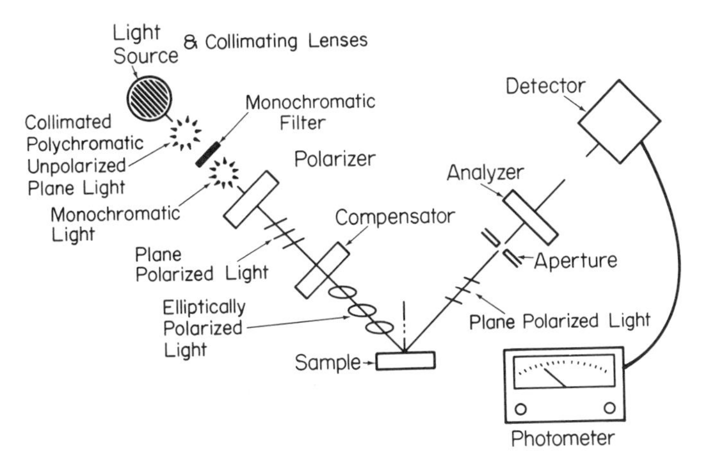

Fig. 7.14. The component parts of an ellipsometer.

additional equations and to solve for more than two unknowns. If the film absorption is significant, such an approach is mandatory.

Now let us turn our attention to the ellipsometer apparatus and the mechanics of ellipsometric measurement. A schematic of a typical ellipsometer is shown in fig. 7.14. Monochromatic light from a lamp or a laser is collimated and polarized. The polarized beam passes through a phase retarder, which introduces a phase shift between the p and s projections of the polarized wave. The emerging beam is elliptically polarized. The relative retardation of the p and s waves is changed by rotation of the polarizer. The reflected beam is passed through a second polarizer (called the analyzer). The intensity of the beam that emerges from the analyzer is measured with a photometer. The state of beam polarization is altered as the beam passes through the film. The p and s wave phase difference is altered, and the possibility exists for the reflected beam to become linearly polarized once more. If the optical axis of the polarizer was normal to the light polarization vector, the photometer would see no optical signal.

In making an ellipsometric measurement, the polarizer and analyzer are rotated randomly until a photometer null is reached. When the null is observed, the following relations hold:

$$\tan \Delta = \sin(\delta) \tan(90^\circ - 2\mathrm{P}_0) \tag{7.57}$$

$$\tan \psi = \cot(L) \tan(-\mathrm{A}_0) \tag{7.58}$$

$$\cos(2L) = \cos(\delta) \cos(2\mathrm{P}_0), \tag{7.59}$$

where P_0 and A_0 are the azimuth angles of the electric field vectors emerging from and incident on the polarizer and analyzer (respectively) when the null condition is observed. As a result of the cyclic nature of these functions, there are two sets of P_0 and A_0 readings at which the null is observed. For one set,

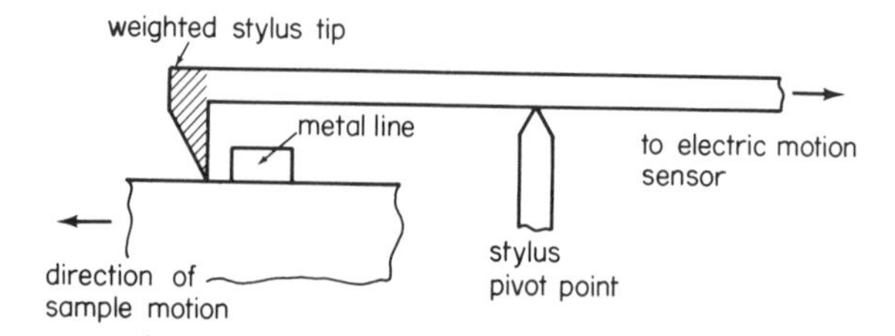

Fig. 7.15. Stylus profilometer measurement of thin film metal line thickness (not to scale). Motion of stylus arm induced by line step is sensed electronically and displayed on a strip recorder.

the analyzer is always in the first quadrant (0° to 90°); for the second set, the analyzer reading lies in the fourth quadrant (270° to 360°). Equations (7.57) to (7.59) hold for the first quadrant results. The fourth quadrant results do not give any additional data on the film properties. However, many of the commercially available data-reduction computer programs require the second data to improve measurement accuracy.

Optical interference techniques are also used in thickness determination. They are not as accurate as ellipsometry, and they involve a prior determination of the film's optical indices. They are, however, nondestructive and they can measure thicknesses in very small areas. "Microscopes" are now available that can measure the thickness of thin residual oxide layers in contact windows of near-micron sizes. Since silicon becomes transparent at infrared (IR) wavelengths longer than a micron, IR reflection is important in measuring silicon epitaxial layers.

For optically opaque films, stylus profilometers are useful (see fig. 7.15). The stylus mass per unit contact area is less than a gram per square centimeter, and it is *usually* nondestructive (this may not be true for very delicate films, though). Films as thin as 20Å have been measured with this technique. Finally, optically opaque films (such as epilayers) can be lapped and stained, that is, the layers can be ground down at some angle (as shown in fig. 7.16). A variety of stains can be used to reveal the layer in question under a light microscope. The angle-lapping "magnifies" the thickness of film presented to the eye under the microscope. The stained film thickness is measured and the true thickness is determined using the geometric correction factor as shown in fig. 7.16.

7.6.2 Electrical Properties

Electrical characterization is required for films that will remain as active and passive elements in the completed device. The type of characterization employed depends on whether the film is to be a conductor, a semiconductor, or an insulator. Consider the case of insulators first. Deposited silicon nitride is a good insulator. However, the current voltage ($I - V$)

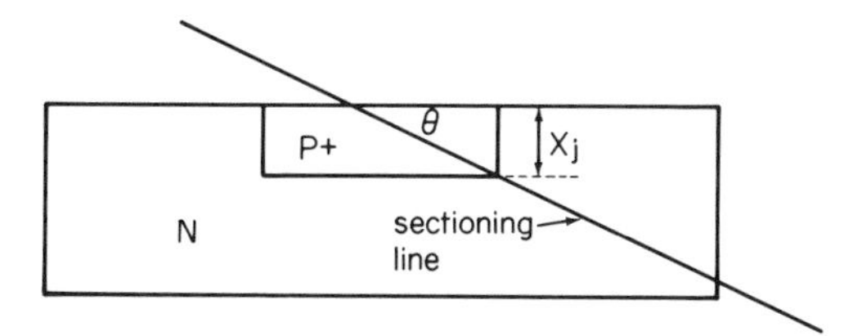

Fig. 7.16. Junction is sectioned at angle θ and stained to reveal p+ layer. Junction *appears* to be $(X_J/\sin\theta$ deep. Making θ shallower, makes junction depth appear larger.

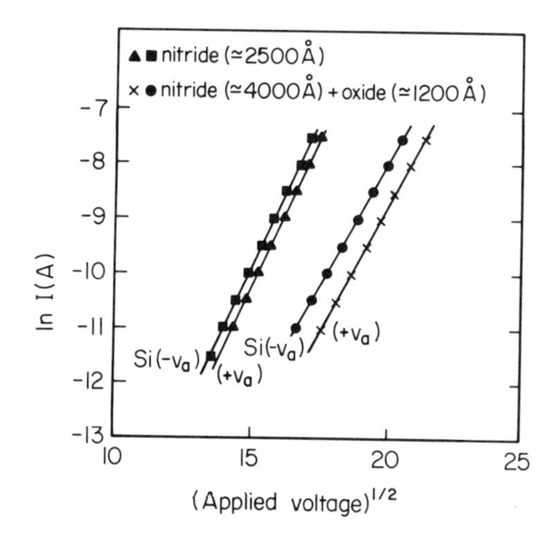

Fig. 7.17. Nitride current as a function of applied field. Nitride was deposited on silicon (as indicated). The applied voltage divided by film thickness gives (approximately) the electric field in the nitride. After reference 19.

characteristics of these films can easily be measured with the typical 30 mil diameter $C-V$-dot structure previously described in Chapter 3. Biases on the order of 10 volts give rise to currents near a microamp through films in the 500 to 1000Å thickness range. Plots of the logarithm of current against the square-root of the insulator electric field yield straight lines. The current is transported through the film by a field-assisted trap-to-trap hopping mechanism [18] (Poole-Frenkel effect). A typical nitride $I-V$ plot is shown in fig. 7.17. An electrometer used as a constant current source can be used to make these measurements. The current setting is dialed in and the voltage at which this current appears is recorded.

Deposited silicon dioxide, on the other hand, is an even better insulator. Current transport occurs by means of tunneling into the conduction band of the insulator (usually Fowler-Nordheim tunneling). The $C-V$ structure draws less than picoamps in a well-prepared film. Special guard-ringed structures (designed to minimize edge and surface leakages) may be needed to make an acccurate determination of film conductivity. However, gross problems associated with reactor contamination or poor gas flow can give rise to conducting films. As such, the electrometer measurement described above will serve as a go/no-go test of film quality.

Typically, four-point and spreading resistance probes are used to assess epi-silicon resistivities [20]. The spreading resistance probe can be applied to angle lapped layers, as discussed in depth in Chapter 4. Similarly, polysilicon layer resistivity can be probed using these techniques. Polysilicon layer resistivities are usually in the 10-30 $\Omega/\square$ range. Epilayer resistivities can be anywhere from hundreds of ohm centimeters to tens of ohm centimeters. Deposited metal line resistances, like evaporated or sputtered metal line resistances, are rarely a problem. Metal failures tend to be catastrophic (line shorts or opens). Contacts between metal and semiconductors (and between metal levels) tend to be much more of a problem than high line resistance. Thin films and contacts are discussed in depth in Chapter 6.

7.6.3 Chemical Properties

Departures from ideal film stoichiometry and the presence of extrinsic contaminants make insulating films electrically "leaky". Such problems can also lead to enhanced diffusion coefficients. This might render the deposited film useless as a diffusion barrier. Stoichiometry can be assessed using secondary ion mass spectroscopy (SIMS) [20]. Small amounts of extrinsic contamination (10 part-per-billion levels) can also be detected in this way. Large amounts of contamination (part-per-million levels) can be analyzed using the e-beam-excited x-ray fluorescence spectrometers frequently mounted on scanning electron microscopes. Rutherford backscattering [20] can be used to analyze multilayer structures. Here, the thickness of submerged layers can be determined and some information on stoichiometry is obtained. All of these techniques can be viewed as destructive. SIMS sputters away the layers in the course of the analysis. X-ray spectroscopy creates radiation damage in the insulating films. Rutherford backscattering is least damaging. However, it causes some radiation damage and can leave residual helium in the analyzed layers.

Ellipsometry can provide some chemical information in a nondestructive way. The index of refraction can be related to the chemical composition of the film. For example, stoichiometric silicon nitride has a refractive index close to 2. The most common extrinsic chemical contaminant in the nitride is oxygen. Oxygen lowers the refractive index. In the limit, the nitride would become silicon dioxide, with a refractive index of 1.45.

We can also view the doping profile of an epilayer as a chemical property. The profile can be intentionally nonuniform (as discussed in the applications section below). The film can also be "autodoped" during the growth process, that is, dopant atoms from the host substrate can diffuse into the grown

Fig. 7.18. Texture plot of niobium thin film on sapphire. Pattern on left is "almost"ringlike (as in a powder diffraction pattern). Regions of high intensity (texture) appear along ring. Region on right shows sharp spots characteristic of crystal sapphire. Courtesy of S. Qadri.

layer (and vice versa). In addition to spreading resistance probes, CV dopant profiling is useful here (see Chapter 4).

7.6.4 Structural and Mechanical Properties

The basic crystalline structure of thin films can be revealed through Laue backscattered x-ray diffraction [22]. Laue spot patterns can be used for crystal identification and for orientation purposes. Amorphous films will show ring patterns whose radii are characteristic of the range parameters of the atoms in the network (see fig. 2.22). Polycrystalline silicon films tend to form with a common grain axis perpendicular to the film surface. The individual grains are rotated about this axis with some degree of randomness. The resulting backscattered pattern, called a fiber pattern, is shown in fig. 7.18. Individual spots are swept out as arcs.

Defect structure of crystalline thin films of silicon can be ascertained in a number of ways. Defect-delineating etches (such as Wright and Sirtl etches) were discussed in Chapter 2. Such etches reveal the number of line dislocations penetrating the crystal surface as well as stacking faults. Similarly, x-

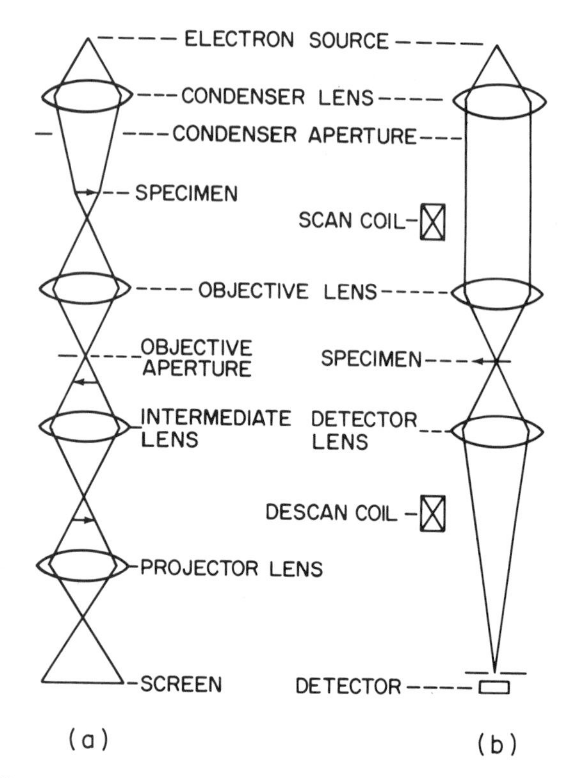

Fig. 7.19. Electron optic configurations for (a) TEM and (b) STEM. After reference 24.

ray topography [22] can also reveal the spatial distribution of surface defects in epitaxial layers (see fig. 2.23). In addition, x-ray topography can be used to measure local strain, as the crystal lattice constant is changed in these strained regions. This leads to a slightly different Bragg angle and a change in reflected x-ray intensity; this is also evident in fig. 2.23. For this same reason, the amount of strain present in a crystal can also be measured by determining the apparent broadening of diffraction spots.

Transmission electron microscopy (TEM) has emerged, over recent years, as an important technique for analyzing the layer and defect structure of IC films [24]. In TEM, a high energy flood electron beam is passed through a sample and the transmitted intensity pattern is brought to focus on a screen (or on a piece of film). In scanning transmission microscopy (STEM) the electrons form a beam that is passed through the sample, and an image is built up by scanning the sample point-by-point and recording transmitted intensities. Both methods are illustrated in fig. 7.19. Contrast is achieved by diffraction effects and is not simply related to the beam-absorption in the sample. To achieve TEM, the sample must be thinned to about 10-100 nm, or a replica of the sample must be made. Thinning is achieved by

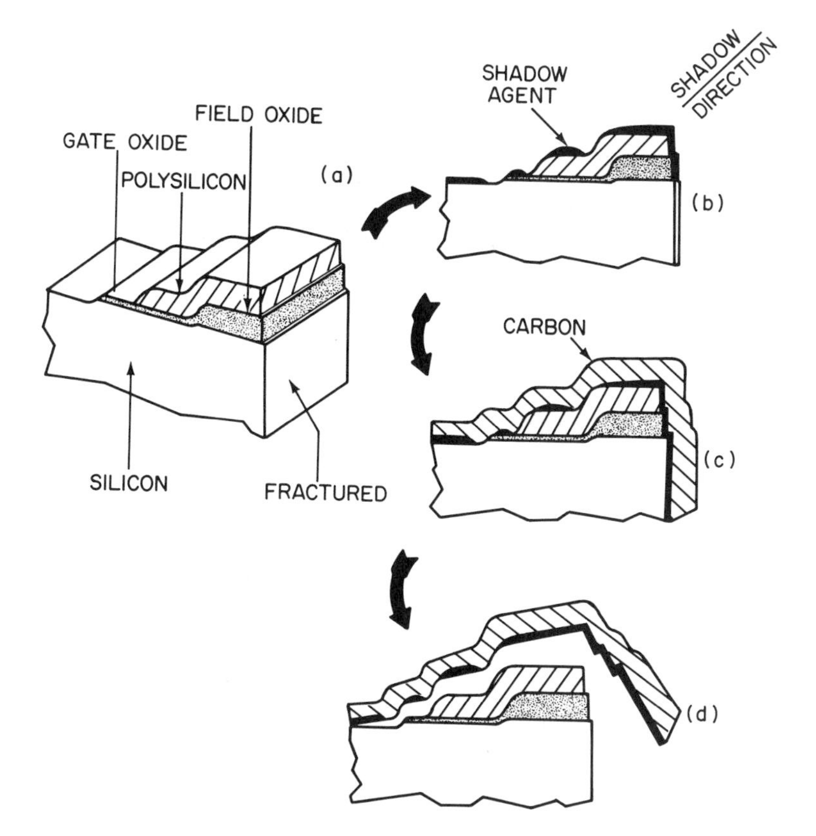

Fig. 7.20. Illustration of a method for preparing a shadow replica of a fractured VLSI chip for TEM study: (a) The fractured sample. (b) Deposition of a heavy metal shadowing agent. (c) Deposition of carbon film. (d) Stripping the replica that contains the shadowing agent. After reference 24.

a combination of grinding and ion-beam milling processes. A typical replica process is shown in fig. 7.20. Contrast in a deposited carbon replica is achieved through the metal shadow process or by steps left in the replica by differences in the cleared-region thicknesses. An application of the process is shown in fig. 7.21 in which amorphous and polycrystalline silicon film structures are compared.

Lattice constant mismatch in deposited thin films causes these films to be stressed. The film stress gives rise to strain in the underlying host lattice. Such strains can bow silicon wafers. The stress can be either tensile (contracting the lattice beneath the film) or compressive (expanding the lattice beneath the film). Stress is defined as the force per unit area (dynes/cm^2). The material is strained (i.e., suffers compression or dilation) as a result of stress. In metal thin films, the typical stresses encountered range from 10^8 dynes/cm^2 to 10^{10} dynes/cm^2. Refractory metals (such as platinum

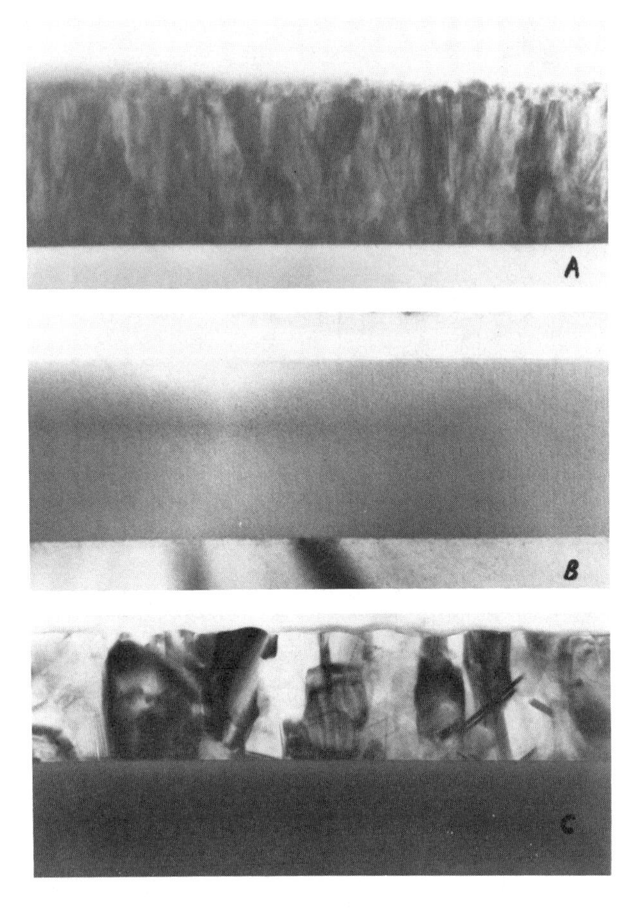

Fig. 7.21. Different states of "polycrystal"silicon as a function of doping and deposition technique: (a) As-deposited at $\sim$630°C, (b) As-deposited at $<$500°C, and (c) after PBr_3 doping at 950°C.

and molybdenum) tend to have high stress, while compliant metals like gold have less stress. Metal stresses tend to be tensile, while dielectrics can be compressed on deposition. Film stress can cause significant bowing in the wafer. This is discussed in greater detail in Chapter 6.

7.7 Applications and Special Techniques

In this section, we cover the major applications of CVD processes in integrated-circuit technology. The organization of the section is by structure type: glasses first, polycrystalline materials next, and epitaxial films last. The section concludes with some discussion of less standard techniques and the roles these techniques are playing in advancing semiconductor technology. Specifically, methods and applications of recrystallization, beam deposition, and selective growth are described.

Silicon dioxide and silicon nitride are deposited as glasses in CVD reactors. CVD SiO_2 is frequently called silox. Silox films can be grown thicker at lower temperature, in less time, than an oxidized silicon film. The deposited film is usually of lower quality than the grown film and is rarely used in the active device regions, that is, it would not be used as a gate oxide. One reason for this is that there are many more charge-trapping sites present in the silox layers. This leads to threshold shifts and instabilities in MOSFETs. This does not preclude the use of this film as a thick-field isolation insulator between the first-level interconnect layer and the substrate. The low deposition temperature creates less spreading of shallow junctions and less wafer warpage. In addition, silox can be used as an insulating layer between metal layers or between metal and polysilicon. Similarly, silox is a popular "scratch protection" deposited over the whole wafer after fabrication. Only the bonding pads are exposed in a subsequent etch step.

Another advantage of silox as an interlayer dielectric is the fact that it can be reflowed. For example, assume that we desire a via hole cut from metal-layer 1 to poly. In VLSI technology this contact might be a square 1-2 μm on a side. Some anisotropic etch technique would be used to define it. The side wall or the via would be extremely thick. Metal 1 would be put down using a conventional planetary deposition technique (as described in Chapter 6). For reasons described in Chapter 6, it is very difficult to cover steep sidewalls with evaporated or sputtered metal. It is desirable to "slope" the sidewall somewhat. One way to do this is to heat the silox after via definition. This softens the glass surface and tension causes it to reflow into a smoother configuration. This is shown in fig. 7.22. Addition of phosphorous to the glass during the deposition process, or combinations of boron and phosphorous cause the reflow temperature to lower. Reflow temperatures as low as 750°C have been reported for borophosphosilicate glasses [25].

Phosphosilicate and arsenosilicate glasses can be produced by introducing phosphine or arsene into the feed-gas mixture [1]. Borosilicate glasses are made (particularly in low-pressure reactors) using mixtures of tetrapropoxysilane, tripropyl borate, oxygen, and nitrogen. Borophosphosilicate glasses can be made by starting with a phosphosilicate glass and further doping it with boron in a separate doping step.

Both silox and silicon nitride are used as diffusion barriers. Of the two, silicon nitride is the best diffusion barrier. The use of deposited films as a barrier for various impurities of importance in IC processing is discussed in Chapter 4. The fact that nitride is a better barrier means that thinner films of nitride can be used to achieve the same level of protection. This

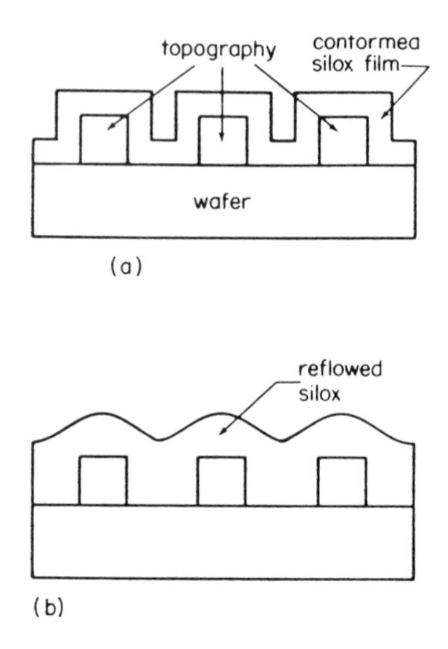

Fig. 7.22. Silox reflow-process: (a) Prior to reflow. (b) Sharp edges round after reflow.

is important in high resolution pattern definition. As a rule, dimensional control varies inversely with the thickness of the layer defined. Thus, nitrides are the barriers of choice when lateral dimension control of the diffusion is important.

Nitride also acts as a diffusion barrier for oxygen. This is the basis of many of the "local" oxidation of silicon (LOCOS) techniques used in VLSI field isolations. A number of variants of the LOCOS process are used in VLSI circuits today. Three of the most common are presented here: straight LOCOS [26], ROX [27], and SILOS [28]. We begin with straight LOCOS, whose process flow is shown in fig. 7.23.

The purpose of LOCOS is to provide a self-aligned field region. The field region provides the boundary between the active channel and the isolation region. As such, it defines the MOSFET channel width, as shown in fig. 1.13. The field is comprised of two elements: a thick field oxide and an inversion-preventing implant. In nMOS technology, the field contains a heavy p-type implant. In older nMOS technology, it is necessary to create the thick field oxidation and the implant in two separate masking steps. The field oxide process sequence shown in fig. 7.23 indicates a technique for doing this in a single mask step. A nitride/oxide dual-dielectric layer is deposited and patterned on the silicon. The oxide is called a "stress-relief" layer. It absorbs some of the stresses generated by the nitride and prevents defect generation and wafer bowing. The patterning resist is used as an implant mask for the p+ layer. After implant, the structure is oxidized to form the thick field region. The nitride film prevents oxidation in the active gate regions of the device. The p+ implant is naturally self-aligned to the field oxide in this way.

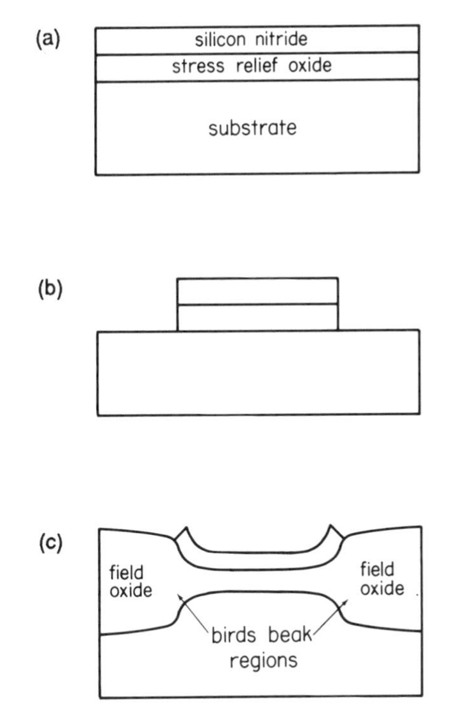

Fig. 7.23. Self-aligned field process: (a) Deposit stress-relief oxide and nitride layer. (b) Define oxide/nitride layers. (c) Grow field oxide.

There are two drawbacks to this process. First, it is not planar. The field oxide projects above the gate oxide and can create metal step coverage problems in subsequent process steps. Also, there is an apparent "birds-beak" effect (encroachment of the oxide under the nitride passivation) due to oxygen diffusion during field oxidation evident in fig. 7.23. Lack of planarity can be overcome with the recessed oxidation (ROX) process (fig. 7.24). In ROX, the nitride and stress relief oxide are deposited and defined. At this point, the process differs from LOCOS. The silicon is etched to half the depth of the field oxide. The structure is then oxidized. The oxidized field swells to become level with the active device window. This eliminates the lack of planarity of LOCOS field oxides, while maintaining the self-aligned character.

A number of techniques have been proposed to eliminate birds beak. In the SILOS (sealed interface local oxidation of silicon) process, this is accomplished through removal of any interfacial oxide between the nitride and the semiconductor (including the silicon native oxide). The justification for this is as follows. The diffusion coefficient for oxygen or water vapor in oxide is greater than that of silicon or nitride. The stress-relief oxide and the native oxide can be viewed as a conduit, bringing oxidizing species under

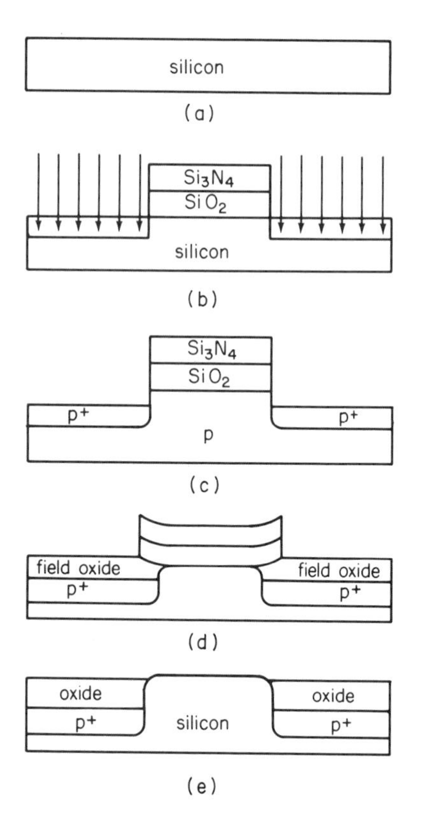

Fig. 7.24. ROX process: (a) Starting slice. (b) Oxide/nitride self-alignment mask and field implant. (c) Resulting structure. (d) Field oxidation. (e) Resulting planar process.

the passivating nitride far from the field boundary. Removing this conduit limits the oxide bird's-beak encroachment. This is illustrated in fig. 7.25. The native oxide can be removed by introducing hydrogen into the reaction tube in a nitrogen carrier gas. The hydrogen reduces the silicon surface, the oxygen blowing away as water vapor. If the nitride is too thick, wafer bowing and silicon damage can result from nitride stresses since there is no stress relief layer. In modern VLSI, the field oxides are not as thick, and nitride films as thin as 750 Å can be used to passivate the gate region. Such nitrides are too thin to damage the silicon.

An interesting problem is associated with the use of nitride in a self-aligned field process. The field oxide is usually grown in steam. The water vapor diffuses under the stress-relief oxide and oxidizes the nitride. This forms an oxide at the old nitride oxide/nitride interfaces. Free ammonia is liberated, which then diffuses to the oxide/silicon interface. This process is shown in fig. 7.26. Oxynitride is then formed at this interface. The nitride and stress-relief oxide are removed, but the oxynitride is not stripped in the buffered

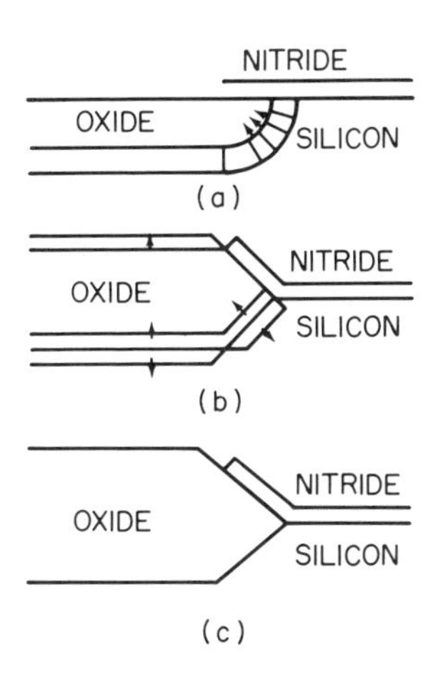

Fig. 7.25. The SILOS process: (a) Nitride is deposited on hydrogen cleaned silicon surface and oxidation commences. (b) there is some diffusion of the oxygen under the nitride, but the amount of oxygen diffusion is reduced over the case in which the native oxide is present. (c) Final structure.

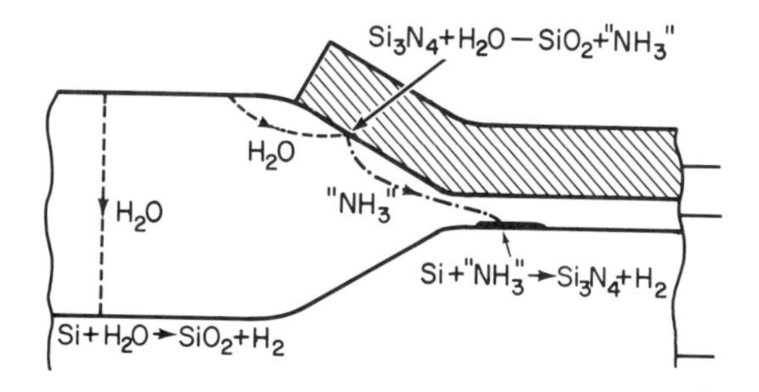

Fig. 7.26. A model for the formation of oxynitride during LOCOS. Water diffusing through the field oxide reduces the nitride, forming ammonia. The ammonia diffuses to the oxide semiconductor/oxide interface, forming oxynitride. After reference 29.

HF stress-oxide removal step. Next, the gate oxide is grown. The thin oxynitride prevents gate-oxide formation. The oxynitride does not insulate the gate from the channel regions and a short occurs. This is called the *white ribbon effect* [29], since the oxynitride appears as a white band surrounding the field oxide boundary. SILOS processes minimize the white ribbon width. Fluoroboric acid dips also helps to remove the oxynitride.

Just as in the case of silox, nitrides are usually too full of electrically charged defects to be used in the active gate regions of MOSFETs. However, the trap structure can be exploited to create an information storage device. This is the basis of the electrically alterable nonvolatile memory, a cross section of which is shown in fig. 7.27. A high positive bias (about 25 V) is applied to the gate and holes are trapped at the native-oxide nitride interface. This trapped charge shifts the threshold of the MOSFET transistor. The

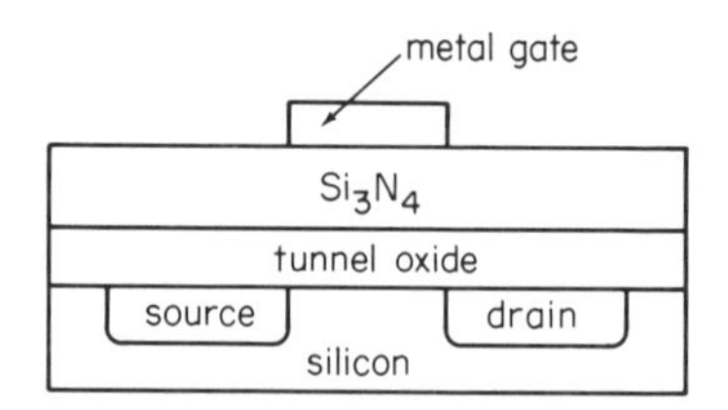

Fig. 7.27. MNOS memory transistor.

shift in threshold can be sensed, and memory storage is accomplished in this way.

Polycrystalline silicon is the usual first level interconnect used in MOS devices, as seen in the CMOS process flow of Chapter 1. The polygate is essential to the modern self-aligned gate used in practically all of today's MOS device processes. In addition, the polygate dimensions determine the transit time and gate capacitance for MOSFET devices. As a result, the smallest feature on an IC chip is, usually, the polygate. An example of the process used to make the gate is shown in fig. 7.28. The polygate shields the channel region, while the source and drain regions are simultaneously formed through implantation. The polygate edge lines up almost exactly with the source drain boundary. This eliminates parasitic overlap capacitances, as shown in the figure. Of course, no process can provide a "completely"self-aligned source/drain to gate boundary. There is some lateral spreading of the implant under the gate, which gives rise to some overlap capacitance (as explained in greater depth in Chapter 5).

The polysilicon interconnect lines must be good conductors. As-deposited films are near intrinsic and have very high sheet resistivities (certainly greater than 1000 $\Omega/\square$). Attempts to dope the layers in situ by introducing arsenic-, phosporous-, or boron-containing gases into the CVD reactor during the growth phase usually fail. The dopant atom is strongly bound to surface nucleation sites. This inhibits silicon layer growth and leads to a highly inhomogeneous film. Typically, the films are doped after deposition and 10-30 $\Omega/\square$ resistivities are achieved. More heavily doped layers have lower sheet resistances, but they present problems in pattern definition. "Clusters"of dopant atoms form that inhibit etching. The most severe problems in this regard appear in boron-doped films. Here, the doped polysilicon layers withstand even a wet chemical etch. As a result, boron-doped poly is rare in IC processes. When resistances of n-type poly approach 10 $\Omega/\square$, there is a degradation both of etch profiles and of selectivity in plasma-etching processes.

The conductivity of the polysilicon line is intimately related to its grain structure. The grain boundaries can localize charge and act as scattering centers. This lowers mobility and reduces bulk conductivity. As mentioned above, grain size depends on the number of embryos that reach critical size

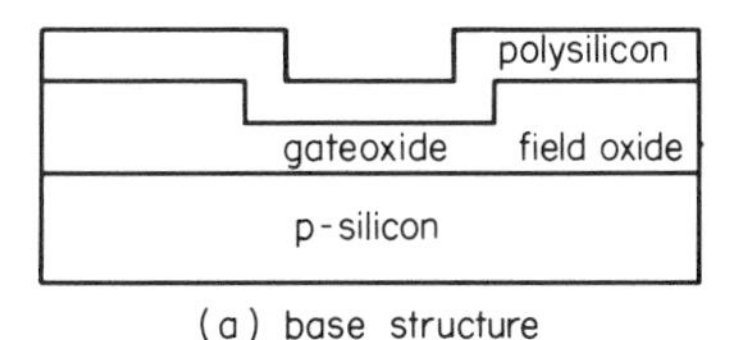

(a) base structure

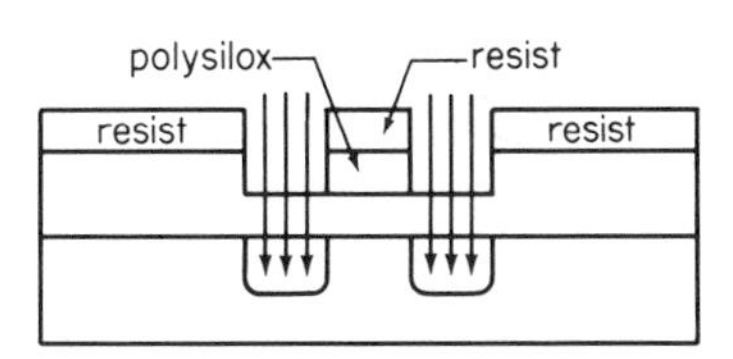

(b) poly definition and implant

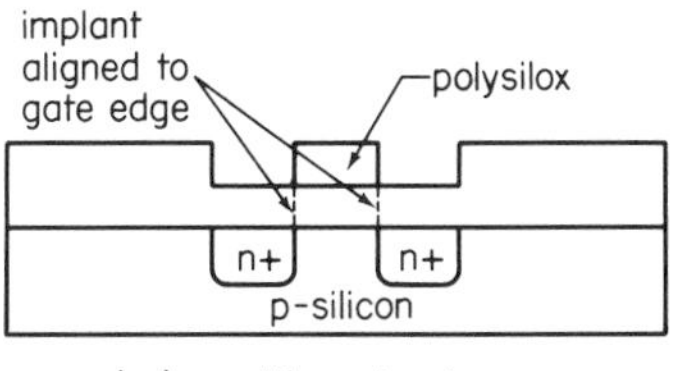

(c) resulting structure

Fig. 7.28. The self-aligned gate process.

in a unit time and the growth rate of these critical nuclei. Subsequent heat treatment tends to make the average grain size bigger. To see why this is the case, consider the "soap-bubble froth" model of the grain structure [30]. Each bubble is characterized by a radius and an internal pressure. Gas molecules will diffuse from bubbles of high pressure to bubbles of lower pressure. A bubble's internal pressure is inversely proportional to its diameter. The small bubbles shrink and the large bubbles grow. To first approximation, we can write the time-rate-of-change of the bubble diamater as inversely proportional to the bubble diameter, or

$$\frac{dD}{dt} = \frac{K}{D}, \tag{7.60}$$

where K is some constant. Integrating this equation yields

$$D^2 = Kt + D_0^2, \tag{7.61}$$

where D_0 is the initial bubble diameter before heat treatment. If the average initial bubble diameter is much smaller than the final bubble diameter, the bubble diameter grows in proportion to the square root of annealing time.

If the process is diffusion driven, the transfer of atoms across the grain boundary would be a rate-activated process. The temperature dependence

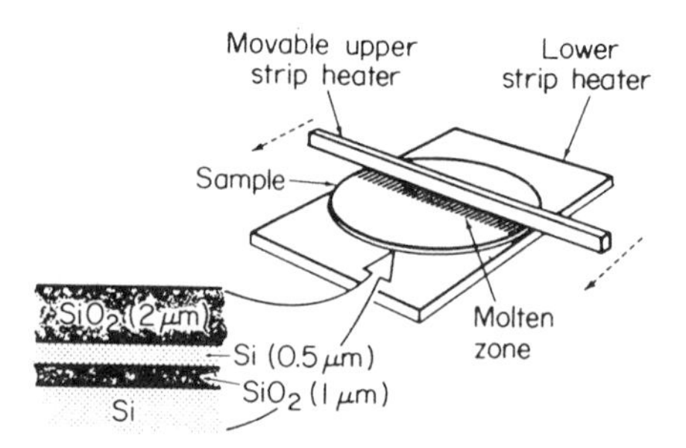

Fig. 7.29. Schematic diagram of sample and graphite strip heaters used in zone-melting recrystallization of SOI films.

of the process can be derived by writing

$$K = K_0 \exp\left(-\frac{Q}{kT}\right), \tag{7.62}$$

where Q is some empirically derived activation energy. For some metal films, eq. (7.61) and (7.62) provide a good fit to experimental data. This is not always the case. The presence of impurities localized in the grain boundaries creates a "drag" effect and damps grain growth. The effect of surface and bulk strains in the solid state contribute differently to facililate the diffusion process. Most experimental data can be fit to a relationship of the form

$$D = K\, t^n, \tag{7.63}$$

where n is a number less than 1.

The simple annealing steps described above do increase grain size. More aggressive techniques are currently being developed to create near-perfect crystal films. One such technique is shown in fig. 7.29. In the first, we begin with a window cut in an oxide layer. Polysilicon is then deposited over the whole surface. An oxide or a nitride capping layer is then deposited over the polysilicon. This layer helps provide a smooth surface after recrystallization. The surface of the wafer is locally heated either by a graphite strip heater or by a laser. In the case of laser heating, the capping layer helps couple optical energy to the polycrystalline film. The region of local heating is scanned across the wafer surface. Melting of the polycrystalline layer occurs. The exposed silicon in the oxide cut acts as a seed for crystal growth. In this way, device-quality films have been created. The resulting material has superior isolation capability. In addition, devices can be fabricated in the underlying silicon substrate. This creates a three-dimensional IC. The resulting material is called SOI (silicon-on-insulator) and is the subject of a large amount of recent research [31].

Crystalline layers can be grown directly on crystal substrates. Epitaxially grown layers are widely used in both MOS and bipolar technology. Historically, the first application of epilayers was in the bipolar area, where it was

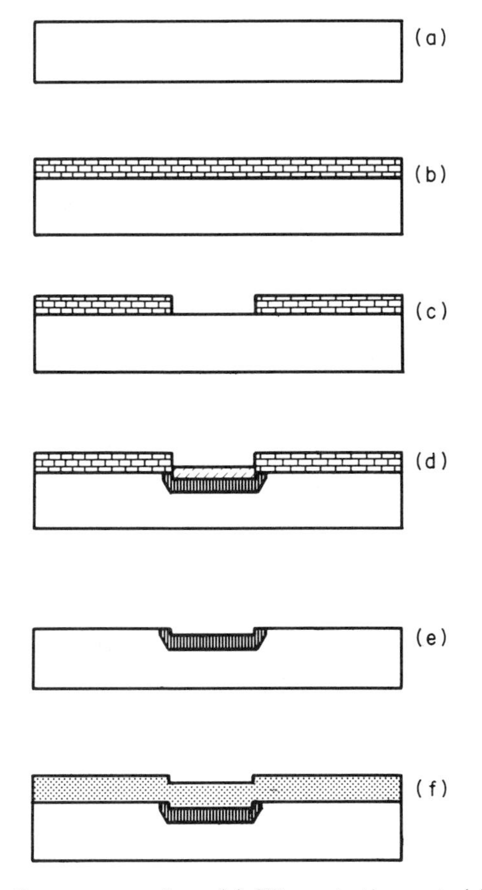

Fig. 7.30. Buried collector process flow. (a) Silicon starting material. (b) Oxidized wafer. (c) Open window for buried collector diffusion. (d) Buried collector diffusion and concurrent oxidation. (e) Removal of oxide layers. (f) Epilayer growth.

used to provide junction isolation between devices in an IC. In addition, the epiprocess also provides the buried collector, a heavily doped region used to minimize collector series resistance. A diagram illustrating the buried collector and a process used to fabricate it is included in fig. 7.30. The heavily doped region is diffused into the silicon substrate through an oxide mask. The doped wafer is reoxidized before the mask oxide is removed. The oxide thickness in the mask region does not change very much (as it is fairly thick already). About 1000 Å of new oxide is formed over the doped region. Next, all the oxides are stripped. This creates a step in the silicon that is visible after the epilayer is grown. The step is necessary, otherwise the position of the buried layer would be unknown after the epigrowth. It would then be impossible to align the transistors to the buried layers.

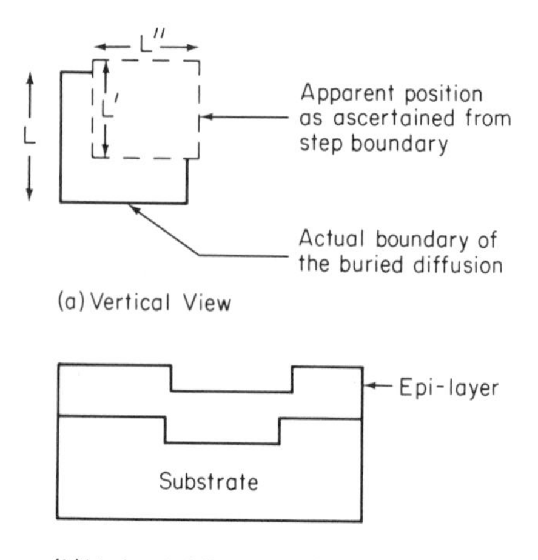

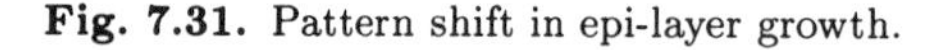
Fig. 7.31. Pattern shift in epi-layer growth.

One problem associated with the process is that the step-boundary pattern apparently shifts after epigrowth. In addition, there is some pattern distortion. These are illustrated in fig. 7.31. The shift and distortion effects are a function of the process used to create the epilayer, as well as the substrate orientation. Subsequent mask layers must be intentionally misaligned with respect to the observed buried layer step to account for this effect. The misalignment and distortion must be characterized by lapping and staining the wafer and by comparing the stained doped layer position and dimensions with the step that appears in the epitaxial layer.

Epilayers are used for a variety of reasons in both n-MOS and CMOS technologies. Consider, as an example, the dynamic RAM. Here information is stored on a MOS capacitor storage node. Single event upset (SEU) is a major problem in such a device [32]. The SEU phenomenon is shown in fig. 7.32. Here, a cosmic ray is incident on the capacitor memory cell. Electron-hole pairs are created all along the particle track. Some mobile charges from all along the track are "funneled" back to the capacitor, erasing the storage. The errors that arise are called soft errors in that they do not represent hard, or permanent, damage to the device. SEU could interrupt the functioning of a core program in a computer, it certainly leads to erroneous information storage. If the device is fabricated in an epilayer deposited on a heavily doped (0.1 Ω-cm) substrate, this effect is reduced. The recombination lifetime in such a heavily doped substrate is extremely small. Many of the electron-hole pairs created by the cosmic ray recombine before they can be funneled to the storage node.

Epilayers are popular in CMOS for a few reasons [33]. First, a heavily doped substrate provides a good ground plane for the device. Minority charges generated in the active device layer can be channeled to ground,

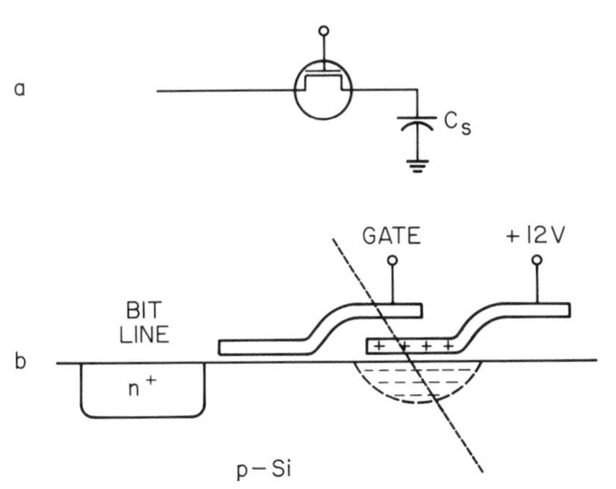

Fig. 7.32. Soft-error in a 1-T RAM. (a) The basic circuit, (b) device cross-section indicating cell-discharge due to cosmic ray hit.

rather than to the parasitic bipolar transistor collector terminals (see Chapter 1) where they can cause latch-up. The substrate resistance, R_s, is also reduced, further impeding latch-up. Soft errors and cosmic-ray induced latch-up can be avoided for the reasons described in the above paragraph. For p-tub processes, n epi on n+ starting materials are used. For n-tub CMOS, p epi on p+ layers are used.

The ultimate in device isolation is achieved using heteroepitaxy. A thin film of device-quality silicon can be grown on sapphire (Al_2O_3). This is called the silicon-on-sapphire (SOS) process [34]. Device "islands"can be made by etching away surrounding silicon (fig. 7.33). Individual transistors can be made on these islands. There is no silicon connecting these transistors. There is a device operating-speed advantage created by the effective removal of the silicon substrate. This occurs because the interconnect-to-substrate capacitance is removed as is the bottom-capacitance of diffused junctions. The SEU and latch-up problems are eliminated since the electron-hole pair generation and minority charge generation rates in the substrate are much smaller than they are for bulk silicon.

The drawback to the implementation of SOS technology is low device yields. There is a defect structure associated with the sapphire/silicon interface that can give rise to unacceptable transistor leakage. In addition, aluminum from the substrate can diffuse into the device islands during processing and autodope the silicon p-type. While VLSI circuits have been implemented in SOS, the yield limitations make these devices very expensive. In high-density VLSI technology, intraline capacitance, rather than interconnect-substrate capacitance, dominates in determining the device speed. As such, much of the SOS speed advantage over bulk devices is

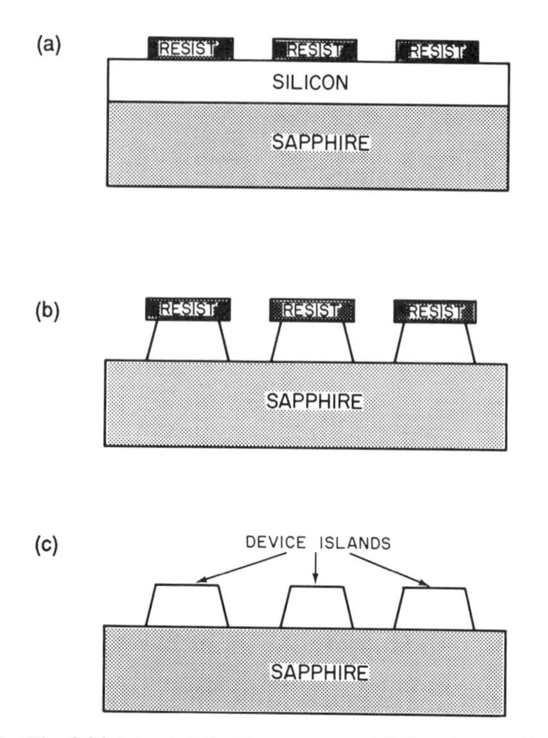

Fig. 7.33. The SOS island definition process: (a) Starting configuration. (b) Island-defining etch. (c) Completed structure.

lost. When latch-up and SEU problems are of prime concern, SOS becomes the technology of choice, despite the expense.

One further point should be stressed about CVD films in general: They tend to be conformal. That is, they create a relatively uniform blanket over underlying topography. This occurs for two reasons. Even though there is a preferred flow direction for the gas, molecular motion is fairly random due to the large kinetic energy of the individual particles. As a result, we do not observe the shadowing and projection effects that occur in directional deposition systems (such as metal evaporators). The kinetic energy of physisorbed species is also relatively high. There can be considerable redistribution of atomic or molecular species after they are captured by the surface. This has a smoothing effect on the film.

Although the films tend to be conformal, this is not always the case. Consider CVD deposition into a deep trench (fig. 7.34) of length L, width W, and depth d. The depositing species along the side wall of the trench have their source in the gas above the trench opening. These species are depleted as we move to the bottom of the trench. The trench itself tends to collimate the momentum vectors of atoms or molecules in the depositing gas. If the trench perfectly collimates the depositing gas, there would be

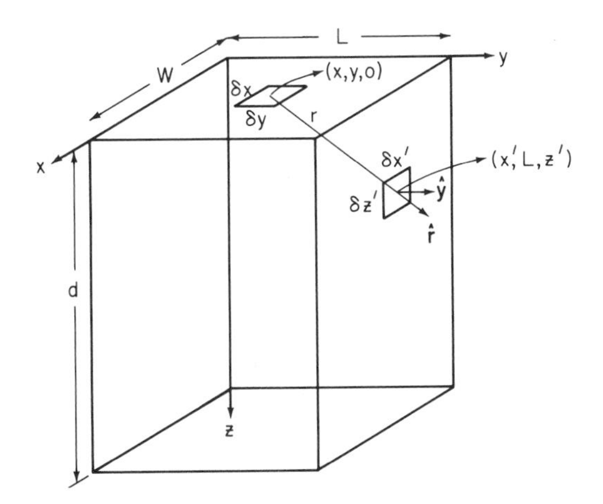

Fig. 7.34. Deposition geometry for a trench.

no film grown on the sidewall. The situation is similar to the rain falling over a building. If there was no wind and the rain came straight down, it would land on the roof and on the ground, but the building walls would stay dry. These effects are accounted for in the calculation of deposition side-wall profile given below.

Once again, consider fig. 7.34. Consider the trench opening to be composed of many infinitesimal sources of area $dA = dxdy$ at a point (x, y, o). Each source radiates isotropically into a hemisphere of 2π steradians. The area $dA' = dx'dy'$, located on the side wall at (x, L, z), intercepts a fraction of the total emission. This fraction, f, is given as

$$f = \frac{d\Omega}{2\pi} , \tag{7.64}$$

where $d\Omega$ is the solid angle subtended by dA' when viewed from (x, y, o). If we assume that each area of source emits I atoms or molecules per unit area per second, the source area emits a total of $Idxdy$ particles per second. Thus, the total number of particles incident on dA' is $fIdA'$. The flux incident, F, is the number of particles incident per unit area per unit time. We can then write the following expression for flux at (x', L, z')

$$F = \frac{I}{2\pi} \int_o^L \int_o^w \frac{\hat{\mathbf{r}} \cdot \hat{\mathbf{y}}}{r^2} \, dxdy , \tag{7.65}$$

where $r =$ the distance from (x, y, o) to (x^1, L, z^1), $\hat{\mathbf{r}} =$ a unit vector pointing to the direction of $\mathbf{r}$, and $\hat{\mathbf{y}}$ = a unit vector in the y direction (i.e., 0,1,0) The vector, $\mathbf{r}$, can be written

$$\mathbf{r} = (x' - x, L - y, z'), \tag{7.66}$$

and the unit vector, $\hat{\mathbf{r}}$, is

$$\hat{\mathbf{r}} = \frac{(x' - x, L - y, z')}{|\mathbf{r}|}. \quad (7.67)$$

The collimating effect is accounted for by the area projector $(\hat{\mathbf{r}} \cdot \hat{\mathbf{y}})$. The depletion of source material is accounted for in the $1/r^2$ term. Please note that this approach does not account for momentum redirection through gas-phase collisions in the trench. Thus, it works best when travel distances are less than the mean free path for depositing species. The model does not account for redistribution or surface desorption of depositing species. Both of these effects tend to smooth the profile and make the deposition more conformal.

7.8 CVD of Metals

CVD is a high throughput process for providing nearly conformal coatings. As a result, deposition of metals like Al, W, Mo, and silicides has been explored extensively [35]. Availability of a suitable gaseous source for CVD is the most important factor in such cases. For example, the only suitable aluminum sources are the trialkyl aluminums, which are extremely explosive materials. In addition, carbon-to-aluminum bonds in such gases are strong, leading to deposits that contain large amounts of carbon. Oxygen is also easily picked up during Al depostion. Thus, in spite of repeated attempts, CVD Al films useful for IC applications have not been successfully deposited. Tungsten and tungsten disilicide, on the other hand, have been successfully deposited by CVD using WF_6, and WF_6 and SiH_4, respectively. Both WF_6 and SiH_4 are readily available gases that easily decompose in CVD reactors. Tungsten is deposited by reducing WF_6 with hydrogen. WSi_2 is produced by the reaction between WF_6 and SiH_4. Gaseous SiF_4 and H_2 are reaction products.

Under certain experimental conditions W metal can be selectively deposited on silicon or polysilicon surfaces only, with no deposit on neighboring oxide surfaces. This eliminates lithographic patterning steps and allows via filling in the simplest way. This technique is still being perfected at this writing.

A number of techniques can be used to form tungsten films selectively [35]. The basis of the selective deposition process for WF_6/Ar gas mixtures is the following reaction:

$$3Si(solid) + 2\ WF_6(gas) \rightarrow 2\ W(solid) + 3\ SiF_4(gas). \quad (7.68)$$

The SiO_2 surface cannot be reduced by the WF_6 molecule, and free tungsten is not released. For the process to continue after the surface is coated with tungsten, WF_6 must diffuse through the film to react at the Si/W interface or free silicon must be present on the W surface. As a result, it is not possible to grow layers thicker than 100 nm in this way. Tungsten deposition can also proceed as follows:

$$WF_6(\text{gas}) + 3H_2(\text{gas}) \rightarrow W(\text{solid}) + 6HF(\text{gas}). \tag{7.69}$$

Thus, in WF_6 and H_2 mixtures, W can deposit over *all* surfaces (oxide and silicon.) However, nucleation sites on the deposited W surface cause reaction (7.69) to proceed much more rapidly over deposited tungsten than over SiO_2. Thus, addition of H_2 to the gas stream still causes enhanced deposition over the already-deposited-tungsten.

In practice, these selective processes are difficult to run. SiF_4, a reaction byproduct of the process expressed in eq. 7.68, can be absorbed on oxide surfaces. The adsorbed SiF_4 can nucleate the growth of free tungsten over the oxide. In addition, WF_6 can diffuse along the Si/SiO_2 interface and cause tungsten encroachment into adjacent device areas. This can cause undesired shorting. All the processes listed above are extremely temperature dependent (normally occurring anywhere from 350°C to 550°C). The selectivity of deposition (and tungsten everachment) are also temperature dependent. Other metals that have been deposited successfully using the CVD method are Mo, Ta, Ti, and their silicides.

7.9 Summary

In this chapter we have covered the fundamentals of chemical vapor deposition. The different types of CVD reactors were described. Reactors differ as to whether the reaction is carried out at atmospheric pressure or at low pressure. In addition, a distinction was made as to whether the gas flow was perpendicular to the wafer surface (a vertical reactor) or parallel to the wafer surface (a horizontal reactor). The reacting gases could be broken down in plasma to reduce the chemical reaction barriers and allow for low-temperature deposition. Low-pressure chemical deposition was discussed as a method for creating a surface-reaction-limited deposition process (as opposed to a mass-transport-limited process.)

The CVD process was modeled by breaking it up into three parts:

- Introduction and equilibrium of the deposition gases into the reactor.
- Transport of reactive species to the deposition surface.
- Nucleation and growth of the thin film.

Modeling indicated that there are two regimes of reactor operation: gas-phase-transport limited growth and surface-reaction-limited growth. Gas-phase-mass-transport-limited growth gave the least uniform deposition due to variations in stagnant-layer thickness and other flow inhomogenieties. Gas-phase transport could be speeded up, moving the reaction into the surface-reaction-rate-limited-regime. If the flow was too fast, the flow would become chaotic. This occurs when the Reynolds number exceeds 2000.

Amorphous, polycrystalline, and crystalline materials could be grown by CVD processes. Growth techniques and IC applications for each type of film were discussed. Glasses could be used as diffusion barriers, as field isolations, and as insulating spacers between layers of interconnect. Polysilicon is widely used as a gate material in MOSFETs. Epilayers are used to provide isolation and buried collectors in bipolar technology. In MOS technology, epilayers are used to supress latch-up and soft error.

References for Chapter 7

1. W. Kern, V. Ban, Chemical vapor deposition of inorganic thin films, in "Thin Film Processes," J.L. Vossen, W. Kern, eds., Academic Press, New York, pp. 257-331 (1978).
2. E.W. Sabin, Laser activated CVD of silicon dioxide and silicon nitride, in "Semiconductor Silicon 1986," H.R. Huff, T. Abe, B. Kolbesen, eds., Proc. Vol. 86-4, Electrochemical Soc. Press, Pennington, NJ, pp. 284-294 (1986).
3. J. Dieleman, R.G. Frieser, eds., "Plasma Processing," Proc. Vol. 82-6, Electrochem. Soc. Press, Pennington, NJ, pp. 478-506 (1982).
4. A.J. Grove, "Physics and technology of semiconductor Devices," John Wiley and Sons, New York, pp. 13-18 (1967).
5. D.R. Stall, H. Prophet, "JANAF Thermochemical Tables," (2nd edition) NSRDS reference Vol. 37, National Bureau of Standards Press, Gaithersburg, MD, (1971).
6. B. Carnahan, H.A. Luther, J.O. Wilks, "Applied numerical analysis," John Wiley and Sons, New York, Ch. 5 (1969).
7. L.P. Hunt, E. Sirtl, A thorough thermodynamic evaluation of the silicon-hydrogen-chlorine systems, J. Electrochem. Soc. 19, 1741-1744 (1972).
8. E. Sirtl, L.P. Hunt, D.H. Sawyer, High temperature reactions in the silicon-hydrogen-chlorine system, J. Electrochem. Soc. 121, 919-924 (1974).
9. V.S. Ban, S.C. Gilbert, Chemical processes in vapor deposition of silicon I. Deposition from $SiCL_3H$ and $SiCl_4$, J. Electrochem. Soc. 122, 1382-1388 (1975).
10. V.S. Ban, Chemical processes in vapor deposition of silicon II. Deposition from $SiCl_3H$ and $SiCl_4$, J. Elctrochem. Soc. 122, 1389-1390 (1975).
11. D.E. Rosner, "Transport Processes in Chemically Reacting System," Butterworth Publishers, Boston (1986).

12. B. Lewis, J.C. Anderson, "Nucleation and Growth of Thin Films," Academic Press, London (1978).
13. R.E. Reed-Hill, "Physical Metallurgy Principals,"2nd ed., Brooks/Cole Engineering Division, Monterey, CA (1973).
14. J.W. Christian, "The Theory of Transformations in Metals and Alloys,"Pergamon Press, London (1965).
15. Y.W. Mathews, ed., "Epitaxial Growth,"parts A and B, Academic Press, New York (1975).
16. J.H. Van Der Merwe, Phil. Mag. 22, 269 (1970).
17. R.M.A. Azzam, N.M. Bashara, "Ellipsometry and Polarized Light," North Holland Personal Library, Amsterdam (1986).
18. D.E. Lamb, "Electronic Conduction Mechanisms in Thin Insulating Films,"Methuen Co., London (1965).
19. B. Swaroop, P.S. Schaffer, Conduction in silicon nitride and silicon nitride-oxide films,"J. Phys. D 3, 803-806 (1970).
20. P.F. Kane, G.B. Larrabee, Characterization of Semiconductor Materials, McGraw-Hill, New York, p. 91 (1970).
21. L.C. Feldman, J.W. Mayer, "Fundamentals of Surface and Thin Film Analysis,"North-Holland, New York (19).
22. B.D. Cullity, "Elements of X-Ray Diffraction,"2nd ed., Addison-Wesley, Reading, MA (1978).
23. B.K. Tanner, "X-Ray Diffraction Topography,"Pergamon Press, London (1976).
24. R.B. Marcus, J.T. Sheng, "Transmission Electron Microscopy of Silicon VLSI Circuits and Structures,"Wiley-Interscience, New York (1983).
25. C. Fu, A novel borophosphosilicate glass process, in Proc. IEEE Int'l. Electr. Dev. Mtg., IEEE Press, Piscataway, NJ, Cat. No. 85CH2252-5, pp. 602-605 (1985).
26. E. Bassous, H.N. Yu, V. Maniscalco, J. Electrochem. Soc. 123, 1729 (1976).
27. I. Magdo, A. Bohg, Framed recessed oxide scheme for dislocation-free planar S. structures, J. Electrochem. Soc. 125, (6), 932-936 (1978).
28. J. Hui, T.Y. Chiu, S.S. Wong, W.H. Oldham, Sealed-interface local oxidation technology, IEEE Trans. ED., ED-29, (4), 554-561 (1982).
29. E. Kooi, J.G. Vanlierop, J.A. Appels, Formation of silicon nitride at a Si-SiO_2 interface during local oxidation of silicon and during heat-treatment of oxidized silicon in NH_3 gas, J. Electrochem. Soc. 123, (7), 1117-1120 (1976).
30. L.A. Chadwick, D.A. Smith, eds., "Grain Boundary Structure and Properties,"Academic Press, London (1976).
31. IEEE Circuits and Devices Magazine, Vol. 3, Issues 4 and 5 are devoted to SOI (1987).
32. T.C. May, M.H. Woods, IEEE Trans. ED., p. 1 (1979).

33. D.S. Yaney, C.W. Pearce, The use of thin epitaxial layers for MOS VLSI, Proc. IEEE Int'l. Electron Dev. Mtg., IEEE Press, Piscataway, NJ, p. 236 (1981).
34. G.W. Cullen, C.C. Wang, "Heteroepitaxial Semiconductors for Electronic Devices,"ed., Pringer, New York (1978).
35. E.K. Broadbent, "Tungsten and Other Refractory Metals for VLSI Applications,"Vols. 1 and 2, Materials Research Society Press, Pittsburgh, PA (1987).

Problem Set for Chapter 7

7.1 Using the concepts developed in this chapter, define the relevant terms necessary for formulating a comprehensive model of CVD deposition rate.

7.2 Consider the desorption of atoms (or molecules) from a depositing surface to be a "rate-activated" (Arrhenius) process.

a. Derive an expression for the fractional coverage of the surface in terms of the activation energy for desorption, temperature, and the "attempt-to-escape" frequency of the adsorbed atom.
b. Derive an expression for the mean residence time, τ_a, for an adatom prior to desorption.
c. Assume that the nuclei grow by attachment of adatoms or molecules that diffuse along the surface and attach to a growing cluster. To within a multiplicative factor, derive an expression for the number of supercritical nuclei that form per unit time per unit area. The expression should contain terms involving the number of adsorbed atoms (N_a), the free energy of activation for surface diffusion (ΔG_s), and the surface diffusion jump frequency (ν).

7.3 Denote the number of supercritical nuclei that form per second per unit area as I^*. Associated with each cluster is a region, frequently called the *capture-zone*, which is relatively free of adatoms as a result of the cluster attachment processes. The capture zone is taken to be circular, of radius $\sqrt{D\tau}$, where D is the surface diffusion coefficient and τ_a is the mean adatom residence time.

a. Write an expression for the fraction of the surface covered by the adatoms and their associated capture zones.
b. Assume that nucleation can take place only on surfaces populated by adatoms. Write a differential equation that expresses the time dependence of I^*.

c. Derive an expression for the substrate area covered by critical-sized nucleii at time t.

7.4 Consider the case of nucleation and growth of volume precipitates. An example of such precipitates would be the SiO_2 precipitates discussed in conjunction with internal gettering (Chapter 2). As in problem 7.3, the concentration of the incorporating species is reduced in the vicinity of the precipitate. The resulting concentration gradient drives the diffusive transport of atoms to the precipitate, thus determining the adatom impingement rate. Assuming a spherical precipitate:

a. Show that the concentration as a function of distance from the center of the sphere can be written

$$C(r) = a + b/r$$

where a and b are constants determined by the boundary conditions.

b. Assume that the relevant boundary conditions are: $C(\infty, t) = C\infty$ and $C(R, t) = C_o$, where $C\infty$ is the concentration some distance from the precipitate, and C_o is the adatom concentration on the precipitate surface, at radius R. Furthermore, assume that the volume of an adatom in the precipitate is v_o. Write a differential equation for the rate of precipitate volume increase.

c. Use this equation to derive an equation for the radius, R, at some time, t.

7.5 Derive an expression for the temperature dependence of the deposition rate of a thin film in a CVD process and specify the range of applicability of the expression derived. What factors make the process more or less temperature sensitive?

7.6 Consider the CVD tube shown in the figure below. As shown in the text, the stagnant-layer thickness changes as a function of position along the susceptor. This will lead to a position-dependent growth rate, even if the depositing gases are well mixed.

a. Explain why this is so.

b. For the conditions shown in the figure, assume a 1 liter/min. flow rate and a 1200°C ambient temperature, estimate the change in growth rate along the susceptor.

7.7 As depositing gas blows down a CVD tube operating in the horizontal configuration, the reacting species can become depleted, leading to a non-uniform growth rate along the susceptor.

a. Define the factors that determine the growth rate non-uniformity.

b. Assuming that the gas is well mixed in the vertical direction and the concentration gradient diffusive mixing in the horizontal direction is

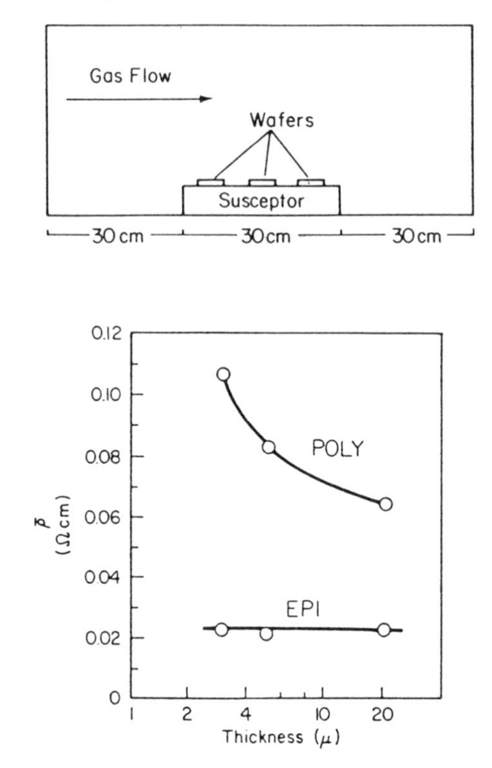

Fig. 7.35.

Fig. 7.36. Resistivity vs. thickness for boron-doped polycrystalline and epitaxial layers deposited at the same time.

small, derive an expression for the growth-rate change along the CVD tube.

7.8 List the advantages and disadvantages of using polysilicon layers thicker than 1 μm in silicon-gate IC processing, rather than thinner layers.

7.9 From the figure below, determine the thickness of the boron-doped polysilicon that would produce a sheet resistance of 130 $\Omega/\square$.

7.10 A doped polycrystalline line is 80 μm long and 3 μm wide and has a resistance of 9.333 kΩ.

a. What is the sheet resistance?
b. Assume the line is 2500 Å thick, what is its resistivity?
c. Assume the polysilicon had been implanted with phosphorous and fully activated, what would have been the implant dose?

7.11 Consider the deposition profile for CVD coverage of a trench wall. Use eqs. (7.64–7.67) to derive this profile for a very narrow trench ($W = dx <<$

$L = 10\ \mu\text{m}, d = 10\ \mu\text{m}$).

7.12 Consider two reacting gases, A and B, in a chamber of constant volume. The dominant reactions taking place are:

Ia.
$$2A_2 + B_2 \longrightarrow 2A_2B$$

and

Ib.
$$2A_2B + B_2 \longrightarrow 2A_2B_2.$$

In addition, assume the reaction takes place at constant temperature. Derive the necessary equations to calculate the equilibrium partial pressures of all relevant species in the reacted mixture from free-energy data and Boyle's Law.

8 Lithographic Tools

8.1 Introduction

Lithography is the technique of reproducing predetermined patterns on arbitrary surfaces. In semiconductor work, the patterns are those of the devices to be fabricated on semiconductor wafers: diffusion tubs, metallization lines, oxide cuts, etc. We can list the goals of lithography for microelectronics as follows:

- To place high-resolution, defect-free patterns over as large an area as possible while controlling the critical dimensions (CDs) over the types of image-plane topography normally found in IC processing.
- To align the image with underlying features on the wafer surface.
- To do the above tasks without damage to the devices fabricated.

Basically, these goals boil down to the following: The success of a lithographic process is defined as our ability to place a material boundary wherever we would like on a semiconductor wafer. This boundary could be the edge of a diffusion tub, the center point of a contact window, or any of a number of material delimiters needed to make an IC. Currently, the IC industry demands a capability to resolve submicron-sized minimum features. These submicron images must be placed over the surface of a 100-mm-diameter silicon wafer with 0.1-micron precision.

Our ability to set boundary positions is a major factor for consideration in determining *design rules.* Design rules are IC construction guidelines that tell us: a) the minimum feature sizes for a given material or process, b) how close we can reproducibly bring two features together without intolerable interactions occurring, and c) with what degree of accuracy we can place a given feature with respect to features already present on the substrate. We refer to this relative accuracy as "precision."These rules are important for a number of reasons. First, they determine the maximum device density of the IC. Obviously, if the minimum MOSFET gate dimensions are $10\mu\text{m} \times 10\mu\text{m}$, we would have trouble fitting two such FETs in an area of $100\mu\text{m}^2$. In addition, if we make the design rules "tight"(near the maximum performance limits of our processing tools), final production yields may be reduced.

Minimum feature sizes are *not* set simply by the optimum resolving power of the lithographic tool used. We have to take into account the surface on which the image is to be defined. If the surface is not flat, the image may be out of focus. If the underlying surface is reflective, standing-wave patterns can form, changing the image dimensions. Two distinct images placed close together may interfere with each other. This is an example of a proximity effect. So far, the concepts of resolution, focus, and alignment accuracy have been used rather loosely. In succeeding sections these concepts will be given some mathematical substance.

We begin with an overview of current lithographic practice. This includes some discussion of mask making and pattern generation. Ultraviolet (UV) and x-ray optics are reviewed. Various types of UV and x-ray lithographic systems are described. This is followed by a description of currently used e-beam lithography tools. All elements of the lithographic system are examined. These elements include exposure sources, optics, and alignment subsystems. The chapter concludes with an overview of techniques for pattern inspection of mask or of printed wafers. A discussion of resist systems is given in Chapter 9. Exposure modeling of various resist systems using the tools described below is also deferred until Chapter 9.

8.2 Lithography: An Overview

The process of optical lithography for microelectronics is similar to the somewhat more familiar process of photography. In lithography, semiconductor wafers are, essentially, turned into photoplates by the application of photoresist (a thin, light-sensitive plastic film). In some ways lithography is simpler than photography. In photography, light of many colors and intensities is used to create an image. In lithography, the range of exposing wavelengths is often extremely narrow. Every attempt is made to ensure uniform intensity over the exposure field. The image is a shadow image of the desired set of shapes, created by a *mask*. Some regions of the mask are opaque, stopping the incident light, while others are transparent. Light passing through the transparent regions falls on the photoresist (PR), exposing it (as illustrated in fig. 8.1). In some cases, where very high resolution is desired, the PR is exposed by a tightly focused electron or ion beam. This beam "writes" the pattern directly on the workpiece, filling in the shapes to be exposed much like the familiar ink-jet printer [1].

There are a number of variations on the optical or beam process, as summarized in fig. 8.2. Entries along the top of the figure indicate three separate exposure techniques: focused beam, pattern projection, and simple shadow mask. Entries along the left-hand column indicate three different types of exposure source. This leads to nine distinctly different types of microcircuit lithography, eight of which have been tried with some degree of success. The ninth entry, e-beam flood of a stencil mask held close to the workpiece (with no intervening electron optics) was attempted by Westinghouse Corporation in their "ELIPS" system development. This was subsequently

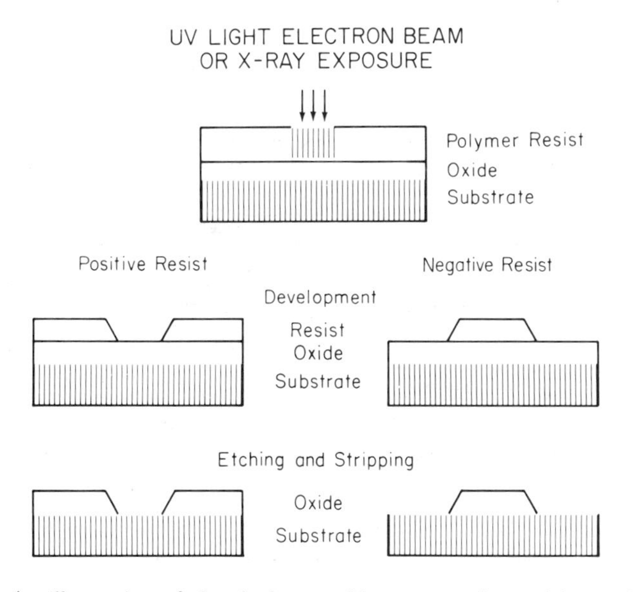

Fig. 8.1. An illustration of the shadow-masking process for positive and negative resists. For positive resists, the exposed region washes away in developer. For negative resist, the unexposed region washes away.

replaced by a patterned photo-cathode exposed by UV light. However, as of this writing, technical difficulties associated with these approaches have not been surmounted. If we read the figure from right to left, we get a historical survey of the field. The first lithography tools employed masks held in contact with the work piece. Each time the mask came in contact with the wafer, though, there was a possibility of mask-feature damage. To remedy this, proximity-printers were used. Here, the mask was separated from the wafer by about $25\mu m$. This helped to reduce mask damage, but it severely degraded resolution (for reasons discussed below).

The next major development in the field occurred in the mid-70s when Perkin-Elmer Corporation introduced the full-wafer projection system. Here, a high-quality image of the mask is projected, like a slide photo onto the wafer. Resolution was better than that obtained with a proximity printer (but not as good as that seen with contact printers!). The optical field of full wafer systems is relatively large. The IC industry, at this writing, works with wafers as large as 150 mm in diameter. It is difficult to maintain resolution over fields this large for a number of reasons. Even though the optical exposure source is near-monochromatic, a number of spectral lines are admitted to the system. For transmission lenses chromatic aberrations become a problem, as does astigmatism. Furthermore, as discussed in Chapter 2, the wafer deforms as a result of processing. This may cause portions of the wafer to be out of focus in a full-wafer projection system (out-of-plane distortion). In-plane distortion also occurs, making registration difficult.

	FOCUSED BEAM (DIRECT-WRITE)	PATTERN PROJECTION	MASK (PROXIMITY OR CONTACT)
PHOTONS	LASER CHEMISTRY	UV SCANNING OR STEP & REPEAT	UV & X-RAY
ELECTRONS	RASTER OR VECTOR SCAN	PATTERN ON CATHODE	—
IONS	FIELD-ION SOURCES	OPEN MASK	CHANNELING IN MASK

Fig. 8.2. Summary of exposure techniques for pattern definition. Courtesy of Dr. D.J. Nagel.

To remedy this problem, the exposure field size is reduced so that it fills only a part of the wafer. A system using this reduced-field-size technique is called a *wafer stepper*, since the image of a single chip is "stepped" across the wafer, one chip at a time. Smaller fields partially overcome the wafer flatness problem and produce a higher resolution image through relatively aberration-free optics. The step-and-repeat machine produces fewer wafers per hour than a full field system, but it allows for smaller linewidths to be made with better registration.

For the highest resolution and overlay accuracy, beam-forming lithographies are currently available. These frequently employ electrons or ions, achieving extremely high resolution because of the very short effective wavelength of these particles. The photon beam, however, occupies a central position in most conventional lithography schemes. The shadow masks used by all of the optical lithographies are frequently made with an optical-beam pattern generator. Here, a focused light beam is shot onto an unpatterned, resist-covered photoplate. The beam is flashed on and off, delivering precise amounts of energy to expose the resist. The table on which the plate sits is moved under computer control to sweep out the desired exposure pattern. Electron and ion beams work in a similar way. These beams, though, are usually moved off the center line to expose a small field before the table is moved. Electron and ion beams are used to expose wafers directly as well as to make mask plates. In the next section the fundamentals of mask making are presented.

8.3 Mask Making

Most lithographic processes involve pattern replication through shadow masks. These masks are usually made by the flash technique described above. Electron-beam generated masks, however, are becoming more and more prevalent. Let us first consider the optical pattern generation process. The light flash is, at most times, a rectangle. The size of the rectangle defines the minimum feature size on the mask plate. The plate itself is glass (or quartz for deep-UV printing) coated with a thin film such as chromium that is opaque to the radiation. When glass plates are used, low-thermal-expansion (LE) borosilicate glasses are usually specified. The light-sensitive photoresist is placed on top of the opaque film. The pattern is first transferred to the photoresist through the flash technique. After developing, the remaining resist serves to protect the underlying thin film from subsequent etches. The unwanted part of the opaque film is etched away and the desired pattern remains. The remaining protective resist is stripped. The resulting plate is called the *reticle.*

In order to make the reticle, the pattern generator must be driven by a computer-originated control tape [2]. There are a number of steps to be taken in the generation of this tape. First, a drawing of the desired pattern must be made by a layout specialist. This drawing is stored in a computer as an *intermediate file* (or I file, for short). In making the I file, the pattern is broken down into a series of smaller geometric units (usually rectangles). Each entry in an I file contains, at a very minimum, the coordinates of the lower-left- and upper-right-hand corners of the rectangle (although a variety of formats have been used to encode this information!). These coordinates are specified with respect to a user-defined origin. The coordinates are expressed in integers known as *grid units.* The user must specify what physical dimension the arbitrary grid unit actually corresponds to. The I file entry can also include instructions, such as "step and repeat the rectangle n times in x, m times in y, each time following some displacement Δx and Δy." Once the I file is prepared the whole file must be recompiled into a format that can be read by the pattern generator. Once this has been done, the pattern-generation (PG) tape is prepared. The pattern generator can then recompile the tape to supply additional machine control information to the PG file. For example, if an e-beam pattern generator is used, information on how much time the beam should spend in a given area can be added to the file. Appending machine control information to the PG tape creates a *job file.* After this machine recompilation, the reticle is written.

Feature dimensions on the reticle are usually larger than the feature dimensions desired on the wafer. The reticle image is demagnified by some factor and stepped across a second photoplate called the *master.* The demagnification process allows the generation of minimum feature sizes smaller than flash dimensions. Some of the rough edges that can result from imperfectly butted flashes also wash out in the demagnification. In the past, 10x reductions were standard. Currently 5x and 1x reductions are used for wafer steppers. The master plate is then reproduced either by contact or by projection printing. In current stepper practice, working plate copies are usually 1x and the stepper projects a demagnified image directly on the wafer.

The issue of 1x versus *n*x reduction lithography is currently a topic of intense debate in the IC fabrication community. It is easier to make projection lenses for 1x systems because of system symmetry and the relatively small size of the plate-imaging field. As a result, various lens aberrations encountered in large-area images are not important in this case. However, there is a significant burden placed on the primary pattern generator to create masks of good edge acuity. In *n*x systems, the rough edges are smoothed out in the demagnification process. In addition, it is much harder to inspect the small features on 1x plates for defects.

There are two types of opaque thin films used in plate making: those that are opaque throughout the visible spectrum and those passing wavelengths that do not expose photoresist. Chrome films 30-50 nm thick are generally used as actinically opaque masks. The material is very hard and is relatively easy to etch without loss of edge acuity. Chrome holds up well in use. "See-through" plates can be made using iron or chrome oxides. Chrome oxide is now the most popular because of its toughness.

To understand the need for the two types of plate we must first understand what is meant by a bright-field or a dark-field mask. In a light-field mask, most of the light-absorbing thin film has been etched away, causing most of the mask to be transmissive. In the dark-field case, the opposite is true. There is great difficulty associated with aligning dark-field masks to features already present on the wafer. To see this, refer to fig. 8.3. Here an attempt to align two crosses is illustrated. One cross is on the wafer, another is on the mask. If the mask was totally opaque, the wafer cross would be totally obscured and coarse alignment would be prohibited. The solution is to use a material that stops the 400-nm light, which exposes resist yet passes longer wavelength light that is visible to the eye and to other electronic sensors.

The concept of *mirror sense* is also important in mask making. In creating contact or proximity prints, the mirror sense of the original image is obtained. This is because when making such copies, the resist/opaque film composites of both master and image must be placed in contact, as shown in fig. 8.4. The copied image, then, is a mirror image of the mask pattern. Thus, the master must be a mirror image of the working plate. Both the mirror sense and the field sense can be set by commands found in the PG tape.

Once the working plates have been made, the actual wafer exposures can commence. There are a number of possible tools available for this purpose. These tools and their limitations are described in the next section.

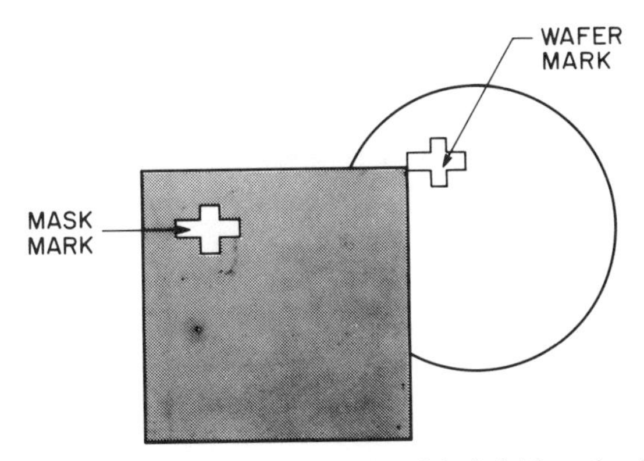

Fig. 8.3. Problems associated with alignment of dark-field masks: The mark on the wafer is hidden by the mask field.

8.4 Lithographic Systems

There are four major parts to every lithographic system. These parts include:

- The radiation source used to expose the resist
- The optical system used to create the desired image
- The wafer handler
- The alignment subsystem

Over time, many varieties of these four elements have appeared. In this chapter, combinations of subsystems currently in major use are reviewed. These include contact and proximity printers, full-field and subfield projection systems, and e-beam and x-ray tools. In addition, the fundamentals of manual and automatic alignment of mask to wafer are presented.

8.4.1 Contact and Proximity Printing

The first lithographic system used in the mass production of semiconductor devices was the contact printer [3]. Contact printers are still widely used in laboratories and in small businesses where high yield and high throughput are not primary goals. Surprisingly, the resolution of contact printers can be quite good. Micron line and space patterns can be easily reproduced with such equipment. However, the reproducibility of contact printing leaves much to be desired. Since the mask comes in contact with the wafer each time it is used, masks must be cleaned frequently. In addition, the masks can be damaged in the contact process. Still, the low cost of the equipment and ease of use and maintenance makes the technique attractive in a number of applications.

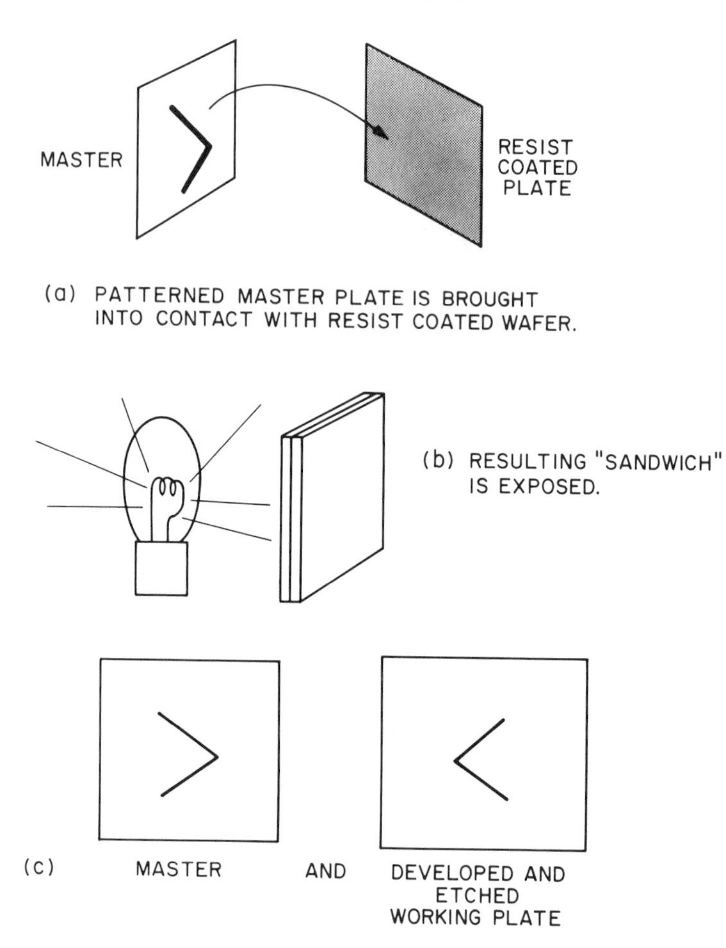

Fig. 8.4. Creation of working plate from a master.

It should be emphasized that there is really no such thing as a true contact printer. Even if the mask and the wafer were held in intimate contact, the exposing light (usually the emission from a mercury lamp, near 400 nm in the UV part of the spectrum) still has to travel through a resist of finite thickness. The wafer can also suffer out-of-plane distortions in high-temperature processes, preventing uniform contact across the exposure field. The presence of extraneous particles can create large mask-to-wafer separations. Resist edge build-up (discussed in Chapter 10) further complicates matters. The inability of the mask to reproducibly contact the wafer creates difficulty in linewidth control and in resolution. The reasons for this are discussed below.

In the absence of diffraction, the resolution limit of the shadow masking technique would be set by our ability to create a collimated beam of light of uniform spatial intensity. However, we know from quantum mechanics that light has both particle and wave aspects. The wave aspects give rise

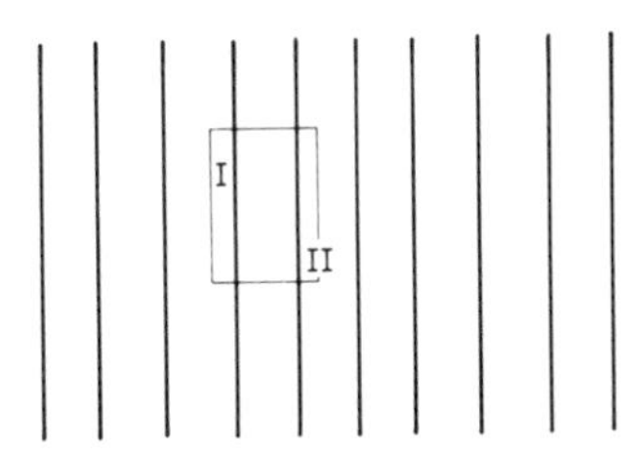

(a) Array of Parallel Wavefronts

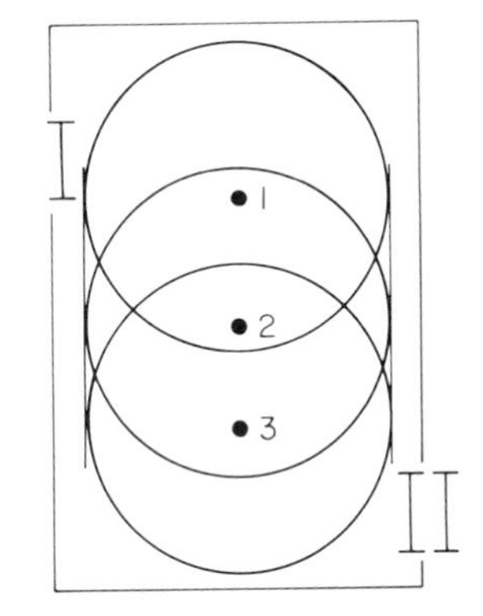

(b) Reconstruction of Wavefront Segment II From Wavefront Segment I Using Three Huygens Wavelets Centered at 1, 2, 3.

Fig. 8.5. Huygens wavelets summing to produce a coherent wavefront.

to diffraction effects. We can view the diffraction problem most easily using the Huygens wavelet picture (fig. 8.5) [4]. In this framework, a plane wavefront is seen as being composed of a lot of tiny, closely spaced sources of spherical wavelets. These wavelets expand and interfere with each other to give rise to the next wavefront. In a sense, the wavelet picture is always an approximation. This is because it would take an infinite number of sources on the first wavefront to exactly produce the second wavefront. The more sources you include, the better the approximation.

Now let us consider the case in which there is an opaque region blocking a portion of the wave train. The opaque region acts as a spatial filter on some of the little wavelets. Thus a smooth, constant wavefront cannot be reconstructed in succeeding fronts on the train. As a result of the spherical wavelet hypothesis, some light actually "escapes" the clear region and appears under the opaque portion. Mathematically, we say that each clear region can be broken into a number of infinitesimal source areas. Spherical Huygens wavelets travel out from each infinitesimal element of the source. For each uniformly illuminated aperture, area contributes an amplitude

$$dA = \left(\frac{A_0}{\lambda r}\right) \exp[j(\omega t - kr)]dS, \tag{8.1}$$

where A_0 = incident amplitude, t = time, k = wave number of the incident light, r = separation from source, dS = source area, $j = \sqrt{-1}$, ω = angular frequency, and λ = wavelength of incident light. The total amplitude at any point on the observing screen is found by summing all the contributions from each differential in eq. (8.1). The intensity is obtained by multiplying the amplitude by its complex conjugate. The pre-factor, λ^{-1}, in (8.1) is required to achieve the proper units and accounts for the vector nature of the light's electromagnetic field.

A convenient coordinate system to solve the problem of intensity beneath an aperture is shown in fig. 8.6. In this coordinate system, r can be expressed

$$r = \sqrt{(x_{\rm p} - x_{\rm p}^1)^2 + (y_{\rm p} - y_{\rm p}^1)^2 + r_0^2}, \tag{8.2}$$

where r_0 is the separation of the image plane from the aperture plane. Other coordinate designators are defined in the figure. The integral of eq. (8.1) is usually done in one of two limits. When r_0 is much larger than $(x_p - x_{\rm p}^1)$ or $(y_p - y_{\rm p}^1)$, $r \simeq r_0$ and a binomial expansion of (8.2) can be considered. This takes the form:

$$r \simeq r_0 + \frac{1}{2r_0}(x_p - x_{\rm p}^1)^2 + \frac{1}{2}(y_p - y_{\rm p}^1)^2 \tag{8.3}$$

Since the time varying part of (8.1), $\exp(j\omega t)$, disappears when intensity is computed ($I = A^*A$), 8.1 can be rewritten as:

$$dA = \frac{A_o}{\lambda r_0}\exp(-jkr_0)\exp\{-\frac{1}{2r_0}jk[(x_p - x_{\rm p}^1)^2 + (y_p - y_{\rm p}^1)^2] \tag{8.4a}$$

If we go one step farther and assume:

$$r_0 >> \frac{k(x_{\rm p}^2 + y_{\rm p}^2)}{2},$$

(8.4a) becomes:

$$dA = \frac{A_0}{\lambda r_0}\exp(-jkr_0)\exp\left[+\left(\frac{jk}{r_0}\right)(x_{\rm p}x_{\rm p}^1 + y_{\rm p}y_{\rm p}^1)\right] \tag{8.4b}$$

The first approximate form for this wavelet contribution, eq. (8.4a), is known as the Fresnel approximation. Here, the *phase* of the wavelet incident at a point on the observing screen varies quadratically as the wavelet's point of origin is moved about the aperture. Eq. (8.4b) is known as the Fraunhofer approximation. Here, the wavelet phase varies linearly as the wavelet's point of origin is moved about the aperture.

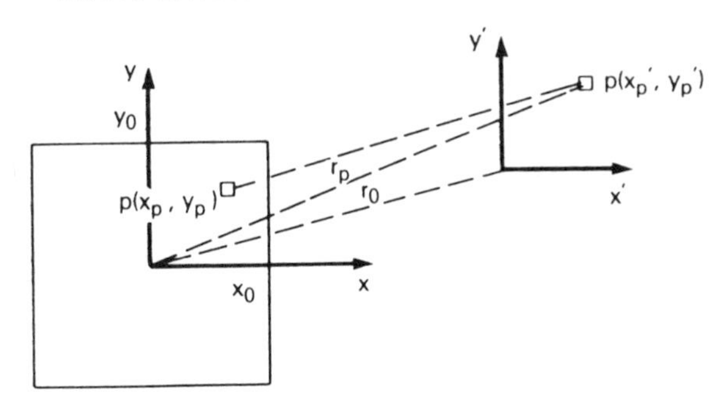

Fig. 8.6. Two-plane coordinate system for solving the diffraction problem.

Mathematically, the Fraunhofer case is much more easily handled. For the case of an infinitely long slit of width b, the intensity on an observing screen is:

$$I = I_0 \left[\frac{\sin\left(\frac{b \sin\theta}{2\lambda}\right)}{\frac{b \sin\theta}{2\lambda}} \right]^2 \tag{8.5}$$

where θ is the angle between a line drawn from the center of the slit to the observation point and a line from the center of the slit drawn normal to the observing screen. Here, λ is the wavelength of the illuminating light. This familiar sine variation is shown in fig. 8.7.

The Fraunhofer approximation is interesting in that it shows the main features of all diffraction intensity profiles, namely, the large central maximum surrounded by less intense side lobes. However, Fraunhofer diffraction is rarely valid in IC lithography. In contact printing, the observing plane is close to the aperture plane. Under these conditions the linear phase variation discussed above is not valid. The mathematically more complex Fresnel approximation gives better results. To perform a Fresnel calculation, (8.4a) is integrated, separating x and y components:

$$A = \int dA = XYZ \tag{8.6}$$

where:

$$X = \int \exp\left[\frac{-jk}{2r_0}(x_p - x_{\mathrm{p}}^1)^2\right] dx \tag{8.7a}$$

$$Y = \int \exp\left[\frac{-jk}{2r_0}(y_p - y_{\mathrm{p}}^1)^2\right] dy \tag{8.7b}$$

$$Z = \frac{A_0}{\lambda r_0} \exp(-jkr_0) \tag{8.7c}$$

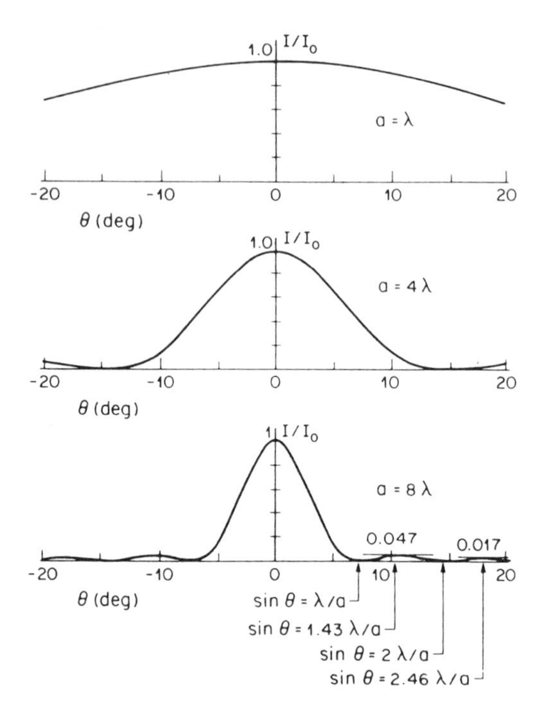

Fig. 8.7. Frauenhofer diffraction intensity pattern. After reference 6.

The X and Y integrals break up into the tabulated Fresnel sine (S) and Fresnel cosine (C) integrals:

$$S = \int \sin \frac{[k(x_p - x_p^1)^2]}{2r_0} dx \tag{8.8a}$$

$$C = \int \cos \frac{[k(x_p - x_p^1)^2]}{2r_0} dx \tag{8.8b}$$

Once A is evaluated, the intensity as a function of position is obtained by taking its complex square product.

Techniques for applying eqs. (8.6) – (8.8) in a number of different aperture configurations are summarized in many standard optics texts [4,5]. Results for a one-micron slit are graphically depicted in fig. 8.8. As in the Fraunhofer case, the large central maximum is evident, as are the less intense sidelobes.

This figure is particularly informative. It shows the tremendous effect of mask-to-wafer separation on resolution in proximity printing. It illustrates an additional problem in lensless systems: the difficulty of critical dimension (CD) control. It is interesting to note that even a fairly "washed-out" exposure intensity profile (shown in fig. 8.8a) can give rise to a reasonable sidewall sloped resist profile (fig. 8.8b). The problem is in the placement

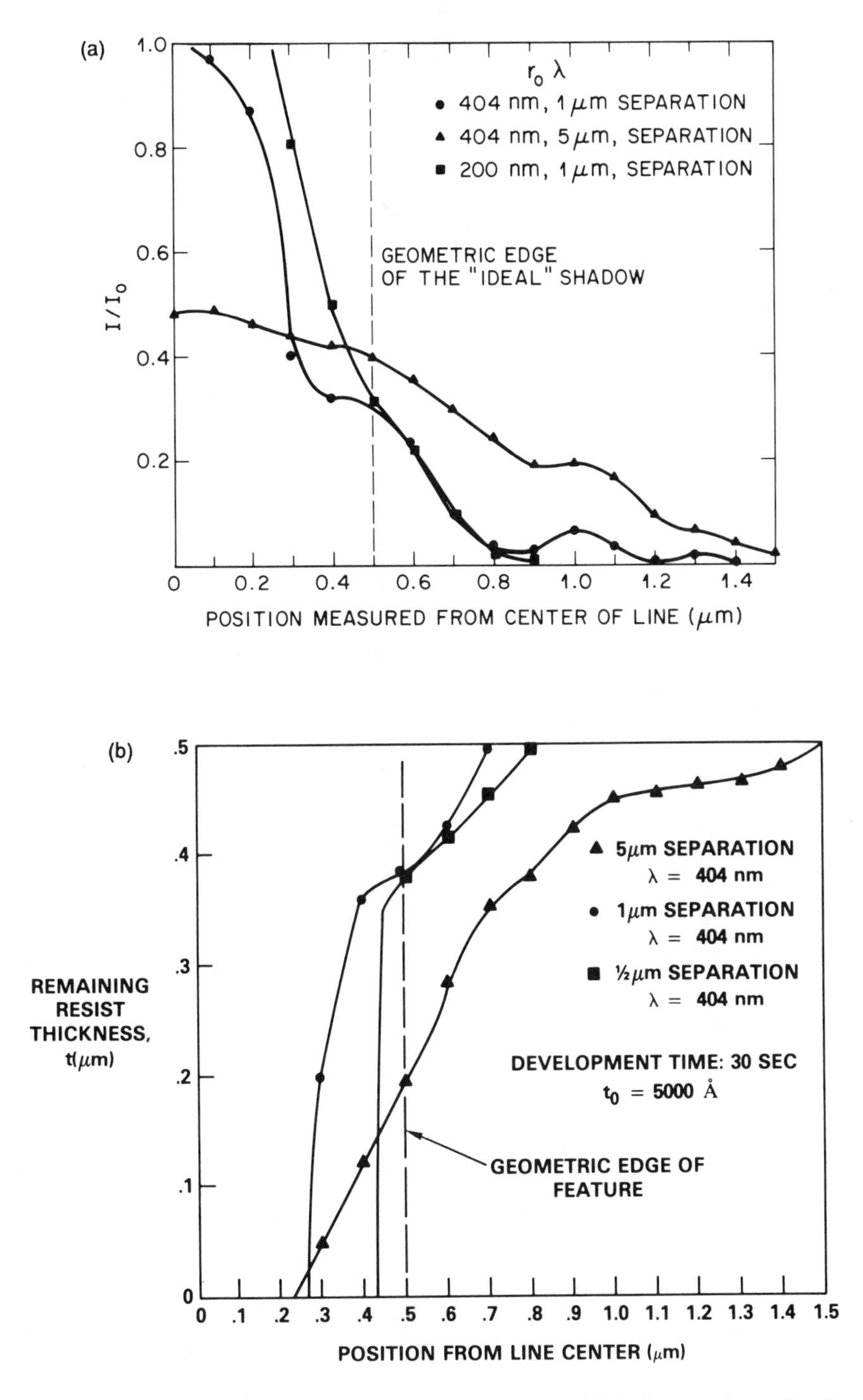

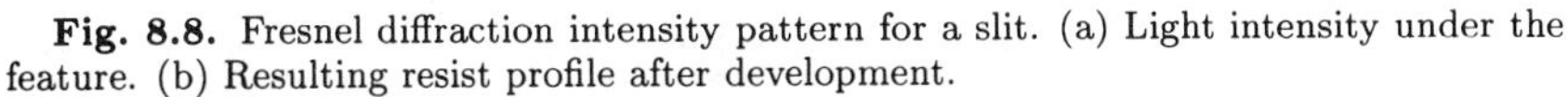

Fig. 8.8. Fresnel diffraction intensity pattern for a slit. (a) Light intensity under the feature. (b) Resulting resist profile after development.

of the resist boundary. All the resist lines in fig. 8.8b can probably be etched into underlying materials, but the boundary positions of these lines will vary greatly.

Many of the commonly used definitions of resolution are based on the appearance of the intensity distribution shown in figs. 8.7 and 8.8. For example, the Rayleigh resolution criterion is based on the Fraunhofer approximation [eq. (8.3)]. In this approximation, Rayleigh observed that the minimum discernible angular separation of two point sources occurs when the central maximum of the intensity pattern of one source falls on the first minimum of the second source. This is shown in fig. 8.7. For a circular aperture, this angular separation is given as [6]:

$$d\phi = 1.22\left(\frac{\lambda}{b}\right), \tag{8.9}$$

where b is the diameter of the aperture. A similar expression is obtained for an infinitely long slit (as shown in the figure), only 1.22 is replaced by 1.43. These formulae provide the basis for most approximations used to estimate resolution.

Now consider proximity printing of an array of closely spaced lines whose period is p. We make use of the observation that in all the Fresnel formulas for the intensity distribution, we find the *product* of λ and r_0. These terms never occur separately. A rule of thumb here is that the minimum printable period is:

$$p_{\min} \simeq 3\sqrt{\lambda r_0}. \tag{8.10}$$

The r_0 term here refers to the mask-to-wafer separation. The fact that p_{min} is proportional to $\sqrt{\lambda r_0}$ is derivable from 8.9 (see problem 8.1).

In addition to diffraction effects, light scattering in the resist can limit resolution. Each time light is incident on a surface that bounds media of different refractive indexes, reflection can occur. This creates a standing wave in the resist. If the light is not perfectly collimated, multiple reflections can carry the light far from the desired shadow boundary. This is illustrated in fig. 8.9. The extent of the light scattering under the masked feature and the amplitude of the standing wave can be computed by summing the incident and reflected wave amplitudes. The reflection coefficient at a boundary (for near-normal incidence) is given by

$$R = \frac{(\eta_1 - \eta_2)}{(\eta_1 + \eta_2)}, \tag{8.11}$$

where η_1 is the refractive index of the medium through which the beam is incident and η_2 is the index of refraction of the medium through which the beam is transmitted. Second, the internally reflected beam is attenuated as it travels through the resist. The amount of attenuation is given by

$$\frac{I}{I_0} = \exp(-ax), \tag{8.12}$$

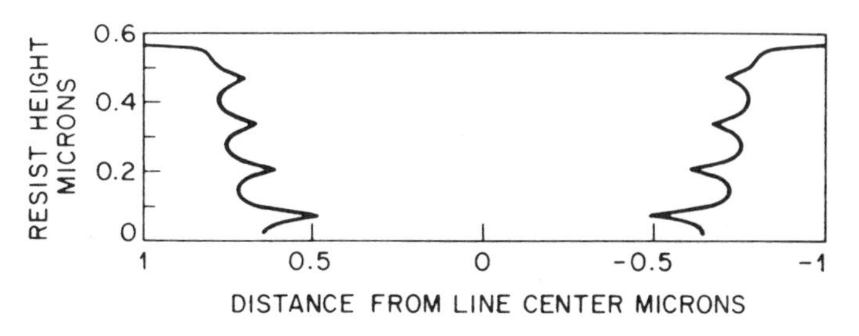

Fig. 8.9. Standing-wave pattern in photoresist.

where a is the absorption coefficient of the resist and x is the path length traveled. This is covered in greater depth in Chapter 9.

8.4.2 Projection Lithography

As mentioned above, contact printing damages masks and proximity printing has relatively poor resolution. The technique of projection printing was developed to remedy these problems [7]. Here, a lens system is used to create a focused image on a desired surface. Diffraction limits and standing waves are still encountered. However, the mask is undamaged in the exposure process since it never contacts the wafer. Even if there are particles or high spots on the wafer, the position of the focal plane is maintained over the bulk of the exposure field. In the section below, the fundamentals of projection printing are presented. Three classes of projection tools — full wafer, wafer scanner, and wafer stepper systems — are analyzed and contrasted.

To begin, we review some concepts from basic optics. Consider the simple lens system shown in fig. 8.10. Light is bent by refraction through a focal point to form an image on a plane (the image plane). The position of the focal point is a function of how the lens is constructed. Changing the position of the focal point changes the size of the object on the image plane. Thus, we can change the magnifying power of the lens. The figure also illustrates another important optical parameter: the cone angle, α_c. If we multiply the sine of the cone angle by the index of refraction, η, of the medium surrounding the image plane, we arrive at a term called the *numerical aperture* (abbreviated NA):

$$\mathrm{NA} = \eta \sin(\alpha_c). \tag{8.13}$$

The NA is proportional to the size of the largest incoming wavefront the lens can bring to focus. The larger the wavefront, the more light can be brought to focus. This makes the image plane brighter and exposes film faster. Thus, the NA is related to the "speed," or photographic f-number of the lens:

$$f\text{-number} = \frac{1}{2\mathrm{NA}}. \tag{8.14}$$

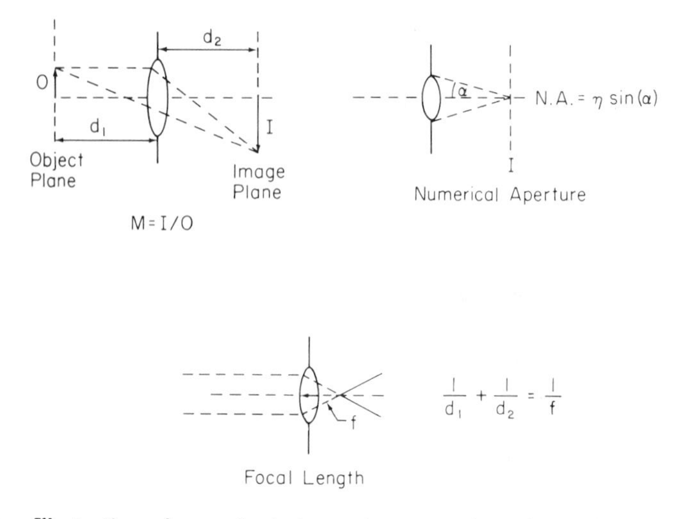

Fig. 8.10. Illustration of some basic terms in geometric optics. Magnifications, numerical aperture, and focal length are illustrated.

Despite the fact that a lens intervenes, eq. (8.9) is still valid in estimating system resolution capabilities. In order to faithfully reproduce an image of the object, we must sum many Huygens wavelets originating at the object. The aperture filters some of those wavelets, degrading the image. The amount of degradation is proportional to the number of wavelets filtered. This is evident in fig. 8.5. It takes many wavelets summed to form a smooth wavefront parallel to the source wavefront. In all cases, the aperture reduces the range of the summation. NAs for today's high-resolution lithographic lenses can be as large as 0.42 [8]. Many microscope lenses have NAs of 1.

Another quantity related to the numerical aperture is the depth of focus (DF). The DF is the minimum distance the image plane can be displaced before the resolution is noticeably degraded. Why does the image quality decline when we move the image plane? For a lens system the reason lies in the fact that all the light forming the image is bent through a common focal point. Light rays diverge as they move through the focus. The closer the focal point is to the lens plane, the more pronounced the divergence becomes. The separation of the focal point from the lens plane is related to the lens numerical aperture. These observations allow us to make an estimate of the DF. Consider a set of parallel wavefronts moving through the lens aperture, as shown in fig. 8.11. Assume these fronts are brought to an ideal focus at f. Displacing the plane of this focus a distance d causes the spot focus to grow. When the spot size reaches the Rayleigh resolution limit, we say (rather arbitrarily) that the displacement is one DF. As can be shown from data provided in the figure

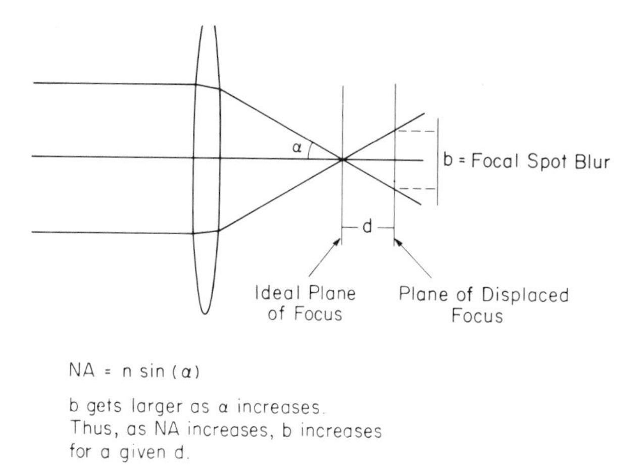

Fig. 8.11. Numerical aperture and depth of focus.

$$\mathrm{DF} \propto \frac{\lambda}{(\mathrm{NA})^2}. \tag{8.15}$$

As resolution, which is proportional to NA, increases, DF decreases. This is a major difficulty in current-day optical lithography. As the lens NA is increased to provide higher resolution, greater packing density, and improved performance, the DF shrinks drastically. It may become impossible to resolve images over the normal wafer topography, over or, out-of-plane wafer distortions. This has given rise to a host of "planarizing" techniques. These are discussed in Chapter 9, and elsewhere in this text. As a rule of thumb, for optical systems in the 1 μm resolution range, the DF approximately equals the NA times one micron.

Throughout this discussion we have assumed that the light is coherent, that is, the amplitude and phase of the electromagnetic field vary in a well-established fashion across a given wavefront. When coherent waves interact, their amplitudes are summed, and we observe the constructive and destructive interference patterns characteristic of diffraction. Intensity is the square of the resulting amplitude. If waves of many frequencies are mixed, this may not be the case. A condition of temporal incoherence exists. Usually, though, the light source is filtered to provide near monochromatic illumination. Despite this, no illumination system used in lithography is completely coherent in a spatial sense. This is because no light source is a true point source. We cannot hope that all the light-emitting oscillators over the face of a source are in phase. This results in a condition called *spatial incoherence.*

Associated with the concept of spatial coherence is the concept of partial coherence [9]. At first glance, it might be thought that the illumination is either totally coherent or totally incoherent. This is not the case. To see this, consider a thought experiment in which two parallel slits are illuminated by a broad area source (fig. 8.12). The resulting diffraction pattern is imaged on an observing screen at some distance from the slits. As the slits are moved farther and farther apart, two things happen. As we might anticipate, the positions of the maximum and minimum intensities shift. As the slits are separated more and more, however, a somewhat more subtle effect also comes into play. The fringe pattern on the observing screen becomes washed out. We say that the visibility of the fringe pattern is reduced. We define visibility as

$$V = \frac{(I_{\max} - I_{\min})}{(I_{\max} + I_{\min})}, \tag{8.16}$$

where $I_{\max}$ is the maximum observed intensity and $I_{\min}$ is the minimum observed intensity. When slit separations are greater than about a millimeter, the visibility sinks to zero. What is happening here? As a larger and larger area of the source is sampled, the coherence of the illumination declines. This is because the wavefronts from the many emitters that make up the source cannot interfere in a predictable manner over sensible regions. The visibility declines in a smooth fashion. Visibility should be greater than 0.5 (preferably ~0.7) for good imaging. This requirement on visibility is used to set the design rules for lines and spaces.

The degree of coherence is a measure of how well the given source approximates a point source. Optical systems can be designed to provide a predetermined degree of coherence. Consider the two-lens system shown in fig. 8.13. This is called the Köhler configuration, and it is frequently used in projection steppers. Here, the image of the source in the final aperture is not a point, but it does not fill the final aperture completely. This "underfilling" provides an adjustable degree of coherence. When the source appears as a point ($d_s^1/d_o = 0$), we have complete coherence as the optical disturbance appears to move out in a spherical wave from one source. If the source appears infinite, it is completely incoherent. The coherence factor, σ is the ratio of the apparent source diameter to the diameter of the final aperture (as shown in the figure).

Since coherence influences visibility, we can anticipate that coherence influences image quality. The nature of this influence is not as straightforward as it might seem. To see this, let us investigate the nature of image formation in somewhat greater depth. *We can consider an image as produced by the interaction of a main light beam (which passes through an object) and the beams diffracted by the object* [11]. For example, consider imaging a grating with a coherent source (as shown in fig. 8.14). The line-and-space period of the grating is $2L$. The spatial frequency of the grating is $f = 1/2L$. For the example worked here, we take only the first-order diffracted beams as significant. The diffracted beams travel in a direction, θ, given by the Bragg

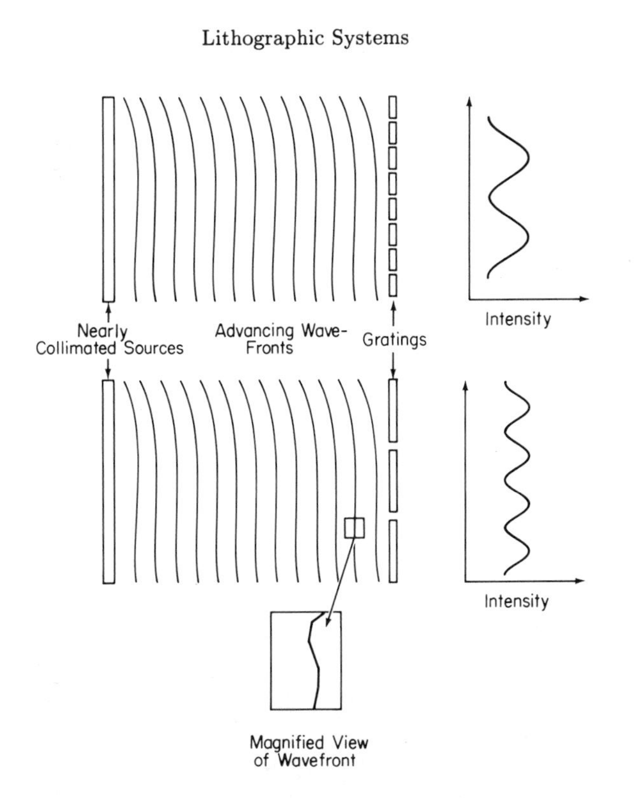

Fig. 8.12. Effect of slit separation on fringe visibility. The microscopic "roughness" in the wave-front causes a small reduction in visibility for closely spaced openings. The uncertainty in wave-front position increases as we travel larger distances along the wave-front. Thus, widely spaced slits have poorer visibility images.

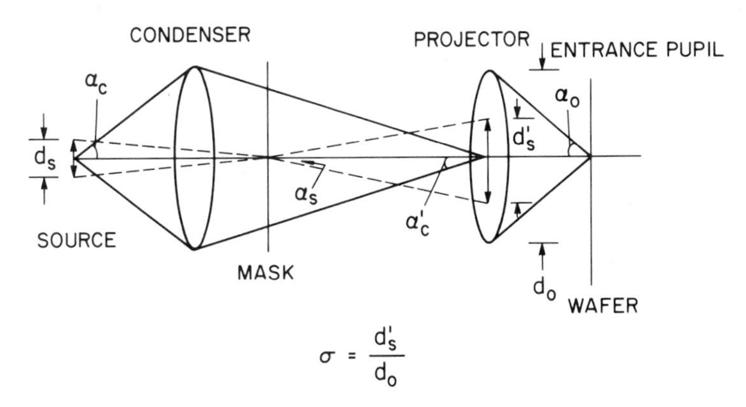

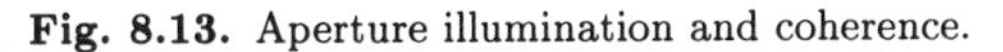

Fig. 8.13. Aperture illumination and coherence.

formula

$$\eta\, 2L \sin(\theta) = N\lambda, \tag{8.17}$$

where N is an integer and λ is the wavelength of the illuminating light.

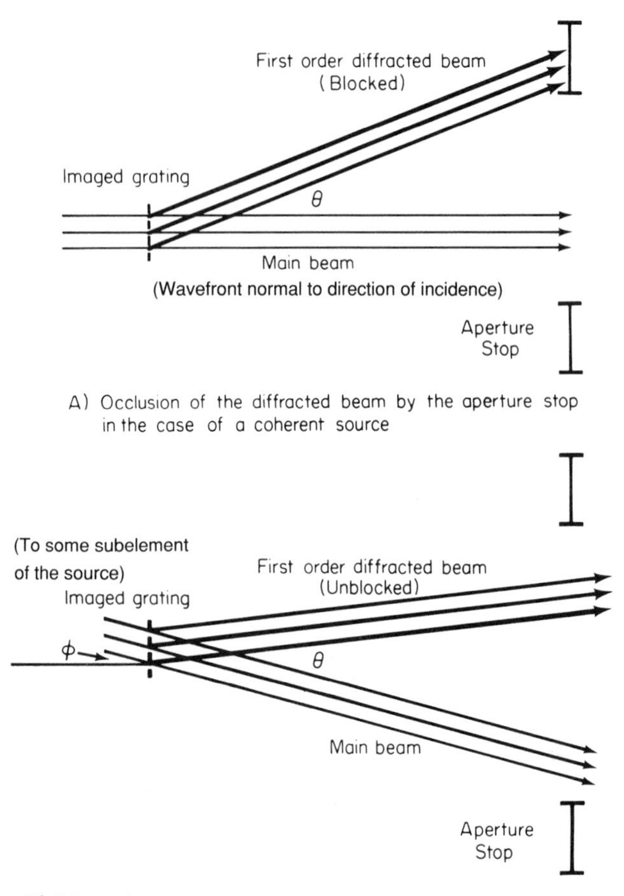

Fig. 8.14. The effect of coherence and the minimally resolved period of a grating.

If the aperture prevents the first-order beams from passing through to the imaging plane, the image cannot be formed. This occurs when $\eta \sin(\theta) =$ NA. Thus, for coherent illumination, the maximum grating frequency that can be imaged is

$$f_{\max} = \frac{\mathrm{NA}}{\lambda}. \tag{8.18}$$

Now take the case of illumination by an extended incoherent source. This is shown in fig. 8.14b. Here, the diffracted peaks occur at the angle, θ, given by

$$\eta\, 2L[\sin(\phi) + \sin(\theta)] = N\lambda, \tag{8.19}$$

where ϕ is the angle between the source point and the normal to the grating plane. As can be seen from the figure, a diffracted wavefront that is

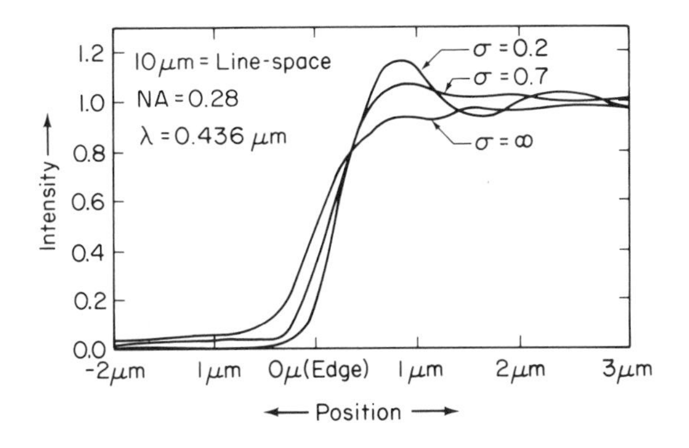

Fig. 8.15. Influence of coherence on resolution [10].

"clipped" in the coherent case is admitted to the image plane in the incoherent case. The maximum grating frequency that can be imaged without the clipping of the first-order diffraction is

$$f_{\max} = \frac{2\text{NA}}{\lambda}. \tag{8.20}$$

From this analysis, it appears that imaging with incoherent light improves resolution by a factor of 2. However, as mentioned above, the visibility declines as the source becomes incoherent. Even though the image-forming information (expressed in the main and diffracted beams) is present, the contrast necessary to form a good image is lacking.

The best compromise illumination source is neither completely coherent nor completely incoherent. The ideal σ (for projection systems using light at a wavelength near 400 nm and lenses with NAs near 0.35) is about 0.7. Here we have the benefit of the system transmitting diffracted beams for relatively fine-pitched gratings combined with the superior visibility of the coherent source. In addition, the sharp changes in intensity near feature boundaries, caused by the coherent-wave amplitude interference, are attenuated. This is seen in fig. 8.15. Here, the intensity near a boundary is shown for three coherence conditions. For complete coherence, there is a steep intensity drop-off near the boundary edge. However, intensity variation near the boundary is dramatic. This makes it difficult to arrive at the ideal exposure level. Incoherent exposure gives a more gradual intensity drop-off. A value of σ close to 1 gives a relatively sharp intensity drop-off and a flat intensity in the clear portion of the mask.

Generally, modeling studies of the type summarized in fig. 8.15 are performed using the transfer function concept [12,13]. The approach taken here is aimed at familiarizing the reader with the computational apparatus of the concept. The mathematics is more rigorously derived in the cited references. To make use of this concept, the image is broken down into its spatial frequency harmonics, that is, we do a Fourier series expansion of the mask image:

$$D(x_0) = a_0 + \Sigma[a_n \sin(2\pi n\nu x) + b_n \cos(2\pi n\nu x)], \tag{8.21}$$

where the a and b coefficients are the Fourier sine and cosine coefficients and n is an integer and $D(x)$ is either an amplitude or an intensity representation of the image. We then multiply each of the spatial frequency components (the individual a and b terms) by a system transfer function and discover what the intervening optical system does to the amplitude of each of the spatial harmonics. The MTF is the magnitude of the optical system transfer function. We then retransform the result to get the aerial intensity distribution in the image plane. The MTF is simply the frequency-dependent visibility of the image divided by the frequency-dependent visibility of the mask.

For analysis, the mask is a simple period grating of perfectly opaque and perfectly transmissive regions. In this case the mask visibility is 1 for all frequencies. For symmetric gratings (equal light and dark areas, with the opaque region just left of the origin), the image intensity distribution is odd. This means the series expansion will be a sine series (b_n =0 for all n). In addition, for most cases, taking only the first term in the series provides an adequate representation of the image-light distribution

$$D_0(x) = a_0 + a_1 \sin(2\pi\nu x). \tag{8.22}$$

For a grating pattern of equal light and dark areas, (8.22) becomes

$$D_0(x) = 0.5D_{\max}[1 + \left(\frac{4}{\pi}\right) \sin(2\pi\nu x)], \tag{8.23}$$

where $D_{\max}$ is the light amplitude or intensity that would be present on the image plane if the mask was completely transparent. Mask visibility here is $4/\pi$.

The effect of the intervening optics is to multiply the sinusoidal part of this expression by the space-frequency-dependent transfer function, $M(\nu)$:

$$D(x) = 0.5D_{\max}[1 + M(\nu)\left(\frac{4}{\pi}\right) \sin(2\pi\nu x)]. \tag{8.23a}$$

The unsubscripted $D(x)$ represent the light-distribution on the image plane through the intervening optics. Whether D is an amplitude or intensity representation, as well as the type of transfer function required, depends on the coherence of the light used to create the image. We consider two limiting extremes here: perfectly coherent illumination and perfectly incoherent illumination. For the partially coherent case, the reader is referred to reference 9, Chapter 10. First, consider the coherent case.

When coherent illumination is used, M_c is referred to as the *coherent transfer function.* D, here, is an *amplitude* representation. To find the intensity distribution, we must take the complex product, D^*D of the result. In this case, all spatial frequencies up to some critical frequency are admitted to the imaging plane without distortion. The origin of this critical frequency can be seen in light of our previous discussion about the role of the diffracted beam in creating the image. The higher the spatial frequency, the larger the angle the diffracted beam makes with respect to the incident beam. When the frequency gets high enough, the aperture intercepts it, and it would not contribute to the formation of the image. Consider the geometry in fig. 8.16. For a magnification factor of 1, $d_o = d_i$ (d_o and d_i are defined in the figure), and the critical frequency ν_c is given by:

$$\nu_c = \frac{L}{2\lambda d_i}, \tag{8.24}$$

where λ is the wavelength of this incident light, and L is the dimension of the aperture. The coherent transfer function is equal to unity for all spatial wavelengths longer than ν_c, and zero for all wavelengths shorter.

To help in calculating the two-dimensional image, we can specify the following functions [13]. In Cartesian coordinates, we define $\mathrm{rect}(x)$ to be unity whenever x is less than one and to be zero for all other values of the argument. Similarly, we define $\mathrm{circ}(r)$ in cylindrical coordinates to be unity when r is less than one and zero otherwise. We can now write the coherent optical transfer function for square and circular apertures as

$$M_c(\nu_x, \nu_y) = \mathrm{rect}\left(\frac{\nu_x}{\nu_c}\right)\mathrm{rect}\left(\frac{\nu_y}{\nu_c}\right) \tag{8.25a}$$

and

$$M_c(\nu_x, \nu_y) = \mathrm{circ}\,\frac{\sqrt{\nu_x^2 + \nu_y^2}}{\nu_c}, \tag{8.25b}$$

where ν_x and ν_y represent the x and y spatial frequency components.

The transfer function approach can be generalized to the case of incoherent illumination. Here, though, the coherent transfer function is replaced by the optical transfer function (OTF). Furthermore, $D(x)$ is given as an *intensity* representation. To define the OTF, we must redraw the aperture using a coordinate system in which x and y are expressed in terms of the space frequencies that impinge on these points from the object plane (i.e., x is written as $\lambda d_i \nu_x$, and y is written as $\lambda d_i \nu_x$). The OTF is the normalized autocorrelation of the aperture with itself in this coordinate system, multiplied by a phase factor, $e^{i\phi}$. This phase term accounts for the effect of the optical components on the light waves transmitted through the system. The normalized autocorrelation is fairly easy to compute for simple geometries. It is just the area of overlap of two copies of the aperture displaced with

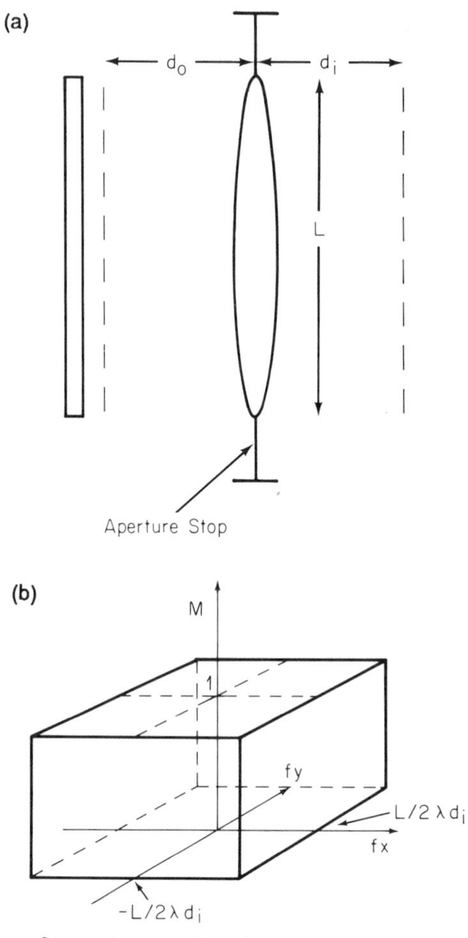

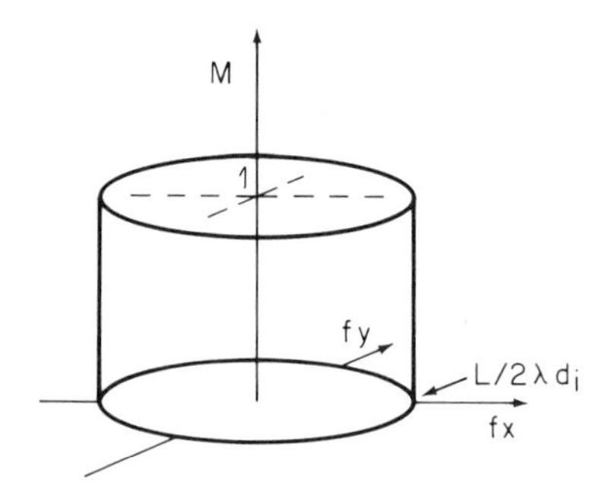

Fig. 8.16. (a) Definition of parameters for the transfer function calculation. (b) The rect and circ functions. After reference 13.

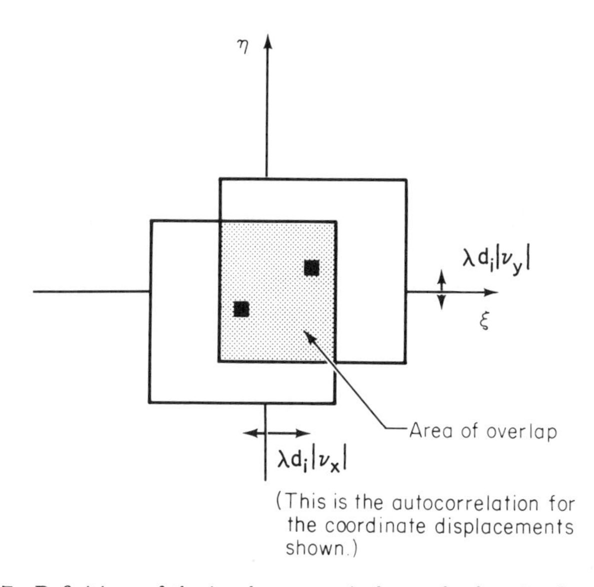

Fig. 8.17. Definitions of the incoherent optical transfer function by autocorrelation. The OTF is the normalized overlap area. The argument of the OTF is the coordinate displacement as shown.

respect to each other by $(\lambda d_i \nu_x, \lambda d_i \nu_y)$. The normalization is accomplished by dividing by the area of the aperture. Thus

$$M(\nu_x, \nu_y) = \frac{\text{area of overlap}}{\text{total aperture area}}\, e^{i\phi}. \tag{8.26}$$

This is illustrated in fig. 8.17. It is left as an exercise for the reader to compute some simple OTFs for slits, squares, and circles. The phase independent magnitude of the OTF is called the *modulation transfer function.*

It is also interesting to note that the transfer function approach can be used to compute the effects of lens aberrations on the image. This is accomplished by applying a phase factor to the aperture function. How the phase is specified determines the type of aberration. The most straightforward aberration to handle is that of a defect of focus. That is, the case in which the image plane is not at the point of ideal focus. For a pattern composed of equal lines and spaces which is coherently illuminated, the following amplitude and intensity formulae become valid:

$$A(x) = \frac{A_{\max}}{2}\{1 + \frac{4}{\pi}\exp[i\phi(z)]\sin(2\pi\nu x)\} \tag{8.27a}$$

and:

$$I(x) = 0.25 I_{\max}[1 + \left(\frac{8}{\pi}\right)\cos(\phi)\,\sin(2\pi\nu x) + \left(\frac{4}{\pi}\right)^2 \sin^2(\pi\nu x)], \tag{8.27b}$$

where

$$\phi = \pi z \frac{\lambda}{p^2}. \qquad (8.27a)$$

Here, z is the separation of the image plane from the plane of ideal focus, and p is the period of the line and space pattern (i.e., the reciprocal of the line and space frequency.) Eq. (8.25) also allows for the computation of the change in the intensity pattern as a function of displacement of the actual image plane from the plane of ideal focus.

An understanding of the optical fundamentals described above leads us to an appreciation of current trends in optical systems design. The desirability of projection optical systems was evident in the late sixties. Projection systems could supply high resolution without mask damage or poor CD control, but this was not as easy to achieve as it seemed. High numerical aperture lenses were needed that had minimum aberration over larger area projection fields. Such lenses could not be made at that time.

To remedy this problem, Markle, Offner, and King proposed the ring field scanner [14]. Here, a collimated beam of light is passed through a semicircular slit (see fig. 8.18a). The semicircular arc is about 1 mm wide, and its length depends on the maximum wafer diameter scanned. The light then passes through a mask and is brought to focus on the wafer with the help of two spherical reflecting mirrors. The reflecting mirrors have centers of curvature along the same axis. The exposure field is swept over the wafer by synchronously moving both the mask and the wafer, as shown in fig. 8.18b. Only a small fraction of the lens area is used for imaging in the ring-scan scheme described above. The limited lens area allows the possibility of a high degree of geometric aberration correction. In addition, the reflecting optics employed alleviates spectral dispersion problems (chromatic aberration effects) encountered in transmission optics.

Such scanning systems are usually used as full-wafer exposure tools. As a result of the depth-of-focus problems discussed above, the stepper approach is usually used in near-micron lithography. Typical exposure fields are relatively small (<1 cm^2). Even if 5x reduction systems are used, the mask image is less than 25 cm^2. It is currently possible to make lenses capable of micron resolution for such imaging areas without the need for scanning. It may be possible to extend these systems to below 0.5-μm resolution using deep-UV exposure tools. Whether such systems will be 1x or nx in their final form remains to be seen.

It also should be pointed out that the illumination source for such a system remains in doubt. Mercury lamps are the traditional illuminators in most lithographic systems. There are three primary mercury excitation lines used in lithography. They are the G, H, and I lines. Most lens systems are designed to focus on the G line (436 nm in wavelength). More recently, I-line (365-nm) systems are being marketed. To go below a half micron in resolution, it will probably be necessary to use sources that emit at wave-

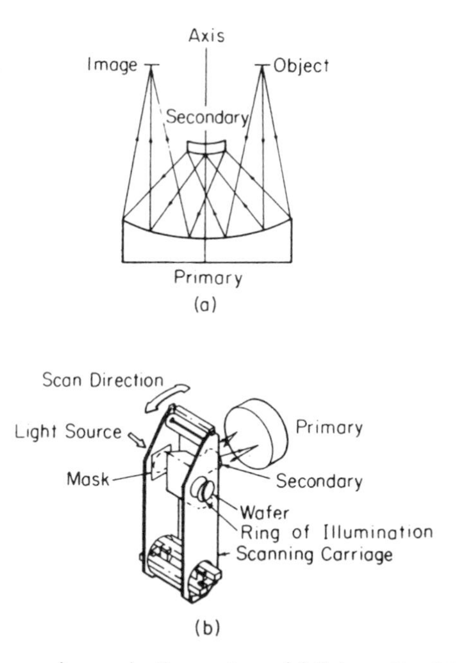

Fig. 8.18. The ring-scanning projection system: (a) Schematic of the optical system. (b) Scan mechanism.

lengths deeper in the ultraviolet. For example, KrF and ArF excimer lasers emit at 248 nm and 193 nm wavelengths, respectively. These sources are intense enough to expose photoresist in times on the order of 100 ms (using currently prototypical deep-UV systems [15]), but exposure uniformity and laser reliability remain a problem.

In going to wavelengths below 200 nm, radical changes must be made in the stepper optical system. Glass will not efficiently transmit these wavelengths. Furthermore, combinations of positive and negative (convex and concave) lenses of appropriate refractive index to provide correction for chromatic aberration are not available. Combinations of reflecting and transmissive or solely transmissive optics must be used. Deep-UV resists are also in a state of development. The highest resolution resist sensitive in the deep UV (polymethyl methacrylate) will not hold up in VLSI etch processes. Other materials absorb too little of the beam and are not exposed. Other materials absorb too strongly and give poor side-wall profiles. More is said on this in Chapter 9.

To conclude this section we have used a Fourier-sum approach to study image formation in optical systems. The space-frequency filtering properties of apertures on most features were shown to give rise to diffraction blurring. A system transfer-function approach was developed to provide a mathematical formalism to deal with these concepts. Geometrical optics was used to analyze the depth-of-focus principle. The section concluded with a discussion of how these principles applied to modern lithographic systems.

8.4.3 E-Beam Lithography

Current e-beam machines are an outgrowth of the work done on scanning electron microscopes (SEMs) [16]. Much of the literature on SEM beam-forming equipment is of immediate relevance to e-beam lithography. Most e-beam systems can be divided into three parts: the beam-forming element (or column), computers for lens control and data handling, and the mechanical components. The column consists of the electron source (or gun), beam blanking coils, electron lenses, scanning coils, and astigmation correctors. The mechanical components consist of the vacuum system and stage translation equipment. We begin our discussion by considering some of the physics of beam formation in the column.

Figure 8.19 is a schematic representation of a typical e-beam column. Free electrons are created and accelerated in the gun. Electrons are boiled off the filament (as in a standard vacuum tube). The filaments are usually a tungsten hairpins. Sharpened rods of lanthanum hexaboride (LaB_6) are also used as thermal-emission cathodes. Most systems used today rely on LaB_6 rods. The LaB_6 provides the highest brightness and smallest spot size. Brightness, B, of an electron source is defined as the current density per unit of solid angle

$$B = \frac{J}{\Omega}\left(\frac{\text{amps}}{\text{cm}^2 - \text{Sr}}\right). \tag{8.28}$$

The electrons reach their final velocity on passing through the anode, which is biased positively with respect to the filament. In the Wehnelt geometry gun shown in the figure, an electrode is placed around the filament. This electrode is negatively biased, causing electrons to be bent through a crossover point. It is a demagnified image of this crossover that is focused by the lens system at the workpiece.

The lenses serve to focus and to demagnify the electron beam emerging from the gun. There are two ways to do this, either electrostatically or by magnetic coils. Small amounts of contamination and charging severely affect the uniformity of the electrostatic focusing. To avoid this problem, magnetic lenses are usually used. A cross section of a magnetic lens is shown in fig. 8.20. If a divergent beam of electrons is incident on the lens entrance, electrons will have a velocity component normal to the plane of the cross section. The **v** x **B** force of the magnetic field on the electron will force the beam to convergence. In addition, the electron is caused to spiral around the magnetic field line. This causes a rotation of the beam shape about

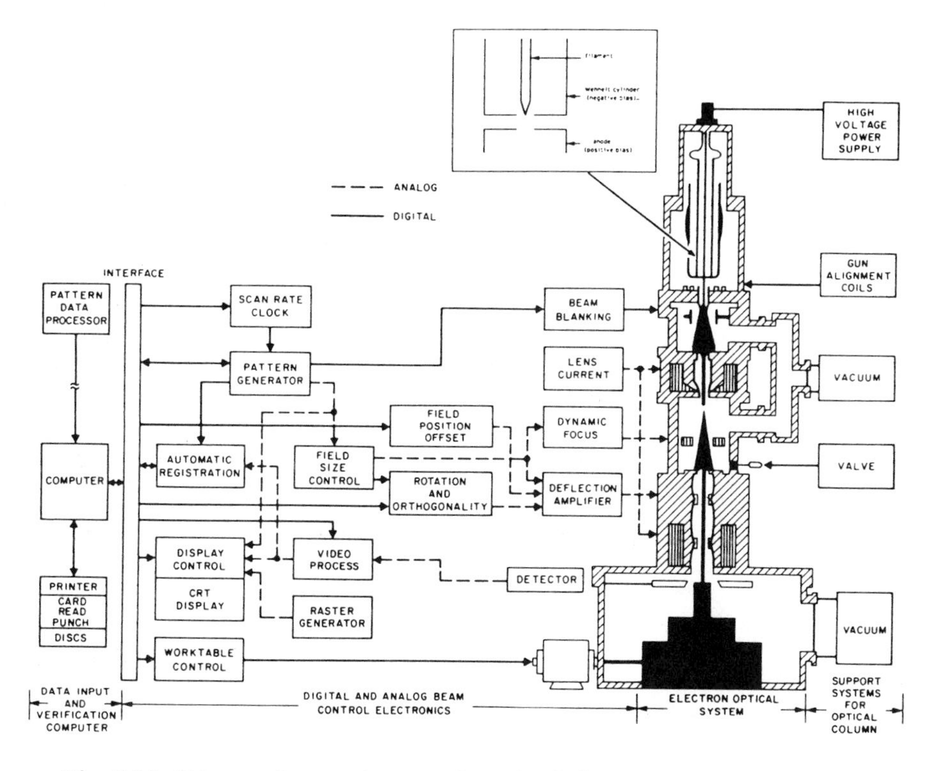

Fig. 8.19. E-beam column and gun configuration [16].

the central axis of the column. For Gaussian "round"beams, this is not an issue. For shaped apertures (discussed below), this could influence image formation. All magnetic lenses are convergent.

Most magnetic lens systems are stable enough to allow some deflection of the beam to the writing area. Typically, the area over which the beam can be electronically scanned (i.e., the writing field) can range from 0.1 x 0.1 mm^2 to about 5 mm^2. This scanning is accomplished, once again, with magnetic coils. The current in the coils is ramped to create the scan. There are two ways in which features can be written in the field. The first technique employed is called *raster scan.* In the raster mode, the beam is scanned television fashion through the whole writing field by the deflection coils. The writing field is broken up into squares of minimum resolution size called *pixels.* The scan coils would cause the beam to address every pixel. However, a blanking coil is employed to cancel the beam completely when the address loaded on the primary deflection coils correspond to an unwritten pixel. The stage is moved after the field is written.

Some e-beams work on a *vector scan* principle [17], that is, the beam is vectored to the shape to be written and the shape is written by the beam.

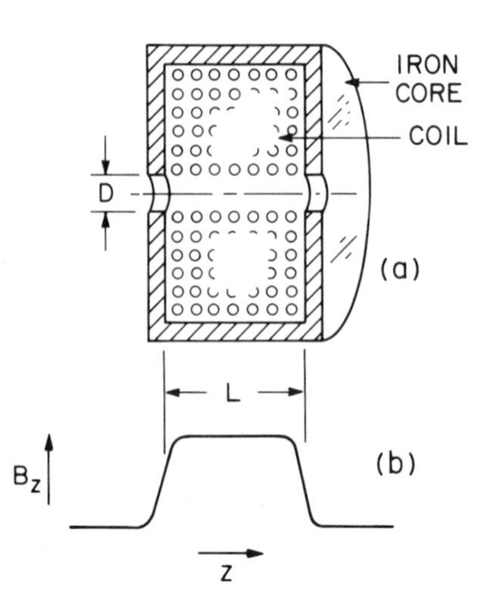

(a) A TYPICAL MAGNETIC LENS

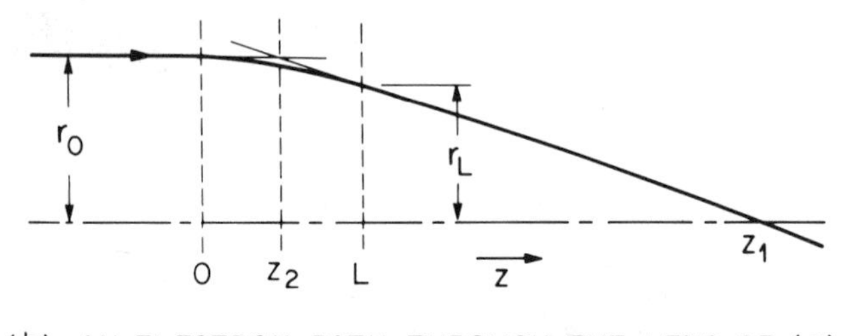

(b) AN ELECTRON PATH THROUGH THE LENS OF (a)

Fig. 8.20. Magnetic focusing lens cross section.

Only the addresses corresponding to the written regions are loaded onto the deflection coils. A first-order analysis indicates that there would be a significant speed advantage in a vector scan machine. To a first approximation, the plate write time is equal to the number of pixels addressed times the system clock speed (the time needed to load and to stabilize an address on the deflection coils). The larger the pixel-to-pixel deflection distance, though, the more time is required for the beam to stabilize. For raster scan, the pixel-to-pixel deflections are small and uniform. In the vector-scan case, the written regions may be widely separated, requiring long settling times. In addition, there is a bandwidth problem in the vector scan case. In raster scan, the range of signal frequencies on the deflection and on the blanking coils is limited. In the vector scan case, these signal frequencies have a wider range. This requires design of coils with a larger bandwidth of response. Such wide-bandwidth coils are frequently more sensitive to noise pickup.

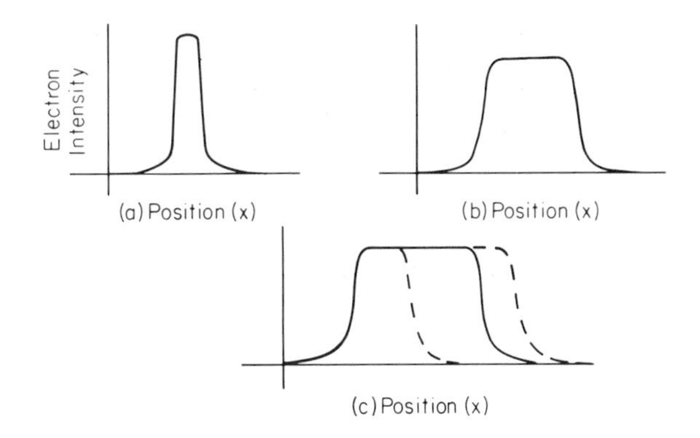

Electron Beam Probes (a)Gaussian Round Beam (b)Fixed Shaped Aperture (c)Variable Shape Aperture

Fig. 8.21. Electron-beam spatial energy distributions.

It is also important to consider the spatial and energy distribution of electrons in the beam. The discussion that follows on beam-energy distributions and aberration parallels that of [17]. There are three types of spatial distributions that can be obtained with commercially available machines [18]. These distributions are shown in fig. 8.21. The first and most common distribution is the Gaussian round beam (fig. 8.21a). The following expression is usually used to describe the spatial distribution of current density in such a beam

$$J = J_p \exp - \left(\frac{r}{s}\right)^2, \tag{8.29}$$

where J_p = peak beam current density, r = distance off central axis, and s = standard deviation. The $1/e$ point depends on the temperature of the electron source and the accelerating voltage:

$$s = \left(\frac{\mathrm{k}T}{eV}\right)^{1/2} \times \text{Gauss shape factor}, \tag{8.30}$$

where e = charge the electron, V = accelerating voltage, T = cathode temperature, and k = Boltzmann constant. Using the Wehnelt gun and magnetic lens configurations to form a minimum diameter probe, a Gaussian round beam is obtained. A variation on the round beam is shown in fig. 8.21b. This is called the fixed-shaped beam configuration. Here, the beam is expanded to yield a broader area exposure. The exposed area is a square. This is achieved using an aperture that clips the "skirts"of the Gaussian. The increased beam dimension improves throughput (by expanding the pixel size) but also sacrifices resolution. In some machines it is possible to vary the aperture dimension, creating the variable aperture distribution shown in fig. 8.21c. This allows improved throughput without sacrifice of resolution. In highly sophisticated machines, a combination of apertures is used to create a variety of shapes in the writing field (fig. 8.22).

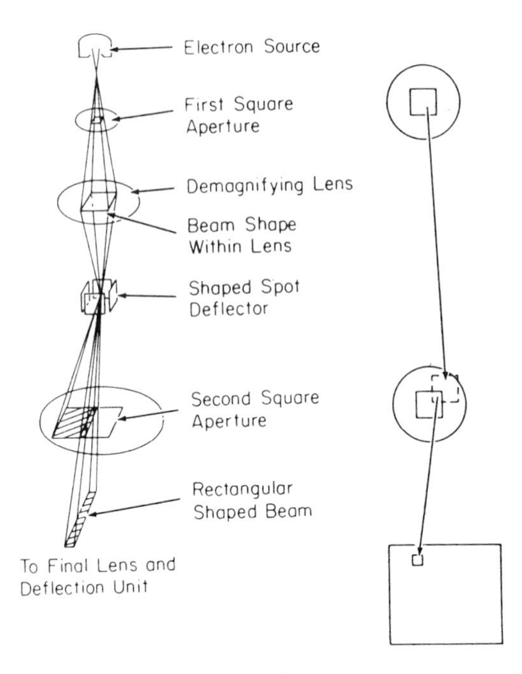

Fig. 8.22. Two apertures in the column give rise to a variety of "shapes in the field."

The Gaussian intensity profile is an immediate consequence of the energy distribution of the electrons emitted by the gun. The trajectory of an electron through the lens is a function of the incident direction and the velocity. If there is a distribution of velocities and incidence angles, electrons cannot be brought to a point focus. The width of the electron energy distribution is a function of the source temperature. Higher temperatures mean broader energy distributions. It can be shown that there is a maximum beam current density, $J_{\rm m}$, that can be brought to focus through a convergence angle, α, (for a given source temperature and accelerating voltage). This maximum is given by the Langmuir equation

$$J_{\rm m} = J_0 \left[1 + \left(\frac{eV}{{\rm k}T}\right)\right] \sin^2(\alpha) \approx \frac{J_0 V \alpha^2}{(T/11,600)}, \tag{8.31}$$

where J_0 = current density emerging from the gun cathode.

For a Gaussian round beam, the total beam current is given by

$$I = S^2 \pi^{3/2} J_p \tag{8.32}$$

In practice, roughly 80% of the total beam current is included within a diameter, d_g (which equals $2s$) of the beam center. Thus, a relationship is obtained that expresses the minimum radius of the focused spot

$$J = \frac{0.8I}{(1/4)\pi\, d_{\rm g}^2} \leq \frac{J_0 V \alpha^2}{(T/11,600)} \tag{8.33}$$

or (since $B \approx J/\pi\alpha^2$)

$$d_{\rm g,min} = \frac{1}{\alpha}\left(\frac{I}{3.08B}\right)^{1/2} \tag{8.33a}$$

This minimum radius spot is frequently called the disc of least confusion and is independent of the electron-optical system employed.

The above paragraphs define the physical limitations on minimum spot size encountered in an ideal electron-optical system. There are defects in real lens systems that give rise to astigmatism and to aberration that also cause an increase in the minimum-diameter spot size. Astigmatism results from small misalignments in the optical system components. If we consider the beam as a cylinder, the projection plane then intersects the beam axis at an angle that differs from 90°. The result is an elliptical spot on the workpiece. The astigmatism thus induced is corrected using *stigmators*. These are magnetic or electrostatic pole pieces that force the spot into a circular symmetry.

There are two types of aberrations that are important in e-beam lithography [18], chromatic and spherical aberrations. First, consider spherical aberration. All electrons do not enter the lens along the center line. Magnetic fields in the lens are weaker at the centerline of the lens. The crossover point is a function of the separation of the electron entry point from the lens centerline. This gives rise to spherical aberration. Both of these effects are illustrated in fig. 8.23. Both of these aberrations are a function of the convergence angle, α, at the position of the disc of least confusion. Spherical aberration is illustrated in fig. 8.23b. We can write an expression for the disc of least confusion that results from each of these processes as a constant times some power of the crossover angle. For spherical aberration we can write

$$d_{\rm s} = \left(\frac{1}{2}\right) C_{\rm s}\alpha^3, \tag{8.34}$$

if α is expressed in radians and $d_{\rm s}$ is expressed in centimeters. $C_{\rm s}$, the spherical aberration constant, is also expressed in centimeters.

Chromatic aberration (fig. 8.23a) occurs as a result of the energy spread of electrons in the probe beam. Slow electrons are brought to crossover closer to the lens than are fast electrons. For chromatic aberration, we have

$$d_{\rm c} = C_{\rm c}\left(\frac{\Delta V}{V}\right)\alpha, \tag{8.35}$$

where ΔV represents the energy spread in the electron beam. The chromatic aberration constant, $C_{\rm c}$ (also expressed in centimeters), is a function of lens

(a)

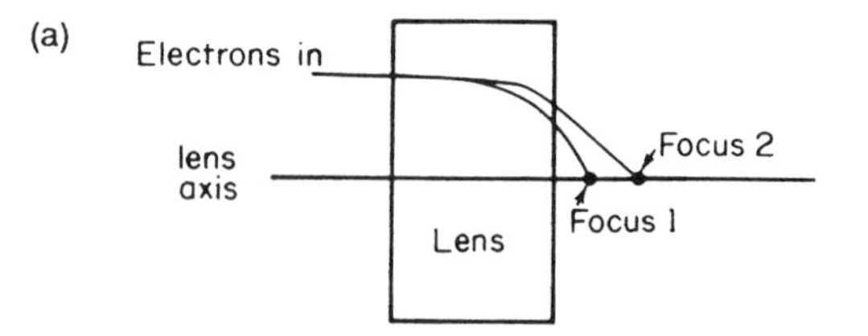

Magnetic lenses bring electrons of different energy to different points of focus

(b)

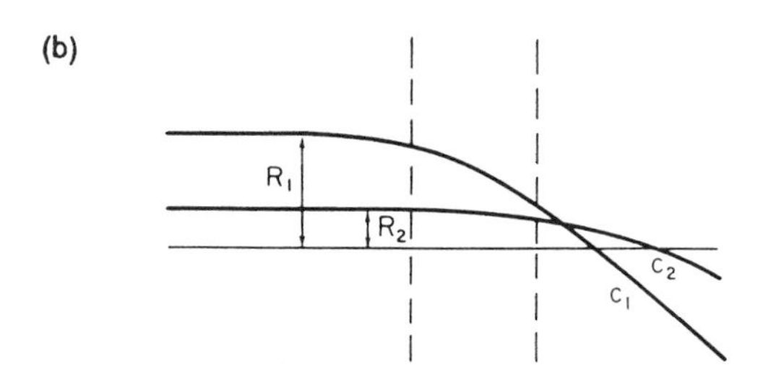

Origins of Spherical Aberration in a Magnetic Lens

Fig. 8.23. Spherical and chromatic aberrations in magnetic lens: (a) chromatic aberration, (b) spherical aberration.

focal length, f. Frequently, we write $C_c = K_c f$, where K_c is constant. Chromatic aberration increases as focal length increases.

As e-beam speed improves, or as novel approaches (such as the variable shape technique) become available, new resolution and throughput limits are reached. At very high electron-gun current, the electrons emitted at the source give rise to a negative space-charge surrounding the filament. This space-charge impedes further emission and limits the maximum current that can be drawn from the gun. This is called the *space-charge limit* [20]. The Coulomb interaction of the beam in the crossover region (the beam's focus) broadens the energy distribution and enhances the chromatic aberration in subsequent lens stages [21]. Kramer [22] has shown that the energy spread induced by this effect is

$$\Delta V_b \simeq \left[\left(\frac{J r_0}{\alpha^2}\right) / e\, V_0^{1/2}\right]^{1/2} \{1 + \left[\frac{kT_b}{(e\, V_0 \alpha^2)}\right]\}, \tag{8.36}$$

where T_b = temperature of the electron source, r_0 = radius of the beam at crossover, and V_0 = beam energy. When the second term in the bracket is ignored, we have (since $B \approx J/\pi\alpha^2$)

$$\Delta V_b \sim \frac{(B r_0)^{1/2}}{V_0^{1/4}}, \tag{8.37}$$

where B is the source brightness. In the next section we consider other advances in lithographic technique designed to surmount the throughput problems without loss of resolution.

While the wavelength of the electron is small, it is not zero. There is some diffraction limit to the smallest focused spot possible. This is expressed as

$$d_{\mathrm{d}} = 0.6\frac{\lambda}{\alpha}, \tag{8.38}$$

where λ is given as $1.225 \times 10^{-7}/\sqrt{V}$ (the effective wavelength of the electron). Since all of these effects are statistically uncorrelated, the total disc broadening goes as the sum of the squares of the individual contributions. The total disc diameter can be expressed, using equations 8.33, 8.34, 8.35, and 8.38, as

$$d_{\mathrm{t}}^2 = \frac{1}{\alpha^2}\left[\frac{5.4 \times 10^{-15}}{V_0} + \frac{I}{3.08B}\right] + \alpha^6\left(\frac{1}{4}C_{\mathrm{s}}^2\right) + \alpha^2\left(C_{\mathrm{c}}\frac{\Delta V}{V}\right)^2. \tag{8.39}$$

It is interesting to note that this equation implies that there is an optimum crossover angle for a given set of optics and beam currents. Eq. (8.39) expresses the influence of the optical system on the minimum resolved spot size. It should also be noted that the beam diameter is only one of the factors defining minimum resolvable feature size. The way the electron interacts with the resist usually has an equal influence. This is discussed in greater depth in Chapter 9.

The paragraphs above largely deal with resolution. It must be emphasized that one of the major drawbacks to production-line use of e-beam machines is throughput. E-beams are usually *serial* exposure tools, exposing one pixel at a time. If the pixel size is small, it will take a long time to print a pattern of reasonable density. Actual throughput limits are a result of what we can term *machine-overhead* penalities, that is, the inherent throughput limit rests in the settling times of the various mechanical, electronic, and electron optic components and *not* on the exposure sensitivity of the resist.

To summarize these machine-overheads, we have:

1. Alignment time (i.e., time required for the beam to seize on an alignment mark and make the necessary pattern-placement corrections)
2. Mechanical settling time of the stage (~100 ms)
3. Settling time required for the digital-to-analog converters (DACs), which take digital pixel address data from memory and convert this information to lens currents.

Recent advances in VLSI technology have created vast improvements in (3). In the past, typical "clock-speeds"(minimum address loading frequencies) were 10-20 Mhz. Newer machines tout 100-200 Mhz clocks. It should be noted that these improvements affect only (3) and not (1) or (2).

In addition, the mechanism of how the e-beam partitions its pattern generation data fields impacts throughout. For example, consider the data structure of the Cambridge EBMF system [19]. Each major writing field is divided into subfields. The machine writes all data within a subfield using addresses loaded from a single DAC. Subfields are changed by a second DAC. The fastest "clocks"occur in the writing of a single subfield. In vectoring fast subfield to subfield, from DACs must be reset. The setting time of the second DAC takes into account the longer lens reset time required for larger vector shifts of the writing beam. Of course, calibration of the two DACs is critical in minimizing subfield misalignment (i.e., the so-called subfield-butting error).

At the time of this writing, it appears that throughput limitations prevent the introduction of e-beam tools to the production line. It must be emphasized, though, that e-beams have had a central position in microlithography by virtue of their capabilities as a mask maker. They will continue to hold this central position as a mask maker for the high-throughput deep-UV lithographies or for x-ray lithography, discussed in the next section.

It should also be pointed out that other charged-particle-based lithographies are possible. Ion beams have been employed as wafer-patterning and mask-patterning tools. Typically, ion beams are still pixel-by-pixel exposure tools and are slow. Ion flood beams can also be used in conjunction with a mask, but these tools have not provided good submicron pattern replication due to scattering effects in ion masks. However, the ion-beam writer is still an attractive tool because it does not suffer from the proximity effect. Electrons released in the ion-bombardment process are of very low energy and do not expose adjacent pixels. As shown in Chapter 9, this leads to very high resolution. Chapter 11 is devoted to beam-processing techniques.

8.4.4 X-Ray Lithography

Based on the above considerations, the reader may ask, is it possible to expose all the pixels in a writing field *in parallel* with radiation of wavelength so short that diffraction no longer provides the resolution limit? The answer is, at present, a tentative yes. Some improvement in resolution is obtained by moving into the deeper ultraviolet for our exposure wavelength. This can be done in a number of ways. As mentioned above, commercial steppers for handling mercury I-line radiation (rather than G-line) are becoming available. There is also significant radiation from these lamps near 220 nm. These lines can be selected by optical filtering techniques. Excimer laser sources can be used to obtain even shorter wavelengths. Shortening the wavelength also improves depth of focus, as proved above. For the ultimate in resolution, x-ray lithography is possible. In this section, x-ray sources suitable for lithographic work are described. X-ray mask technology

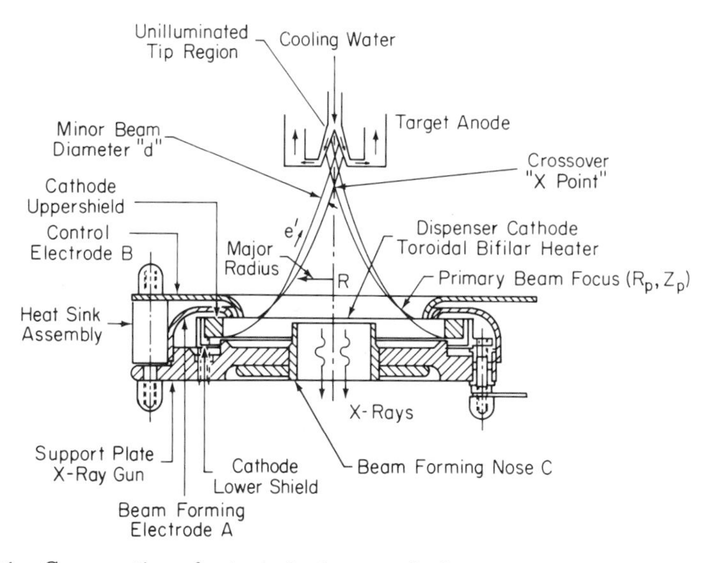

Fig. 8.24. Cross section of a typical e-beam-excited x-ray source: an x-ray tube. Courtesy of Perkin-Elmer Corp.

is reviewed and the fundamental resolution limits of x-ray lithography are presented.

X-rays of wavelengths from 0.1 to $\sim$ 4.5 nm are useful for lithography. X-rays can be obtained in an x-ray tube by focusing a beam of electrons on a piece of metal or silicon (called the anode). A cross section of such a tube is shown in in fig. 8.24. The radiation output spectrum from such a tube is shown in fig. 8.25. There are two components to such a spectrum: sharp lines and a broad-band continuum. The continuum, or Bremstrahlung, (braking radiation) component arises as a result of the deceleration of incident electrons in the anode. The lines are characteristic of the anode material and are due to interactions between the incident beam and the cores of the atoms that make up the anode. An incident electron knocks an electron out of the atomic core orbitals. Electrons in higher energy states fill the resulting "hole,"giving up their extra energy as x-rays. It is usually the line radiation that is most important in exposing x-ray resist. Anodes of aluminum, tungsten, palladium copper, and diamond have been successfully used in lithography. These materials have line radiation at 0.834 nm, 0.630 nm, 0.437 nm, 1.33 nm, and 4.5 nm, respectively.

A simple stationary anode x-ray source is fairly inexpensive, but it is also slow. Even relatively high-speed x-ray resists are exposed at a rate of 0.25 cm^2/minute exposure times with stationary anodes. This is hardly faster than e-beam lithography machines. To achieve higher power, the anode can be made as a disc that can be rotated at high speed. The disc is water cooled during operation. In this way, the beam's energy is distributed over

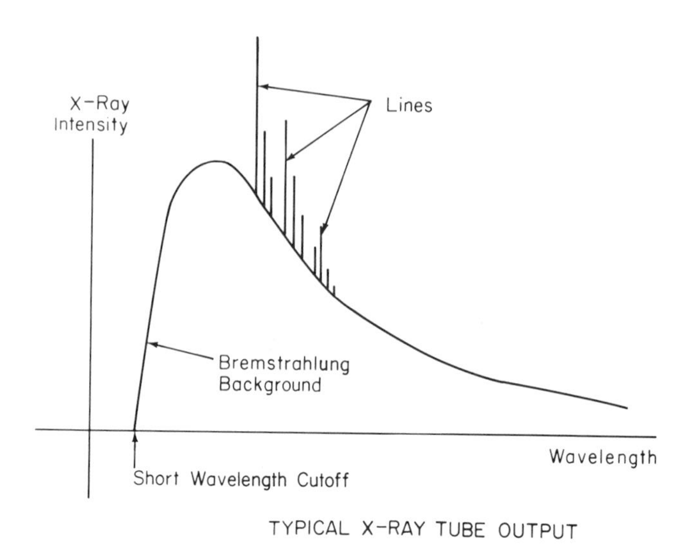

Fig. 8.25. Spectral output from the source shown in fig. 8.24.

the disc (rather than in a single spot.) A 2-3 cm^2/min exposure rate for a resist of 400 mJ/cm^2 incident sensitivity is today's benchmark for such equipment. The reason for this slow throughput is the low x-ray conversion efficiency, that is, only about one part in ten thousand of the electron-beam energy incident on the anode is converted into x-rays. For aluminum anodes operated at 20 keV, 60 microwatts of x-rays are emitted per steradian for every watt of electron energy incident.

Brighter sources can be achieved using more exotic means. Promising high-brightness x-ray sources include the x-ray synchrotron, the plasma puff, and the laser-plasma source. The synchrotron is a type of particle accelerator that causes electrons to race about in a circle [25]. Since the direction of the electron velocity vector is continually changing, the electron is in a continuous state of acceleration. A very bright pencil-like beam of x-rays is emitted, sweeping out from the plane of the electron orbit. This is shown in fig. 8.26. For such a source to be useful in lithography, the beam must either be spread or translated over the workpiece. In more aggressive approaches, the electron's plane of motion can be raised or lowered. It is very difficult to accomplish any of these types of scanning goals with conventional optical techniques. The index of refraction for x-rays in most materials is unity. Thus, x-ray light cannot be bent in an x-ray lens. Typically, grazing incidence mirrors are used to translate the beam over the workpiece. The synchrotron beam is incident at an angle more shallow than the critical reflection angle, creating an efficient reflector that can be mechanically driven to sweep the beam. Small amounts of contamination severely affect exposure uniformity.

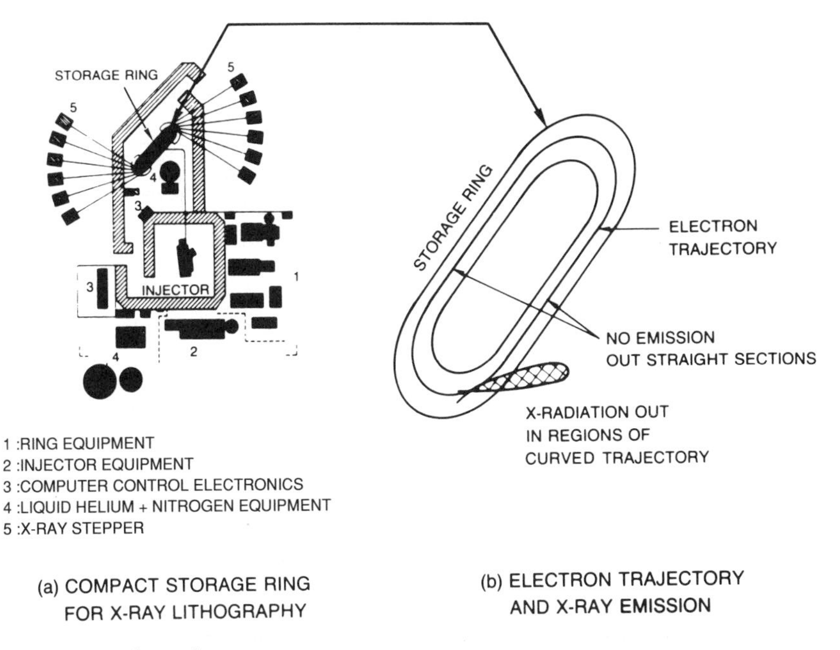

Fig. 8.26. A synchrotron x-ray source.

Figure 8.27 shows a schematic of the plasma puff [26]. In this system, a nozzle injects a puff of gas into a chamber. The charge output from a high-energy storage capacitor system is used to create a discharge in the chamber. The discharge contains an ionized gas, or plasma. The stored energy dumped into the resulting plasma heats the plasma. The magnetic field created in the discharge causes the hot gas to implode. This creates a tremendously high local energy density. Heated electrons have sufficient energy to interact with the gas atom cores to create x-rays. The resulting emission is largely line emission (although some continuum is also observed). Once again, the wavelength of the line emission is a characteristic of the puff gas used. The energy conversion efficiency is high. Up to 10% of the energy stored in the capacitor bank can be converted into usable x-rays. One disadvantage of this source is the fact that the high-energy plasma erodes the chamber walls, hurtling debris out toward the workpiece. This can damage the x-ray mask or the workpiece itself. Thin plastic debris shields are helpful in eliminating this problem, as is placing the workpiece at some angle off the cylinder axis of the puff. Both of these solutions reduce the x-ray intensity on the target.

Another attractive x-ray source is the laser plasma x-ray generator [27] (fig. 8.28). Here, an intense laser light beam is incident on a solid target. The target ablates and the ablated material is further heated by the laser beam.

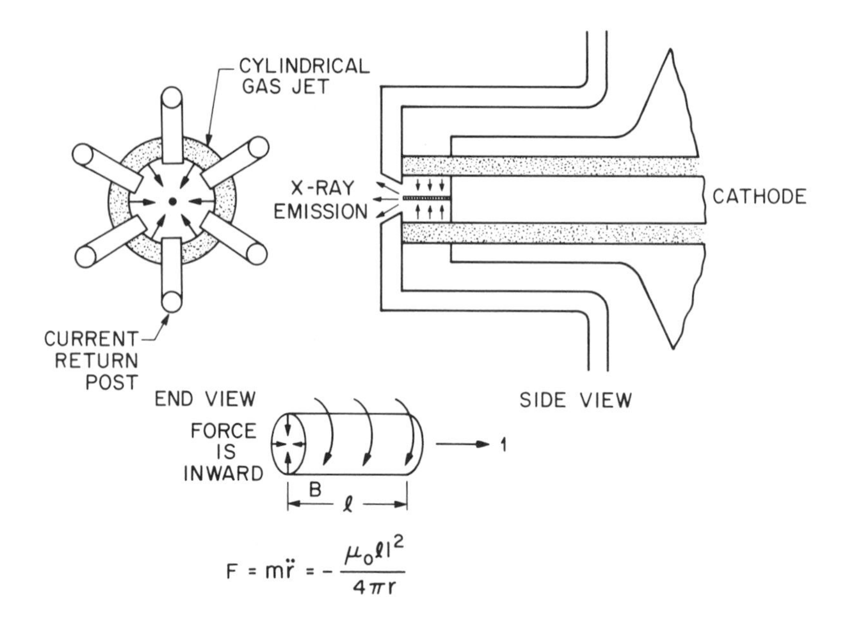

Fig. 8.27. A schematic of a plasma-puff x-ray source (courtesy of Maxwell Corp.).

A high-temperature plasma is formed. The free electron energy in the plasma is high enough to interact with the ion cores. X-ray line emission is excited in this process. Conversion efficiencies of 1% to 10% have been obtained. Optimization of the beam to the x-ray conversion process is difficult, as many parameters influence conversion efficiency. The wavelength and time duration of the laser pulse significantly influence the x-ray output intensity, as does the reflectivity of the material on which the laser beam is incident. Still, it is currently believed that such sources can provide order of magnitude throughput improvement over the brightest rotating anode sources. A comparison of the various types of x-ray sources currently available is given as table 8.1.

It should also be noted that the plasma sources are pulsed. That is, the x-ray output is delivered either in a single pulse or as a group of pulses. This can present a problem. The large amount of x-ray energy incident on the mask and the workpiece may damage either one or both of these items. Current trends in source design are aimed at spreading the exposure out over a few seconds using a number of pulses (anywhere from 10 to 100 pulses). Such systems are, currently, very hard to build. A reliable gas injection nozzle capable of opening and closing 100 times a second is required for the plasma puff. A high-energy laser capable of recharging every hundredth of a second is needed for the laser-plasma x-ray generator.

What are the resolution limitations in x-ray lithographic systems? Obviously, the near-angstrom wavelengths of the exposing radiation will not suffer

TABLE 8.1: Source characteristics, sizes and costs of x-ray devices used for exposure of photoresists.[a] (courtesy of Dr. D.J. Nagel)

	Electron Impact Sources		Synchrotron Radiation Sources	Plasma Sources	
	Static Anode	Rotary Anode		Nd Lasers	Electrical Discharge
Spectral Character	Lines and Continuum		Continuum Only	Strong Lines and Weak Continuum	
Photon energy range	3 > 10 keV (Sealed Source) <0.1 to > 10 keV (windowless)		UV to > 50 keV	Most intense < 1keV	
Source size (mm^2)		1–20	5	0.2	0.2
Emission solid angle (sr)	10^{-2}		10^{-5}	4π	0.1
Exposure area (cm^2)	<50		0.1–5	<200	<20
Emission time	Continuous emission		Steady, pulsed emission $<10^{-6}$ sec/pulse		
Time between Pulses (sec)	—	—	10^{-8}–10^{-6}	1	0.1
PMMA exposure Time (sec) [b]	10^4		< 100 (present sources) < 10 (new sources)	100	100
Device size (m)	1 × 1	2 × 2	[c]	2 × 5	1 × 1
Device cost ($)	2×10^4	10^5	10^6–10^7	2×10^5	2×10^4
Other factors	Vacuum or helium atmosphere used in the exposure regions			Exposures in vacuum to date	

[a] Approximate or typical values are given. Data are from references cited in text.
[b] Times for electron-impact and synchrotron radiation sources usually produce relief structure in PMMA in a single shot.
[c] Storage ring radii are in the 1- to 100-m range, with machines adequate for resist exposure having 1- to 10-m radii.

from diffraction limits as severely as optical systems with wavelengths in the thousands of angstrom's range. This does not mean that angstrom resolution is possible. To see this, the x-ray optics of the system must be studied in depth. The major resolution limit encountered here is due to *penumbral blur*. Penumbral blur occurs due to the fact that most x-ray sources are not exactly point sources. The e-beam incident on the anode creates an extended area of x-ray emission. The puff creates a cylinder of emission that

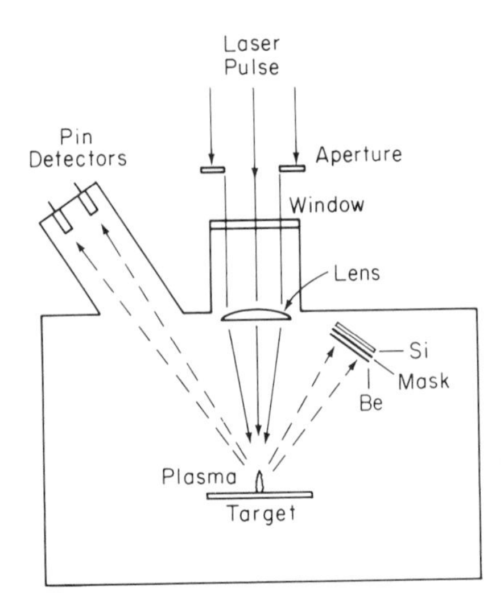

Fig. 8.28. The laser-plasma x-ray generator.

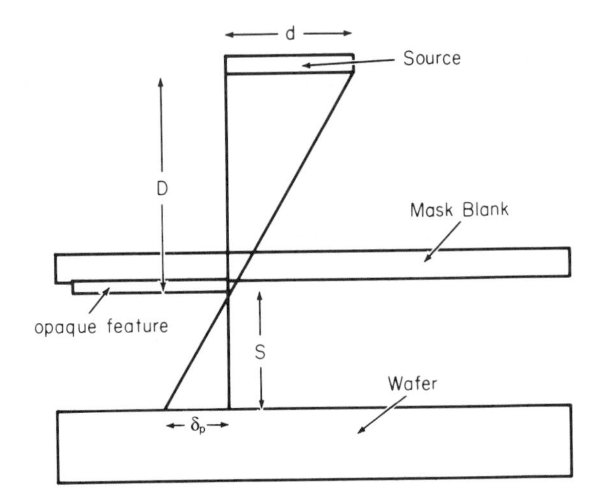

Fig. 8.29. Penumbral blurring of x-ray exposed features.

projects to a spot along the cylinder axis. The laser plasma source is also a cylindrical plume. As shown in fig. 8.29, such a source geometry undercuts the mask feature. The maximum amount of blur is given by

$$\delta_{\rm p} = S\left(\frac{d}{D}\right), \tag{8.40}$$

where $\delta_{\rm p}$ = extent of the penumbral shadow, S = mask-to-wafer separation, D = source-to-mask separation, and d = source diameter.

"Point"sources such as rotating anode tubes or plasma PUFFs have emission regions larger than 1 mm in extent. Laser plasma emission regions are usually less than 100 μm. For this case, the resolution approaches the diffraction limit expressed in eq. (8.10). From this equation, it can be seen that resolutions better than 0.5 μm are obtainable, but the large mask-to-water spacing creates significant diffraction spreading (see problem 8.10).

The intensity drops off fairly linearly from the feature edge as we move into the penumbral shadow. To first approximation, the penumbra causes a blurring of the feature edge. The extent of the blurring is δ_{p}. A more complete evaluation would involve a detailed analysis of how exposed resist behaves through the development cycle as a function of exposure dose. Details of this analysis are given in the next chapter.

Exposure with a point source also gives rise to feature displacement problems. These problems, frequently called run-out, get progressively worse at the edge of the exposure field (see fig. 8.30). In addition, point sources give rise to an apparent feature magnification, also illustrated in this figure. If the mask and the wafer remain parallel, with a constant separation, these problems can be eliminated in the CAD data. Features can be intentionally displaced on the mask to compensate for the edge run-out problem. If, however, the wafer becomes distorted through processing, feature displacements become impossible to deal with. This is one of the prime motivations for using low-temperature processing steps and low-stress-deposited films. These precautions minimize wafer distortion.

It can be argued that the major roadblock to the introduction of x-ray lithography into production is the x-ray mask technology. Extremely tight placement accuracy is required for mask feature (as summarized in Section 8.5). The fabrication of x-ray masks is considerably more difficult than the fabrication of optical masks [24]. The soft x-rays used in lithography are completely absorbed over a short distance in most materials. Typical $1/e$ absorption distances for a number of materials are shown in table 8.2. The short absorption length means that the transparent portion of the mask must be very thin. Typical mask materials are thin silicon membranes or chemical-vapor-deposited boron-nitride thin films. Other materials such as diamond film or SiC are in evaluation. The absorption characteristics of these materials are listed in table 8.2. Patterned heavy-metal absorber thin films provide the opaque regions of the mask. The x-ray transparent substrate must meet the following criteria:

- Stiffness: The substrate material must be stiff enough to resist distortion and deformation through the fabrication process and the absorber must not create strain in the underlying substrate membrane.
- Residual stress: The absorber film should not contain residual stress that might be relieved over time, causing distortion. The substrate base membrane should be in tension.

TABLE 8.2: Absorption coefficients (μ) at various x-ray energies for materials of importance in lithography. ρ = absorber density.

Hydrogen		Boron		Carbon		Oxygen	
Energy (keV)	μ/ρ (cm^2/g)	Energy (keV)	μ/ρ (cm^2/g)	Energy (keV)	μ/ρ (cm^2/g)	Energy (keV)	μ/ρ (cm^2/g)
0.1	1.16 + 4	0.1	1.49 + 4	0.1	2.65 + 4	0.1	6.23 + 4
0.12	6.55 + 3	0.15	6.17 + 3	0.15	1.08 + 4	0.15	2.56 + 4
0.13	5.10 + 3	0.1873	3.673 + 3	0.2	5.58 + 3	0.2	1.32 + 4
0.15	3.26 + 3	0.1887^{K}	0.2381	2.53 + 3	0.3	5.01 + 3	
0.17	2.21 + 3	0.2	6.68 + 4	0.2845^{K}	4.90 + 4	0.4	2.49 + 3
0.2	1.32 + 3	0.3	2.88 + 4	0.3	4.20 + 4	0.5	1.44 + 3
0.25	6.39 + 2	0.4	1.2 + 4	0.4	2.29 + 4	0.5313	1.23 + 3
0.3	3.61 + 2	0.5	7.98 + 3	0.5	1.34 + 4	0.5327^{K}	1.90 + 4
		0.6	4.91 + 3	0.6	8.44 + 3	0.6	1.60 + 4
0.35	2.17 + 2	0.7	3.24 + 3	0.7	5.64 + 3	0.7	1.09 + 4
0.4	1.39 + 2	0.8	2.24 + 3	0.8	3.95 + 3	0.8	7.94 + 3
0.45	9.65 + 1						
0.5	6.80 + 1	0.9	1.61 + 3	0.9	2.87 + 3	0.9	5.89 + 3
0.55	4.89 + 1	1.0	1.19 + 3	1.0	2.15 + 3	1.0	4.50 + 3
0.6	3.71 + 1	1.2	7.38 + 2	1.2	1.32 + 3	1.2	2.91 + 3
0.65	2.96 + 11.5	3.78 + 2	1.5	6.93 + 2	1.5	1.57 + 3	
0.7	2.33 + 1	1.7	2.59 + 2	1.7	4.81 + 2	1.7	1.11 + 3
		2.0	1.58 + 2	2.0	2.97 + 2	2.0	7.00 + 2
0.75	1.82 + 1	2.5	7.98 + 1	2.5	1.53 + 2	2.5	3.69 + 2
0.8	1.45 + 1	3.0	4.55 + 1	3.0	8.81 + 1		
0.85	1.22 + 1						
0.9	1.03 + 1						
0.95	8.41 + 0						
1.0	6.90 + 0						
1.2	4.11 + 0						
1.5	2.15 + 0						
1.7	1.54 + 0						
2.0	1.06 + 6						
2.5	7.06 + 1						
3.0	5.61 + 1						

TABLE 8.2: (*continued*) Absorption coefficients (μ) at various x-ray energies for materials of importance in lithography. ρ = absorber density

Nitrogen		Silicon		Tungsten		Gold	
Energy (keV)	μ/ρ (cm^2/g)	Energy (keV)	μ/ρ (cm^2/g)	Energy (keV)	μ/ρ (cm^2/g)	Energy (keV)	μ/ρ (cm^2/g)
0.1	4.45 + 4	0.1	5.70 + 4	1.17 + 4	0.1	6.27 + 3	
0.15	1.76 + 4	0.12	1.03 + 5	0.15	1.81 + 4	0.1071_{O_I}	5.20 + 3
0.2	9.02 + 3	0.14	9.68 + 4	0.2	1.98 + 4	0.1085	5.75 + 3
0.3	3.42 + 3	0.148_{L_I}	9.13 + 4	$0.2501_{N_{IV,V}}$	1.86 + 3	0.118	4.92 + 3
0.4	1.70 + 3	1.1494	1.01 + 5	0.2515	1.99 + 4	0.15	5.37 + 3
0.4009	1.69 + 3	0.15	1.00 + 5	0.3	0.73 + 4	0.2	1.10 + 4
0.4023^K	2.98 + 4	0.17	8.40 + 4	0.4	1.34 + 4	0.3	1.61 + 4
0.5	1.90 + 4	0.18	7.69 + 4	$0.4467_{N_{II,III}}$	1.18 + 4	$0.3404_{N_{IV,V}}$	1.55 + 4
0.6	1.22 + 4	0.2	6.43 + 4	0.4481	1.29 + 4	0.3418	1.63 + 4
0.7	8.32 + 3			0.5	1.15 + 4	0.4	1.50 + 4
0.8	5.90 + 3	0.24	4.52 + 4				
		0.3	2.82 + 4	0.5943_{N_I}	9.07 + 3	0.5	1.23 + 4
0.9	4.33 + 3	0.4	1.48 + 4	0.5957	9.63 + 3	$0.5775_{N_{II,III}}$	1.02 + 4
1.0	3.27 + 3	0.5	8.71 + 3	0.6	9.48 + 3	1.5879	1.10 + 4
1.2	2.08 + 3	0.6	5.59 + 3	0.7	7.42 + 3	0.6	1.09 + 4
1.5	1.10 + 3	0.7	3.80 + 3	0.8	5.91 + 3	0.7	8.72 + 3
1.7	7.68 + 2	0.8	2.71 + 3	0.9	4.78 + 3	0.7581_{N_I}	7.67 + 3
2.0	4.79 + 2	0.9	2.00 + 3	1.0	3.92 + 3	0.7595	8.12 + 3
2.5	2.49 + 2	1.0	1.53 + 3	1.3	2.33 + 3	0.8	7.40 + 3
3.0	1.45 + 2			1.7	1.32 + 3	0.9	6.04 + 3
		1.2	9.72 + 2	1.809_{M_V}	1.11 + 3	1.0	4.98 + 3
		1.5	5.33 + 2	1.809	3.74 + 3		
		1.7	3.80 + 2			1.2	3.51 + 3
		1.839^K	3.08 + 2	$1.872_{M_{IV}}$	3.42 + 3	1.5	2.223 + 3
		1.839	3.74 + 3	1.872	5.12 + 3	2.0	1.21 + 3
		2.0	3.00 + 3	2.0	4.31 + 3	2.206_{M_V}	9.16 + 2
		2.5	1.66 + 3	$2.281_{M_{III}}$	3.07 + 3	2.206	2.61 + 3
		3.0	1.01 + 3	2.281	3.68 + 3	$2.291_{M_{IV}}$	2.37 + 3
				2.5	2.90 + 3	2.291	3.55 + 3
				$2.575_{M_{II}}$	2.68 + 3	2.5	2.84 + 3
				2.575	2.95 + 3	$2.743_{M_{III}}$	2.243
				2.82_{M_I}	2.33 + 3	2.743	2.68 + 3
				2.82	2.56 + 3		
				3.0	2.18 + 3	3.0	2.13 + 3

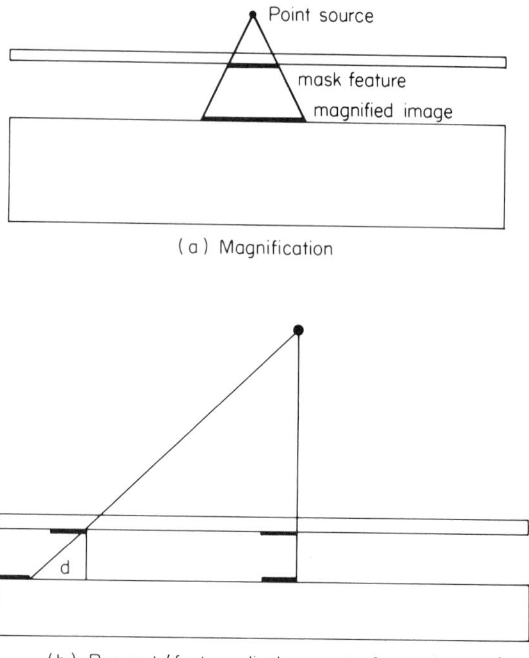

Fig. 8.30. Run out and magnification effects occurring in imaging systems using a point source illuminator.

- X-ray transparency: At least 90% of the incident radiation should be passed.
- Freedom from defects: The membrane should be pinhole free and free of bumps and hillocks.

Some of the criteria appear to be mutually exclusive. For the substrate to be x-ray transparent it usually has to be thin. Thin films frequently have pinholes and/or growth spikes. Thin films tend to be compliant and may not support the tight feature placement accuracy required for today's VLSI.

Silicon membranes and boron-nitride films have shown the best performance to date as x-ray masks. A typical x-ray mask cross section is shown in fig. 8.31. Gold is used as the absorber material for two reasons. First, it absorbs strongly those x-rays most useful for lithography. Second, it can be low stress and thus would not deform the substrate on deposition. The gold itself can be excited by the incident beam and made to emit characteristic x-rays. Typically, the gold M line (0.585 nm) is excited. Current x-ray resists are very sensitive to radiation of this wavelength. As a rule, most materials are fairly transparent to their own characteristic radiation. Thus, the gold features can, on excitation, expose photoresist. For this reason,

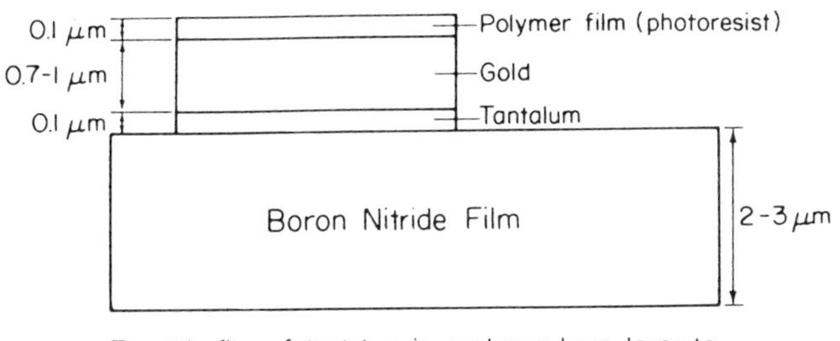

Fig. 8.31. A typical x-ray mask cross section.

a thin polymer absorber is used over the gold. A hard x-ray continuum (the Bremstrahlung component) can give rise to a considerable number of photoemitted electrons that can expose resist. The thin plastic overcoat (in this case, unremoved photoresist) can also absorb this unwanted radiation.

Two severe problems in x-ray lithography remain to be discussed: resists and alignment. Most x-ray resists are fairly "slow" (i.e., they require a long exposure time). In addition, to make proper use of the ability to pattern finer geometry features provided by x-ray lithography, we must have a matching ability to align these finer features within the desired pattern — to place these features where we want them. Resists are discussed in the following chapter. Alignment is discussed in this section below.

8.5 Alignment

In most cases alignment accuracy is as important as resolution in creating high-density integrated circuits. Assume for a moment that it were possible to routinely make an 0.05-μm gate-length transistor. The impact of such an achievement would be lost if we had to follow a 1-μm-per-gate design rule to allow for feature placement tolerance. In this section, various techniques available for the performance of accurate alignments are presented.

Stages that move under micrometer control can provide the mechanical stability to reposition a wafer with 50-nm accuracy. It may be possible to improve this to 10-nm accuracy in the future. It is quite a challenge to build an optical system capable of making full use of this positioning accuracy. Consider the "traditional" method of alignment. A series of alignment crosses are etched into the substrate. Under each cross a number appears. This number refers to the mask level. A corresponding cross is placed on the mask level itself. The mask cross and the wafer cross are viewed through an optical microscope. Micrometers that move the wafer under the mask are then used to bring the workpiece into proper alignment with the wafer. The micrometers move the wafer with three degrees of freedom: X, Y, and θ. Since the Rayleigh resolution of most optical microscopes is about 0.5 μm, it is frequently thought that this is the limit of repositioning accuracy in the cross-on-cross system. This is not the case. The issue of whether or not two lines are viewed as distinct is different from the problem of symmetrically placing one cross within another. In practice, optical repositioning with 1/3-μm accuracy is routinely achieved.

A number of variations of the cross-on-cross method appear in all of the types of lithography systems discussed so far. In advanced wafer steppers, the alignment mark is sighted through the stepper lens itself and viewed with a TV camera. Optical pattern recognition schemes are sometimes used to remove the manual component from the alignment process. Alignment can be global or die-by-die in nature. In the global alignment scheme, the alignment marks are picked up at two or three places on the wafer, and the mask and wafer alignment is accomplished by getting the best alignment using those two or three points. In the die-by-die scheme, each die is aligned separately after global alignment is done. Global alignment can be done manually. The die-by-die scheme requires some computer alignment mark recognition scheme. The die-by-die method also takes time, cutting wafer throughput by as much as 80%.

Still, the 1/3-μm repositioning accuracy does not stress the capability of the mechanical system. To achieve the ultimate in repositioning accuracy with an optical system, the wave-interference properties of light must be exploited. For example, consider the alignment system shown in fig. 8.32. Here, a beam of light is brought to a tight focus using a zone plate on the mask. A zone plate is an array of opaque and transparent lines. The opaque lines subtract the portions of the diffracted wavefront that would cause destructive interference at some focal point of the lens. This creates a submicron line of high intensity incident on the wafer surface. The line strikes a linear diffraction grating printed on the wafer surface. The grating can either be etched into the silicon or made of some thin-film material that will survive the wafer-processing steps. When the whole line is in perfect alignment, the diffracted beam reaches maximum intensity. Each linear grating is capable of creating alignment in one direction only. Thus, two gratings are needed for X and Y alignment. It is possible to get repositioning accuracies better than 10 nm with such a procedure [28]. Alignments of this type are suitable for both x-ray and optical lithography.

In addition to the X, Y, and θ degrees of freedom, the mask-to-wafer gap and tilt must be set. This can be done either optically or electrically. There are a number of optical approaches possible. In the simplest case, the best possible focus is achieved for a small spot on the workpiece. Next, a small spot on the mask is focused on. The change in the focal position of the alignment microscope is measured. This is the wafer-to-mask separation

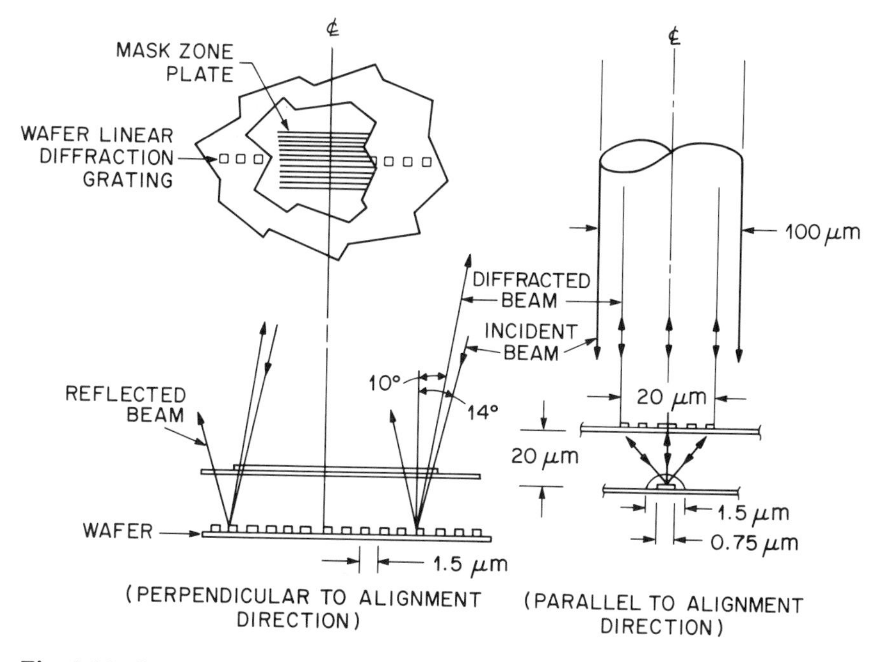

Fig. 8.32. Zone-plate internal alignment.

for the spot being analyzed. The accuracy of the measurement is just the microscope depth of focus (usually a few microns). In addition, the zone-plate measurement also gives a determination of gap. Here, maximum reflected intensity is sought after course alignment. To assure wafer-to-mask parallelism, these procedures are repeated at a number of places across the exposure field. The wafer-to-mask separation can also be sensed capacitively. Conducting discs must be accessed on the plane of the workpiece and on the mask.

Automatic alignment systems are also possible. One such scheme is shown in fig. 8.33 [29]. In this figure, we see a combination of lenses, disc-shaped figures, apertures (squares with circles cut in them), and miscellaneous optical components. The reticle, R, is normally imaged onto the wafer, W, through lenses L_1 and L_2. Light from the illuminator enters through apertures and is collimated with lens L_c. In the alignment mode, mirror M blocks the main illuminator. The wafer and the reticle are illuminated with a low-intensity source, S. Illumination of the wafer alignment mark is accomplished with the beam-splitter, B. Lens L_2 forms a narrow diffraction pattern of the alignment mark in aperture A. Clear regions on the wafer produce no diffraction. Reflected light from these clear portions is stopped by the opaque block at S. This forms a dark-field image of the mask on the reticle. The dark-field image of the wafer-alignment mask passes through the reticle-alignment mark and is reflected by M into the detector, D. The wafer is translated until

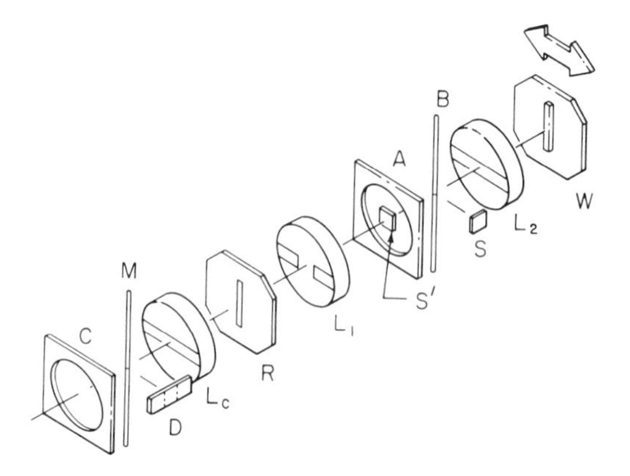

Fig. 8.33. Automatic through-the-lens alignment [28].

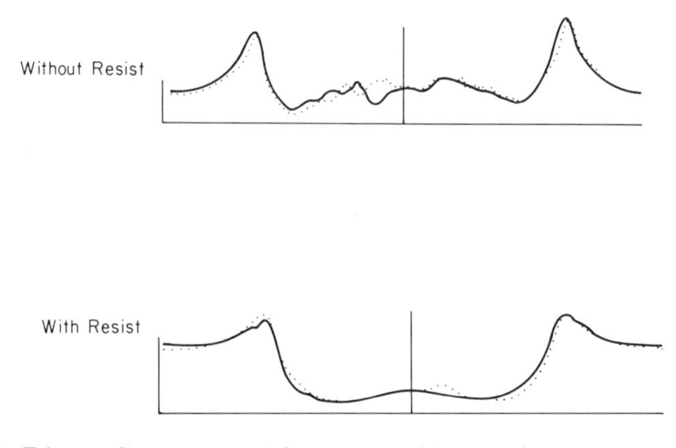

Fig. 8.34. E-beam alignment signal (courtesy D. Oleniewski).

the signal into the detector is maximized. Alignment usually occurs in two steps. First a "global"alignment is accomplished, which brings the wafer and mask alignment marks in close proximity. This is followed by a fine alignment. This is but one scheme in current use. Recently, fine-focused laser-scanning alignment has been reported [30]. The standard deviation of the displacements from the ideal repositioning point for all of these methods is less than 150 nm.

For e-beam lithography, a small portion of the writing field is scanned with the e-beam itself. Backscattered electrons reflected from the wafer are collected and measured. The e-beam alignment mark is either a pit mark etched into the silicon or a heavy-metal square defined in previous lithographic operations [31]. A critical change in the backscattered electron intensity is taken as an indication that the beam has swept over an alignment mark. A picture of such a mark and the backscattered electron intensity profile is shown in fig. 8.34. Typically, such techniques can provide better

TABLE 8.3: E-beam pattern generator specifications for mask making

Parameter \ Minimum Feature Size	1.25μm	0.5μm	0.25μm
Edge Acuity (Line Edge Roughness) [nm]	125	50	25
Repositioning Accuracy (within a field) [nm]	100	50	10
Field-to-Field Butting Accuracy [nm]	125	60	30
Minimum Address Size [nm]	125	100	50

than 100-nm repositioning accuracies. It is a significant problem to create an e-beam alignment mark that holds up in processing and that can be visualized with the e-beam probe under the various thin-film layers of an integrated circuit. In all cases, optical as well as e-beam, the visibility of the alignment mark below the resist is a key limiter in achieving good alignment.

The feature positioning and realignment tolerances possible for today's VLSI technology are shown is table 8.3. Along with this, we see requirements for 0.5-μm and 0.25-μm lithography. It can be seen from this table that going from current capability to 0.5 μm is not as large a step as in going from 0.5 μm to 0.25 μm. In addition to achieving such tight specifications, it is also necessary to demonstrate that these specifications have been met. Techniques for mask and wafer inspection are discussed in the next section.

8.6 Inspection and Repair

The goal of inspection at the wafer or at the reticle level is to ascertain critical dimension (CD) tolerances, registration accuracy, and defect levels. In sophisticated systems, defects can also be repaired when detected. The basics of inspection and repair technology are reviewed here.

There are a number of techniques available for linewidth determination [32]. Of course, optical microscopes with Filar measurement gratings impressed on the visual field have been used for years. Typically, these tools are limited to near-micron line measurements. Focused laser beams are also used to scan over a line while the reflected beam intensity is measured. Of course, the reflected light intensity is not a step function of beam position. The beam intensity is not uniform across its diameter. Diffraction effects contribute to rounding of the signal or position plot. One can arbitrarily set the line boundary at some position (usually close to the half-maximum position) on the plot. Intensity profiles based on diffraction theory that account for the degree of coherence can help to assign boundary positions and make the technique more accurate. Such techniques could be useful to the half-micron level. Scanning micrographs are accurate to below 1/2 μm, but they are not easily automated.

Critical dimensions and defects can be assessed using either die-to-die or die-to-data-base inspection techniques [33]. Here, an image of the chip is projected onto the imaging plane of a position-sensitive detector (PSD), such as a charge-coupled device. The chip image can be that of the pattern on the wafer or that of the pattern on the reticle. In die-by-die alignment, a second chip image is simultaneously projected onto a second PSD. The image can be that of an adjacent chip or that of a reference image. The PSDs are read out a pixel at a time and the outputs are compared. A defect or pattern-sizing error occurs at spots where the PSD pixel outputs from two sensors disagree. In die-to-data-base inspection, the pixel outputs from a single PSD are directly compared with the PG data base. Such a technique eliminates the need for "perfect" reference images or the need to use extra work to determine which of the two inspected chips the flaw occurred in. Such techniques can spot ~0.4-μm defects. While this seems impressive, even a "relaxed" specification for 0.25-μm linewidths requires visualizing 50-nm defects. If we are working with 5x optical reticles, this specification relaxes to 250 nm (1/4 micron.) This is somewhat easier, but still a challenge!

Once defects have been located, they can sometimes be repaired. In the past, opaque defects in reticles or in masks could be ablated ("zapped") with a tightly focused laser beam. Such repairs could be performed on defects no smaller than 1-2 μm. Laser techniques could also be used for the inverse operation — covering a hole or in bridging an open line. In the plugging operation, the mask or reticle is covered with thick photoresist. The laser blows a hole in the photoresist over the defect. A thin chrome film is deposited over the whole workpiece. The remaining resist is soaked off (usually in acetone). The metal over the remaining resist washes away also. This is a variant of the "lift-off" process discussed in Chapter 9. The only metal that remains after the soak is the metal over the formerly clear defect. The metal blocks light transmission thus "healing" the defect. Repair of smaller features found in today's VLSI requires finer tools to drill away submicron defects and to run a submicron variant of the laser-plug operation described above [34]. Advanced beam techniques for submicron lithographic repair tools are discussed in Chapter 11.

8.7 Conclusion

In this chapter we have reviewed the major systems used in current lithographic practice.

- Contact printers
- Proximity printers
- Full-field projection systems
- Wafer steppers
- E-beam systems
- X-ray systems

To achieve high resolution, all the optical systems surveyed attempt to achieve high numerical apertures. This causes depth-of-focus limitations that may cause more problems than poor resolution. Such problems can be remedied by using shorter wavelengths in the exposing radiation. By such means, both depth-of-focus and diffraction limitations can be reduced. X-ray lithography has superior resolution and minimal depth-of-focus limitation. Faster resists, brighter sources, and more durable masks must be developed to make this technology commercially feasible.

E-beam systems do not suffer from diffraction limited resolution or from depth-of-focus problems. However, beam-resist interactions cause the beam to spread out and hence resolution can be degraded. To fully comprehend how this takes place, the resist exposure process must be reviewed. This is accomplished in the next chapter. In the next chapter it is also shown how resists can be used to provide an optically flat surface to minimize depth-of-focus effects in conventional optical lithography.

Resolution is not the only concern for good lithographic practice. Feature positioning on glass masks and repositioning on the workpiece must be accurate to better than 100 nm for 0.5-μm minimum feature-sized lithography. On going to devices requiring 1/4-μm minimum feature size, alignment accuracies better than 50 nm will be required. This will severely stress the limits of mechanical and optical systems. In addition, inspection to verify the meeting of these goals is a problem for current-day research.

Acknowledgment

This chapter is dedicated to C.J. Taylor, head of microlithography at the Westinghouse Advanced Technology Laboratory (retired).

References for Chapter 8

1. O.G. Folberth, W.D. Grobman, eds., "VLSI: Technology And Design," IEEE Press, New York pp. 25-56 (1984).
2. C.A. Mead, L. Conway, "Introduction To VLSI Systems,"Addison-Wesley, Reading, MA pp. 91-144 (1980).
3. G.I. Geikas, B.D. Ables, Contact printing and associated problems, Kodak Photoresist Seminar Proceedings II, p. 47 (1968).
4. J. Strong, "Concepts of Classical Optics,"W.H. Freeman, San Francisco (1958).
5. F.A. Jenkins, H.E. White, "Fundamentals Of Optics,"4th ed., McGraw Hill, New York (1976).
6. J.W. Ford, "Classical And Modern Physics,"Vols I & II, Ch. 18 (1973).
7. B.J. Lin, Optical methods for fine line lithography, in "Fine Line Lithography,"R. Newman, ed., North Holland, Amsterdam (1980).
8. P. Burggraaf, Wafer steppers and lens options, Semicond. Internat'l 9 (3), 56 (1986).
9. M. Born, E. Wolf, "Principles of Optics,"6th ed., Pergamon Press, Oxford (1980), (particularly chapters 9 and 10).
10. L.F. Thompson, M.J. Bowden, The lithographic process: the physics, in "Introduction To Microlithography,"L.F. Thompson, C.G. Willson, M.J. Bowden, eds., Washington, D.C., ACS Symposium Series #219, pp. 16-85 (1983).
11. F. Zernike, The wave theory of microscopic image formation, in "Concepts of Classical Optics,"J. Strong, e.d., W.H. Freeman, San Francisco pp. 525-536 (1958).
12. M.C. King, Principles of optical lithography, in "VLSI Electronics: Microstructure Science,"Vol. 1, N.G. Einspruch, ed., Academic Press, New York pp. 42-81 (1981).
13. J.W. Goodman, "Introduction to Fourier Optics,"McGraw-Hill, New York (1968).
14. D.A. Markle, Solid State Technology 17(6), 50 (1974).
15. Bennewitz et al., Excimer laser-based lithography for 0.5-μm device technology, Proc. IEEE Int'l Elect. Dev. Mtg., p. 312 (1986).
16. H.R. Brewer, ed., "Electron-Beam Technology in Microelectronic Fabrication,"Academic Press, New York (1980).
17. D.R. Herriott, G.R. Brewer, Electron-beam lithography machines, in "Electronic-Beam Technology in Microelectronic Fabrication,"Academic Press, New York (1980).
18. H.C. Pfeiffer, Recent advances in electron-beam lithography for high volume production of VLSI devices, IEEE Trans. Elect. Dev. ED-26 (2), 663-674 (1979).
19. Further information on this system is available from Cambridge Instrument North America Field Service, Chicago, IL.

20. D.R. Corson, P. Lorraine, "Introduction To Electromagnetic Field And Waves," W.H. Freeman, San Francisco pp. 168-170 (1962).
21. H. Boersch, Zeitschrift Fur Physik, 139, p. 15 (1954).
22. W. Kramer, Optik (Stuttgart), 54 (3), pp. 211-234 (1979).
23. A.P. Neukermans, Current status of X-ray lithography, Sol. St. Technol. 27, (9), pp. 185-188 (1984).
24. A.R. Shimkunas, Advances in X-ray mask technology, Sol. St. Technology 27, (9), pp. 192-199 (1984).
25. A. Betz, F.Y. Fey, A. Heuberger, P. Tischer, Calculation of optimation electron energy of a dedicated storage ring for X-ray lithography, IEEE Trans. Elect. Dev. ED-26, (4), pp. 693-697 (1979).
26. J. Pearlman, T. Riordan, Voc. Sci. Tech. 19, p. 4 (1981).
27. Peckerar et al., X-ray lithography with laser produced plasmas, Proc. Symp. Elect. Ion Beam Sci. and Tech., p. 432 (1978).
28. J. Itoh, T. Kanayama, "Optical-heterodyne detection of mask-to-wafer displacement for fine alignment, Jpn. J. Appl. Phys. 25 (8), L684-L686 (1986).
29. R.S. Herschel, Autoalignment in step-and-repeat wafer printing, Proc. SPIE, Vol. 174, "Developments In Semiconductor Microlithography," IV, pp. 54-62 (1979).
30. A. Suzuki, Double telecentric wafer stepper using laser scanning methods, Proc. SPIE, Vol. 538, Optical Microlithography, IV, pp. 2-8 (1985).
31. A. Neureuther, Visibility of electron beam alignment mark, Solid St. Elect. 26, 27 (1979).
32. W.M. Bullis, D. Nyysonen, Optical linewidth measurements, in "VLSI Electronics: Microstructure Science," Vol. 3, N. Einspruch, ed. pp. 301-346 (1982).
33. E. Jozefoz, S. Follis, W.E. Luch, Die-to-database inspection — An effective method of detecting and locating defects on reticles, Sol. St. Technol. 30, (1), 79-82 (1987).
34. B.W. Ward, D.C. Shaver, M.L. Ward, Repair of photomasks with focused ion beams, SPIE Vol. 537, Electron-Beam, X-ray and Ion Beam Techniques for Submicron Lithographies IV, pp. 110-116 (1985).

Problem Set for Chapter 8

8.1

a. Use eq. (8.9) to show that $p_{\min} = c\sqrt{\lambda r_0}$, where c is some constant.
b. What factors influence the magnitude of c?

8.2 Use the Fresnel diffraction formulae [eqs. (8.6–8.8)] to calculate the intensity pattern on wafer for contact printing under the following sets of circumstances:

a. 1/2-μm opaque slit,
b. 1-μm mask-to-wafer separation, $\lambda = 200$ nm.
c. 1/2-μm opaque slit, 5-μm mask-to-wafer separation, $\lambda = 404$ nm.
d. 1/2-μm opaque slit, 2-μm mask-to-wafer separation, $\lambda = 404$ nm. Reference 4, Chapter 9 is particularly informative with regard to this problem.

8.3 Use the geometry shown in fig. 8.11 to demonstrate the validity of eq. (8.15).

(The reader may wish to consult refs. 9 and 13 while working on problems 8.4, 8.5, and 8.6.)

8.4 Consider the transfer function of coherent optical systems as expressed in eqs. (8.24) and (8.25). Show that eq. (8.24) is just a restatement of eq. (8.18).

8.5 Consider a clear region on a mask between two opaque regions. Does the "printability" of an opaque defect in the clear space depend on the position of the defect in the space? If so, why does this occur?

8.6

a. Derive an expression for the MTF of an incoherently illuminated system employing (1) a slit of width a, and (2) a circular aperture of radius a. The autocorrelation displacement represents the spatial frequency at which the MTF is evaluated.
b. Compare and contrast the slit and circular aperture MTFs derived in this problem with the coherent transfer functions for the same geometry apertures.

8.7 Assume the MTF must exceed 0.5 for a feature to be printed, what is the minimum printable defect size for the two MTFs discussed in problem 8.6?

8.8 Using the coherent transfer function and the OFT for an infinitely long slit aperture, derive the area-intensity profiles for a 1-μm line in both cases.

8.9 Consider the case of e-beam lithography. As stated in the text, there is an optimum crossover angle for e-beam focus.

a. Derive an expression for that angle.
b. What is the minimum spot size for this focus?
c. How much must the focal plane be displaced to get an 0.1-μm growth in the minimum spot diameter

8.10 Assume that C_S = 40 cm, C_c = 10 cm, V = 20 KeV, and $B = 1$ x 10^6. Calculate the minimum spot size for the following conditions:

a. $I = 1$ x 10^{-6}, $\Delta V = 7V$.
b. $I = 1$ x 10^{-8} $\Delta V = 2V$.

What is the major factor influencing spot size in both cases?

8.11 Estimate the minimum resolvable line-and-space pattern period for a mask-to-wafer spacing of 20 μm and an exposure wavelength of 10 Å, using the results of problem 8.1.

8.12 In the case of x-ray lithography, consider a 25-μm ideal mask-to-wafer spacing. Assume a 3-cm exposure field and a wafer taper of 0.5 μm per cm. How does this taper affect feature placement at either end of the exposure field? How does it affect feature shape?

8.13 Consider the fabrication of an x-ray mask. Candidate substrate materials are silicon, carbon, and boron nitride. Candidate absorbers are gold and tungsten. In order to ensure an 80% transmissivity for aluminum Kα emission, what must the maximum substrate thicknesses be for each of the candidates listed? To achieve a 5:1 contrast ratio, what must the absorber thickness be for each of the candidate absorbers on maximum thickness substrates?

8.14 Demonstrate that the point-source magnification factor is the same in the center and at the edge of the exposure field.

8.15 In point-source illlumination, the shadow associated with two closely spaced lines will occlude the space between them. That is, at the edge of the field, there will be no illumination of the "clear" region between two opaque features. Explain this effect using diagrams and derive a maximum field-size limit based on this effect.

9 Resist Systems and Exposure Modeling

9.1 Introduction

The first medium into which lithographic images are transferred is resist. The resist must be capable of reproducing these images with good resolution. It must be a true resist in that it should hold off etches or block implants without falling off the workpiece. The resist is expected to cover relatively sharp edges without rupturing and to be pinhole-free over large areas. Requirements for a successful resist are listed in table 9.1. For many applications, thickness uniformity is also of prime importance. It is one of the miracles of our technology that materials that satisfy most of these criteria have been developed. Resists that exhibit tens-of-nanometer resolution, good adhesion to the work surface, a good degree of process resistance, and thickness variations of less than 10 nm have been obtained.

All resists can be classified as *radiation-sensitive thin films*. A wide variety of materials operating under a number of different physical and chemical principles are used as resists [1]. Organic polymers are the most common materials encountered. The radiation sensitivity of a polymer can arise either directly from the interaction of the incident flux with the base polymer or from the interaction with chemical groups that have been incorporated into the polymer to make them radiation sensitive. Inorganic components, such as chalcogenide glasses [2], have also been tried as resists. The inorganic resists have an advantage in that they can be used as solid, localized doping sources following development. Combinations of different resists have also been used in processing. These "multilayer"systems have been used to provide exceptionally thick, ultra-flat, process resistant composites for high resolution pattern transfer. Multilayer systems have also been used in "lift-off"schemes for metal patterning without etching, as described below.

In this chapter, the basic concepts of resist materials technology are covered. First, the chemical constitution of many of the most commonly used resists is presented. The basic terminology of resist photochemistry is then reviewed. Methods for computing the amount of energy deposited in the resist for a given exposure condition are derived. This is done both for optical and for e-beam lithography. As is shown below, such calculations are particularly complex for e-beam lithography. This is because electron scattering processes in the resist, or in the substrate, cause energy deposition far from the point of beam entry into the solid. Special techniques are developed to "correct"for the problems caused by this effect. Once we know how to

TABLE 9.1: Resist requirements for IC fabrication

Functionality	Manufacturing Processibility
Sticks to relevant surfaces	Adds minimum number of steps (or, at least, "easy" steps)
Can be patterned with existing exposure sources in times consistent with current-day stepper throughput (50 wafers/hr)	Does not require major retooling
Pinhole free	Has reasonable shelf life and minimum toxicity
Covers steps	
Doesn't contaminate devices	
Holds up in reactive ion etch (RIE)	
Has suffiecient contrast to provide good resolving with minimum standing wave effect.	

compute energy-deposition profiles, we proceed to the computation of resist development sidewall profiles. The chapter continues with a description of basic resist handling operations: application, baking cycles, etc. The chapter continues with advanced approaches to resist system development. These advanced system techniques include multilayer resist lift-off processing and contrast enhancement. Resist characterization methods are described at the end of the chapter.

9.2 Resist Chemistry

The history of photosensitive thin films is long. In 1826, J.N. Niepce discovered the light sensitivity of asphalt. Making use of this effect, he was able to etch patterns in pewter. A.E. Becquerel discovered that organic colloids could be made insoluble when exposed to bright light. Talbot developed a technique in 1852 for selectively etching copper with a resist made of gelatin sensitized with bi- or dichromate salt. The history of the photoresist industry is described in somewhat greater depth in the book by W.S. De Forest [1].

Present-day resists fall into two broad classes: those that are *positive acting* and those that are *negative acting*. Positive-acting resists become more soluble in the developing solvent after exposure. Thus, if we expose

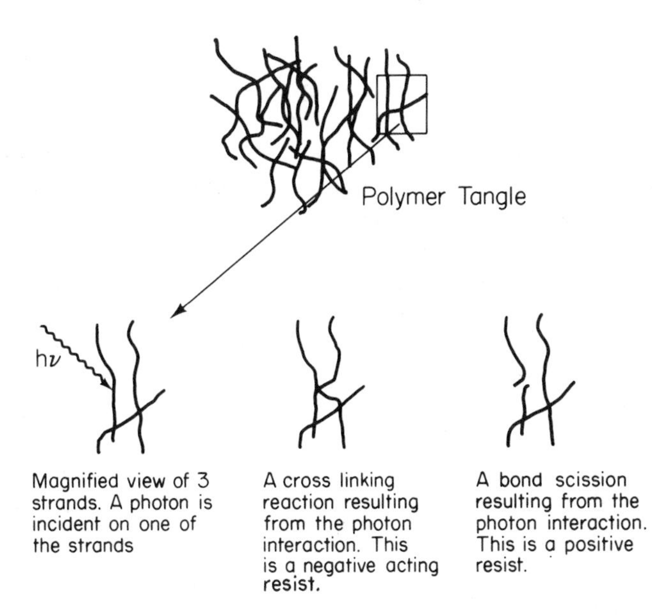

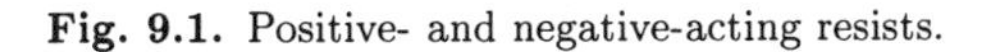

Fig. 9.1. Positive- and negative-acting resists.

a small region of the positive-resist covered film, a hole will appear in the resist in this small region after development. Negative-acting resists become less soluble in solvent after exposure. If a small region of a negative resist-covered film is exposed, only the exposed region will be covered by resist after development. There are a number of mechanisms that can cause changes in solubility after exposure to radiation.

In the case of negative resists, a crosslinking of polymer chains occurs. This is illustrated in fig. 9.1. Two (or more) adjacent polymer strands become linked as a result of radiation-induced chemical changes in one of the strands. The molecular weight of the crosslinked strand is greater than that of a single strand. Solubility is a function of molecular weight in that polymer chains of lower molecular weight usually dissolve more rapidly. Thus, the *unexposed* region of the negative resist washes away faster than the exposed region. For some positive-acting resists, exposure to light causes local bond breakages, also illustrated in fig. 9.1. This process is called *bond scission.* The result is that the exposed region contains material of lower mean molecular weight. The exposed region will wash away during development.

Many times, the base polymer will behave as a positive or negative resist without modification. For example, polymethylmethacrylate (PMMA) will engage in bond scission when exposed to e-beams, x-rays, or to deep ultraviolet radiation. Most resists (both positive and negative) are two-component systems. That is, they contain the basic polymer and a second material incorporated into the polymer, which determines the photochemistry on exposure. This second material can be a "sensitizer"that enhances the crosslinking or bond-scission process. Alternatively, a photoactive compound that undergoes a radiation-driven chemical reaction during exposure can be grafted onto the polymer chain. The resulting compounds are soluble. This, too, gives rise to a positive-acting resist.

Let us consider these processes in somewhat greater depth [3]. The most frequently encountered negative resist is a two-component system. The base polymer is a partially cyclized polyisoprene rubber. Such rubbers are soluble in nonpolar organic solvents such as toluene or xylene. Bis-arylazides are added to the rubber base to create a sensitized resist. Material such as Kodak KTFR and Hunt waycoat or EMerk Selectilux are representatives of this class of two-component resists. The azide component, during exposure to light, engages in a number of chemical reactions with the surrounding polymer units that result in crosslinking. The crosslinked component is insoluble in the organic solvents. Single-component resists, like Mead's COP, crosslink without the aid of a sensitizer.

The main types of positive resists currently in use are made from a Novalac-resin. A photoactive compound, called an *inhibitor*, (usually diazonaptha-quinone) is incorporated into the polymer chain. The inhibitor renders the polymer film insoluble in alkaline solutions. When exposed to light, the inhibitor is converted to a carboxylic-acid-containing compound. This acid is aqueous-base soluble. After exposure to light, the resist film becomes soluble in alkaline solutions, such as dilute KOH. A positive-acting resist is obtained by this process. This is shown schematically in fig. 9.2.

Positive-acting resists have demonstrated the best resolution capabilities to date. Negative-acting resists tend to swell on development, causing feature distortion. In addition, negative resists may form filaments, or strings, at the feature boundaries. These strings mask underlying material from etch, giving poor edge definition. As a result, negative resists are less frequently used in optical lithography. For reasons described in Chapter 8, negative resists are still attractive in e-beam lithography. Electron-beam writing time for a light-field pattern can be reduced with such a resist.

9.3 Resist Exposure Characterization

In characterizing resist exposure, we are interested in two things: how long it takes to expose a given resist and what kind of boundary definition the exposed resist provides. The minimum amount of energy deposited in the resist that is required to create an exposure is called the *resist sensitivity.* There are two factors that affect boundary definition in the resist itself: light undercutting the mask feature boundary, and the slope of the resist. Sidewall

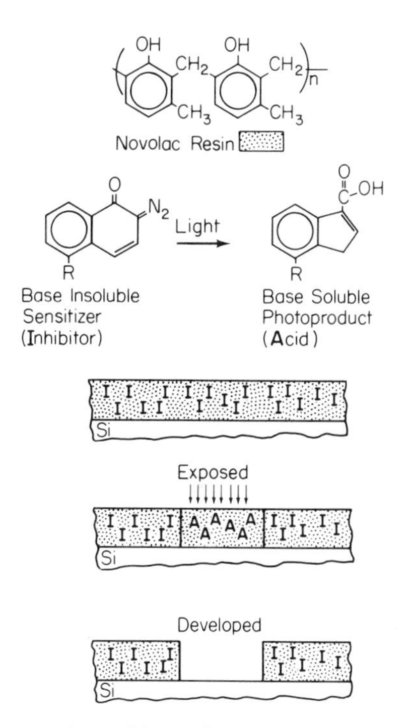

Fig. 9.2. Sensitizer action in positive resists.

slope can cause problems when the thin region at the resist termination point becomes too thin to hold up in subsequent etch processes. This creates edge roughness and boundary shifts.

Let us first consider the issue of resist sensitivity. Incident radiation causes chemical changes in the resist. The photochemical quantum efficiency, ϕ, is defined as the ratio of the number of molecules transformed to the number of photons incident:

$$\phi = \frac{\#\text{ of molecules transformed}}{\#\text{ of photons incident}}. \tag{9.1}$$

For a positive resist, the molecular transformations referred to are bond scissions (S). For a negative resist, the reference is to crosslinks (X). The ionizing radiation efficiency, the G value of the resist, is related to ϕ. For a positive resist, $G(S)$ is defined as the number of bond scissions per 100 eV of absorbed energy. For a negative resist, $G(X)$ is defined as the number of crosslinks per 100 eV of absorbed energy. For positive resists with photoactive compound additives, the definition can be changed to read the "number of photoactive sites transformed per 100 eV of absorbed energy."

From a knowledge of the resist G value, the average molecular weight of the resist after absorbing an energy dose, D (eV/g), can be computed. To do this, realize that the average molecular weight prior to exposure, M^0_{ave} is given by the formula

$$M^0_{\text{ave}} = \frac{wN_\text{a}}{N_0}, \tag{9.2}$$

where N_a = Avogadro's number, N_0 = total number of molecules present before exposure, and w = weight of exposed polymer (in grams). Consider the case of the positive-acting resist. After exposure, the total number of bond scissions produced in the sample is given as

$$N^* = \left[\frac{G(S)Dw}{100}\right]. \tag{9.3}$$

Thus, the average molecular weight after exposure is

$$M_{\text{ave}} = \frac{wN_\text{a}}{(N_0 + N^*)}. \tag{9.4}$$

Combining eqs. (9.2)-(9.4) yields

$$\left(\frac{1}{M_{\text{ave}}}\right) = \left(\frac{1}{M^0_{\text{ave}}}\right) + \left[\frac{G(S)}{100N_\text{a}}\right] D. \tag{9.5}$$

In general, the dissolution rate of polymer in solvent is inversely proportional to the molecular weight of the polymer. This consideration leads to the concept of a *development curve.* Such a curve is a plot of the thickness of resist remaining (after exposure to developing solvents for a fixed amount of time) as a function of the log of the absorbed energy dose. Two such curves are shown in fig. 9.3. Figure 9.3a shows the situation for a positive resist, while fig. 9.3b shows the situation for a negative resist. It is interesting to note that there is still film dissolution, even in the case of the totally unexposed positive resist as well as in the case of the fully exposed negative resist.

Both curves exhibit a linear portion as well as an apparent "saturation" level. The slope of the linear portion is known as the resist contrast or gamma value (γ). Examination of these curves demonstrates the ambiguity associated with the notion of sensitivity. How do we define "full" exposure? For a negative resist, do we place the exposure dose at the saturation point on the plot? At 10% of the saturation point? Typically, the exposure dose is defined as the D_0 point labeled on the plot. This point is the intercept between the linear extensions of the saturation and the linear regions of the development curve. For a positive resist, it seems logical to place the exposure dose at the zero-thickness intercept. Remember, however, that the zero-thickness intercept dose will vary as a function of development time. "Forced" development can occur by lengthening the development time. This generally reduces the resist gamma, leading to a sloped resist sidewall. These effects can be modeled, as described below.

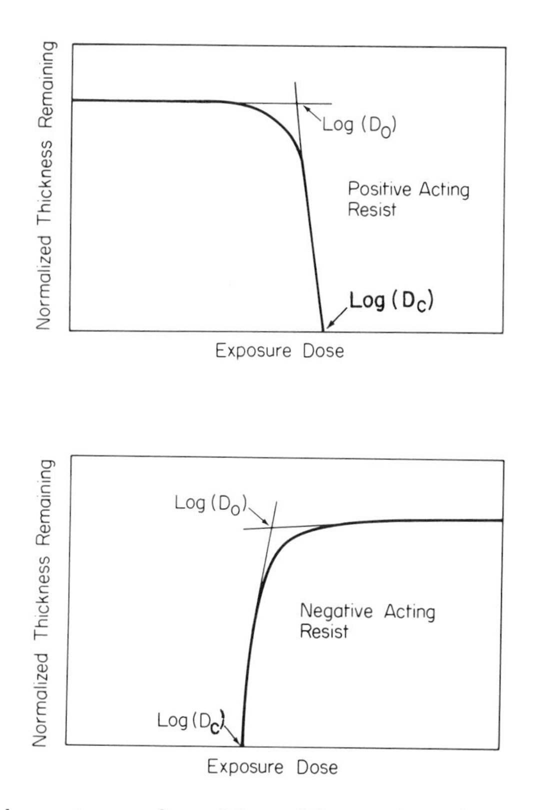

Fig. 9.3. Development curves for positive and for negative resists.

9.4 Optical Exposure Modeling

For modeling resist profiles during development, two pieces of information must be known: the absorbed dose in the region of interest and the resist development curve. For the common case of a two-component resist, the approaches for calculating absorbed dose indicated in Chapter 8 must be expanded. During exposure, the inhibitor in the Novalac base converts to soluble products. These inhibitors are light absorbers. On degradation, the film "bleaches,"becoming more transparent. The rate at which the inhibitor degrades is proportional to the rate at which incident light energy is absorbed. However, the rate at which light energy is absorbed is, in turn, proportional to the amount of inhibitor present. This is further complicated by the fact that the bottom layers of resist are shielded from light by the top layers.

The approach generally used to model these effects was developed by Dill [4]. To begin, we consider the three constituents of the two-component resist: the inhibitor, the resin base, and the reaction products. For such a system, we can write the following equation for the change in light intensity at time t per unit depth at some depth x in the resist:

$$\frac{\partial I(x,t)}{\partial x} = -I(x,t)[a_1 m_1(x,t) + a_2 m_2(x,t) + a_3 m_3(x,t)], \qquad (9.6)$$

where $I(x,t)$ = light intensity at any depth (x) in the film and exposure time (t), a_1 = molar absorption coefficient of inhibitor, a_2 = molar absorption coefficient of base resin, a_3 = molar absorption coefficient of reaction products, $m_1(x,t)$ = molar concentration of inhibitor, $m_2(x,t)$ = molar concentration of base resin, and $m_3(x,t)$ = molar concentration of reaction products. This relationship is an expression of the Lambert-Beer law for optical absorption. For a single-component system (where $m_1(x,t) = 1, m_2(x,t) = m_3(x,t) = 0$) the equation reduces to eq. (8.12).

The rate at which the inhibitor is destroyed is given by

$$\frac{\partial m_1(x,t)}{\partial t} = -m_1(x,t)I(x,t)C, \qquad (9.7)$$

where C is the fractional decay rate of the inhibitor per unit intensity at a depth x into the resist. Dill also makes the following assumptions: the incident intensity at the surface of the resist is constant, I_0; the inhibitor is initially uniform in its concentration throughout the resist, $m_1(x,0) = m_{10}$; the resin concentration is uniform and does not change through the exposure process, $m_2(x,0) = m_{20}$; one mole of inhibitor becomes one mole of reaction product after exposure, $m_3(x,t) = m_{10} - m_1(x,t)$. With these conditions in mind, eq. (9.6) becomes

$$\frac{\partial I(x,t)}{\partial x} = -I(x,t)[m_1(x,t)(a_1 - a_3) + a_2 m_{20} + a_3 m_{10}]. \qquad (9.8)$$

Next, define the following parameters (sometimes called the Dill parameters)

$$M(x,t) = \frac{m_1(x,t)}{m_{10}} \qquad \text{(This is the time–dependent fractional inhibitor concentration.)}$$

$$\begin{aligned} A &= (a_1 - a_3)m_{10} & [A] &= L^{-1}(\mu\text{m}^{-1}) \\ B &= (a_2 m_{20} + a_3 m_{10}) & [B] &= L^{-1}(\mu\text{m}^{-1}) \\ C &= C & [C] &= L^2 E^{-1}(\text{cm}^2/\text{mJ}). \end{aligned}$$

Substitution of these parameters into eqs. (9.7) and (9.8) yields

$$\frac{\partial I(x,t)}{\partial x} = -I(x,t)[AM(x,t) + B],$$

and

$$\frac{\partial M(x,t)}{\partial x} = -I(x,t)M(x,t)C. \tag{9.10}$$

This set of coupled differential equations, combined with the boundary and initial value conditions specified above, can be solved numerically. The question now remains, how are A, B, and C determined?

To make this determination, we begin by defining the internal transmittance of the photoresist as the ratio of the light intensity at the top of the resist to the light intensity at the resist substrate boundary. It can be shown that the Dill parameters obey the following relations:

$$A = \left(\frac{1}{d}\right) ln[T(\infty)/T(0)], \tag{9.11a}$$

$$B = \left(\frac{1}{d}\right) ln[T(\infty)], \tag{9.11b}$$

$$C = \left\{ \frac{(A+B)}{AI_0T(0)} \frac{1}{[1-T(0)]} \right\} \frac{dT(0)}{dt\,|_{t=0}}, \tag{9.11c}$$

where d = resist thickness, I_0 = light intensity at the top of the resist, $T(0)$ = transmittance of the unexposed resist, and $T(\infty)$ = transmittance of fully exposed resist. The value of I_0 is not quite that of the incident intensity. This is because some of the incident light is reflected. The reflection loss can be accounted for by writing

$$I_0 = I_{\text{inc}} \left\{ 1 - \left[\frac{(\eta_1 - \eta_2)}{(\eta_1 + \eta_2)} \right]^2 \right\}, \tag{9.12}$$

where I_{inc} is the incident intensity and η_1 and η_2 are the indices of refraction of air and photoresist, respectively.

Inspection of eqs. (9.11a)–(9.11c) suggests a method of obtaining A, B, and C. We simply measure the initial transmittance and the transmittance after all the photoactive compound sensitizer (PAC) has been transformed. In principle, this is easy. In practice it is a bit harder. Care must be exercised in selecting substrates that have optical indices that are matched to the photoresist (this prevents multiple reflections of light at the substrate/resist boundary). Also, the bottom of the substrate should be coated with an antireflection coating for similar reasons. In measuring the transmitted light, the absorption loss in the substrate *and* in whatever antireflection coatings are present must be accounted for.

Once the Dill parameters have been evaluated, the amount of energy absorbed at any point in the resist can be calculated. Once again, this is easier in principle than in practice. For the case of a large area exposure, the formation of standing waves, due to multiple internal reflections, must be dealt with in the calculation. For the case of a resist of homogeneous refractive index, this is not too difficult. The reflections originate at the resist/substrate interface. The amplitude of the lightwave reflected back into the film is:

$$A_{\text{ref}} = A_{\text{inc}} \left[\frac{(\eta_1 - \eta_2)}{(\eta_1 + \eta_2)} \right], \tag{9.13}$$

where A_{inc} is the amplitude of the wave incident on the interface and the η's are the resist and substrate refractive indices. Assume for a moment that absorption in the resist is small, that the dominant reflection comes from the resist/substrate interface, and that the incident beam direction is normal with respect to the resist surface. The reflected wave will interfere with the incident wave, giving rise to a periodic light intensity variation in the resist from top to bottom. The amplitude of the light as a function of position x in the resist is (calling the top of the resist the $x = 0$ point)

$$A(x,t) = \text{Re}\{A_{\text{inc}} \exp[i(\omega t - kx)] + A_{\text{ref}} \exp[i(\omega t - k(2d - x))]\}, \tag{9.14}$$

where Re means we take the real part of the complex expression. The time-independent intensity in the resist is found by multiplying $A(x,t)$ by its complex conjugate.

Equation (9.14) does not account for the fact that reflections can occur within the resist itself. This is because, as shown above, there is a position-dependent resist refractive index. To account for this effect, the resist must be considered to be composed of a series of thin layers. The reflection at each of these layers must be calculated and the resulting wave amplitudes must be summed [5]. Diffraction effects at boundaries also are not treated in the above analysis. This is usually handled by extending the calculated standing-wave intensity pattern under the feature. The intensity under the feature is attenuated by multiplication with a Fresnel diffraction envelope (such as computed from eqs. 8.5–8.8). The results of one such calculation are shown in fig. 9.4. Once the resist energy absorption has been calculated, the resist sidewall profile in the vicinity of a boundary can be computer modeled. Such modeling is discussed in Section 9.6.

9.5 E-Beam Exposure Modeling

Particle beams interact with matter in fundamentally different ways than light does. In modeling optical exposure, the wave nature of light had to be called into specific account. In the case of electron beams (in the range of energy and feature size we currently consider), the particle nature of the incident radiation dominates the resist response. Particle electrons are scattered by atoms in the resist polymer network or in the workpiece. The amount of energy deposited in the solid traversed by the beam depends on

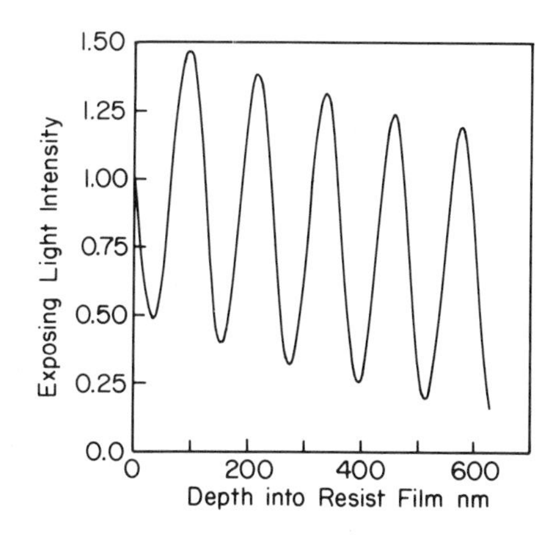

Fig. 9.4. Standing-wave effects in resist exposure.

the energy of the particles in the beam and the atomic constituency of the solid. The current most popular method of analyzing energy distribution in e-beam lithography is called the Monte Carlo method [6]. This is described below.

In the Monte Carlo analysis, the trajectory of a single electron is traced through the solid using a technique of statistical analysis. There are a number of ways this can be done. All of these methods rely, to some extent, on the probabilistic nature of the particles-solid interaction. For example, let us consider a rather "brute-force" approach. A given configuration of scattering centers is chosen. The electrostatic energy surface associated with the configuration is then fixed. The classical equations of motion for the particle trajectory are solved. The random variable is the particle's entry point into the energy surface. In a somewhat simpler approach, the Rutherford scattering probability formula is used, as described below. Either of these approaches would describe interaction of the electron with the scattering network. The electron can lose energy to the network in a number of ways. It can a) create secondary ionization, b) cause vibrational motion in the network, or c) create optical excitations. There is some probability that each of these interactions occur, just as there is some probability that the electron will pass through the region without losing energy.

The probabilistic analysis proceeds as follows. The real-number line is partitioned in such a way that it assigns weights to each of the processes under consideration (see fig. 9.5). For example, let us assume that there is only one inelastic interaction mechanism available and that there is a 30% chance that this process occurs in the time interval under consideration. Further, assume that the process causes a 0.01-keV energy loss. A random number from 1 to 100 is chosen. If the number is in the range from 1 to 30 the particle is assumed to have lost energy. The particle leaves the differential scattering volume with 0.01 keV less energy than it had when

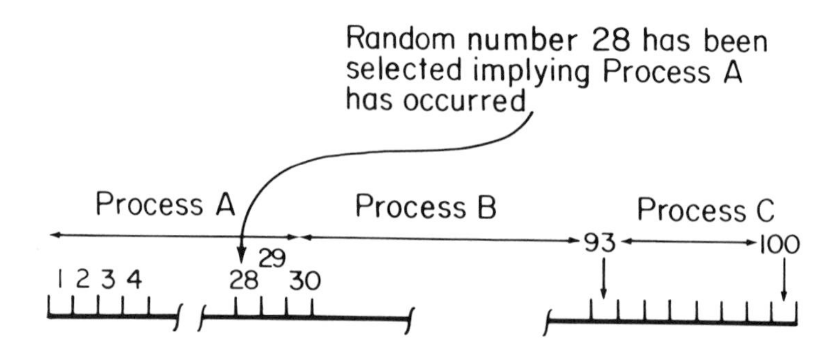

Fig. 9.5. Partitioning of the real number line for Monte Carlo analysis.

it entered. Selection of numbers 31-100 implies that the particle passes the volume without losing energy. The scattering volume is broken up into cubes and the amount of energy deposited in each cube is recorded. This analysis is done again and again for as many as 10,000 trajectories.

Next, energy deposition profiles must be made. This is done by moving down into the resist along an axis normal to its surface. For a round beam, the energy deposition has cylindrical symmetry about this axis. Denote the cylindrical radial distance by r. The mean energy deposition as a function of r is then evaluated for each x. The radial energy distribution function can be fit to some smooth distribution curve for use in subsequent calculation. Once the energy deposition has been evaluated, the resist exposure profiles can be obtained as they were in the case of optical exposure. The procedure is complicated a bit by the fact that the central axis of the beam moves from pixel to pixel during the exposure.

Most attempts at e-beam exposure modeling use an even simpler approach than the one described above. Rather than looking in detail at the different inelastic interactions that can occur, a "continuous slow-down" approximation is used. That is, the electron loses energy to the scattering network in an amount that depends only on the differential path length traveled and the average energy of the electron over that path length. For electrons, we write

$$\frac{dE}{ds} = -\left(\frac{2\pi e^4 n_e}{E}\right) ln\left(\frac{1.166E}{I}\right) \tag{9.15}$$

where n_e = the density of atomic electrons over the path ds, E = the average energy of the electron over ds, and I = the mean ionization energy for atoms encountered.

We must then find the path length traveled. Rather than solving the equations of motion precisely, the Rutherford scattering formula [7] is adapted for use in the Monte Carlo approach. We assume that the particle trajectory is a straight line for each path length considered. On arriving at the end of each path length, the electron is scattered at some angle, ϕ, with respect to the cylinder axis. The probability, p, of observing a given scattering angle, ϕ, is a function of the electron energy:

$$p(\phi\ , E) = \frac{e^4 n_e Z \Delta s}{16E^2} \sin^4\left(\frac{\phi}{2}\right), \tag{9.16}$$

where Δs is the path length and Z is the mean atomic number of the atoms in the scattering field. Again, scattering is assumed to be isotropic about the path length axis. Thus, the aximuthal scattering angle is chosen by picking a random number from 1 to 360.

Results obtained by Monte Carlo analysis are usually put into some analytic form. To do this, we assume that the incident beam is normal to the resist surface. At any depth in the resist, the energy deposition profile along a radius, r, perpendicular to the beam axis is frequently taken to be a double Gaussian [8, 9]

$$\varepsilon(r) = K_n \left[\exp\left(-\frac{r^2}{\beta_f^2}\right) + \eta_e \left(\frac{\beta_f}{\beta_b}\right)^2 \exp\left(-\frac{r^2}{\beta_b^2}\right)\right], \tag{9.17}$$

where $\varepsilon(r)$ = energy deposition (J/cm^3) at point r, β_f, β_b = half widths of the forwardscattered and backscattered electron distributions (cm), η_e = ratio of the number of backscattered to forwardscattered electrons at the depth of interest, and K_n = a normalization factor (J/cm^3). The incident beam is taken to be of zero diameter.

The notion of "forwardscattered" and "backscattered" electrons may, at first, seem puzzling. But the breakdown of the total electron ensemble into two groups makes good physical sense. The forwardscattered distribution contains those electrons that have interacted very little with the resist with the underlying substrate. Their trajectories are still relatively perpendicular to the beam axis. The backscattered electrons have interacted with the material so strongly that their velocity vectors are reversed. These electrons can deposit energy in layers that they have previously traversed at a distance from the beam axis. This is shown in fig. 9.6. The term η_e refers to the ratio of the total number of backscattered electrons to the total number of forwardscattered electrons at some depth beneath the resist surface.

Equation (9.16) can be modified to take into account the fact that the incident beam is *not* of zero diameter. We do this as follows: Assume that the e-beam creates a "Gaussian-round-beam," of half width β_*. The β_f and β_b half widths are now defined as

$$\beta_f = \sqrt{\beta_*^2 + \beta_f^0}, \tag{9.18a}$$

and

$$\beta_b = \sqrt{\beta_*^2 + \beta_b^0}, \tag{9.18b}$$

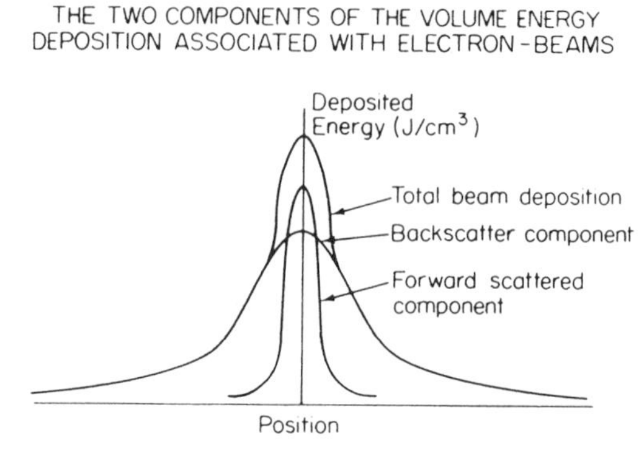

Fig. 9.6. Forwardscattered and backscattered contributions to electron energy deposition.

where β_f^0 and β_b^0 are the characteristic half widths of the forward and backscattered distributions of a point probe.

Monte Carlo analysis is used to provide estimates of β_f, β_b, η_e, and K_n. Analytic expressions have also emerged that enable computation of K_n as a function of resist penetration depth, z. The average energy dissipation, dv, per unit distance, dz, in the z direction for a single electron is given as [10]

$$\frac{dv}{dz} = \frac{E_0}{R_G} \Lambda(f) \left[\frac{\text{eV}}{\text{cm}}\right], \tag{9.19}$$

where E_0 = incident beam energy (eV), R_G = characteristic electron pentration range, the so-called Grün range (cm), $f = z/R_G$ = normalized penetration depth, and $\Lambda(f)$ = a depth-dose function, given below.

Multiplying this expression by the number of electrons/cm^2 incident gives the energy absorbed in the resist per unit depth at depth z. The Grün range can be evaluated using the expression, in which E_0 is given in keV

$$R_G = \left(\frac{0.046}{\rho}\right) E_0^{1.75} \,[\text{cm}] \tag{9.20}$$

where ρ is the density ($\sim$1 for most resists). The depth-dose function is given as a polynomial [11]

$$\Lambda(f) = 0.74 + 4.7f + 8.9f^2 + 3.5f^3. \tag{9.21}$$

Using eq. (9.19) in conjunction with eqs. (9.20) and (9.21), we can get the energy, dV, deposited in thickness interval, dz. The integral of eq. (9.17) over r must equal dV for the layer thickness (dz) chosen. K_n is adjusted to make this occur, keeping in mind that the units in eq. (9.19) must be converted to J/cm^3 to match the units in eq. (9.17).

TABLE 9.2: A tabular listing of parameters for double-Gaussian fitting of Monte Carlo data [12]

E_0 (keV)	t (μm)	β_f (μm)	n_E	β_b (μm)	A (μm)	β_b/R
10	0.5	0.22	0.51	0.65	1.58	0.41
15	0.5	0.13	0.51	1.14		
15	1.0	0.44	0.52	1.41	3.12	0.45
25	0.5	0.06	0.51	2.6		
25	1.0	0.22	0.49	2.9	7.36	0.39
25	1.5	0.43	0.52	2.9		
40	0.5	0.04	0.42	6.0		
40	1.0	0.11	0.45	6.0	16.22	0.37
40	1.5	0.22	0.44	6.2		

Next, we must consider evaluation of β_f, β_b, and η_e. As mentioned above, these terms are fitting parameters used to match eq. (9.17) to the Monte Carlo results. Most sophisticated modeling takes into account the fact that η_e, β_f, and β_b vary as a function of depth z into the resist film. Tabular listings of these data are shown in table 9.2 for e-beams of various energies incident on polymethyl methacrylate (PMMA) resist [12]. The resist thickness is given as t, and the resist is applied on a silicon substrate. Discussions of the assignment of parameter values are given in references 6, 12 and 13.

As seen in table 9.2, β_b is a weak function of beam penetration depth. This is not true for β_f. Greeneich [13] has related the value of β_f to the number of elastic scattering events, P_e, an electron undergoes while moving through the resist. This relation is shown graphically in fig. 9.7. Furthermore, P_e is related to the depth of penetration z

$$P_e = \frac{400z(\mu m)}{E_0(\text{keV})}. \tag{9.22}$$

It should also be noted that η_e varies weakly with film thickness and beam incident energy. The backscatter distribution depends strongly on the atomic weight of the material on which the resist sits. Large differences in the backscatter distributions are observed on going from silicon or aluminum substrates to copper or to GaAs [12,13].

Using the techniques described above, it is possible to derive two-dimensional equal-energy deposition contours in the resist. Such contours are shown in fig. 9.8. A comparison of these contours with the Monte Carlo

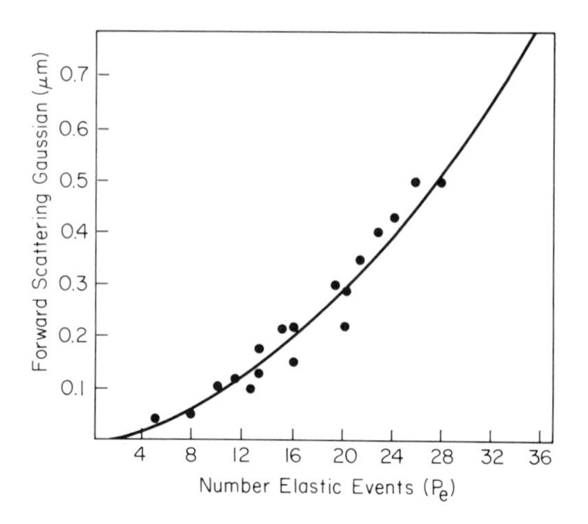

Fig. 9.7. Characteristic Gaussian half-widths for forwardscattering as a function of the number of elastic events. After [13].

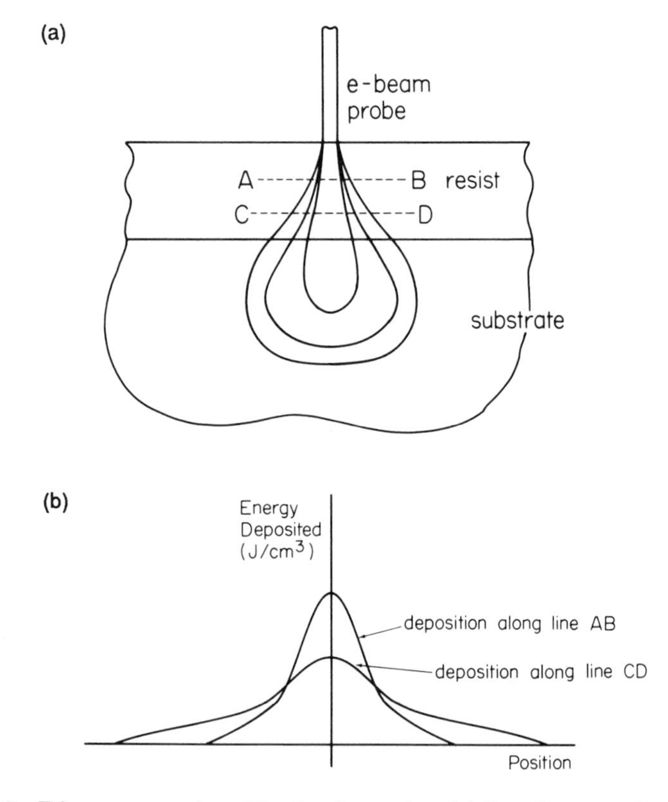

Fig. 9.8. E-beam energy deposition in photoresist: (a) Two-dimensional energy deposition contours for e-beam energy dissipation in resist. (b) Deposition as a function of position at two fixed depths.

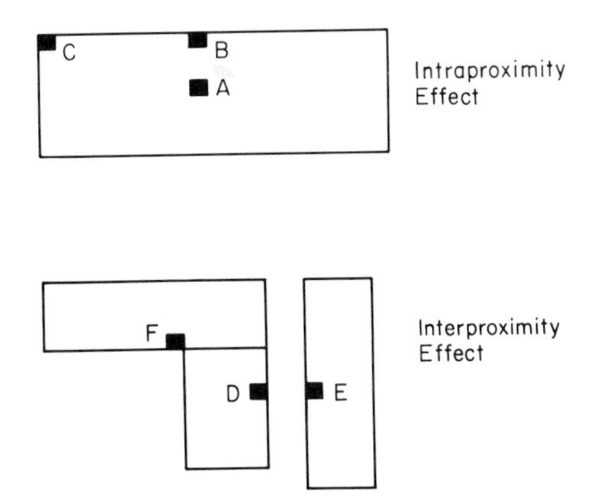

Fig. 9.9. Interproximity and intraproximity effects.

simulation is shown in the figure. These data can be converted into resist sidewall profiles using the methods described.

9.6 Electron-Beam Proximity Correction

As a result of the scattering effects described above, energy can be deposited in the resist by electrons traveling far from the axis defined by the incident beam. Such scattering gives rise to *proximity effects.* We distinguish two types of proximity effects: interproximity and intraproximity effects. Both are illustrated in fig. 9.9. First consider fig. 9.9a. All points in the boxed area are exposed. The pixel at point A receives the highest exposure dose. Backscattered electrons (and some forwardscattered electrons) from above, below and from either side will enter pixel A causing exposure. Pixel B will get less exposure, drawing backscattering and forwardscattering electrons only from the region beneath the top feature boundary. Pixel C gets the lowest dose, drawing extra exposure only from the lower right-hand quadrant of the feature.

Interproximity effect refers to the interaction of the resist with scattered electrons originating from differently exposed shapes in the pattern. This is illustrated in fig. 9.9b. Points D and E get more exposure than they would if the two shapes were very far apart. Thus pixels D and E are more exposed than pixel B. Similarly, pixel F is more exposed then pixel B. The excess exposure of F is also called the interproximity effect, even though it is largely due to electrons scattered from a continuous feature. In pattern generation, this L-shaped feature is viewed as resulting from two distinct rectangles. Both of these proximity effects give rise to an inhomogeneous exposure dose across the pattern. This translates into dimensional uncertainly in the developed resist pattern. There are three basic approaches to

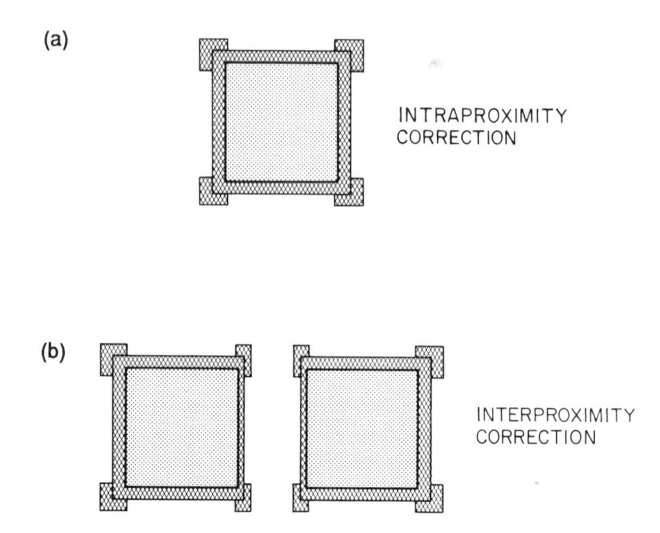

Fig. 9.10. Proximity correction by shape manipulation.

proximity effect correction: shape manipulation [14], dose manipulation [15], and background equalization [16]. Each of these are discussed below.

Shape manipulation is the easiest technique to comprehend and the easiest to implement. To get a feeling for the basis of shape manipulative correction, consider the effects of intraproximity and interproximity. In the intraproximity effect isolated shapes of small dimension appear underdosed as a result of electrons "escaping" the feature into the unexposed field. In order to develop these features in positive resist, stronger solvents and longer time development procedures are required. This alters the dimensions of larger features. One way to correct this is to set the development strength and the development time to provide proper feature sizes for the large-area patterns. The small patterns must be made larger in the pattern generation data base. In addition, the small shapes are further altered to make the dose uniform at the corners. All these modifications are shown in fig. 9.10a.

In the interproximity effect, electrons escaping one pattern lead to exposure in another. If the shape-altering procedure outlined above did not account for the presence of nearby features, overexposure would occur. Thus, the writing boundaries between two adjacent features are not extended as far as they would be if the adjacent features were far apart, as illustrated in fig. 9.10b.

To summarize, in shape manipulation, the writing-field boundaries of small features are pushed out to correct for intraproximity and interproximity. The amount the feature boundary displacement depends on feature size, nearness to a corner, and on the distance to nearby features. Simulations, such as those described above, can be used to develop the rules for boundary placement. Frequently, though, cut-and-try techniques are used. Once the shape manipulation rules are decided, the original pattern-generation data base is resorted and each primitive pattern is classified. The necessary boarder correction is then applied.

Now let us turn our attention to dose manipulation as a method of proximity correction. As indicated in Chapter 8, most e-beam machines can alter the time the beam dwells on a given pixel within a shape. This is referred to as altering the system clock. Thus, the incident e-beam charge delivered to a given pixel can be altered. Parikh has shown [15] that it is possible to achieve a uniform energy deposition within a feature by altering the exposure dose to each pixel.

This is done using a matrix algebraic technique. First, the e-beam writing field is broken down into pixels. Next, an interaction matrix, f_{ij}, is defined. Each individual matrix entry provides the following information: every joule of energy incident on pixel j leads to f_{ij}. Joules absorbed in pixel i. Under these circumstances, we can get the total energy deposited in the ith pixel as follows:

$$\varepsilon_{\text{tot},i} = \sum_j f_{ij} d_j, \tag{9.23}$$

where d_j is the total energy taken out of the e-beam by the jth pixel. The d_j^s can be obtained directly from eq. (9.19).

In Parikh's method, the goal is to make the $\varepsilon_{\text{tot},i}^s$ all equal by adjusting the individual $d_j s$. Take the required exposure dose to be ε_0. In algebraic notation, the problem is expressed as the solution to the series of equations:

$$\begin{pmatrix} \varepsilon_1 \\ \varepsilon_2 \\ \vdots \\ \varepsilon_n \end{pmatrix} = \begin{pmatrix} f_{11} & f_{12} & \cdots & f_{1n} \\ f_{21} & f_{22} & & \\ \vdots & & \ddots & \\ f_{n1} & & & f_{nn} \end{pmatrix} \begin{pmatrix} d_1 \\ d_2 \\ \vdots \\ d_n \end{pmatrix}. \tag{9.24}$$

There are a couple of ways to evaluate f_{ij}. In the "easy"method, we assume that the individual pixels are so small that the energy deposition contributions from nearby volumes are constant across the area of the receiving pixel. We then assign one double-Gaussian energy-deposition profile for each emitting pixel. Here, η_e, β_f, and β_b are taken as the average values for these terms over resist depth. Next, take the normalization constant, K_n, of the emitting pixel to be such as to make the magnitude of the double-Gaussian equal unity at the center of this pixel (i.e., when $r = 0$). Then, the f_{ij} is the magnitude of the double-Gaussian where r is the separation of the i and j pixels. Parikh [15] has given an in-depth discussion of advanced methods for solving eq. (9.24) and for evaluating the f_{ij} even when the deposition is not constant across a pixel.

While the dose-manipulation method probably gives the best results, it is the hardest to implement. The f_{ij} matrix is large. Even a 100 x 100 pixel feature has 10,000 f_{ij} entries. It is difficult to invert such a matrix. In addition, the problem may be mathematically "ill-posed." That is, there may not be a true inverse to 9.24. This is related to the fact that we are writing "square" images with "round" beams. Special mathematical techniques are required to get a "best" inverse.

Obviously, the whole pattern-generation data base cannot be treated at once. Recurrent shape combinations must be dealt with separately. Extensive use must be made of geometric symmetry. The matrices may be be made sparse by setting $f_{ij} = 0$ for points separated by more than a small number of pixels. Once the analysis has been done, the data base must be scanned to find the shapes for which a given type of dose correction applies. The individual shapes must be partitioned and appropriate clocks assigned.

The final proximity correction to be considered is called *background equalization* (or *GHOSTing*), as developed by Owen and Rissman [16,17]. According to these authors, the major difficulty associated with proximity effect is the fact that the background exposure level is wildly varying over the exposure field. This is illustrated in fig. 9.11. Here, e-beam exposure of seven closely spaced lines is shown along with the energy deposition contour of the uncorrected exposure. The Gaussian backscatter components associated with each exposed line sum to give a broad, position-dependent background on which the main exposure peaks ride. Thus, the background exposure levels are position dependent *and* the main peak exposures levels in the exposed regions vary. Peak 1 gets *much* less exposure than peak 4, for example.

In a background equalization scheme, the primary exposure is accomplished first. This is followed by exposing the complementary field (the normally unexposed region) with a lower current, slightly broadened beam. This creates a "GHOST" image of the original pattern. The GHOST exposure sums with the primary pattern exposure to create the uniform exposure shown in fig. 9.12.

If Q_e is the nominal exposure dose, Owen and Rissman recommend that the GHOST exposure dose, Q_c should be

$$Q_c = Q_e\left(\frac{\eta_e^*}{1 + \eta_e^*}\right), \tag{9.25}$$

where η_e^* is the ratio of energies deposited in the resist by the ackscattered electrons to that deposited by the forward scattered electrons. This η_e^* term is related to η_e, defined above, through the relationship

$$\eta_e^* = \eta_e \left(\frac{< \Delta V_b >}{< \Delta V_f >}\right), \tag{9.26}$$

where $< \Delta V_b >$ and $< \Delta V_f >$ are the mean energies deposited by backscattered and forwardscattered electrons. These terms are evaluated by Monte

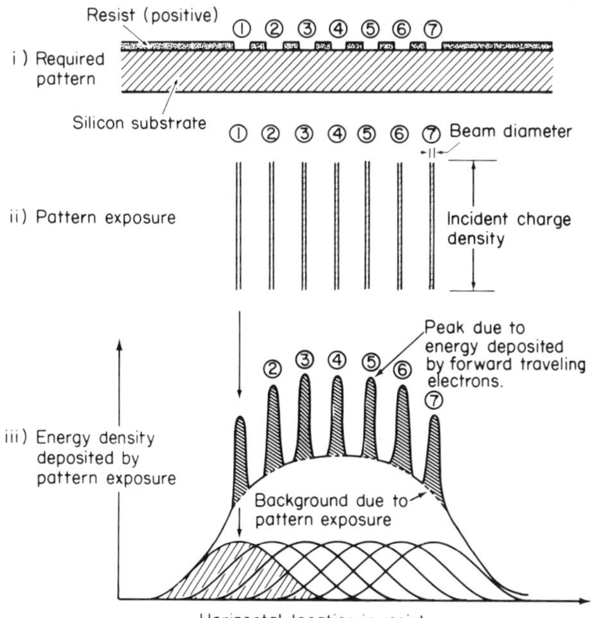

Fig. 9.11. Exposure of seven closely spaced lines and the resulting resist energy deposition contour. After [16]. Note the main peaks, superimposed over the backscatter distribution, give rise to spikes on a more gently varying backrground dose.

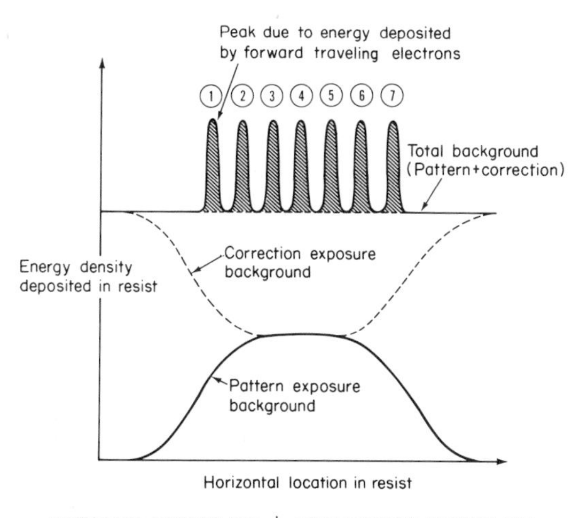

Fig. 9.12. Dose equalization using a GHOST corrector. If the gently varying background in fig. 9.11 was flat, there would be no linewidth variation. To accomplish this "flattening,"the exposure field (normally unexposed) is written with a defocused beam. After [16].

TABLE 9.3: Development rate curve–fit parameters [ep. (9.27)][a]

		PMMA Etch Rate Model				
PMMA	DEV	R_1	C_m	D_0	α	Comment
2010	Conc	8.33	1.0	325	1.303	Data fit
2041	Conc	8.33	.309	325	1.404	M_N Scaled
2010	1:1	1.0	1.0	199	2.0	Data fit

[a] Taken from reference 18.

Carlo analysis. The recommended exposure spot size, σ_c, is

$$\sigma_c = \frac{\sigma_b}{(1 + \eta_e^*)^{1/4}} , \tag{9.27}$$

where σ_b is the width of the *backscattered* beam of the focused exposure probe.

GHOSTing has the advantage of ease of implementation. No extensive sorting and rearrangement of the original data set is required. No extensive computation of dose in individual pixels is required. Background equalization has been applied successfully for 0.5 μm lithography. However, since *all* the resist gets some exposure, there is significant thinning of the protecting resist during development. This can lead to pinholes or to resist decomposition in etching. In addition, since all pixels are addressed, writing time goes up.

9.7 Sidewall Profile Models

To be of true use the energy deposition calculations described above must be converted into resist-development models. That is, we want to take energy deposition data and use it to predict sidewall profiles and feature dimensions for development resist. As is generally the case in engineering modeling, there are a number of ways to do this. In the simplest approach, we assume an infinite gamma for the resist-development curve (fig. 9.3); that is, when energy absorption in the resist is below some critical value, the resist dissolution rate (DR) is small. Exposure at levels above this critical value give DRs that are large and dose independent. Thus, the resist behaves as a switch: off below some exposure level, on above it. To predict resist development profiles for such a system, we must simply provide equal-dose contours. Resist will develop out to the dose contour of the critical energy.

At the next level of sophistication, a "string-model" is employed [18]. Here, the developer/resist interface is modeled as a "string" with points, or "knots," marked off at uniform intervals. Since we know the energy absorption at each knot, we also know the resist dissolution rate at each of these points.

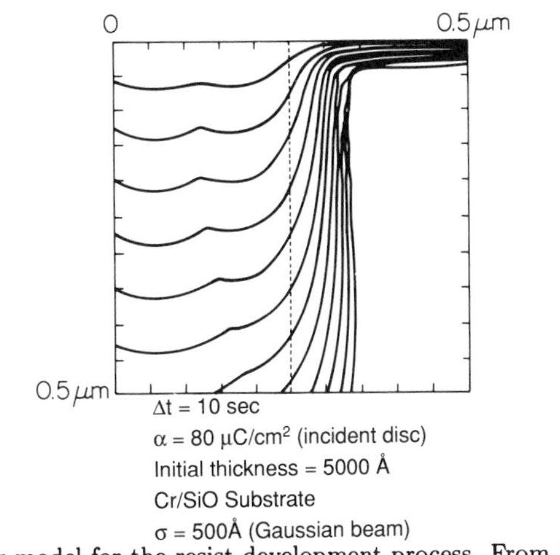

Fig. 9.13. String model for the resist development process. From [18].

The knots originally lie on a straight line. In the first step of the profile calculation, we assume the development has proceeded for some short time interval Δt. We move each knot a distance equal to the local development-rate times Δt in a direction perpendicular to the original line. We now fit a smooth curve through the new knot positions. We thus have an estimate of the resist profile at time Δt into the process.

The process continues by moving each knot a distance equal to the local development-rate times Δt along a local normal to the modeled resist/developer interface. We repeat this process n times until $n\Delta t$ equals the development time. (It is sometimes necessary to add more knots at steep spots in the string profile.) In this way we are able to get a "motion-picture"of the resist sidewall through the development process.

In a computer model, we can use a "look-up"table for the development curve or, we can use a formula that represents the curve. One such formula for positive resist is given here [18]:

$$R(D) = R_1(C_{\rm m} + \frac{D}{D_0})^{\alpha} \, [\text{cm/sec}], \tag{9.28}$$

where $R(D)$ = dose-dependent dissolution rate, R_1, $C_{\rm m}^{\alpha}$ = background (unexposed) etch rate, D = absorbed dose, D_0 = knee energy of the development curve (see fig. 9.3), and α = high-dose slope of the development-rate dose curve. Values for these constant terms are given for three different types of PMMA e-beam resist and for various developer concentrations in table 9.3. Results of a typical string model sidewall computation are shown in fig. 9.13.

As a supplement to these computation intensive techniques, Blais has presented a method for estimating resist sidewall slope [19]. Let us call the sidewall slope-angle, Θ, the angle made by the resist sidewall and the

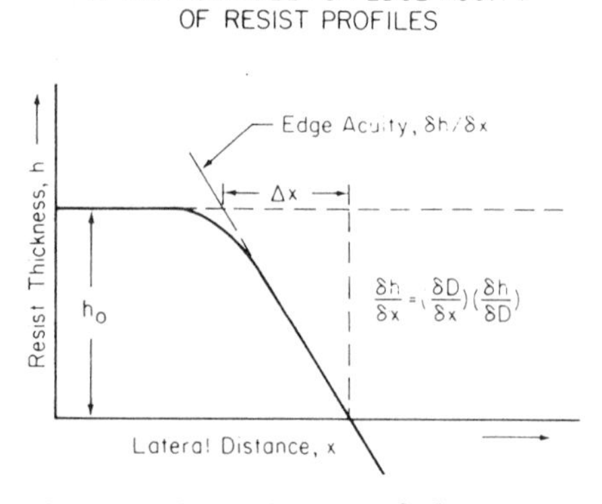

Fig. 9.14. The Blais sidewall profile model. From [19].

substrate. If $T(x)$ is the position-dependent thickness of the resist,

$$\frac{dT}{dx} = \tan\Theta = \left(\frac{\partial T}{\partial D}\right)\left(\frac{\partial D}{\partial x}\right), \tag{9.29}$$

where x is the position along the substrate and D is the absorbed dose, as illustrated in fig. 9.14.

$T(D)$ is determined from the development curve. For a resist development curve of slope γ:

$$T(D) = T_0\gamma\ell n(\frac{D_c}{D}), \tag{9.30}$$

where T_0 is the thickness of the unexposed resist and D_c is the threshold exposure dose (the dose needed to create a zero-thickness remaining film after development.) Thus, we have

$$\frac{\partial T(D)}{\partial D} = -\frac{T_0\gamma}{D}. \tag{9.31}$$

Absorbed dose as a function of position can be calculated using eq. (8.23) of the last chapter:

$$D(x) = \frac{1}{2}D_m[1 + \frac{4}{\pi}M_I(\nu)\sin(2\pi\nu x)]. \tag{9.32}$$

Here, we assume incoherent illumination. Once again, M is the space-frequency dependent modulation-transfer function and D_m is the dose achieved in a large clear area of the mask. For a grating stucture, ν is the frequency of the grating. Applying eqs. (9.32) to (9.31) we get

$$\mathrm{Tan_I}(\Theta) = 4\gamma T_0\left(\frac{D_\mathrm{m}}{D(x)}\right)\nu M_\mathrm{I}(\nu)\,\cos(2\pi\nu\, x), \tag{9.33}$$

where the I subscript refers to "incoherent" illumination.

Thus, we have presented a variety of techniques of varying sophistication to calculate resist exposure profiles. We now move on to the basics of resist handling in production-line practice.

9.8 Resist Application

Now that the fundamentals of radiation-matter interaction in photoresist processing have been presented, we turn to even more basic questions. How are resists "installed" on the wafer surface. What are the relevant issues of resist handling for a successful process? These questions are addressed in this section. There are a number of ways to apply resist to wafers. Three methods are currently in widespread use. The most frequently employed method is *spinning*. In this method, the wafer is placed on a vacuum chuck capable of holding it securely while the wafer is spun at high speed (1-10 kRPM). Resist is spilled on the wafer surface prior to the spin. During the spin, excess resist is flung off the wafer and surface tension pulls the remaining film flat. Resist can also be sprayed or the workpiece can be dip-coated. Prior to any resist application step, the wafers usually receive a dehydration bake at 200°C to remove residual water vapor.

The final thickness of the spun resist layer is determined by the spin speed, as well as by the polymer composition, mean molecular weight, and viscosity of the resists

$$d = KC^{\beta}\frac{[\eta]^{\gamma}}{\omega^{\alpha}}, \tag{9.34}$$

where d = thickness of the spun film, C = concentration of the polymer in the solution (g/100 ml of solution), $[\eta]$ = viscosity of the resist, ω = spin speed, and K, α, β, γ = constants. The powers in eq. (9.34) must be determined for a given polymer/solvent system. The spinning technique provides films of exceptional uniformity. This is indicated in fig. 9.15 for resist spun on a 50-mm wafer. Better than 10-nm thickness control was obtained, as verified by ellipsometry on a flat wafer.

If the wafer is not flat, the resist will conform, to some degree, to the topography of the underlying features. This is illustrated in fig. 9.16. Assume that the underlying topography provides a peak-to-trough separation of distance b. Once the resist is spun, the peak-to-trough distance of the overlying resist is some distance a. We can speak of the degree of planarization, p, provided by the resist as

$$p = 1 - \frac{a}{b}. \tag{9.35}$$

What factors affect the degree of planarization? Immediately after spinning, the resist is fairly conformal (i.e., $p = 0$). There is some surface tension associated with the air/resist interface. The system can lower its free energy by minimizing top-surface area. The minimum surface area possible occurs when the top surface is planar. The resist cannot respond instantaneously to the planarizing force supplied by the surface tension. The more viscous the resist, the more sluggish the response. The peak-to-trough height, $h(t)$,

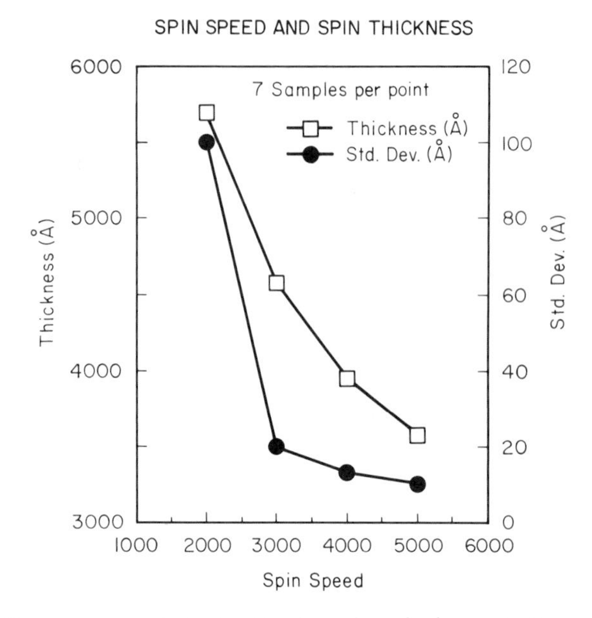

Fig. 9.15. Thickness control obtained in the resist spinning process.

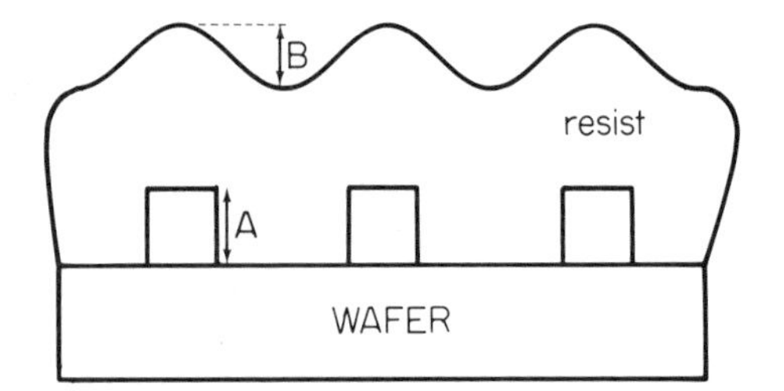

Fig. 9.16. Nonconformal nature of resist overcoats.

decays exponentially

$$h(t) = h_o \exp\left(-\frac{t}{\tau}\right), \tag{9.36}$$

where h_o = original peak to trough distance, τ = time constant of the decay, and t = time between spinning and the bake process in which the film is set (i.e., spinning solvents driven off).

Assume, for a moment, that the original contour variation of the resist surface is sinusoidal, of spatial period L. From the Navier-Stokes equation of fluid dynamics, the following expression for τ is derived

$$\tau = \frac{3\mu L^4}{16\pi\sigma h_o^3}, \tag{9.37}$$

where μ = dynamic viscosity of the resist (poise) and σ = surface tension (ergs/cm^2). According to eq. (9.36), the system should, given enough time, planarize completely (i.e., reach the $p = 1$ point). In practice, this does not happen because the spinning solvent dries and the resist no longer flows. This can be modeled by assuming some planarization cut-off time, t_c, in eq. (9.36). The t_c can be increased by allowing the viscous material to planarize in an ambient containing a vapor of its solvent. For reasons cited in Chapter 8, surface planarization is of primary importance in IC manufacture. As a result of the limited depth-of-focus of high resolution optical systems, a flat, well-defined imaging plane is critical.

After the resist is spun, its spinning solvent must be completely removed. Small amounts of residual spinning solvent severely inhibit the photosensitivity of the resist. Solvent removal is accomplished in a pre-exposure heat treatment (usually called prebake or soft-bake) [20]. In addition to rendering the resist photosensitive, prebake also improves adhesion.

Just after all the spinning solvent is driven off, resist is most sensitive. Frequently, the resist is too sensitive for controlled processing. Small variations in exposure intensity across a wafer, or changes in substrate reflectivity severely change the shape of the development curve. This leads to uncontrolled sidewall profiles and to wildly varying linewidths. Thus, the prebake sometimes extends beyond the time needed to just remove all the spinning solvents. This destroys some of the photoactive compound and leads to a better-controlled process.

While spinning is the method of choice for resist applications in the IC field, there are many areas of electronics technology in which spinning is not practical. Frequently, the work piece cannot fit on a spinning chuck. In addition, as a result of surface tension effects, "edge beads" form (see fig. 9.17). This is a resist build-up at the edge of the wafer. This "bead" prevents the mask from coming in contact with the wafer in a contact aligner. Many devices, such as surface acoustic wave devices and discrete ICs, use contact printing to achieve near-micron printing capability. Thus, the edge bead is a problem because it seriously degrades resolution. In such cases, the workpiece can either be dipped in resist or resist can be sprayed on with an airbrush. Both dipping and spraying operations should be done in an ambient containing the evaporated resist solvent. This allows the resist to flow for a greater period of time and achieve a high degree of planarity.

Once the resist has been exposed, it must be developed. This is done by spraying or by dipping the resist with solutions of the polymer solvent. Frequently, the solvents are diluted with a fluid that cannot dissolve exposed resist. This slows the development process and creates a more uniform development. The ratio of solvent to nonsolvent volumes in the developer is referred to as the strength of the developer. Current trends in stepper-based lithography call for strong developers. This forces an "overdevelopment" of the resist and allows for some degree of underexposure in the exposure tool.

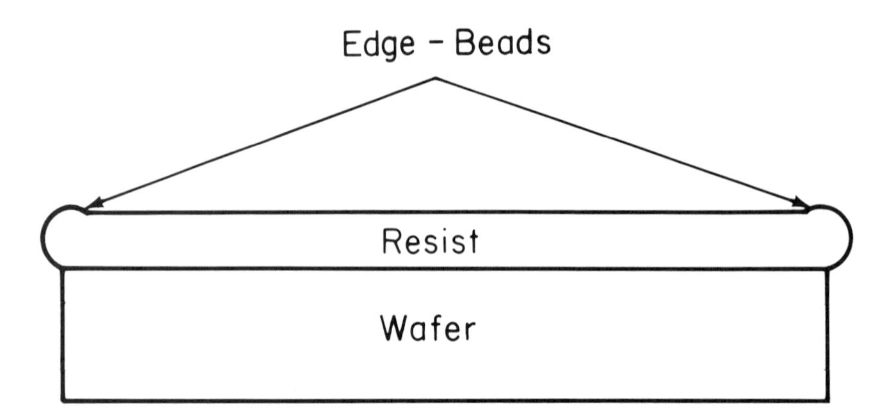

Fig. 9.17. Edge bead.

This improves stepper throughput without causing the adhesion problems that would be caused by a short prebake. Again, the rapid development procedure could cause loss of critical-dimension control. Exposure levels, development temperature, and developer strength must be rigidly controlled to avoid this problem.

Following the development step, the resist is "postbaked." This drives out residual developer that could cause resist swelling. In addition, the postbake improves resist adhesion and limits film breakdown during subsequent etches. Sometimes, resist is exposed to a temperature cycle prior to development or postbake. This causes some redistribution of the photoactive compound, washing out the standing-wave pattern in the resist sidewall [21].

9.9 Resist Systems

In addition to the single-layer resists just discussed, it is possible to use different resists in combination to form a multilayer sandwich. Resists can be used in conjunction with other polymeric layers. The surfaces of resists can be chemically treated to create a surface layer with markedly different exposure and chemical characteristics than the bulk. Such layer combinations are referred to as *resist systems.* The theory and application of resist systems are discussed below.

The obvious reason behind using two or more different resists in the same photomasking step is to obtain the main benefit of each component without the drawbacks associated with each when used alone. For example, thick resists are useful because they are extremely durable and hold up well in strenuous processes. Such resists usually do not demonstrate good resolving power, however. They are so thick that significant diffraction spreading

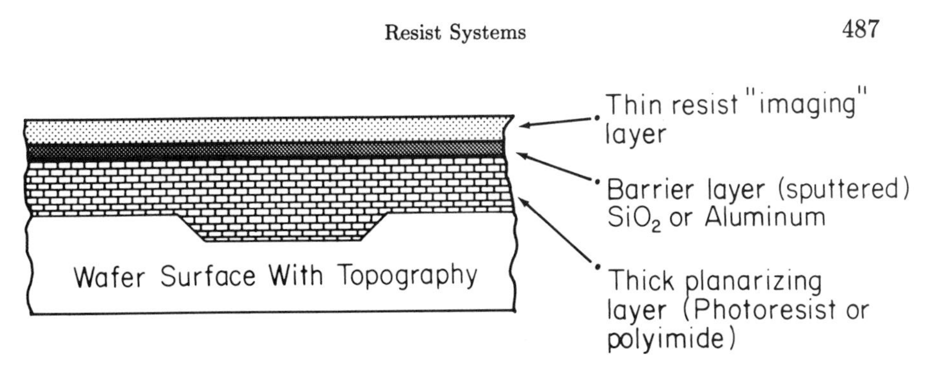

Fig. 9.18. The trilevel resist system.

occurs within them. Multiple reflection effects are also usually more severe. Thin resists exhibit good resolution but are not very process resistant.

One way to get around this is to spin a thick, process-stable resist base. A thin, high-resolution resist of relatively high sensitivity is used on top of this. The top layer is exposed and patterned. After the patterning takes place, the whole wafer is then flood exposed without a mask. This exposes the less sensitive base, which is subsequently developed. The top-layer resist can be matched to absorb, or at least attenuate, a key wavelength necessary to expose the bottom layer. Thus, the top layer becomes a "conformal" mask in perfect contact with the bottom layer. This scheme suffers from the following problem: The bottom layer must be insoluble in the solvent of the top layer. Otherwise, a mixed layer of unknown lithographic capability forms. It is very difficult to control this transition layer.

In order to solve this problem, a third layer is frequently added [22]. This layer is a chemical barrier between the top and bottom layers. Currently, sputtered or low-temperature-plasma-deposited silicon dioxide glass is the favored barrier layer. It is transparent and allows optical alignment with features under the resist. Spin-on glasses (composed of silicate gels) have also been tried. When level-to-level realignment is not necessary, aluminum serves as a good pinhole-free barrier material. A schematic diagram of this "trilevel" process is shown as fig. 9.18.

The trilayer process has proved useful in applications where a high degree of planarity is required. For the ultimate in planarization, the bottom layer can be replaced by very thick, radiation-insensitive material, such as polyimide. Once the image has been transferred into the barrier layer, the thick-base polymer can be etched with a reactive ion etcher using an oxygen plasma. The barrier layer is a good etch mask against the oxygen discharge, and the RIE system anisotropically etches the polymer base. The thickness and viscosity of the base can be adjusted to planarize highly irregular topographies.

Finally, let us consider the types of surface treatments currently employed in resist processing. Chemical modification of the resist/substrate and resist/air interfaces are common in semiconductor processing. Adhesion promoters, such as hexamethyldisilosane (HMDS) are frequently spun on the wafers prior to resist application. This renders the wafer surface hydrophobic. The resist films adhere better to such surfaces. An anti-reflection (AR) coating may also be used on the wafer/resist interface to damp multiple reflections.

A number of air/resist interface treatments are designed to provide specific processing capabilities. Foremost among these is the chlorobenzene treatment for liftoff metallization. In the liftoff process, metal lines can be defined without etching. This is especially useful for gold or aluminum alloys, which are particularly hard to define with good resolution. In the liftoff process [23,24], the resist is exposed and developed. Metal is evaporated over the resulting pattern. The workpiece is then soaked in resist solvent. The metal that overlays the resist is washed away (i.e., "lifted off" by the solvent). This process is illustrated in fig. 9.19. To do liftoff effectively, the metal must break cleanly over the resist sidewall. This is necessary to ensure that solvent can reach the unexposed resist. Bridges may result in shorts or simply tear away during the metal liftoff. To avoid this, the resist is surface-treated to achieve a "reentrant" profile, illustrated in fig. 9.19. The surface treatment consists of a short (less than one minute) soak in chlorobenzene after exposure and prior to development. This hardens the surface to a depth less than 0.1 μm. This hardening creates the reentrant profile.

In recent years, it has been shown that the contrast factor (gamma) of the resist development curve can be improved through surface treatment. A water-soluble bleachable dye, called a *contrast enhancement material* (CEM), is spun on the top surface of the resist [25]. Let us assume that there is some diffraction blurring of the feature. If we look at the exposure intensity as a function of position in the feature, we would see the intense main exposure peak surrounded by the diffraction sidelobes. The intense main peak bleaches rapidly, leaving a transmission window for exposure of the underlying resist. The sidelobes are not intense enough to bleach the dye and, hence, create no exposure. The dye is dissolved prior to resist development. Exceptionally sharp sidewalls are obtained in the resist in this way. The technique also remedies some of the image washout that occurs over stepped topographies. On the negative side, though, the process requires 3-4 times the exposure dose of normal resists. Exceptionally good control of the uniformity and reproducibility of the light exposure field is also needed. Small changes in the thickness of the CEM layer can cause large changes in the exposed feature size. Despite these limitations, the technique has found widespread use in the near-micron resolution range of lithographies.

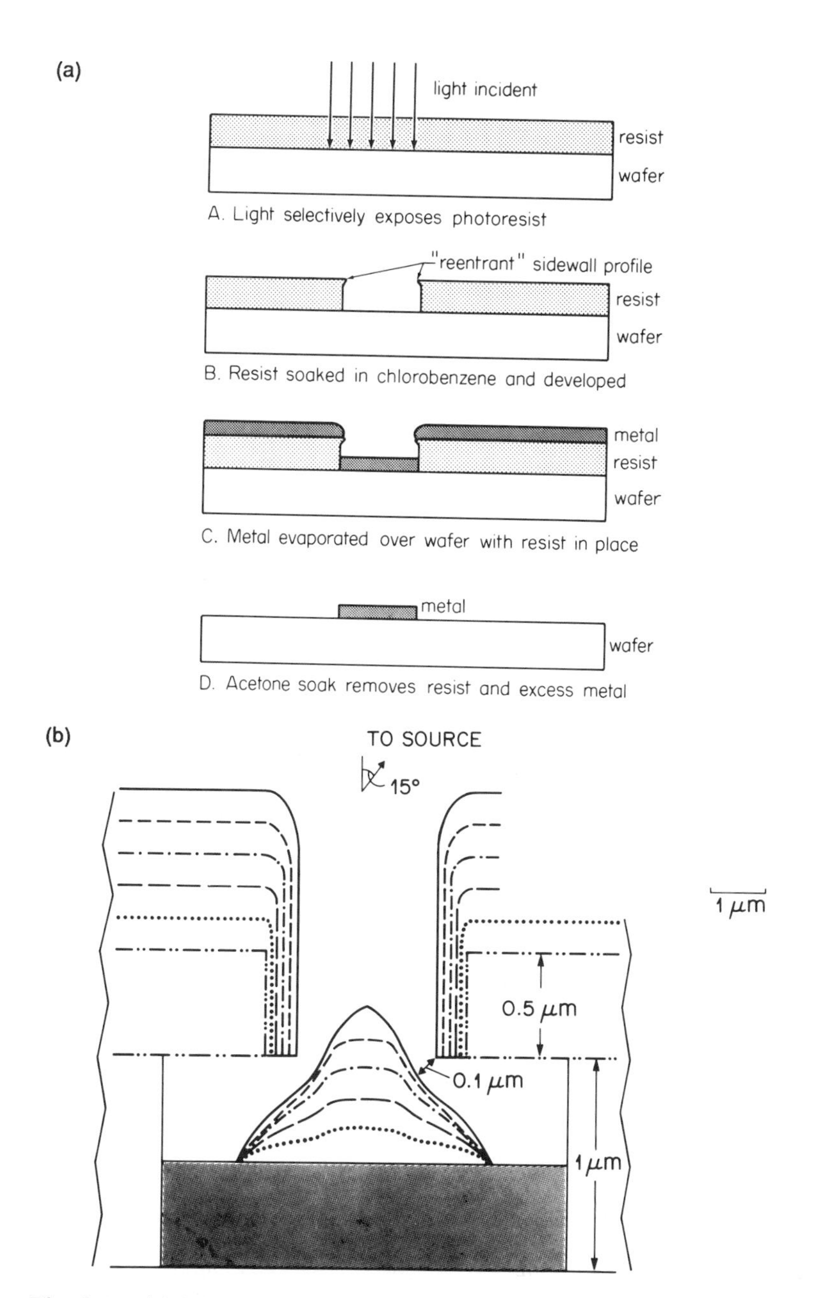

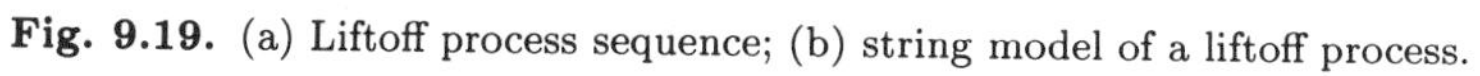

Fig. 9.19. (a) Liftoff process sequence; (b) string model of a liftoff process.

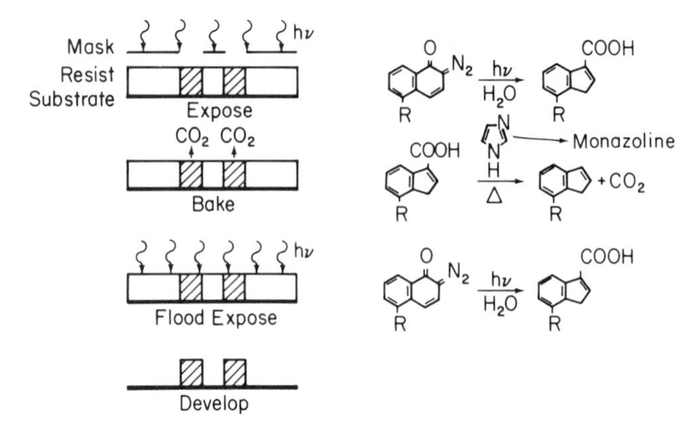

Fig. 9.20. Image reversal of a Novalac based positive resist. Primary exposure creates a "normal"latent image (carboxyl containing photo product). Postbake decarboxylates the acid group (as catalyzed by monazoline). Subsequent re-exposure creates a "normal" latent image in the field.

9.10 Novel Resist Processes

As our understanding of the photochemistry of resist materials has improved, new mechanisms for resist action have been proposed. The resist-handling techniques that have emerged are generally compatible with standard integrated circuit fabrication practices, but the steps involved make for profound changes in the basic chemistry of the photosensitive materials. In this section we review two of these new techniques. The first technique, image reversal [26], addresses the issue of how to obtain a plasma-etch-resistant negative resist that does not swell or string. The second addresses the need for more sensitive resists as stepper or e-beam speeds increase.

An illustration of the image reversal process is shown in fig. 9.20. Note that the resist material base is novalac resin. This is an extremely plasma-etch-resistant material. The major difference in chemistry for the resist shown in the figure is the addition of an imide (such as monazoline or imidazole) to the novalac base. The resist is exposed in the conventional way. Development at this point yields the normal positive image.

If the resist is postbaked prior to development, the imide catalyzes the decarboxylation of the exposed photoactive compound. Remember, the exposed PAC contains an aqueous-hydroxide soluble carboxylic acid group. Decarboxylation (the evolution of CO_2 gas) makes the exposed region less soluble in the developer. When the baked film is flood exposed with UV radiation, the previously unexposed PAC converts to its soluble form containing the carboxylic acid group. This region is now highly soluble and washes away on development. Thus, only the material that received the primary exposure remains after development. This is a negative image, achieved with a normally positive resist, which does not swell or string.

The imide catalyst does not have to be present during the exposure cycle. In fact, baking in ammonia vapors has the same effect. The image-reversal

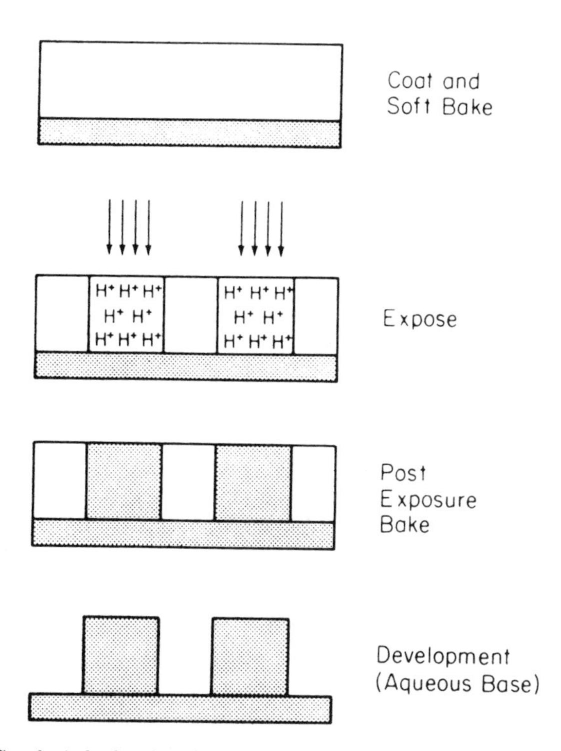

Fig. 9.21. Catalytic hydronium ion produces chemical amplification of the crosslinking process.

process can be accomplished in e-beam machines delivering high doses [27]. Apparently, vacuum-related esters play the catalyzing role. Lack of water in vacuum stops formation of indene carboxylic acid. Instead, the ketene intermediate reacts with the novalac forming an insoluble ester.

The chemical-amplification process is illustrated in fig. 9.21. Once again, a novalac-base resin is used. However, a proprietary material is included that releases hydronium (H^+) ions on exposure to UV and x-ray radiation. The hydronium ions are mobile in the polymer tangle. These ions catalyze crosslinking and make the resist less soluble. Once the catalytically driven reaction is completed, the ion is released and can go on to create more crosslinks. While a single photon can create a single hydronium ion, the ion, once produced, can drive many linking reactions. Thus the chemical effects amplify the effect of the primary exposure. Such resist is currently marketed by Shipley Co. as SAL-601 e-beam resist.

9.11 Characterization

In this section we focus on techniques that are required for resist exposure characterization. Of course, linewidth and pattern fault detection are important in characterizing resist features. Such characterization techniques are reviewed in Chapter 8. Here, we concentrate on sidewall profiles, development curve generation, and on methods for obtaining the Dill parameters.

The preferred method for sidewall profiling is the scanning-electron microscope. Here, the exposed wafer is cleaved along a resist feature. Note, the wafer must be cleaved and not sawed or scribed. These techniques will smear the resist and prevent profile determination. The specimen must be mounted so that the cross section of the resist is perpendicular to the scanning plane. If this is not done, geometric projection effects would complicate linewidth measurement. The results of such an effort are shown in fig. 9.22. For very thick resists, the resist cross section must sometimes be coated with a thin evaporated conducting film (carbon, gold, or gold alloys) to prevent charging effects from destroying the image.

Typically, development curves can be obtained by the tedious technique of exposing the resist, developing the resist for some time, and measuring the remaining thickness either ellipsimetrically or by surface profilometer. If the resist's optical absorption is known, the Lambert-Beer law can be used to determine thickness *during* the development process [28]. Here, the resist is spun on a transparent substrate whose refractive index is matched to that of the resist. A light is shown through the resist/substrate combination. The ratio of incident to transmitted intensity is used to determine resist thickness. This measurement can be done during the development process. In addition, the $T(\infty)$, $T(0)$, and $dT(0)/dt\,|_{t=0}$ terms can be evaluated for a given resist with the method described above. These terms, when applied to eqs. (9.11a) – (9.11c) give the Dill parameters.

9.12 Summary

In this section, we have summarized the behavior of the two types of resist currently in use today: positive- and negative-acting resists. Resists can be positive or negative acting as a result of bond scission (for positive resist) or, as a result of crosslinking (for a negative resist.) Most resists in current use achieve their positive-acting behavior through photoactive compounds (PACs).

The basics of resist chemistry and resist handling were presented. Resist handling techniques include:

- Dehydration bake prior to resist application to remove excess water vapor, which could cause adhesion problems after application.
- The use of adhesion promoters (such as HMDS).
- The spin-application method.
- Prebake for adhesion and to remove excess spinning solvent.

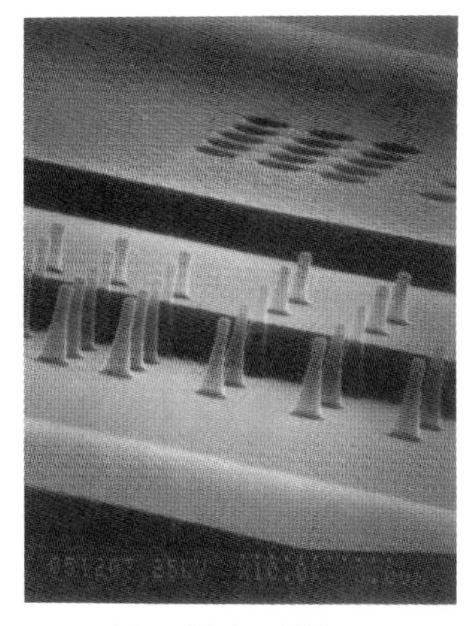

Resist - Shipley 2400
Developer - Shipley 2401
Exposure - 240 mJ/cm^2
Resist Thickness - 1.60 μm

Resist - Shipley 2400
Developer - Shipley 2401
Exposure - 240 mJ/cm^2
Resist Thickness - 1.60 μm

Resist - Shipley 2400
Developer - Shipley 2401
Exposure - 240 mJ/cm^2
Resist Thickness - 1.60 μm

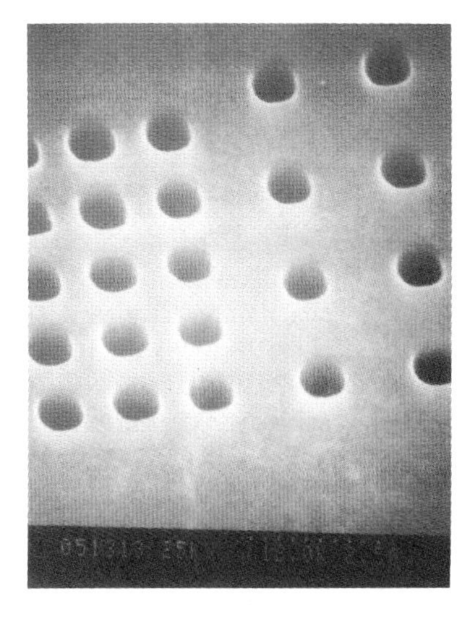

Resist - Shipley 2400
Developer - Shipley 2401
Exposure - 240 mJ/cm^2
Resist Thickness - 1.60 μm

Fig. 9.22. SEM resist sidewall profile. Courtesy of M. Lubin, Hampshire Instruments.

- Exposure and exposure modeling for optical and e-beam lithography.
- Predevelopment bake to aid in adhesion and to redistribute PACs to lessen standing-wave effects.
- Development.
- Postdevelopment bake to improve adhesion through subsequent etches.

The planarizing and resolution enhancing properties of resist and the purposes of multilayer resist systems were also discussed. Novel developments, such as image reversal and chemical amplification were outlined.

In addition, the basic principles of resist exposure modeling were presented. This modeling was done for optical (i.e., the Dill model and the standing-wave model) and for e-beam lithographies. The purpose of this introduction to exposure modeling was to develop correction techniques for feature inteference and proximity effects. Three main e-beam proximity correctors were presented:

- Shape manipulation
- Dose manipulation
- Background equalization

References for Chapter 9

1. W.S. DeForest, "Photoresist: Materials And Processes,"McGraw-Hill, New York (1975).
2. H. Nagai, A. Yushikawa, Y. Togoshima, O. Ochi, Y. Mizushima, A new application of Se-Ge glasses to silicon microfabrication technology, Appl. Phys. Let. 28 (5), 145 (1976).
3. C. G. Willson, Organic resist materisls — Theory and chemistry, in "Introduction To Microlithography,"L.F. Thompson, C.G. Willson, M.J. Bowden, eds., ACS Symposium Series 219, American Chemical Society, Washington, D.C. pp. 87-119 (1983).
4. F.H. Dill, W.P. Hornberger, P.S. Hauge, J.M. Shaw, Characterization of a positive photoresist, IEEE Trans. ED, ED-22 (7), 445-452 (1975).
5. F.H. Dill, A.R. Neureuther, J.A. Tuttle, E.J. Walker, Modeling projection printing of positive photoresists, IEEE Trans. ED, D 22 (7) pp. 456-464 (1975).
6. R.J. Hawryluk, Exposure and development models used in electron-beam lithography, Vac. Sci. Technology 19 (1), 1-17 (1981).
7. R.M. Eisberg, "Fundamentals of Modern Physics,"Wiley, New York pp. 100-106 (1961).
8. T.H.P. Chang, J. Vac. Sci. Tech. 12, 1271 (1975).

9. M. Parikh, D.F. Keyser, IBM Research Report, R.J. 2261 (1978).
10. L.F. Thompson, M.J. Bowden, Lithographic processes, in "Introduction to Microlithography," L.F. Thompson, M.J. Bowden, eds., ACS Symposium Series 219, American Chemical Society, Washington, D.C., pp. 15-85 (1983).
11. T.E. Everhart, P.H. Hoff, J. Appl. Phys., 42, 5837, (1971).
12. M. Parikh, D.F. Keyser, Energy deposition functions in electron resist films on substrates, J. Appl. Phys. 50 (2), 1104-1111 (1979).
13. J.S. Greeneich, Electron beam processes, in "Electron-Beam Technology in Microelectronic Fabrication," G.R. Brewer, ed., Academic Press, New York, pp. 59-140 (1980).
14. N.D. Wittels, G.I. Youngman, Proc. Eighth Int'l, Elect. Ion. Beam Conf., R. Bakish, ed., Electrochem. Soc. Press, New York, p. 361 (1978).
15. M. Parikh, Corrections to proximity effort in electron-beam lithography: I. Theory, II. Implementations, III. Experiments, J. Appl. Phys. 50 (6), 4371-4387 (1979).
16. G. Owen, P. Rissman, Proximity effect correction for electron-beam lithography by evaluation of background dose, J. Appl. Phys. 54 (6), 3573-3581 (1983).
17. U.S. Patent applied for, Geraint Owen, Paul Rissman, Serial No. 389306, June 17, 1982, (Japanese and European patents also applied for).
18. A.R. Neureuther, D.F. Keyser, C.H. Ting, Electron-beam resist edge profile simulation, IEEE Trans. ED-26 (4), 686-693 (1979).
19. P.D. Blais, Sol. St. Technology, 20 (8), 76 (1977).
20. D.J. Elliot, "Integrated Circuit Fabrication Technology," McGraw-Hill, New York (1982).
21. E.J. Walker, Reduction of standing-wave effects by post-exposure bake, IEEE Trans. ED-22 (7), 464-466 (1975).
22. B.J. Lin, E. Bassous, V. Chao, L. Petrillo, Practicing the Novolac deep-UV portable conformable masking techniques, J. Vac. Sci. Technol. 19, p. 1313 (1981).
23. T. Sakurai, T. Serikawa, Lift-off metallization of spattered Al alloy films, J. Electrochem. Soc. 126, 1257 (1979).
24. T. Batchelder, A simple metal lift-off process, Sol. St. Technology 25, 111 (1982).
25. D.R. Strom, Optical lithography and contrast enhancement, Semicond. Internat'l. 9 (5), 162-167 (1986).
26. S.A. MacDonald et al., Image reversal: The production of a negative image in a positive resist, "Proc. of the Kodak Interface Conference," p. 114 (1982).
27. J.M. Shaw, M.A. Frisch, F.H. Dill, IBM J. Res. Develop. 21, p. 219 (1977).
28. K.L. Konnerth, F.H. Dill, In-situ measurement of dielectric thickness during etching or developing processes, IEEE Trans. ED 22, 440 (1975).

Problem Set for Chapter 9

9.1

a. Assume a resist incident light intensity with a spatially Gaussian profile. The width of the profile (i.e., the Gaussian standard deviation σ) is one micron. In addition, assume that the energy absorption in the resist is uniform throughout the thickness of the resist. Consider the cases in which the doses delivered to the resist at the Gaussian peak is D_0, 1.5 D_0, 2 D_0, 3 D_0, and 4 D_0. The development curve of the resists is switchlike; that is, the dissolution rate is zero for absorbed energies less than D_0 and maximum for absorbed energies greater than D_0. Plot the linewidth in the developed resist for the five doses listed.

b. Describe the effect of having a resist development curve of infinite γ on a developed linewidth.

9.2

a. Plot the resist sidewall profiles for the five doses listed above assuming the switchlike development curve. Furthermore, assume the same Gaussian profile as the one described in problem 9.1.

b. Describe, qualitatively, the effect of finite resist γ on sidewall profile.

9.3 Once again, assume the same Gaussian profile as described in problem 9.1. Plot the resist sidewall profiles for development curve γ's of 1 and 10, assuming the dose delivered to the resist at the Gaussian maximum is 1.5 D_0. Here, D_0 is taken to be 500 Å/sec, the development time is 20 sec and the resist starting thickness is 5000 Å.

9.4

a. For a 10-keV incident electron beam, using the equations defined in the text of this chapter, follow the trajectory of a single electron through 20 scattering events using the Monte Carlo simulation method.

b. How would you define the total number of scattering events engaged in by an electron before it comes to rest?

9.5

a. Using the data supplied in table 9.2, calculate the energy deposition profile for a 10-keV electron beam 0.1 μ in diameter as incident on the surface of the resist.

b. Assume that the electron density at the center of the electron probe is 1.5 times enough to create a maximum dissolution rate in the resist on development. Assuming the switchlike development curve described in problem 9.1, calculate the resist sidewall profile after development.

9.6 Discuss the effect of the following parameters on resist sidewall slope: (a) strength of developer, (b) prebake time and temperature, and (c) development time.

9.7

a. Using the Blais model (eq. 9.30), plot the final resist thickness profile for equal lines and spaces, assuming line width (in microns) is 0.5, 1.0, and 1.5; $M_I(\nu) = 0.6$, and $\gamma = 2$.
b. The approximation is not valid near $T = T_0$. Explain why.
c. Propose a better approximation, accounting for the effect described in (b).

(Courtesy C. Steinbruchel)

10 Pattern Transfer

10.1 Introduction

In the two previous chapters we spoke about resist patterning techniques. Rarely does resist remain as an integral part of a finished device structure. The resist pattern must be transferred into underlying materials. Holes must be etched in insulators, and metal lines must be defined to create interconnects. The base semiconductor is frequently patterned to make alignment marks or deep trenches for isolation. There are several requirements that define a successful pattern transfer process. First, the resist layer defining the pattern must survive intact through the process to its completion. Next, the resist image must be transferred into underlying material with minimum distortion. The ability to achieve 1-μm wide metal lines is a technological achievement. However, if we intended those lines to be 0.5 μm wide, our process is a failure. In addition, the pattern-transfer process must be *selective* for the layer defined; that is, if the goal is to etch a polysilicon line over oxide, we do not wish to affect the underlying oxide.

There are a number of different ways to achieve pattern transfer in a given material. Some of the more common techniques are discussed below. The goal of this chapter is to supply a conceptual framework for understanding the processes described. Some amount of "cookbook"description is unavoidable. Wherever possible, the underlying physical and chemical principle, are emphasized.

Perhaps the simplest pattern transfer technique routinely employed is wet chemical etching. Highly selective wet-etch processes are available. Wet etching is amenable to batch processing; that is, large numbers of wafers can be processed simultaneously. Fresh etch bath can be continuously circulated around the workpiece to avoid depletion of the active etchant. Wet etches usually suffer from one major drawback. They are *isotropic*. That is, they etch horizontally at the same rate as they etch vertically. This leads to undercutting of the resist and a shrinkage of etched lines. Undercutting is a particular problem in making near-micron lines. The resist may be completely undercut and float away before the line is fully defined. Completely *anisotropic* (vertical sidewall) etches may be undesirable due to the difficulty of covering the surfaces of vertical groove, with metal (see Chapter 6). However, totally isotropic etches create feature resolution problems in high-density integrated circuits. In today's integrated circuit process technology, a controlled amount of anisotropy is desired.

There are a number of ways to anisotropically etch materials. In solution chemistry, there are potassium hydroxide- (KOH) based etchants for silicon that are selective for various crystallographic planes. Vertical sidewall trenches can be etched into the <110> surface of silicon and pyramids can be etched into the <100> surfaces. Since such etches rely on the crystallographic perfection of the etched substrate, they cannot be used to define polycrystalline or glassy layers.

The main method currently used to achieve anisotropy is plasma etching. Reactive ions and neutral species are created in a glow discharge. A directional etch effect is achieved by accelerating ions along the normal to the work surface. By controlling the ratios of neutrals to ions and by using various gas mixtures, RF powers, and frequencies etches ranging from completely isotropic to completely anisotropic can be achieved. Plasma-etch reactors can be configured in a number of different ways. The basic reactor configuration also influences the quality of the work turned out. While good pattern definition is usually achieved, etch uniformity (both across a wafer and from wafer to wafer in a batch) is a major problem. These issues are discussed in greater depth below.

In this chapter, the fundamentals of wet etching are presented. Both isotropic and plane-dependent etches are discussed. The use of etch stop diffusions to control etch depth is explained. The major configurations of plasma etchers are described. The effect of configuration on performance is emphasized. A model of plasma etching that relates anisotropy to etcher configuration is derived.

10.2 Wet Etching

In wet etching, acids or bases are used to remove unwanted materials not protected by resist. The basic mechanism of wet etching is as follows [1,2]: The etched surface is either oxidized or reduced by the active etchant. The term *oxidation* is here used in the chemical sense: An electron is removed from the etched surface. In reduction, an electron is added. The resulting surface reaction product is soluble in the etch solution. Typical etches used for key materials employed in semiconductor technology are listed in table 10.1. Clean semiconductors, such a silicon or germanium can react directly with halogens like chlorine or flourine to form water soluble flourides or chlorides. If the etch solution's dipole fields are not able to lift the reaction product from the etched surface, complexing agents, such as polyhydric alcohols, can be used. Acids can be used in combination. One acid will form an intermediate compound that is insoluble. The second acid will then form a soluable compound.

TABLE 10.1: Typical wet chemical etches used in semiconductor process technology

Material Etched	Etchants	Comments
1. Silicon	HF: HNO_3: Acetic acid	This is an isotropic etch. Varying the HNO_3 concentration with respect to that of the HF varies surface roughness and changes the dependence of the etch on the substrate doping. Acetic acid enhances the oxidizing power of the HNO_3, making the etch rate more dependent on HF concentration.
2. Silicon	KOH or Ethylene Diamine/ Pyrochatechol/ water	Anisotropic etches (crystallographic plane etch rate dependent). Plane etch rate orders (high rate to low rate): ethylene diamine (100), (110), (111); KOH (110), (100), (111).
3. Silicon dioxide	Buffered HF	Ammonium flouride is the usual buffer. This etch is also used for boro- and phosphosilicate (B and P) glasseș. P-glass etches faster than undoped glass. Buffers make B-glass etching practically composition independent.
4. Silicon nitride	BHF, Hot phosphoric acid (@ 140°–200°C), HF(48%), HNO_3 (70%) 3: 1 @70°C	Unbuffered HF-based etches are very fast. Hot phosphoric acid gives slow controlled etch, stopping at silicon dioxide.
5. Aluminum and aluminum alloys	H_3PO_4:HNO_3: acetic:water (4: 4: 1: 1)	A good polishing etch slow enough to do fine-line definition without bubbling.
6. Aluminum and aluminum alloys	HCL and water (4: 1) @ room temp	Good for fine-line etching without resist lifting.

It is very important that the etch rates are stable and controllable in wet etching. Typically, this means that the pH of the solution must be stable, the temperature of the bath must be controlled, and there should be a constant flow of fresh etchant over the etched surface. The solutions are stirred to keep fresh etchant circulating. Frequently, solutions are buffered to stabilize pH during the etch cycle. From elementary chemistry, we know that acids mixed with their salts will resist changes in hydrogen ion concentration. The

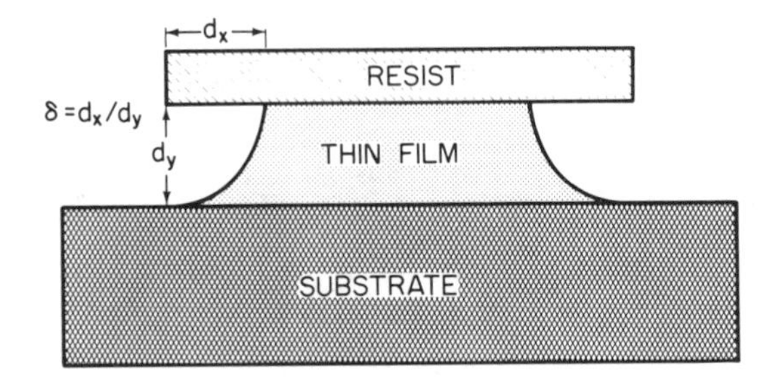

Fig. 10.1. Isotropy factor and resist undercut.

pH of a solution is defined as follows:

$$pH = -\log_{10}[H^+], \tag{10.1}$$

where $[H^+]$ is the hydrogen-ion concentration in moles/liter. Hydrofluoric acid (HF) is usually buffered with NH_4F (ammonium fluoride), for this reason.

Furthermore etchants may be diluted to "cut" the etch rate. This is done to prevent overetching. For example, suppose we try to etch a contact via in a 1/2-μm thick insulating layer between two metal levels. If the oxide etch rate is 10 microns a minute, it may be difficult to quench the etch in 3 seconds. This would lead to severe undercutting and ballooning of the contact window. Thus, the etch rate must be controlled to be consistent with the thickness of the etched film to achieve the desired feature size control. This ballooning is shown schematically in fig. 10.1.

There are three steps to any etch process [3]. Any one of these steps can be rate limiting (i.e., determine the etch rate of a given surface). These steps are:

1. Diffusion of the reacting species to the etched surface.
2. Chemical reaction at the surface.
3. Dissolution of the reaction products.

These classifications are useful in understanding some of the behavior of various etches as a function of temperature, agitation of fluid bath, etc. For example, if we find that steps 1 or 3 are rate limiting, agitation of the solution would, effectively, enhance the arrival or departure rate of chemical species from the surface. Thus, we would observe a pronounced agitation dependence of the etch. We might also anticipate difficulties in achieving etch uniformity, as the flow of material around the wafer is random and uncontrolled.

If step 2 is limiting, we might expect a significant amount of temperature dependence to the etch. To see why this is so, consider the fact that both chemical surface reactions and diffusion are thermal rate-activated processes; that is, the rate at which these processes occur is given by a relationship of the form

$$r = A(T) \exp\left(\frac{E_a}{kT}\right), \tag{10.2}$$

where $A(T)$ is a temperature dependent constant and E_a is an activation energy. Typically, E_a for diffusion in a liquid must be smaller than E_a for surface-chemical reactions. This leads to pronounced temperature dependence for etches dominated by step 2.

Etches for which step 3 is important may also demonstrate especially poor consistency over large areas. The agglomeration of reacted etchant over a given region may prevent fresh etch from reaching the surface. A similar problem is encountered when bubbling occurs as a result of gas evolved during the etch. Such etches may also exhibit anisotropy. That is, they will undercut a resist feature more slowly than they will etch down into the unprotected areas of the film. This occurs because depleted, slow diffusing etchant gets trapped in the channel that forms beneath the undercut resist. The isotropy factor, δ, is also illustrated in fig. 10.1. The expression for δ is

$$\delta = \frac{d_x}{d_y}, \tag{10.3}$$

where d_x and d_y are the horizontal and vertical etch distances, respectively. For completely isotropic etching, δ equals 1. For vertical sidewall or anistropic etches, δ equals 0. Most wet etches approach $\delta \approx 1$.

With these fundamental considerations in hand, we now proceed to a discussion of some practical etch processes. Materials considered fall in three broad classes: dielectrics, semiconductors, and metals. In the area of dielectrics, deposited and grown SiO_2, as well as silicon nitride are covered. Etching techniques for crystalline and polycrystalline silicon are presented. Metals of interest are aluminum and aluminum alloys.

10.3 Wet Etching of Dielectrics

Dielectric etching is usually done either with HF or with H_3PO_4. Silicon oxides are etched almost exclusively with fluorine-based etchants. This is true in both wet and plasma processes. The silicon-oxygen bond is one of the most stable bonds found in inorganic chemistry. Only the Si-F bond strength is comparable. Thus, HF contributes fluorine to form a soluble (or volatile, in the plasma case) silicon fluoride, usually $SiF_6^{=}$. Buffering also appears to impede the undercut of the resist by the acid.

TABLE 10.2: Activation energies and etch rates for $Si0_2$ etching in buffered HF as a function of buffering[a]

Solution Composition (% Flourine)		$[NH_4F]/[HF]$ Ratio	Activation Energy (Kcal/Mole)	Etch Rate (Å/sec @ 30°C)
NH_4	HF			
0	10	0	6.7	36.2
2.0	10	0.2	8.1	60.05
1.0	2.5	0.4	8.2	15
8.0	10	0.8	8.8	112
8.0	2.5	3.2	10.3	19.4
10	3.5	3.0	11	22

[a] After reference 3.

Since all three etch rate-determining steps listed above are thermally rate-activated, we would anticipate that the etch rate, R_e, would be expressed in the form

$$R_e = A_\mathrm{p} \exp\left(\frac{\Delta E_\mathrm{app}}{\mathrm{k}T}\right), \tag{10.4}$$

where ΔE_app is an apparent activation energy and A_p is a prefactor that depends on the molar concentrations of the acid and its buffers. There may be some weak temperature dependence to A_p. Results of such an analysis for thermally grown SiO_2 are given in table 10.2 [4].

More detailed analysis of SiO_2 etching must take into account the nature and concentration of the reacting species. In HF-based SiO_2 etching, the two dominant reactive species are undissociated HF and HF_2^-. It is interesting to note that the etch rate depends on the concentrations of these two species, rather than on the F^- concentration. HF_2^- is formed through the coupled reactions

$$\mathrm{HF} \longrightarrow \mathrm{H}^+ + \mathrm{F}^- \qquad k_1 = \frac{[\mathrm{H}^+][\mathrm{F}^-]}{[\mathrm{HF}]}$$

$$\mathrm{HF} + \mathrm{F}^- \longrightarrow \mathrm{HF}_2^- \qquad k_2 = \frac{[\mathrm{HF}_2^-]}{[\mathrm{HF}][\mathrm{F}^-]}$$

where $k_1 = 1.3x10^{-3}$ @ $25°C$; $6.5x10^{-4}$ @ $60°C$ and $k_2 = 0.104$ @ $25°C$; $3.66x10^{-2}$ @ $60°C$. A more general form for the etch rate is

$$R_e = [\mathrm{HF}]A + [\mathrm{HF}_2^-]B + C. \tag{10.5}$$

A and B contain the rate activation exponentials and prefactors that depend on the concentrations of the various reacting species. C is a temperature-dependent correction factor that accounts for the fact that diffusive transport as well as reaction rate is important in determining etch rate. Eq. (10.5) can be rewritten to account for the exponential terms as

$$R_e(\text{Å/sec}) = 5.0\times10^7[\text{HF}_2^-]\exp\left(\frac{\Delta E_1}{RT}\right)+2.2\times10^6\,[\text{HF}]\exp\left(\frac{\Delta E_2}{RT}\right)+C(T), \tag{10.5a}$$

where $C(T) = 0.02T(T - 292)$, $\Delta E_1 = -9.1 \times 10^3$, $\Delta E_2 = -8.1 \times 10^{-3}$, T = temperature,° K, and R = universal gas constant.

As discussed in chapter 7, CVD glasses are widely used in integrated circuits. They are employed as a dielectric isolation between metal layers and as a scratch protection deposited over the whole chip at the end of the process. As discussed in chapter 7, this glasses are frequently doped with phosphorous to form phosphosilicate glass (PSG) or boron to form borosilicate glass (BSG). These additives immobilize harmful ionic impurities, such as sodium, and they lower the reflow temperature of the material. This makes edge rounding of steep via walls easier, improving metal step coverage.

Phosphorous and high-concentration boron doping tends to *increase* the glass etch-rate in buffered HF. Care must be taken to prevent overetch ballooning of contact windows and vias. Apparently, the increased etch rate comes from the formation of P_2O_5 or B_2O_3 clusters that are water soluble. At low boron concentration, though, BSG etches rather slowly. At these low concentrations, borosilicates are formed, rather than boron oxides. Borosilicates are less soluble in HF than SiO_2 [5].

CVD silicon nitrides are frequently used as diffusion barriers and in self-aligned oxidation processes. These films are etch in buffered HF at room temperature or in H_3PO_4 at 140°-200°C. The etch rate is strongly affected by the presence of oxygen (as silicon oxynitride) in the dielectric film. HF-based etch rates go up with increasing oxygen content; phosphoric acid etch rates go down with increasing oxygen content.

Nitride etch rates in buffered HF exhibit behavior similar to that of SiO_2. Relations for etch rate similar to those expressed in eq. (10.5a) have been determined [6]:

$$R_e(\text{Å/sec}) = A[\text{HF}] + B[\text{HF}_2^-] + C, \tag{10.6}$$

where A = 9.6 Å/sec @ 25°C, 114 Å/sec @ 60°C; B = 10.6 Å/sec @ 25°C, 222 Å/sec @ 60°C; C < 0.001 Å/sec @ 25°C, = −0.12 Å/sec @ 60°C.

10.3.1 Wet Etching of Crystalline and Polycrystalline Silicon

First consider the etching of crystalline silicon. This material can be etched isotropically (δ = 1) or *anisotropically* with wet-chemical techniques. Isotropic etches usually involve mixtures of HF, acetic, and nitric oxides [7,8]. Nitric acid oxidizes the surface of the silicon, which is subsequently etched away by the HF present. Acetic acid contributes protons (H^+), which help keep the HF in an undissociated state, enhancing the oxide etch rate. The acetic acid also keeps the HNO_3 undissociated, enhancing its oxidizing capability.

This type of etch is also suitable as a wet etch for polysilicon. However, it is very rapid and it frequently leads to resist lifting and inhomogeneous etches. Polycrystalline silicon, as a gate material, requires the highest degree of definition control. The gate length is usually the minimum resolved feature size. As a result, polyetches are usually done with anisotropic plasma processes, as discussed in the next section.

Detailed studies of silicon etching using the HF/HNO_3/acetic-acid system have uncovered a number of interesting chemical principles at work [7,9]. For example, there appears to be an autocatalytic component to the etch process. That is, a reaction product is formed (which may or may not be part of the etch reaction) that acts as a catalyst to enhance the etch rate. For the case at hand, surface damage sites appear to serve as nucleation centers for catalyst production. Work-damaged surfaces not only etch faster themselves but also cause an increase in etch rate of surrounding, undamaged material. This leads to some interesting consequences regarding the area uniformity of the etch rate (see problem 10.3). The presence of aluminum also creates local enhanced etching of p^+-silicon surfaces [9]. This may be due to catalytic effects of aluminum, or it may be due to charge transfer effects between the metal and semiconductor (galvanic effects). Apparently, such charge transfer effects are most important in reactions involving HNO_3. Providing sufficient HNO_3 creates a uniform layer of oxide, which is dissolved by HF. HF availability, in this case, provides the rate limit in the reaction. Resulting surfaces are usually shiny and do not exhibit the aluminum-induced etch rate acceleration described above.

A number of integrated-circuit and microstructure-fabrication projects are coming to rely on anisotropic wet-chemical etching of silicon [10]. Trenches many microns deep can etched with vertical sideways and minimal mask undercutting using KOH and water solutions. A 300:600:1 etch-rate ratio is obtained for the <100>, <110> and <111> planes of silicon respectively (44% KOH by weight in H_2O) [11,12]. For all practical purposes, the etch rate in the <111> direction is zero. Ethylene diamine combined with water and pyrochatechol has also been used for such etches [15,14]. We currently are uncertain as to the mechanism that creates this strong plane dependence to the etch rate, even though the overall silicon dissolution kinetics are thought to be understood [14]. Oxidation of the silicon occurs due to the presence of H_2O and OH^-. The resulting oxide is removed from the surface through the formation of an amine-soluble complex (trispyrochatecholato-silicon). The etch ratios for the three-plane sequence <100>, <110>, and <111> is 50:30:3 μm/hr.

Regardless of the mechanism, the plane-dependent etch rate enables the etching of some interesting and useful microstructures on silicon surfaces.

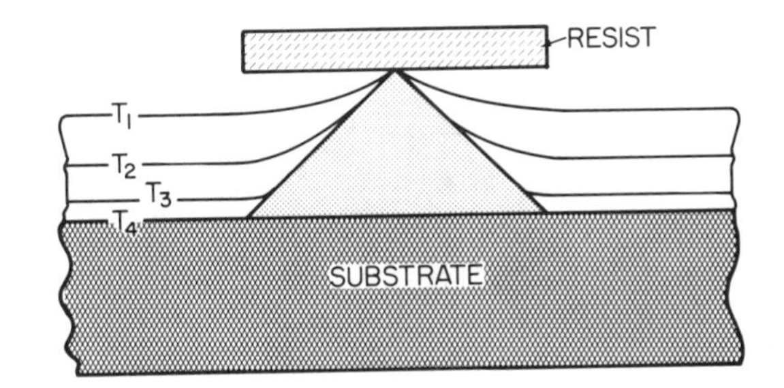

Fig. 10.2. Exposure of <100> planes with anisotropic etches.

For example, consider the following process: Deposit a suitable masking material on the exposed <100> silicon surface. Typically, photoresist will not hold up in such a strong reducing agent as KOH. Oxides or oxide/nitride combinations are frequently used. Pattern a circle in the masking film. Next, etch the silicon in a KOH/water solution. At first the etch will proceed vertically with some undercut. Eventually, as the undercut becomes larger, a <111> face of silicon is exposed. The etching then ceases normal to this newly exposed plane. This is illustrated in fig. 10.2. Faceted etch pits develop in the silicon. The facet faces are the exposed <111> planes. If the radius of the circle is twice the final depth of the etch, a pyramid will develop, as shown in the figure. Such sharp-tipped microstructures could conceivably be used as field emitting cathodes. An "inverse pyramid" or pit can also be made by masking the region around a circular hole. This technique can be used to create apertures in silicon with tight dimensional control since the etching ceases when the <111> plane is fully exposed. Such apertures are currently used in ink-jet printers. Inverse pyramids are also used as e-beam alignment marks.

If we can find a way to etch parallel to the <111> faces, it would be possible to etch deep trenches in silicon. These structures might find use as device-isolation regions. To make this deep trench, start with the <110> face of silicon exposed. Next, make a groove in a masking layer on the silicon surface. This groove wall must lie parallel to the $< \bar{1}11 >$ or $< 1\bar{1}1 >$ face as shown in fig. 10.3. As Kendall has pointed out [11], the (110) surface can make vertical sidewall trenches in silicon. However, the edges of the trench are not orthogonal structures. The (111) faces exposed are at angles other then 90° with respect to the surface normal (as shown in fig. 10.3). Thus, if one starts with a trench of length L, and one etches to a depth, $D_{\max}$ the rectangle at the bottom of the trench will have length L_{B}. Computation of L_{B} are given in this figure. In addition, the rectangle will not be centered in the trench. The offset, (l - l′), is also computed in the figure. The effect of slight misalignment is shown in fig. 10.4. Usually, alignment is accomplished by first etching a "fan" test pattern of grooves, as shown in the figure. The narrowest groove is the best aligned and is used to align subsequent trenches.

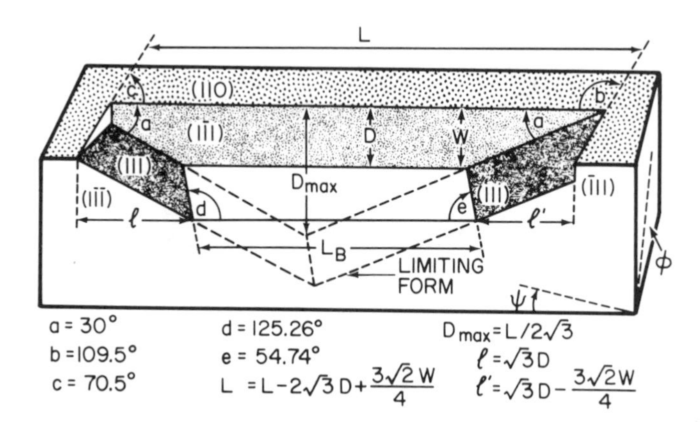

Fig. 10.3. Generation of trenches on <110> surfaces of silicon [11].

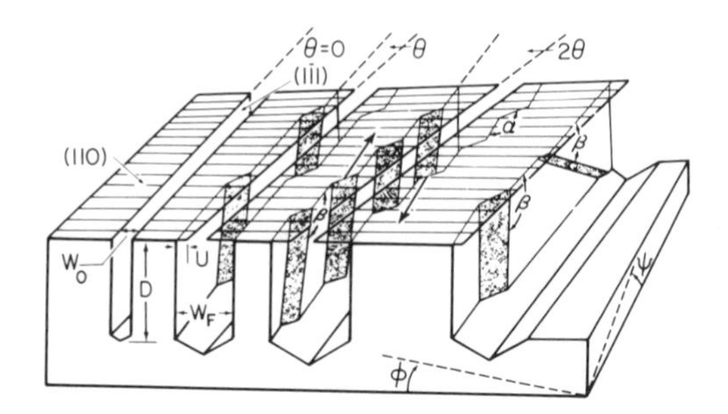

Fig. 10.4. Misalignment effect in groove etching [11].

Examples of pyramids and "slabs" (an inverted groove) are shown in fig. 10.5.

In addition, silicon can be etched to form membranes as thin as 1 μm, using 200- to 300-micron thick wafers as starting material. This is done by heavily diffusing (or implanting) boron into the silicon surface. The boron-diffused layer acts as an etch-stop. Typically, the etch-rate, r, of the doped layer in KOH solutions is characterized by the following proportionality [15]:

$$r \propto \left[1 + \left(\frac{p}{p_\text{c}}\right)^2\right]^{-4}, \tag{10.7}$$

where $p_\text{c} = 5.5 \times 10^{19}$ /cm^3 and p is the mobile hole density in the material. The mobile hole density is usually approximately equal to the boron doping density. Apparently, KOH etching of silicon involves both oxidation and reduction processes [16]. Both H_2O and OH^- are needed to produce soluble silicates of the form $Si(OH)_2(O^-)_2$. As the hole doping level increases, the Si Fermi level lowers toward the valence band edge, inducing spontaneous passivation (oxidation) of the Si surface. This oxide is a slow etcher that reduces the net Si etch rate by about two orders of magnitude. In this way,

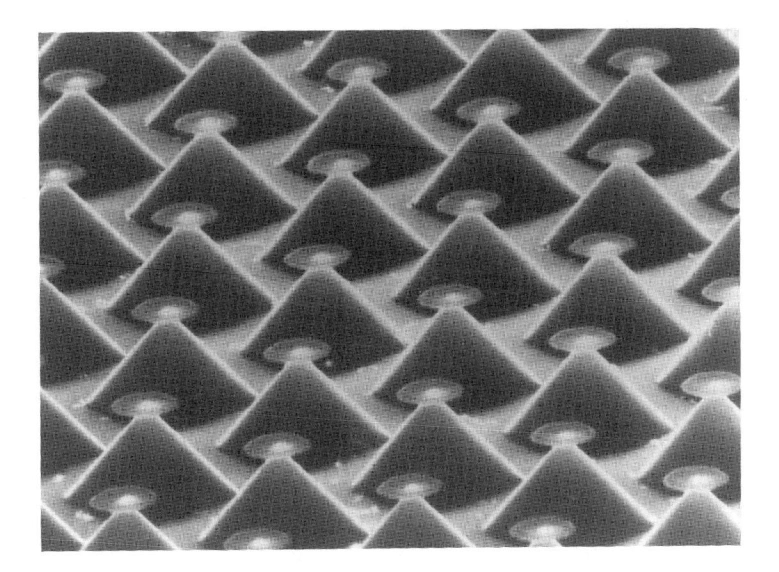

Fig. 10.5. Examples of anisotropically etched silicon [11].

the undoped silicon can be dissolved, leaving the doped silicon membrane. X-ray lithographic masks are made using this technique.

10.3.2 Metal Etching

Solvation of metals generally proceeds by oxidation. Water soluble metal compounds are formed by the oxidizing action of acids on the metal surface. The term oxidation is used in the chemical sense: removal of electrons from the material. Cations, anions or nonionic species can be formed. Cation production and dissolution proceed via the reaction

$$\mathrm{M} \longrightarrow \mathrm{M}^+ + e^- \tag{10.8a}$$

$$\mathrm{M}^+ \overrightarrow{\text{solvent}}\, \mathrm{M}^+(\text{solution}), \tag{10.8b}$$

Here, M stands for some metal atom. Free electrons are rapidly dissipated via the reaction

$$2(\mathrm{H}^+ + e^-) \longrightarrow \mathrm{H}_2(\text{gas}). \tag{10.9}$$

Hydrogen is evolved in the form of bubbles through the etch process. Alternate processes are possible, such as

$$H^+ + HNO_3 + 2e^- \longrightarrow H_2O + NO_2^- \qquad (10.10)$$

for nitric acid etches. More complex reactions are responsible for the production of anion and nonionic species

$$M + H_2O \longrightarrow MO + H_2 \qquad (10.11a)$$

$$M + H_2O_2 \longrightarrow MO + H_2O \qquad (10.11b)$$

$$MO + OH^- \longrightarrow MO_2^= + H^+. \qquad (10.11c)$$

When the oxide is not water soluble, *passivation* can occur. That is, the surface oxide can quench the etch process. In such cases *complexing agents* are required to convert the oxide into a soluble anion. Similar processes were discussed in the case of silicon etching. Here, HF forms a hexafluoride complex that is soluble. Polyhedric alcohols and organic acids have a similar effect in metals etching. Table I indicates that aluminum and aluminum compounds (Al:Si or Al:Cu) can be etched in HCl or NaOH. Somewhat slower etches that produce small amounts of bubbling are used for fine-line pattern definition. Usually phosphoric acid etches are used for such a task.

To summarize the results of this section, we see that wet chemistry can be used to define patterns in a variety of materials. Usually, wet etching is isotropic. However, anisotropic wet etches known as *metallographic etches* have been developed that attack crystallographic planes preferentially. Many of the materials used in semiconductor processing, though are not crystals. Rather, they are glasses, and new techniques must be used to achieve fine-line pattern definition. Plasma etch methods have been developed to accomplish this in a wide variety of materials. Plasma etching is discussed in the remainder of this chapter.

10.4 Plasma Etching Apparatus

Current plasma-etch techniques are an outgrowth of the work on plasma ashing of organic films that was done in the late sixties. Wet chemical strippers are frequently useless in removing hardened photoresist. Resist is largely carbon and hydrogen. These films can be burned to produce CO_2 and water. However, the idea of setting fire to a wafer is unattractive, to say the least (even though it has been attempted in the past!). Oxygen can be made to react with a variety of hydrocarbon thin films without combustion if the oxygen is activated in a glow discharge. In the discharge, an electric field causes ionization of the gas species. A variety of unstable atomic and

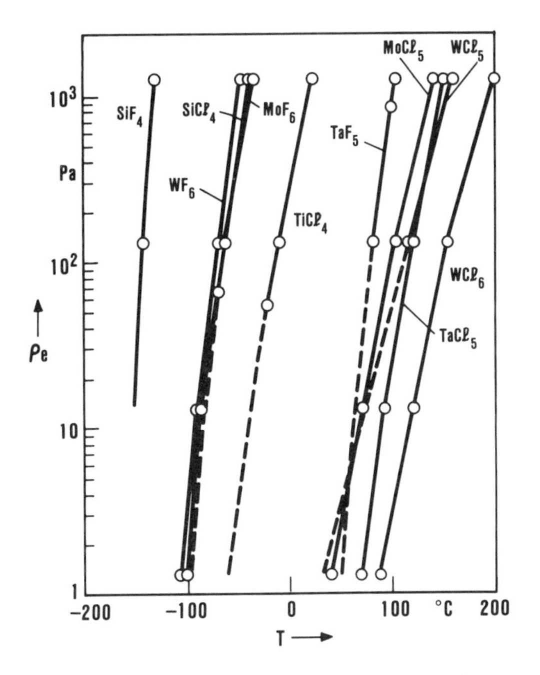

Fig. 10.6. Vapor pressure of various compounds formed in plasma etching.

molecular species appear. These species include O, O^+, and O^-, as well as ionized and neutral molecular oxygen. The charged and neutral atomic oxygen atoms are much more reactive than the oxygen molecule. They can attack the hydrocarbon surface at low temperature at a controllable rate. The rate depends on the densities of reacting species and on the power supplied to the discharge. It was subsequently discovered that such a process can be used to etch polysilicon, silicon dioxide, silicon nitride, or even metals when chlorinated or fluorinated gas compounds are used. These gases form volatile reaction byproducts which are easily removed from the etched surfaces. This is seen by looking at the vapor pressure of the formed compounds (fig. 10.6). Early etchers were particularly useful in etching polysilicon interconnects with a degree of control superior to wet etches. In this section, the most common plasma etchers currently in use are described. Models of the plasma etch process are supplied in the next section.

The discharge can be sustained either by a DC or an RF supply. Typically, RF sources are used because they can attract both positive and negative charges from the plasma. This reduces charging effects in insulators. Insulator charging can prevent reacting species from impinging on the workpiece. The frequency of the RF supply has been shown to be a variable affecting etch rate and isotropy factor. Care must be taken in all cases not to violate government RF emission regulations.

There are primarily three configurations of etcher in current use [17].

TABLE 10.3

Etcher	Configuration	Salient Feature
Barrel	Wafers sit in boat. RF supplied to cylinder surrounding boat.	Isotropic etch, high background pressure > 1 torr)
Parallel Plate	Two parallel plates. RF supply to top electrode.	Isotropic/anisotropic etches possible at background pressures 0.1 – 0.3 torr
Reactive Ion Etcher	Top electrode removed, base pressure lowered. Walls grounded RF capacitivity coupled to electrode on which wafers sit	Often anisotropic. Background pressure < 0.1 torr

These are shown in fig. 10.7. The salient features of each of these configurations is listed in table 10.3. The earliest plasma etcher marketed is the barrel etcher (fig. 10.7a). This etcher is just a quartz barrel surrounded by two hemicylindrical electrodes or inserted into an RF coil. RF power is applied to the system. Reacting gas is bled in from the rear of the barrel, and the workpieces sit in a boat on the barrel wall. The gas pressure is commonly 0.5-2.0 torr, and the RF field configuration does not favor directional etching. Such machines are still used in resist stripping. In addition, brief exposures to oxygen plasma will remove underexposed resist to improve pattern transfer. This is called *descumming.*

The next generation of plasma etchers is known as the planar or parallel plate etcher (fig. 10.7b). Here, the RF discharge is struck between two parallel plates. The RF supply is attached to the top electrode, and the bottom electrode is grounded. The operating pressure is lower than the barrel etcher and the electric field lines are more uniform. These field lines are generally perpendicular to the electrode surfaces. The lower operating pressure and the more regular field configuration makes for some degree of anisotropy in the etch.

The final configuration considered (fig. 10.7c) is the reactive ion etch (RIE) configuration. Here, the RF supply is capacitively coupled to the lower plate. The workpieces are placed on the lower electrode. The counterelectrode is removed. The operating pressure is extremely low. As shown below, removal

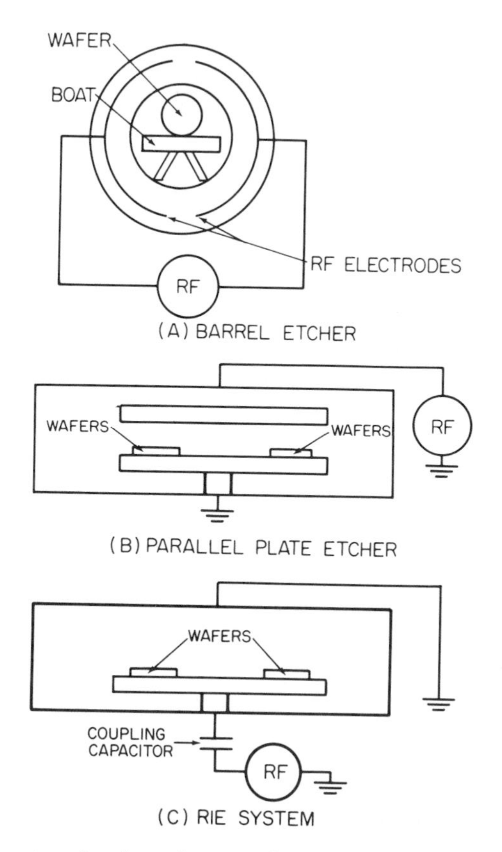

Fig. 10.7. Configuraton of various plasma etchers.

of the counterelectrode causes a greater amount of ion acceleration toward the etched surface. The capacitive coupling aids in development of a negative DC potential, further enhancing the acceleration. The reduced pressure reduces ion scattering in the gas. All of these effects combine to create a highly anisotropic etch. Some typical gas etchants used in etching materials of interest in semiconductor technology are shown in table 10.4 [18]. Some justification for the selection of these gases is given below. As a general trend, though, note that silicon and silicon-based compounds are usually etched in fluorine or chlorine etchants, while aluminum and aluminum compounds are etched only in chlorine-based systems.

It should be pointed out that many of the etch processes listed in table 10.4 do not have the impressive *selectivity* that wet etches usually have. For example, if we are etching polysilicon lines over thin oxides, the oxide may etch at a rate which is only 5 or 10 times slower than the polysilicon. This etch rate differential is about an order of magnitude below that achievable with wet chemistry. Newer equipment and chlorine etchants, however, have

TABLE 10.4: Plasma etch processes [18]

Summary of Etching with Unsaturated Fluorocarbon Plasmas			
Materials	Gases	Anisotropy	Selectivity
Si	CF_4, C_2F_6	Partially anisotropic at low-pressure, high ion energy	—
Si O_2, Si_xN_y	CF_4, C_2F_6, C_3F_8 CHF_3, CF_4, or C_2F_6 with H_2 CH_4, C_2H_2 or C_2H_4 additions	Anisotropic at high ion energies	High over Si and III-V compounds
TiO_2			High over Ti
V_2O_5			High over V

Summary of Plasma Etching in Chlorine- and Bromine-Containing Plasmas			
Materials	Gases	Anisotropy	Selectivity
Si	Cl_2, CCl_4, CF_2Cl_2 CF_3Cl, Cl_2/ C_2F_6, Cl_2/CCl_4 Br_2, CF_3Br	Anisotropic at high and low ion energies; isotropic for doped Si under high pressure conditions with pure Cl_2.	High over SiO_2
III-V Compounds	Cl_2CCl_4, CF_2Cl_2 CCl_4/O_2, Cl_2/ O_2, Br_2	Anisotropic at high and low ion energies (for chlorocarbon and BCl_3 mixtures); isotropic for Br_2 and Cl_2 under some conditions.	High over SiO_2, Al_2O_3, Cr MgO
Al	Cl_2, CCl_4, $SiCl_4$, BCl_3, Cl_2/CCl_4 Cl_2/BCl_3Cl_2/ CH_3Cl, $SiCl_4$/ Cl_2	Anisotropic at high and low (for chlorocarbon mixtures and BCl_3) ion energies; isotropic for Cl_2 without surface inhibitor	High over Al_2O_3 and some photoresists
Ti	Br_2		
Cr	Cl_2/O_2Ar $CCl_4/O_2/Ar$	Anisotropic at low ion energies	1: 1 vs. CrO_2 at >20% O_2
CrO_2	Cl_2/Ar, CCl_4/Ar		High vs. Cr
Au	$C_2Cl_2F_4$, Cl_2		

TABLE 10.4 (continued)

Summary of Plasma Etching with F-atom Source Plasmas			
Materials	Gases	Anisotropy	Selectivity
Si	F_2, CF_4-O_2 $C_2F_6-O_2$, $C_3F_8-O_2$,SF_6-O_2, SiF_4/O_2, NF_3, Clf_3	Isotropic	High over SiO_2. Si_xN_y, metals, silicides
SiO_2		Isotropic at low ion energy; anisotropic at high energies	Very high over III-V compounds
Si_xN_y		Similar to SiO_2	Selective over SiO_2 in isotropic range
$TiSi_2$, $TaSi_2$, $MoSi_2$, WSi_2		Partially anisotropic at high ion energies	High over SiO_2, Si_xN_y~1, 1 over Si
Ti, Ta, Mo, W, Nb		Anisotropic at high ion energies	High over SiO_2 at high powers; selective over TaO_2
Ta_2N		Isotropic at low ion energy	Selective over TaO_2, SiO_2, Al_2O_3

demonstrated much greater selectivity. To prevent overetching, some method of endpoint detection is frequently used. This may involve the addition of an optical spectrometer to the etch chamber. The spectrometer is tuned to look for the light emission associated with the desired etch process. When this emission ceases in the course of the process, we can assume that the process is finished. The laser interferometric method can also be used. Here, the thickness of the test piece is measured throughout the etch process. When the full thickness etch point is reached, the process terminates. When etching various low-conductivity films, the sample conductivity can also be measured with respect to the plasma. When the conductivity increases, the sample is etched. It must be emphasized that as a result of variations in the etch rate across the sample and from sample to sample in the etch batch, endpoint detection is of limited value. Highly selective processes must be developed in the plasma area.

10.5 Plasma Etching: Mechanisms and Models

Plasmas are in a constant state of interaction with the walls of their confinement chamber and with objects placed in their midst. Usually, electrons are far more mobile than positively charged ions. Electrons collide with the walls much more frequently than the ions, and the plasma becomes ion-rich in the vicinity of the wall. Electrons also charge insulating surfaces. Thus the plasma's bulk potential, V_p, is usually more positive than the wall potential, V_w (which is usually grounded) or the workpiece surface potential. An unconnected object assumes some (negative) floating potential, V_f, in the plasma as referenced to ground. A space charge of positive ions thus surrounds the walls and negatively charged objects in the plasma. This space charge is called a *sheath.* Positively charged ions are accelerated by the electric field in the sheath. It is this acceleration that gives rise to the etch anisotropy, as shown below.

In reading the discussion that follows, the reader is asked to keep the following facts in mind: The plasma-etch process is one that depends on many variables. It is difficult to alter these variables independently to ascertain their individual effects. In addition, the analytic techniques frequently used to ascertain the chemical state of the etched surface (such as Auger on photoelectron spectroscopy) are not applicable here. In addition, temperature, a key factor influencing any chemical reaction, is frequently not specified. As a result, this important parameter is missing from the chemical rate equations. While powerful statistical techniques have been employed [19], the reader must view the following analyses as approximate, providing rules-of-thumb, explaining gross trends. It should be pointed out that this is a fertile area for research and future development. As minimum feature sizes are reduced, more burden is placed on plasma etch processes to provide faithful pattern transfer.

The following discussion parallels the description of plasma etch anisotropy given by C.B. Zarowin [20]. The goal of this work is to provide a framework for understanding how the parameters associated with the plasma etcher itself influence the etch process. It is this parameter set (RF power, frequency, pressure, etc.) that, together with gas composition, is usually the most readily accessible to the process engineer. Small variation in gas composition or material parameters (e.g., doping concentration of polysilicon) frequently makes a major impact on rate and anisotropy. Some of the reasons for this are summarized below.

Consider the etching of solid, S, by a reactive species, A. In the plasma A can either be neutral or charged positively (symbolized by A^+). The following reactions can take place:

$$\mathrm{S + A \rightarrow B},\ k_{\mathrm{f}} = \frac{[\mathrm{B}]}{[\mathrm{A}]} \tag{10.12a}$$

and

$$S + A^+ + e^- \rightarrow B,\ k_f^+ = \frac{[B]}{[A^+][e^-]}, \tag{10.12b}$$

where B is a volatile reaction product and the k_f's are the forward rate constants. We assume that the forward reactions dominate and that the total reaction rate is the sum of the ion and neutral driven reaction rates:

$$R_0 = k_f[A] \tag{10.13a}$$

and

$$R^+ = k_f^+[A], \tag{10.13b}$$

where the square brackets indicate the concentration (i.e., in number per cm^3) of the species in question.

The total etch rate, R is just the sum of eqs. (10.13a) and (10.13b). Etching is a thermally rate-activated process. Thus, the forward rate constant for the two types of etch can be written

$$k_f = k_f^0 \exp\left(-\frac{\Delta}{kT}\right), \tag{10.14a}$$

and

$$k_f^+ = k_f^0 \exp\left[-\frac{(\Delta - U)}{kT}\right], \tag{10.14b}$$

where Δ is the neutral rate activation barrier, k is the Boltzmann constant, T is the temperature, and U is the amount of barrier lowering achieved by the RF field heating the plasma and k_f^0 is the exponential prefactor constant. We usually take U to be proportional to the kinetic energy of the ions. Thus, we have

$$R = k_f^0[A] \exp\left(-\frac{\Delta}{kT}\right)\left[1 + \alpha^+ \exp\left(\frac{U}{kT}\right)\right], \tag{10.15}$$

where α^+ is the degree of ionization of the reaction ($[A^+]/[A]$).

As a result of ion acceleration in the sheath, the mean kinetic energy of the plasma in the y direction (normal to the etch surface) is greater than the mean kinetic energy in the x direction (parallel to the surface). The isotropy factor will be equal to the ratio of the etch rate in the x direction to that in the y direction

$$\delta = \left[1 + \alpha^+ \exp\left(\frac{U_x}{kT}\right)\right] / \left[1 + \alpha^+ \exp\left(\frac{U_y}{kT}\right)\right]. \tag{10.16}$$

When $U_y >> U_x$ (as it would be when ions are strongly accelerated in the sheath), δ is close to zero. The etch is highly anisotropic. When U_x and U_y are about equal or near zero (as in the case of weak sheath acceleration)

δ is nearly 1. The etch is isotropic. We define the following parameter: $\rho^2 = U_x/U_y$, the directionality factor of the etch.

Next, we would like to gain some quantitative insight into the effect of the sheath field and gas pressure on etch isotropy. To do this, we must first realize that in a steady state, the force on the charge as it plummets through the sheath is balanced by the momentum lost per unit time through collisions. Even when the steady state is not reached, the momentum lost per collision will be proportional to this force. The most likely scattering event is the interaction of an ion and a neutral (there are many more neutrals than ions). The momentum lost per unit time is given by

$$\Delta p = \bar{u}\mu\nu\,, \tag{10.17}$$

where $\bar{u}$ = mean directed velocity (i.e., average velocity of an ion due to acceleration in the sheath field), μ = mass of the ion-neutral colliding pair, ($1/\mu = 1/m_{\text{ion}} + 1/m_{\text{neutral}}$), and ν = collision frequency. Equating this expression to the force due to the mean electric field in the sheath, E_{s}, gives:

$$\bar{u} = \frac{eE_{\text{s}}}{\mu\nu}\,. \tag{10.18}$$

Thus, the mean velocity in the direction of the sheath accelerating field depends linearly on the sheath electric field.

Next, define the root-mean-square (rms) total ion velocity as $\sqrt{< v^2 >}$. The total ion velocity is the result of the isotropic thermal motion of the ion *and* whatever velocity is imparted by the plasma electric fields. Let us say that w represents the total instantaneous ion velocity outside of the sheath. In that case, both the mean and rms of w are zero, since the ion motion in the bulk plasma is random and isotropic. The total mean velocity is calculated by summing the directed and thermal mean velocities:$< v >= \bar{u}+ < w >$, $< v^2 >= \bar{u}^2+ < w^2 > +2\bar{u} < w >$. While $< w >$ is zero, $< w^2 >$ is not. The squaring operation, done prior to averaging, removes the cancelling negative terms in the average sum. Thus, we have

$$\sqrt{< v^2 >} = \sqrt{\bar{u}^2+ < w^2 >}. \tag{10.19}$$

It is well known from statistical mechanics that the rms velocity of atoms or molecules in a gas is just the mean-free-path-length between collisions, λ, times the collision frequency, ν. So we have

$$\lambda\nu = \sqrt{< w^2 >}\sqrt{1 + \frac{\bar{u}^2}{< w^2 >}}\,. \tag{10.20}$$

Solving this equation for ν and substituting into eq. (10.18) gives

$$\frac{eE_{\text{s}}\lambda}{\mu < w^2 >} = \left(\bar{u}/\sqrt{< w^2 >}\right)\sqrt{\left(1 + \frac{\bar{u}^2}{< w^2 >}\right)} \equiv \alpha, \tag{10.21}$$

where α is a parameter defined by the equation. The kinetic energy of a particle is one-half the mass times the velocity squared. In terms of the U

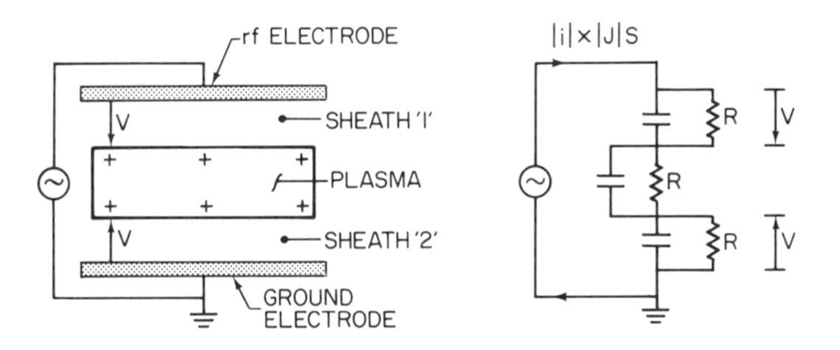

Fig. 10.8. Equivalent circuit for a parallel-plate etcher.

notation used above, U_x corresponds to $< w^2 >$ and U_y to $\bar{u}^2 + < w^2 >$. After some manipulation, we can show that

$$\rho^2 = \frac{2}{(1 + \sqrt{1 + (2\alpha)^2})}. \tag{10.22}$$

We now make the following substitutions based on results derived in statistical mechanics. We equate the mean free path with the reciprocal of the density of neutral scatterers, n_0, times the collision cross section, Q (i.e., $\lambda = 1/n_0 Q$). Next we recognize, from equipartitioning of energy, that $m_+ < w^2 > /2 = 3\text{k}T_+/2$. The m_+ refers to the ion mass and T_+ refers to its temperature. Thus:

$$\alpha = \left(\frac{e}{3\text{k}T_+}\right)\left(\frac{E_\text{s}}{n_0}\right)\left[\frac{1 + m_+/m_0}{Q}\right]. \tag{10.23}$$

These results indicate the importance of the parameter E_s/n_0 in the etching process. When this parameter is large, the directionality factor goes to zero and the etch becomes *anisotropic.* The parameter is large when E_s is large or when n_0 is small. The density of neutrals is, by Boyle's law, proportional to the pressure. Equation (10.23) also indicates that anisotrpic etching can be done at relatively high presssure if the sheath field is high.

The pressure can be set by adjusting the gas flows at the inlet to the etcher. However, what factors influence sheath field? As we might anticipate, RF power density certainly influences E_s. It is somewhat less obvious, but the RF frequency, ω, also influences this field. To see how this comes about, we first consider the equivalent electrical circuit off the etcher. Then, we write Ohm's law for this circuit. The equivalent circuit commonly used to represent a parallel-plate etcher is shown in fig. 10.8. The workpieces are placed on the grounded electrode (labeled with the subscript 2). Generally, this circuit can be further simplified. There are two limiting cases for the circuit. At low frequencies, the sheaths are modeled as resistors. At high frequencies, the sheaths are modeled as capacitors.

Now consider how Ohm's law would work for these systems. Consider the high-frequency case first. There are two components to the conductivity: a DC component and a displacement component (resulting from the shifting space-charge distributions associated with a sinusoidally varying field). Ohm's law, here, is

$$J_{\mathrm{rf}} = (\sigma + i\omega\epsilon_0)E, \tag{10.24}$$

where σ = local DC conductivity, ω = angular frequency of the exciting voltage, ϵ_0 = permitivity of free space, E = average electric field in the region of interest, and $i = \sqrt{-1}$. The conducitvity of the plasma is the sum of the conductiviteis of each of the charged species. The conductivity is given by the expression

$$\sigma_\pm = \frac{n_\pm e^2}{m_\pm \nu_\pm}, \tag{10.25}$$

where + and − refer to positive ion and electron coductivities, respectively (see problem 10.8). In the bulk of the plasma, the conduction is dominated by electrons. In the sheath, conduction is dominated by positively charged ions. When $\omega\epsilon_0 >> \sigma$, we have displacement current-dominated conductivity. When $\omega\epsilon_0 << \sigma$, we have the D.C. resistive transport case. Generally, the bulk conductivity of the plasma (including electron terms) is high and the D.C resistive case applies. Once again, assuming collisions with neutrals dominate, the collision frequency can be expresssed in terms of the neutral density, n_0, the collision cross section Q_+, the ion temperature, T_+, and ion mass m_+

$$\nu_\pm = n_0 Q_\pm \sqrt{\frac{\mathrm{k}T_\pm}{m_\pm}}\,. \tag{10.26}$$

These equations will allow us to solve for the mean sheath field under a wide variety of conditions, as shown below.

First, consider the parallel-plate etcher operating in a high-frequency regime. Here, displacement currents dominate in the sheath

$$E_s = \frac{J_{\mathrm{s}}}{i\epsilon_0\omega}, \tag{10.27}$$

where J_{s} is the RF current through the sheath. The average RF current drawn by the workpiece is generally not measured. The RF power coupled to the plasma, P_{rf}, is usually measured and read out on an equipment panel display. The RF power density is related to the RF current in the bulk plasma

$$\frac{P_{\mathrm{rf}}}{Sl_{\mathrm{p}}} = \frac{J_{\mathrm{p}}^2}{\sigma_-}\,. \tag{10.28}$$

The subscript p refers to plasma. The σ_- conductivity is used, since electron conduction dominates in the bulk plasma. The term on the left-hand side of eq. (10.28) is the RF power density (power per unit volume). The volume in question is just the volume between the parallel plates: Sl_p. S is the area of the plate and l_p is the plate separation. Note that the RF current through the bulk plasma is the same as the current through the sheath. Thus, using the D.C. transport bulk plasma condition:

$$J_{rf} = \sigma_- E_p \,, \tag{10.29}$$

and we can define an effective bulk field, E_p, in terms of the magnitude ratio

$$\frac{E_s}{E_p} = \left(\frac{\sigma_-}{\epsilon_0 \omega}\right) . \tag{10.30}$$

Combining eqs. (10.28) - (10.30) yields

$$E_s = \left(\frac{\sqrt{\sigma_-}}{\omega \epsilon_0}\right) \sqrt{P_{rf}/Sl_p} \, . \tag{10.31}$$

This equation yields the important result that the mean sheath field is proportional to the square root of the ratio of the RF power to the plate area. Thus, higher power densities (P_{rf}/Sl_p) cause more anisotropic etches. The higher sheath field can cause ion-sputtering of the workpiece.

A similar expression is obtained in the case of low-frequency operations. Only here, the DC conductivity dominates in determining the sheath current and $J_s = \sigma_+ E_s$. The magnitude ratio of the high-frequency sheath field to the low-frequency sheath field is given by $\sigma_+/\epsilon_0 \omega$. This number is usually less than 1. Thus, the low-frequency plasmas have higher sheath fields. The low-frequency plasmas are somewhat more anisotropic as a result.

The major trends indicated above hold for the RIE case as well as for the parallel-plate case. However, the equivalent circuit must be modified somewhat. The R_{s1} term is removed and the C_{s1} term is made very large. This is because the new C_{s1} is the result of the sheath associated with the chamber walls. We can view both the high-frequency and low-frequency equivalent circuits as kinds of voltage dividers. In the high-frequency case, we have a capacitive divider. The AC drop across this big capacitance is small. In the low-frequency limit, we remove R_{s1}. By removing the counterelectrode voltage drops, we increase the drop across the sheath surrounding the workpiece. This decreases the isotropy factor. In the limit of small R_p, the sheath field would double in both cases by removal of the counterelectrode.

Converting to the RIE case also entails the use of an RF coupling capacitor (see fig. 10.7c). What is the purpose of this addition? When the RF supply cycle swings positive, electrons will stream to the top plate of the capacitor (common with the workpiece electrode). This capacitor plate can only discharge by attracting positive ions from the bulk plasma. The

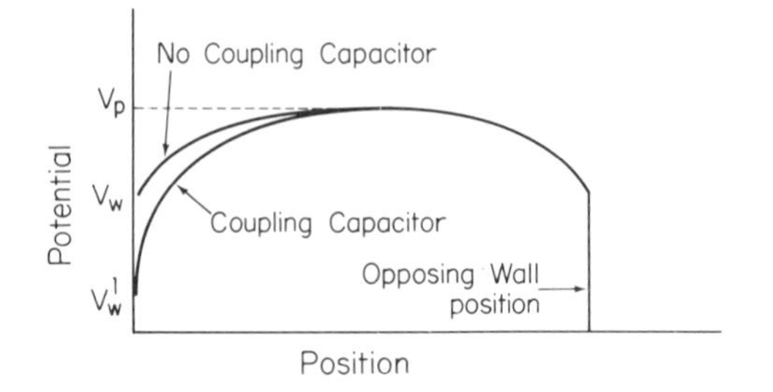

Fig. 10.9. Enhancement of electric field through capacitor coupling in RIE.

mobilities of these ions is much less than that of the electrons. Thus, the plate may retain some negative charge. This creates a workpiece bias effect. The workpiece electrode will assume some level of negative DC bias with respect to the plasma bulk. This is illustrated in fig. 10.9. This added bias enhances the mean electric field in the sheath and lowers the isotropy factor.

The above discussion emphasizes the role of the plasma etching tool in providing ion acceleration through the sheath. As a result, it provides a framework for understanding the impact of frequency, RF power, and background pressure on the vertical etch rate and anisotropy. Of course, the availability of the reacting species and the chemical state of the surface are also important factors. Such factors can and frequently *do* dominate the etch process. Coburn [21] demonstrated that hitting the surface with inert-ion beams activates the surface, increasing the etch rate and anisotropy factor. Other mechanisms for achieving anistropic plasma etches have been proposed. These are now discussed. The first mechanism is termed *side-wall passivation.* Here, either plasma-generated polymer films, or relatively unreactive chemical groups (such as chlorine ions), decorate all the surfaces in the etcher. Ions accelerated through the plasma sheath sputter away the film or unreactive groups, leaving the surface free for attack by more reactive species (such as flourine) or by neutrals. The sidewalls are parallel to the ion bombardment and the ion stream does not erode the passivating polymer. The polymer is continuously sputtered away at the surface perpendicular to ion acceleration. The eroded region etches much faster than the passivated sidewall. This is sometimes called the "peanut-butter" mechanism of anisotropic etching (the polymer, here, is the "peanut-butter".) This is illustrated in fig. 10.10.

A final mechanism that can create anisotropy in plasma etching involves energy transfer from the accelerated ion to the etched surface. Energetic ions incident on the surface can rearrange chemical bonds on the surface. These rearranged bonds react more readily then the untouched surface. This is termed *surface activation.* The sidewalls, once again, are not activated, as they do not receive energy from the ion bombardment. These effects are illustrated in fig. 10.11.

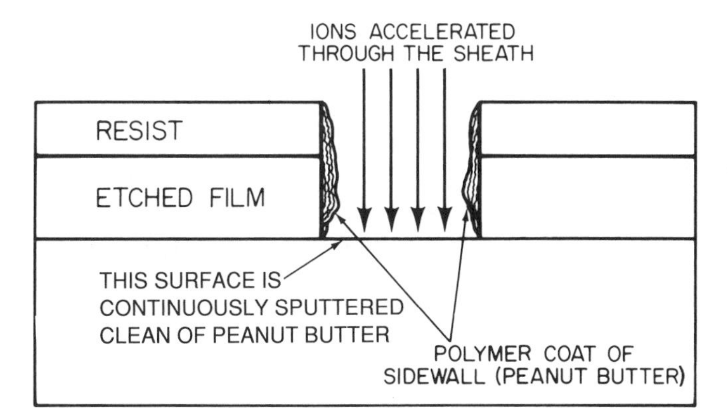

Fig. 10.10. "Peanut-butter" mechanism for anisotropic plasma etching.

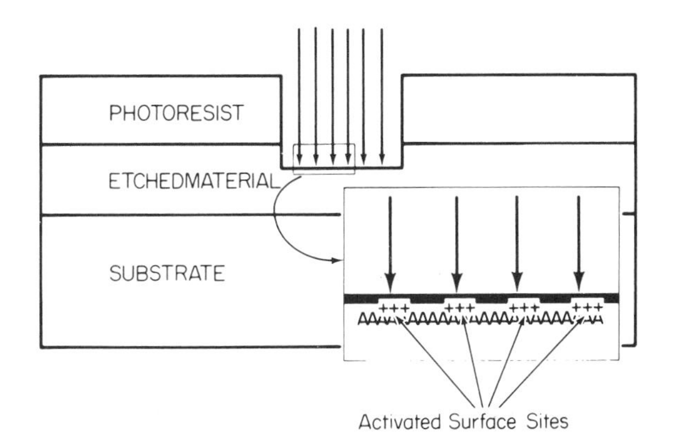

Fig. 10.11. Surface activation and etch anisotropy.

Of course, anisotropy is not the only concern of plasma etching. Etch rate and selectivity are also important factors in judging the success of any etch process. The types of atoms and molecules present in the discharge, their state of excitation, and their charge state are important in determining the etch rate of various materials. For example, increasing the oxygen content of a CF_4/O_2 gas discharge from 0 to 10% causes a fivefold increase in etch rate of Si. Increasing the oxygen concentration to 20% causes a similar increase in etch rate of SiO_2. Increasing oxygen concentration beyond these points causes a marked decrease in etch rate.

Mogab et al. [22] explain this as follows: The dominant species responsible for etching Si and SiO_2 is free atomic fluorine formed from the discharge-induced breakdown of CF_4. As in any reaction, the back reaction is also possible. CF_x molecules combine with free fluorine. Oxygen combines with CF_x to form COF_2, CO, and CO_2. These species cannot reattach free fluoride. As the oxygen concentration increases, the fluorine concentration diminishes due to dilution of the CF_4 in the feed stream. Mogab also points out that for Si etching, the etch-rate maximum does not correspond to the maximum plasma fluorine concentration that occurs when the O_2 level hits 20%. Oxygen can decorate the silicon surface, causing some degree of surface passivation. Larger oxygen concentrations in the feed gas stream cause more surface passivation and lower etch rates. This adds another variable to the process, which shifts the etch-rate maximum. Since the SiO_2 surface is always saturated with oxygen, the etch rate of this material maximizes at 20% (free-flourine maximum point.) Increasing the oxygen concentration from 0 to 10% in the CF_4 feed stream may change the polysilicon etch rate from 1000 to 5000 Å/sec. At the 10% level for such a discharge, the SiO_2 etch rate increases from about 40 to 250 Å/sec. Thus, the selectivity of the etch also improves.

Adding hydrogen to the CF_4 stream will reduce the silicon etch rate without reducing the SiO_2 etch rate significantly [23]. The chemical mechanism for this effect is presently unclear. Possibly, nonvolatile silicon hydrides (SiH_x) form to passivate the surface and prevent etching [18]. In any event, addition of H_2 to CF_4 makes a plasma etch that favors SiO_2 over Si. Such passivating hydrides apparently do not form in this case. Spectroscopic studies of the discharge that determine the type and density of the various atomic and molecular species present are necessary to further evaluate the mechanism. The broad range of possible chemical-based etch-rate and anisotropy enhancement mechanisms should give the reader an indication of the limitations of the Zarowin model.

10.6 Applications of Anisotropic Plasma Etching

In the past, wet chemistry was the workhorse of the industry for pattern-transfer technology. The applications of anisotropic wet etches for silicon were outlined above. In addition, we have already discussed the resist-stripping properties of the barrel etcher and the ability of the barrel etcher to etch polysilicon in a controlled way. In most VLSI applications, the polysilicon self-aligned gate is the feature with the smallest resolved dimension. As of this writing, the industry is attempting to create processes for 1/2-μm-length MOS transistors. Obviously, if the poly is 1/2-μm thick, it would be impossible to etch such a line with an isotropic process. Usually, vertical sidewalls are desired at this phase of the process. Since the poly must be covered with a thin-film insulating layer, it might be thought that a step-coverage problem would result. This does not occur since CVD oxides are conformal. In addition, the steps to be covered are not too high (almost always less than 1/2μm) at this point in fabrication. Thus, the processes that offer the highest degree of anisotropy (well-controlled planar or RIE etching) are selected for gate definition. Furthermore, many novel processes are run using anisotropic plasma etches. Some of these processes are outlined below [24]. Specifically, RIE processes for sidewall spacers, for dielectric planarization, and for trench isolation are presented.

First, we consider the formation of sidewall spacers. As discussed in Chapter 6, self-aligned silicides are used to simultaneously contact shallow junctions and to lower polysilicon sheet resistivity. A silicide former, like cobalt, is deposited over the whole wafer. Silicon is exposed only at the tops of the junctions to be contacted and at the polysilicon lines. The rest of the surface is oxide coated. The cobalt is heated and the silicide forms only where the cobalt contacts with bare silicon. Unreacted cobalt is etched away in cobalt etch. It is desirable to keep silicide from forming over the sidewall of the polysilicon. Silicide, when formed, can grow away from the poly line and silicide can come up from the contact and create a gate-to-drain short. If the side of the polysilicon can be oxidized, without oxidizing the top, an oxide spacer would form, which helps prevent shorting.

Next we ask, how are such sidewall spacers made? A sample process flow making use of anisotropic etching is shown in fig. 10.12. After the polysilicon is deposited and defined, the polygates are oxidized. (The process also can be designed using a conformal CVD-deposited oxide film.) Next, the wafer is placed in an anisotropic etcher. The oxide overcoat is etched. The etcher thins all surfaces perpendicular to the direction of the incident ions at the same rate. This removes the oxide from the top of the polysilicon as well as from the region over the silicon substrate. The vertical height of the oxide in the sidewall region is bigger than the vertical height of the oxide in regions above the poly or silicon substrate (away from the spacer). After etching, the spacers remain. These spacers help contain the region of silicide formation and prevent shorting.

If the initial source-drain implant is a weak-doping step (making an n-layer), the spacer fabrication can be followed by a heavier, n^+ deposition step. This is shown in fig. 10.13. In this way, the active channel under the polygate is contacted by a resistive n^- layer. The potential drop across this layer lowers the electric field in the drain region. This low-doped drain (LDD) structure is used to reduce hot-electron injection from the high-field drain into the gate oxide.

Plasma etching is also used to define small contacts and vias as well as aluminum (or aluminum alloy) metal lines. The contact windows or vias should not have vertical sidewalls, though. This would lead to a metal-step coverage problem. The contact and via etch should contain isotropic as well as anisotropic components. Metal lines create a particular problem. First of all, to etch aluminum with any degree of isotropy requires a heavily chlorinated etchant (as discussed below). Chlorine gas is highly toxic and it corrodes machine parts. In addition, aluminum alloys used in IC processes frequently contain copper. Copper cannot be etched with most gases used in plasma etching. RIE is required to create the highest energy directional ion flux. This flux can sputter away the unwanted copper. Aluminum chlorides

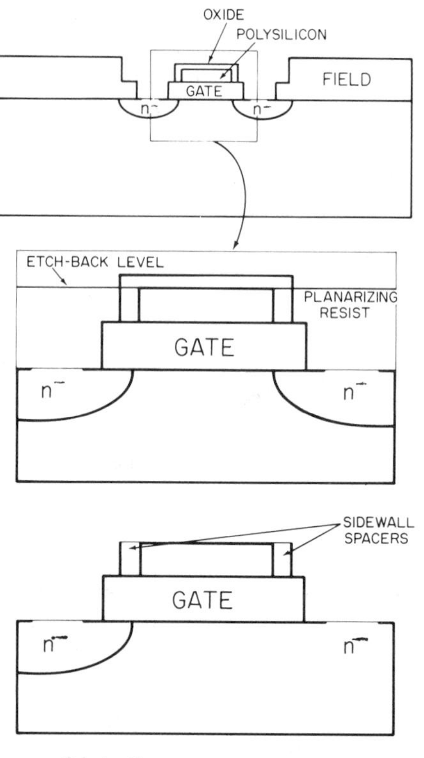

Fig. 10.12. Sidewall-spacer fabrication.

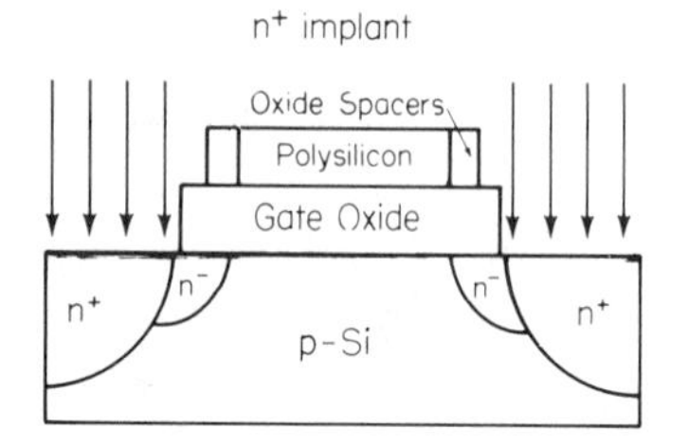

Fig. 10.13. The low-doped drain (LDD) structure.

are not all carried away in the gas phase. Such chlorides create HCl on exposure to moist air. This is both a safety hazard for the machine and a reliability hazard for the device. Both the machine and the device must be cleaned after frequent exposure to chlorine.

Plasma processes can also be used to planarize insulation surfaces. The importance of planarization for high-resolution lithography was stressed in the last chapter. Consider the structure shown in fig. 10.14. A metal line is coated with a thick, conformal silox overcoat. This creates a highly non-planar surface. Resist of planarizing polyimide is spun over the surface. A

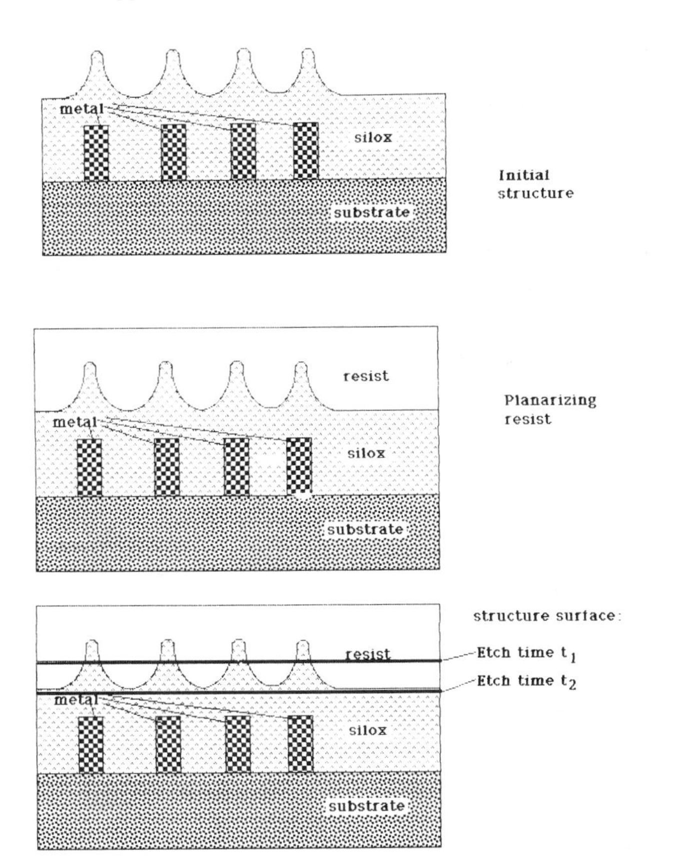

Fig. 10.14. Schematic view of planarization processes using a resist planarizing layer.

plasma etch that etches resist and polyimide at the same rate as silox is then used. The etch proceeds until all the resist or polyimide is gone (as shown in fig. 10.14). This results in a planar surface. As a final application of anisotropic etching, we consider the use of deep trenches in silicon. Such trenches can be more than 2μm deep and less than a micron wide. These structures have obvious application as isolation regions in CMOS processing. Typically, the surface of the etched trench is coated with a thin insulator and the whole trench is refilled with undoped polysilicon. It should be noted that the deep trench creates an effective increase in silicon surface area through "corregation,"or texturing, of the surface. Devices can be fabricated on the sidewalls of the trenches. For example, if the polysilicon deposited to fill the trench is conductive polysilicon, a capacitor can be formed. Such capacitors can be of very high value without increasing chip area. This is extremely useful in dynamic RAM technology [25]. In addition, active devices such as transistors have been fabricated on trench walls [26]. This is one approach to three-dimensional structure technology.

10.7 Summary

To conclude, the key points raised in this chapter were:

- Wet chemical etching is an inexpensive technique for achieving pattern delineation in a variety of materials with good selectivity. The technique is usually isotropic. This leads to unacceptible undercutting in near-micron line definition.
- There are some anisotropic wet etches. These etches will attack different crystallographic planes at different rates. This is the mechanism that gives rise to the anisotropy. Thus, only crystals (such as silicon substrates) can be etched in this way.
- Both parallel-plate and RIE plasma etchers can create anisotropic etches in both crystals and glasses. The main mechanism responsible for the anisotropy is the lowering of the reaction activation barrier in a direction perpendicular to the etched surface. The activation barrier is lowered by the energy the ion receives on acceleration through the plasma sheath.
- Other mechanisms can create anisotropy in plasma etching. Sidewall passivation by unreactive species or by polymerization products of the plasma can create anisotropy. In addition, surface activation caused by ion bomabardment can also create anisotropic etches.
- Plasma etching frequently is not as selective as wet etching. This may necessitate some form of endpoint detection included in the etcher.

References for Chapter 10

1. H.G. Hughes, M.J. Rand, eds., "Etching for Pattern Definition," Electrochemical Society Press, Princeton, NJ (1976).
2. W. Kern, C.A. Deckert, Chemistry etching, in "Thin Film Processes," J.L. Vossen, W. Kern, Academic Press, New York pp. 481-496 (1978).
3. J.S. Judge, The etching of thin film dielectric materials, in "Etching for Pattern Definition," H.G. Hughes, M.J. Rand,. eds., Electrochemical Society Press, Princeton, NJ (1976).
4. J. Dey, M. Lundren, S. Harrell, "Proceedings of the Kodak Photoresist Seminar," Vol. II, p. 4 (1968).
5. A.S. Tenney, M. Ghezzo, J. Electrochem. Soc. 120, 1091 (1973).
6. C.A. Decker, J. Electrochem. Soc. 124, 320 (1970).
7. H. Robbins, B. Schwartz, J. Electrochem. Soc. 106, 505 (1959).
8. H. Robbins, B. Schwartz, J. Electrochem. Soc. 107, 108 (1960).

9. B. Schwartz, H. Robbins, Chemical etching of silicon IV, etch technology, J. Electrochem. Soc. 123, 1903 (1976).
10. G. Kaminsky, Micromachining of silicon mechanical structures, J. Vac. Sci. Technol. B, 3 & 4, 1105 (1985).
11. D.L. Kendall, Vertical etching of silicon at very high aspect ratios, Ann. Rev. Mater. Sci. 9, 373 (1979).
12. K.E. Bean, IEEE Trans. ED-25, 1105 (1978).
13. A. Reisman, M. Berkenblit, S.A. Chan, F.B. Kaufman, D.C. Green, The controlled etching of silicon in catalyzed ethylene diamine-pyrocatechol-water solutions, J. Electrochem. Soc. 126, 1406 (1979).
14. R.M. Finne, P.L. Klein, A water-amine-complexing agent system for etching silicon, J.Electrochem. Soc. 114, 967 (1967).
15. N.F. Raley, Y. Sugiyama, T. Van Duzer, J. Electrochem. Soc. 131, 161 (1984).
16. E.D. Palik, V.M. Bermudez, O.J. Glembocki, Ellipsometric study of etch stop mechanism in heavily doped silicon, J. Electrochem. Soc. 132, 135 (1985).
17. P.D. Chapman, "Glow Discharge Processes," John Wiley and Sons, New York (1980).
18. C.M. Melliar-Smith, C.J. Mogab, Plasma-assisted etching techniques for pattern delineation, in "Thin Film Processes," J.L. Vossen, W. Kern, Academic Press, New York, pp. 497-555 (1978).
19. M. W. Jenkins, M.T. Morella, K.D. Allen, H.H. Sawin, The modeling of plasma etching processes using a response surface methodology, Sol. St. Tech, 29, 175 (1986).
20. C.B. Zarowin, Mechanisms of anisotropic plasma etching, J. Electrochem. Soc. 130, 1144-1152 (1983).
21. J.W. Coburn, H.F. Winters, Ion-and-electron-assisted gas-surface chemistry—An important effect in plasma etching, J. Appl. Phys. 50 (5), 3189 (1979).
22. C.J. Mogab, A.C. Adams, D.C. Flamm, Plasma-etching of Si and SiO_2—The effect of oxygen additions to CF_4 plasmas, J. Appl. Phys. 49, 3769 (1978).
23. L.M. Ephrath, Selective etching of silicon dioxide using reactive ion etching with CF_4/H_2, J. Electrochem. Soc. 126, 1419 (1979).
24. J. Dieleman, R.G. Prieser, eds., "Plasma Processing," Vol. 82-6, Electrochem. Soc. Press, Pennington, NJ (1982).
25. M. Sakamoto, et al., Buried storage electrode (BSE) cell for megabit DRAMS, Proc. IEEE IEDM, IEEE Press, New York, pub. # 85CH2252-5, pp. 710-713 (1985).
26. W.F. Richardson, et al., A trench transistor cross-point DRAM cell, Proc. IEEE IEDM, IEEE Press, New York, pub. # CH2381-2, pp. 714-717 (1986).

Problem Set for Chapter 10

10.1 Consider a string model for determining etch profiles. Such a model is accomplished in the following way: First, graphically (either manually or by computer) create a cross-section plane that will eventually contain the desired profile. Break the surface exposed to the etch up into equidistant points. For an isotropic etch, move each point in a direction normal to the string that connects the individual points. The distance moved should be equal to the local etch rate of the material attacked multiplied by some small increment of time. Repartition the newly exposed surface into equidistant lines and begin the process again. For the case illustrated below

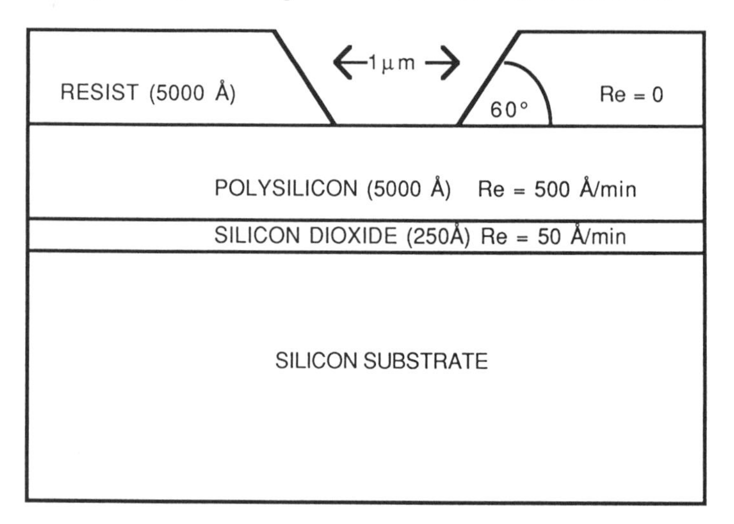

derive the etch profile at $t = 5$ minutes, 10 minutes, and 15 minutes.

10.2

a. Propose a method for modifying the string-model approach described in problem 10.1 to account for etch anisotropy.
b. Again, consider the films depicted in problem 10.1. Assume etching of the resist and oxide proceeds isotropically, but the polysilicon etches with an isotropy factor of 0.2. Furthermore, assume the oxide etches at at a rate of 50 Å/min. Derive etch profiles for the times listed in the problem above.

10.3 Consider the etching of silicon in some isotropic etch. The statement is made that the wafer etches faster near the edge if the reaction rate limit is determined by the diffusion of active species from the etch bath. Conversely, the wafer etches faster near the center if the reaction is autocatalytic. Explain.

10.4

a. Using the Zarowin model as a guide, explain, qualitatively, the effect of RF power density, RF frequency, and background pressure on etch rate and on etch anisotropy.

b. Describe the effect of oxygen and hydrogen on the plasma etching of silicon with freon-based plasma etches.

10.5 How would increasing the surface temperature affect anisotropy a) assuming the Zarowin model, b) assuming the surface passivation process is operating, and c) assuming the surface activation process is operating?

10.6 Frequently, the etch rates (in both wet and dry etch processes) depend on the number of wafers etched at a given time. This is called a "loading effect." Describe, qualitatively, how this effect comes about. Derive a first-order model that quantitatively represents the influence of loading on etch rate.

10.7 As in any chemical reaction, the equilibrium concentration of an active species is the result of a competition between species-creation and species-destruction processes. Explain how the consequences of this concept can affect etch rates and isotropy factor in reactive-ion etching and in wet-etching.

10.8 Many assumptions go into the derivation of equations 10.29–10.31. For example, consider the DC conductivity expression (eq. 10.25). This can be derived from Newton's Law:

$$m\frac{dv}{dt} = eE - m\nu v,$$

where the first term on the right is the standard electric force and the second term is a "collision force" term. Assuming $v = v_o \exp -(iwt)$, derive 10.25 and state the nature of the approximation you use to achieve the form of eq. 10.25.

11 Future Beam Processes and In-Situ Processing

11.1 Introduction

In the last decade and a half, several significant developments were made that led to a continued miniaturazation of semiconductor devices, an increase in the device complexity, and speed. Simultaneously the number of devices on a chip grew from a few thousand to more than a million. The success is associated with the radical developments in processing such as (a) ion-implantation doping (Chapter 5), (b) dry etching (Chapter 10), (c) multilevel resists and new exposure systems (Chapter 9), (d), built-in redundancy and laser repairs (Chapter 7), (e) electron-beam lithography and mask making (Chapter 9), (f) polycide gate and interconnection (Chapter 7), (g) low-pressure chemical vapor deposition (Chapter 6), and (h) planarization (Chapter 10). In spite of these successes, which have led to the manufacturing of devices with near one-micrometer dimensions, further shrinking of dimensions to considerably less than one micrometer and incorporation of clever circuits on a chip are facing an impasse due to both the process and materials limitations. Many of the newly proposed devices and circuits require unconventional processing techniques. Laboratory experiments have been carried out by various researchers to meet the challenge by the use of processes that utilize beams, as is evident from a large number of publications and presentations [1]. In addition, the issues of automation [2], human-free fabrication [3], particle-free fabrication [4], etc. have been raised. All these have led to a serious consideration of new processing concepts and the concept of *in-situ processing*.

The words *in-situ processing* have meant different things to different people. Some explain it as holding wafers in a fixed position and bringing processes or process tools to them. Others link in-situ processing to an ultra-high vacuum system. Still others call it a new name for automation. In-situ processing means all of these and more, and can be defined as:

Processing of devices and integrated circuits in a single particle-free enclosure (or a multiple interconnected chamber) free from direct human involvement.

The in-situ processing concept differs from computer-aided automation of IC processing. The new concept is born out of the thrusts of using new processing ideas in such a way as to minimize the number of processing steps by carrying out localized and direct processing. For example, thin-film deposition followed by lithography and etching sequence could be carried out by direct deposition of the desired pattern, thus obviating the need for the last two steps. The lithography step itself consists of several individual processing steps, namely, the preclean, prebake, application of photoresist, postbake, alignment and exposure, resist development, cleaning, and postbake.

In a typical NMOS VLSI effort, process steps could number as many as 150, whereas a typical CMOS process may have over 200 processing steps. Many of these steps have been introduced to maintain the required cleanliness and minimize particulate contamination. For example, chemical cleaning, using a sulfuric acid and hydrogen peroxide [5] mixture or a RCA clean (which in itself involves about 15 steps) [6] followed by a 100:1 (water:HF) dip and deionized water rinse, is used before and after each major step — doping, lithography and etching, anneal, thin-film deposition, gettering, and even major oxidation. A 100:1 or other HF solution will etch away small amounts of existing silicon oxide on the wafer, causing line-width and thickness loss. In addition, such cleaning steps may introduce harmful effects in thin dielectrics and metallization. Thus far, the processing has tolerated these by allowing for these losses in the design and ignoring less critical effects. For device dimensions less than $0.5\mu m$, such process-induced loss of control may not be tolerated. Also there are new devices and concepts and custom-made devices that will require, by necessity, extremely finer controls in dimensions and in processing.

One must also consider the role of particulates, bacteria, and viruses affecting the device processing. Particulates, generated during any given process or present in the surroundings of the wafers, are the reason for a large fraction of the loss of the yield reduction [7]. They make the fabrication of high-density and larger size circuits with an economically acceptable yield very difficult. This has forced the industry to create clean rooms with as low a particle count as possible. The cleaner the rooms, the higher are the initial capital investment and their day-to-day maintenance cost. Equipment, processes that use gases and liquids, and, most importantly, humans create particulates and contamination of all kinds. Table 11.1 lists the sources of process-induced particles. Humans are perhaps the worst offenders, whereas gases create the least number of particles. A study in 1982 [3] showed that 46% of the particles generated in IC processing were human-related. Liquids, although greatly improved within the last few years, still carry a large number of particles [8]. Recent advances in gas filtration technologies have brought down the particulate contamination in gases to such low levels that the filtration efficiencies of 99.9999997% are not uncommon [9].

In addition, other particulates one must consider are the bacteria and viruses, whose sizes range from tens of angstroms to a few micrometers. Figure 11.1 compares the minimum feature size with the killing defect size [10]. It is apparent that heretofore-ignored defects — bacteria and viruses — could play an important role in determining the yield.

TABLE 11.1: Sources of particulate generation

1. Humans
2. Liquids
3. Mechanical motion
4. Chemical vapor deposition
5. Other thin-film deposition techniques such as evaporation, sputtering, or plasma depositions
6. Reactive ion etching and plasma etching
7. Rubbing of the boats and boat carriers in the furnace tube
8. Turbulence during pumping and backfill
9. Gases

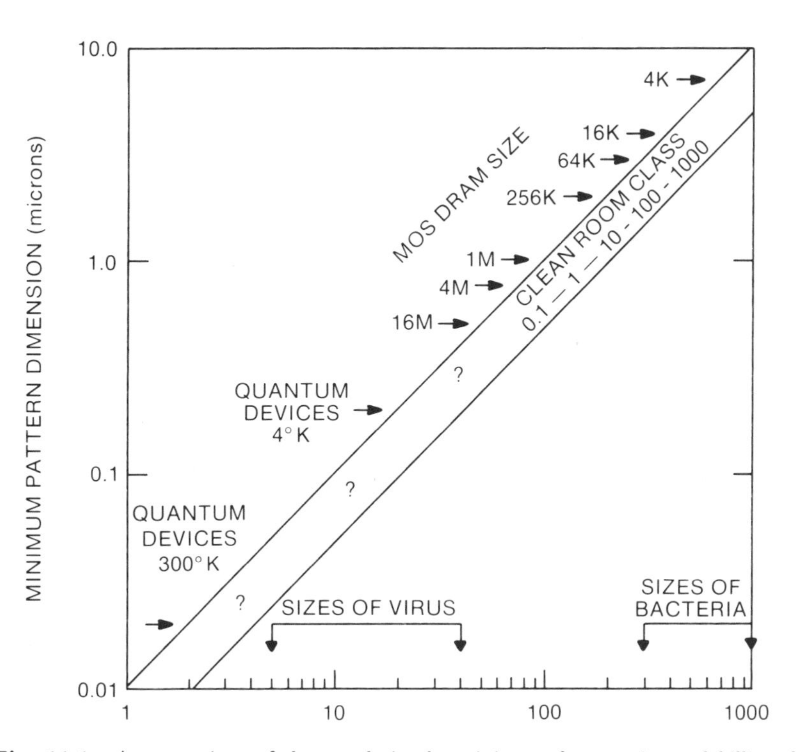

Fig. 11.1. A comparison of the trends in the minimum feature size and killing defect size, showing the importance of organic bacteria and virus as killing defects. From Larrabee [10].

In this chapter we discuss the new processes based on the use of beams of electrons, ions, photons, and neutral atoms and molecules. Some of the processes, e.g., ion implantation, reactive ion etching, rapid thermal annealing, etc. have been discussed in various chapters. Finally, in-situ processing will be examined.

11.2 Beam Processes

Beams of all kinds can be grouped in various ways: charged particle beams versus atomic, molecular, or photon beams; large-area or broad beams versus focused beams, etc. For semiconductor processing, all types of beams may be used. For this type of use two groupings are especially useful.

For charged particles, such as ionized atoms or molecules and electrons, broad or focused, acceleration using desired electric potentials and scanning and focusing using a combination of electrical and magnetic fields are very conveniently accomplished. Neutral particle beams, except for photons, cannot be focused or accelerated. Photons can only be focused by the use of conventional optical means. Scanning by photons has been carried out by oscillating reflective optics or accoustically coupled mechanical movements. Since photons cannot be accelerated, their use as penetrating beams is limited to high-energy photons such as x-rays and γ-rays. Charged beams can be accelerated to energies corresponding to desired penetration depths, limited only by the availability of accelerating potentials. Photon sources such as lasers (see below), on the other hand, can be used to heat the material and, therefore, can be used to cause thermally induced chemical or metallurgical reactions and phase transformations.

11.2.1 Laser-Assisted Processing of Semiconducting (LAPS) Materials

Laser-assisted processing of semiconducting (LAPS) materials has attracted considerable interest in the last few years, particularly because of the increasing complexity and density on the chip. LAPS has been seen as an alternate to existing processes, not only to help implement these complicated integrated circuits, but also to simplify the process [11,12]. LAPS could cover practically all aspects of processing: cleaning, thin-film deposition of insulators, conductors, and semiconductors, annealing, doping, etching, photolithography, etc., and LAPS could be either a large area processing or a focused-beam processing technique. Figure 11.2 lists various processes plotted according to photon fluxes and peak temperatures associated with them. Thus, for example, pure photochemical processes such as laser etching of GaAs in bromomethane or polymerization of PMMA (polymethyl methaacrylate) lead to a temperature rise of about $1 - 10°$ K, whereas for

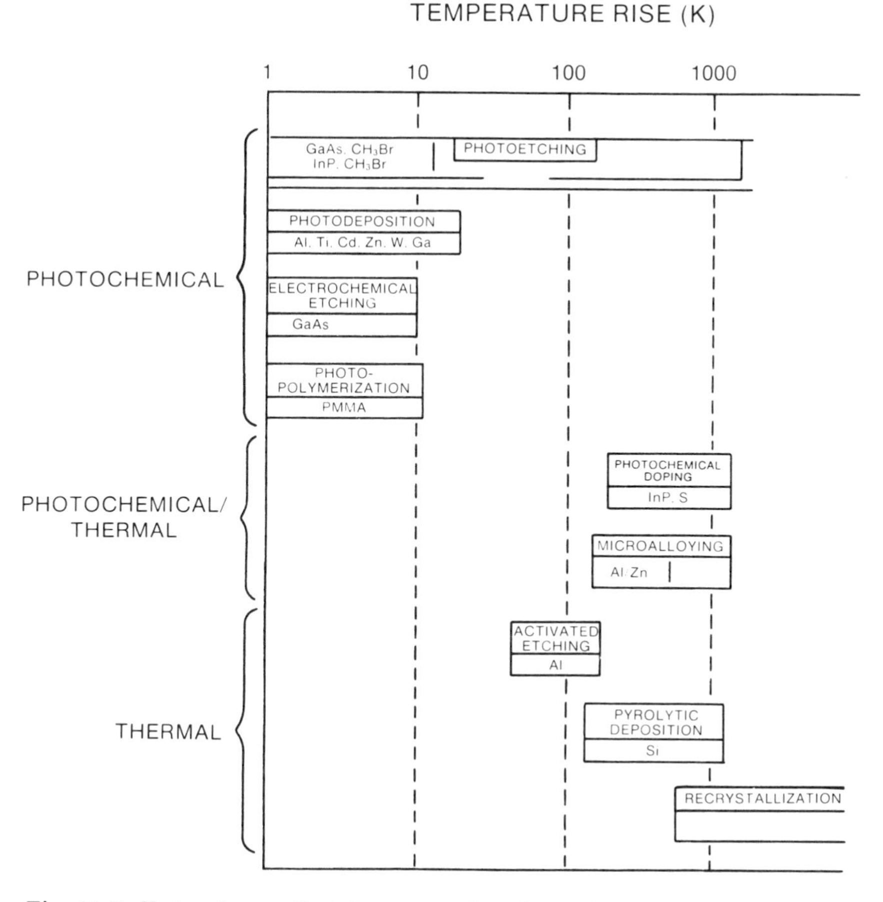

Fig. 11.2. Various laser activated processes plotted according to the rise in temperature associated with such processes. From Ehrlich, Tsao [11].

pure thermal processes temperature rise could be as high as 1000-2000° K. One, therefore, must be very sensitive and careful in selecting a process.

LASER is the acronym for Light Amplification by Stimulated Emission of Radiation. The laser is a device that utilizes the natural oscillations of atoms for amplifying or generating electromagnetic radiation at wavelengths in the range of very long infrared to the deep ultraviolet and even x-rays. Lasers, in a very general sense, generate or amplify light just as vacuum tubes or solid-state devices amplify electronic signals.

Two distinctly different types of lasers have been employed for the processing of semiconductor materials. Pulsed lasers (usually referred to as Q-switched lasers) [13] irradiate samples with very-high-energy, short-duration (5-ns to 500-ns range) pulses. Nd: YAG and Ruby Lasers fall in this category. When irradiating substrates with such lasers, the energy is generally absorbed in surface layers and, in many cases, depending on the energy absorbed, localized melting occurs. This is followed by recrystallization as soon as the laser is switched off or moved away from that location.

For pulsed lasers, with very small exposure times, the temperature rise in the irradiated surface can be calculated by assuming that absorption depth is small compared to the heat diffusion length $(D\tau)^{1/2}$ where D is the thermal diffusivity and τ laser irradiation or dwell time. If ΔT is the temperature rise, C_v and ρ are the specific heat and density of the substrate respectively, then [14]

$$P(1 - R) = \rho C_v \Delta T (D\tau)^{1/2}, \tag{11.1}$$

where P is the laser power that is the product of its intensity and τ, and R is the reflection coefficient of the substrate surface. In deriving eq. 11.1 it is assumed that no other losses of energy occurred. If the substrate temperature was T_0 and the temperature after laser irradiation of duration τ becomes T_R, then $\Delta T = T_R - T_o$ and

$$T_R = T_o + \frac{P(1 - R)}{\rho C_v (D\tau)^{1/2}}. \tag{11.2}$$

If K is the thermal conductivity of the substrate then $D = K/\rho C_v$ and

$$T_R = T_o + \frac{P(1 - R)}{(K\rho C_v \tau)^{1/2}}. \tag{11.3}$$

Since K is a temperature-dependent quantity, substrate temperature T_o will influence the second term in eq. 11.3 as well.

Continuous wave or CW lasers, such as an Argon laser, irradiate substrate with lower energy for a long duration (0.1 ms to as much as a few tens of seconds). In this case the energy absorbed is generally not enough to melt the substrate (silicon), and all transformations occur by solid-state processes. CW lasers are most suitable for semiconductor processing. A typical laser processing setup is shown in fig. 11.3 [14]. As shown, the setup consists of a laser, beam-related optics, and a sample table that is computer driven so that the sample can be scanned in the x-y direction.

Scanning the sample leads to a localized temperature increase in the sample. An accurate knowledge of the temperature is essential in laser processing of materials. For CW lasers with a dwell time of the order of milliseconds, a thermal equlibrium in the substrate can be assumed. This enables the calculation of surface temperature rather simpler. The maximum linear surface temperature at the center of a circular beam of radius W has been calculated to be given as [15]

$$T_{\max} = \frac{1}{2\sqrt{\pi}K(T_0)} \left[\frac{P(1 - R)}{W}\right], \tag{11.4}$$

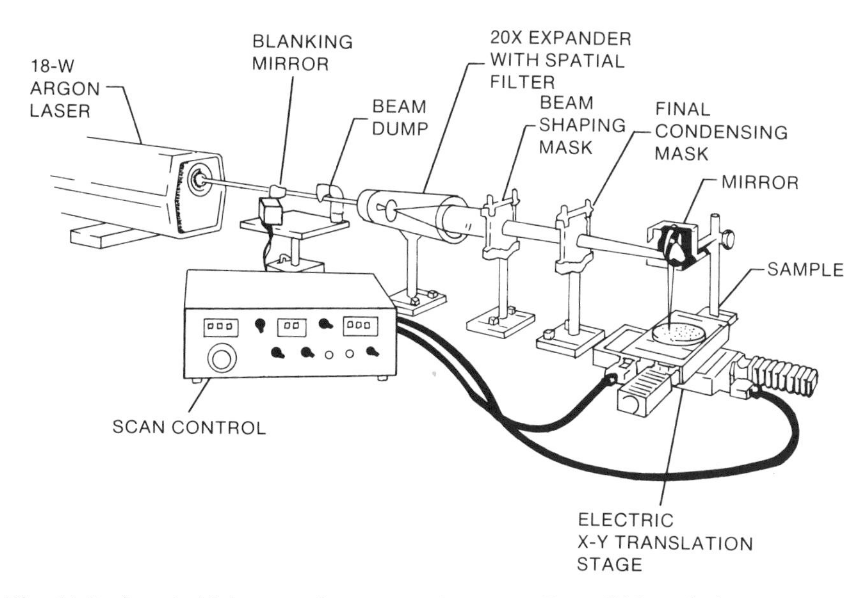

Fig. 11.3. A typical laboratory laser processing setup. From Gibbons [14].

where T_0 is the substrate temperature far away from the irradiated surface (or so-called substrate back-surface temperature) so that $K(T_0)$ is the thermal conductivity at T_0. In deriving this equation several assumptions were made: a) the laser beam is stationary, b) the wafer is semi-infinite, and c) the thermal absorption depth is zero. For most laser annealing experiments these assumptions are satisfactory.

Figure 11.4 shows calculated maximum surface temperatures of various semiconductors as a function of the quantity $[P(1-R)/W]$ and the substrate back-surface temperature T_0. It is apparent that the efficiency of the lasers in heating the substrate is significantly better at higher substrate temperatures. This is because the thermal conductivity of the substrates decrease with increasing temperature, resulting in lower heat losses.

11.2.1.1 Laser Cleaning of Surfaces

The semiconductor surface carries a native oxide (0.5-3nm thick). In addition, there may be surface contaminants before or during processing. Raising the temperature of the silicon wafers to around 900°C or higher in a vacuum has been used to remove the native oxide and organic contaminants. In the absence of oxygen and moisture, the oxide interacts with underlying silicon and produces SiO vapor that is pumped out. A few angstroms of silicon is lost from the surface, leaving the surface clean. Lasers could be used to raise the temperature to induce SiO formation and thus to clean the silicon

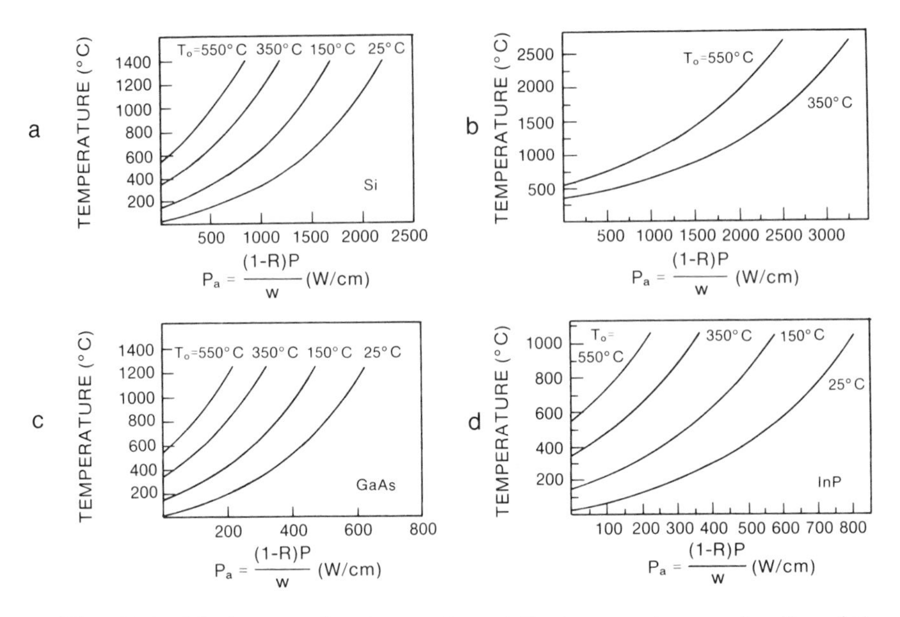

Fig. 11.4. Maximum surface temperature at the beam center as a function of the absorbed power per unit radius for (a) silicon, (b) silicon carbide, (c) gallium arsenide, and (d) indium phosphide. From Lietoila et al. [15].

surface. The process is characterized carefully to eliminate production of defects associated with thermal gradients.

Laser-assisted etching of thicker SiO_2 films, however, has been the most difficult task. It has been concluded, "As a rule, insulators tend to be much more inert than either semiconductors or metals and, therefore, more difficult to etch"[11]. However, this can be used to our advantage, since a high selectivity is required when etching Si in the vicinity of SiO_2.

11.2.1.2 Laser-Assisted Annealing

Annealing of semiconductor materials was perhaps the first application of lasers. Most of the earlier work focused on recrystallization of the polysilicon films, silicide formation, and annealing of ion-implantation damage. A large number of research and review papers have been published in the last 15 years [1,14]. A variety of results are obtained depending on the laser power utilized. For example, fig. 11.5 describes the general behavior in the CW laser-annealed polysilicon films [16]. Low-, medium-, and high-power levels in these experiments correspond approximately to < 5, 7-9, and 11-12 watts, respectively. Thus if only limited dopant activation is desired with no dopant distribution and grain growth in polysilicon, low-power anneal will be appropriate. On the other hand, if an increase in the grain size, uniform dopant distribution, and complete activation are desired, high-power anneal will be

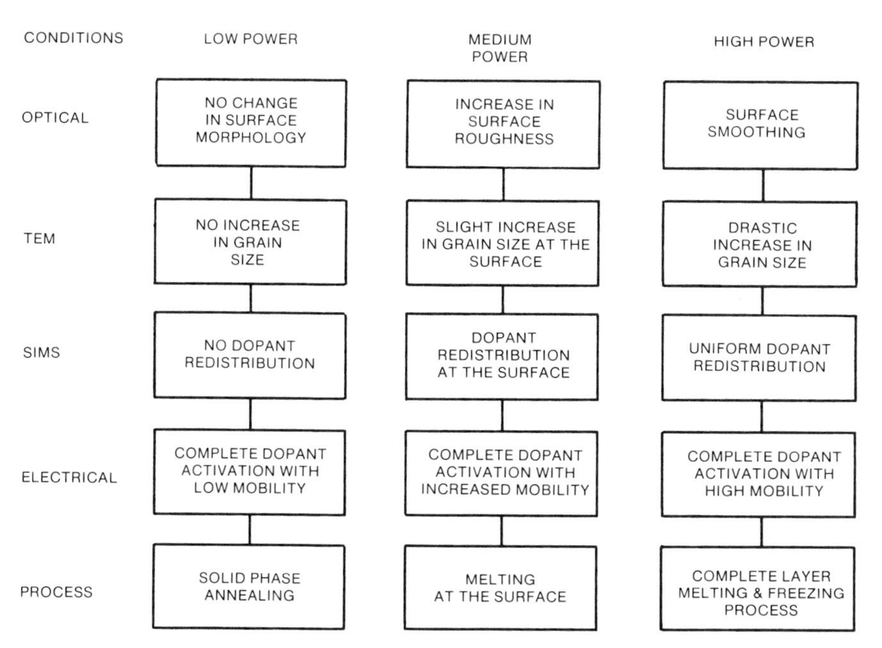

Fig. 11.5. Summary of the effects of laser power on the annealing of polysilicon films and the corresponding measurement methods. From Lee, Stulz, Gibbons [16].

necessary. Similarly, a variety of results can be obtained in laser annealing of ion-implanted silicon. A 100% activation of dopants in silicon can be achieved without affecting a dopant redistribution. This is very important for the formation of shallow junctions.

Ion implantation of material leads to amorphization. Annealing of dopant-implanted semiconductors leads to regrowth of this amorphized region and to dopant activation. Table 11.2 compares the characteristics of ion-implanted silicon annealed in a furnace with those annelaed by pulsed or CW laser [17]. Furnace and CW-laser anneals produce nearly identical results following a solid-phase regrowth mechanism. Pulsed-laser anneals produce different results because of a liquid-phase regrowth mechanism. Figure 11.6 schematically explains the two mechanisms [17]. In the solid-phase regrowth process, the crystalline amorphous interface movement is related to the solid-state diffusion and hence to the point and other defect concentrations, stored energy, temperature, and time. Figure 11.6a shows an amorphized silicon and crystalline silicon interface. Figure 11.6b shows the motion of this interface towards the surface. Solid-phase regrowth often is not ideally perfect and leaves behind a variety of defects whose type and distribution depend on the substrate, initial damage, and annealing conditions. Figure 11.6c shows two types of liquid-phase regrowth. The first one is associated with the anneal when deposited power was not sufficient to melt all (or more) of the amorphized region, so called below threshold. Thus in this case

TABLE 11.2: Comparison of furnace and laser annealing effects in ion-implanted silicon. From Williams [17]

	FURNACE	CW LASER	PULSED LASER
Regrowth Mechanism	Solid-Phase Epi	Solid-Phase Epi	Liquid-Phase Epi
Anneal Parameters	Regrow takes place ar >550°C. Regrowth quality depends on time/ temp.	Regrowth quality depends on power/ dwell time.	Single crystal recovery above threshold power. Poly below threshold.
Dose Dependence	i) High doses regrow poorly. ii) For low doses of dopant, regrowth rate increases with dose.	Same as furnace	Essentially no dose dependence
Orientation Dependence	Poor regrowth for (111): best regrowth for (100)	Same as furnace	No orientation dependence
Microstructure	Never completely defect free: some dislocations (twins in (111)).	Lower dislocation density than furnace	Perfect crystal: extended defect-free
Implant Profile and substitu-tionality	No redistribution below solubility limit (<850°C). Normal diffusion above ~850°*C*. Grain boundary outdiffusion and "push out" above solubility limit	No redistribution below solubility limit. Same as furnace above solubility limit	Always redistribution within melt. Zone refining Substitutionality exceeding equilibrium values
Surface Topography	Featureless	Usually featureless but surface slip and cracking for "overannealing" and (111)	Appearance of frozen liquid: gross surface damage at excessively high laser powers

melt depth does not penetrate into the crystalline silicon and no seeding for epitaxy occurs. This leads to a polycrystalline layer formation on top of the amorphous layer. In the second case, the deposited energy is enough to melt all of the amorphized layer and more. Thus liquid depth penetrates into the crystalline substrate, allowing epitaxial seeding by the substrate during cooling. In both cases, the crystallization proceeds from the melt-substrate interface up towards the surface. The liquid-phase regrowth is, however, very short lived, short time being associated with submicrosecond duration of the pulsed lasers.

Table 11.3 summarizes the laser process parameters and silicides formed by CW laser annealing [18]. A variety of silicides are formed in this manner.

Although laser spot size is large in comparison of silicon integrated circuits being produced these days (or in the future when they will be still smaller in dimension), the laser-assisted annealing (LAA) could still find use in lo-

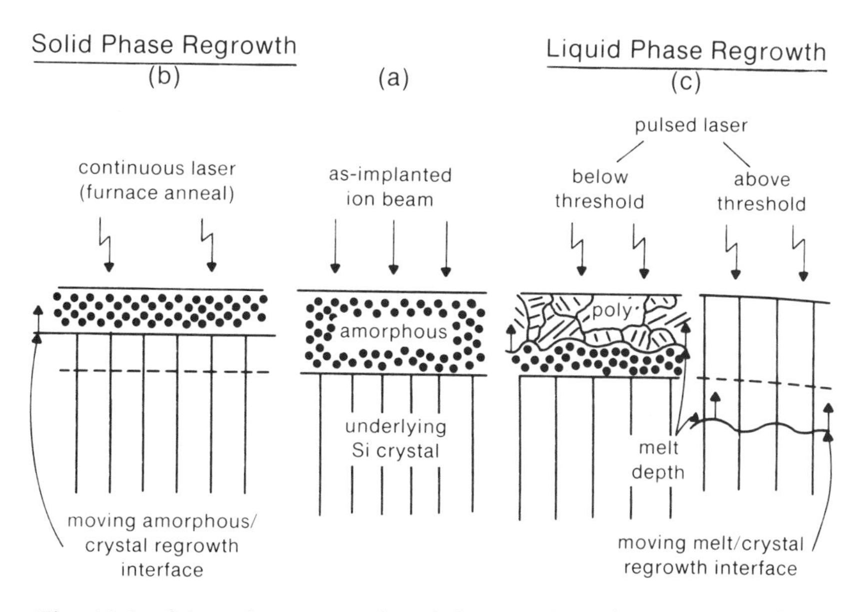

Fig. 11.6. Schematic representation of the amorphized (Ion-Implantation) silicon before and after annealing using continuous or pulsed laser or furnace annealing (Adapted from Williams [17], with permission from the Electrochemical Society).

calized processes. Localized annealing of metals, silicide formation, dopant activation, etc. could be carried out using appropriate-wavelength lasers. Focused IR or visible laser beams have been used in several laboratories to control localized substrate heating and thereby activate micrometer-scale thermal reactions. Theoretically one can envisage a selective optical absorption edge for each material. This absorption edge could be significantly larger for a given material than the surroundings. In such a case, even larger area beams could be used to induce desired heating in small areas without affecting the surroundings.

11.2.1.3 Laser-Assisted Deposition of Metals, Insulators, and Semiconductors

Laser-assisted deposition of the electronic materials has been most extensively investigated. Thin films of various materials have been deposited using one of two deposition techniques: thermochemical or pyrolytical deposition and photochemical or photolytical deposition. In the former, the decomposition of gaseous species is caused by heat, whereas in the latter the decomposition is caused by the radiation absorption by the gaseous species. Pyrolytic processes are a) very fast, a factor of 100-1000 times faster than photolytic processes; b) similar to CVD processes; and c) self-annealing, in that deposited materials do not require additional annealing to attain bulk properties. Photolytic processes, on the other hand, offer

TABLE 11.3: Beam process parameters and resultant reactions for the silicides. From Shibata et al. [18]

Metal Film	Reacted Phase	Reacted Thickness[a] (Å)	P (Normalized Laser Power)	Power Parameters[b] (watts/μcm)	T_{sub} (°C)	Resistivity (μΩ-cm)
Pd	Pd_2Si	1930	0.71	0.154	50(± 0.5)	37
	PdSi	2600	0.9 1.4	0.2 0.3		31
Pt	Pt_2Si[c]	–	0.69	0.148		–
	Pt_2Si[c]	–	0.86	0.185		–
Mo	$MoSi_2$	1450	0.88	0.093	350(± 0.0)	190
W	WSi_2	1200	0.85 (10 scans)	0.087		110
Nb	$NbSi_2$	1350	0.9 (21 scans)	0.089	–	–
Pd	Pd_2Si	2280		0.112	50	39
	PdSi	2730		0.136		20
Pt	PtSi	1670		0.093		28
Nb	$NbSi_2$	1200[d]		0.173	–	–

[a] Calculated from backscattering spectra using the near surface approximation for the energy loss.
[b] Calculated values of absorbed power/beam radius: (1-R)P/w
[c] A mixed phase compound including metastable Pt_2Si_3.
[d] Thickness of a partially reacted film.

the advantage of being independent of local substrate characteristics. In addition, for pyrolytic decomposition there is nonlinearity in deposition due to the exponential dependence of the reaction rate on the reciprocal of the surface temperature. This confines the decomposition to the center of the focused laser beam, leading to the narrowing of the deposited lines. Thus, narrower deposits can be made by using wider laser beams [12]. The selection of the correct deposition method depends on several factors, which include availability of appropriate lasers, gases, and substrates, and the temperature threshold to which the substrate and the surroundings can be taken to.

Pyrolytic decompositions of gases could be substrate dependent and thus may present problems when various different materials form the substrate surface, such as a patterned oxide on silicon substrate. Also heating the substrate to the decomposition temperature may not be desirable. In direct writing the obtainable linewidths will be those determined by the exposure area on the substrate and once again by the type and cleanliness of the substrate. Of several examples of the pyrolytic decomposition and deposition, the deposition of polysilicon lines have been extensively studied [11,12,19].

Silane (SiH_4) or chlorosilanes (SiH_xCl_{4-x}) at reduced pressure in a cavity are exposed to lasers with wavelengths that cause heating in the substrate with no absorption in gas phase. Temperatures in the range of 600-1500°C have been used. Very high deposition rates of narrow lines have been achieved. In addition, the deposited films have been doped with boron or phosphorus by introducing small amounts of dopants in the gas phase (e.g., BCl_3 or PCl_3). Recently demonstrated writing of polysilicon lines points to the success of this method in direct writing of semiconductor materials [19].

Substantial advances are still required in understanding the pyrolytic decomposition (which is very similar to a chemical-vapor deposition process). Reported changes in the linewidths as one moves from silicon to SiO_2 [12] and cusping of the deposited polysilicon surface [19] need to be carefully studied with processes outlined to eliminate such effects. In cases where the substrates were locally heated to temperatures above the melting point of silicon, the stability of the substrate and its surroundings became questionable. Reaction between SiO_2 and molten silicon could lead to SiO vapor formation, as well as stresses that must be carefully evaluated.

Photolytic decomposition of gaseous species leading to deposition on substrates could be very useful for deposition on materials that are sensitive to heat, for example, organic materials, low melting-point metals like Al, etc. Most gaseous species will dissociate in radiations of wavelengths in the range of 250-300nm, such as those of CW laser sources. These are relatively low-power lasers and thus deposition rates are low. Also, because the substrate is practically unheated, deposited films may require post-deposition annealing to obtain desired bulk properties. Photolytic decompositions could be very selective since the decomposition follows an absorption of the radiation of a given wavelength. Thus, if the absorption edge is sharp and there are no other materials or molecules with this absorption band, then a selective decomposition could occur. However, since the decomposition occurs in the gas phase, nucleation on the substrate may bacome necessary. The nucleation should precede the growth and should be confined to the laser-illuminated regions on the surface. Thus a truly photolytic growth proces must follow a surface photochemical process for the nucleation. In the absence of such a nucleation on the surface the vapor-phase decomposition may lead to products that could deposit in and around the laser-illuminated surface.

For all practical purposes, there is no truly photolytic or pyrolytic deposition, although most pyroltic deposition are very close to being so. Table 11.4 gives the list of laser-deposited metals and silicon, together with the dominating process type and gases and lasers employed. In most of such cases the feasibility of writing lines on various substrates has been demonstrated without particularly worrying about the quality of materials. For example, the resistivities of the metals so deposited are higher by a factor of 2 to 100 than the bulk values. The higher resistivities, in general, are

TABLE 11.4: Material systems for photolytic and pyrolytic direct writing. From Osgood, Gilgen [12]

Parent gas	Deposited Material (Me)	Process	Wavelength	Ref.[a]
Alkyls				
$Me(CH_3)_2$	Cd, Zn	Photolytic	257 nm	19
$Me(CH_3)_2$	In, Al	Photolytic	257 nm	34
$Ga(CH_3)_3$	Ga	Photolytic	514 nm	26
Carbonyls				
$Me(CO)_6$	Mo, Cr	Photolytic	257 nm	19
$W(CO)_6$	W	Pyrolytic	10.6 μm	37
$Fe(CO)_3$	Fe	Pyrolytic	10.6 μm	37
$Ni(CO)_4$	Ni	Pyrolytic	530-470 nm	38
Halogenide				
$Mo(F)_6$	Mo	Pyrolytic	10.6 μm	37
$TiCl_4$	Ti	Photolytic	257 nm	7
$TiCl_4$	Ti	Pyrolytic	10.6 μm	12
Hydride				
SiH_4	Si	Pyrolytic	10.6μm,~500 nm	23,59
Acetylacetonate				
MeAcAc	Au, Cu, Pt	Photolytic	257 nm	35,36
	Au, Cu	Pyrolytic	514 nm	36

[a] These reference numbers correspond to those given in Ref. 12.

associated with contamination from the gas phase — oxygen, carbon, and molecular fragments. In some cases, such as Al, it has been very difficult to deposit oxygen- and carbon-free metal. For doped polysilicon films, the resistivities are in the acceptable range. However, this does not exclude the possibility of oxygen, carbon, and other contamination of polysilicon. Thus, a serious effort has to be made to examine and improve the characteristics of the laser-deposited metals and semiconductors.

Compared to the number of experiments carried out to laser-deposit metals and semiconductors, there are only a few experiments that report laser-assisted deposition of insulators, particularly SiO_2 and $Si_3 N_4$. All these depositions have been large-area depositions like those carried out by CVD, plasma, or sputtering methods. Recently reported laser depositions indicate that a high-temperature anneal may be necessary to obtain properties compared to those oxides deposited by other methods and to those thermally grown on silicon. Al_20_3 has also been deposited by this method.

Writing insulator lines in a manner similar to that of metals and polysilicon has not been possible.

11.2.1.4 Laser-Assisted Doping

Laser-assisted doping of semiconductor substrates and of polysilicon during direct writing of the polysilicon lines have been achieved by exposing the semiconductors to gases carrying dopants with a laser beam, causing the necessary decomposition of the gas and heating for the dopants to diffuse in the substrate. Doping is effectively controlled by a) interface-limited transfer of the dopants to the surface and/or b) solid-state diffusion into the semiconductor. At low temperatures diffusion is a slow and rate-limiting process. At high temperatures interface-limited transfer is the rate limiting process.

The laser-assisted doping process is similar to direct-write laser deposition. It can be useful in producing submicron wide shallow diffusions since diffusion times can be controlled accurately to a fraction of a millisecond. Linewidths as narrow as 0.25μm have been reported [20].

11.2.1.5 Laser-Assisted Etching

Patterning, by etching, of the semiconductor materials has been the most critical step in IC processing. Etching has thus evolved considerably, and this evolution is responsible for the rapidly shrinking dimensions of the devices. Laser etching of semiconductors and metals has been experimentally demonstrated. For direct-write etching of silicon, high resolution has been reported using visible or UV lasers that photolyze chlorine to produce high density of reactive chlorine atoms, which then etch silicon. As the temperature of the silicon substrate is raised, the reaction rate and, therefore, the etch rate increases. Extremely high etch rates (thouands of micrometers per second) can be obtained if the surface temperature is raised to the silicon melting point. At such temperatures the reaction with molten silicon is extremely rapid and photodecomposition of chlorine is no longer important. Grooves, 0.3-0.4 μm wide, have been created in silicon by the use of argon ion laser and chlorine as the etchant gas. Etch anisotropy is excellent and large aspect ration structures can be created.

Laser etching of metals and insulators such as SiO_2 and Si_3N_4 has not been very successful, especially using dry gases. This is mainly because of native oxide present on the metals. Thus, an in-situ oxide cleaning step has to be introduced prior to etching metals.

Pulsed-laser etching of organic materials such as photoresists and polymers have been reported [21]. In this technique organic substrate absorbs energy locally at the proper laser wavelengths and ablates, leading to a reasonably well-defined pattern. Etch rates are quite high and holes of nearly $0.5\mu m$ dimensions have been created. The process, however, introduces the fragments that create considerable particulate contamination.

Finally, large-area projection etching, using lasers, has been suggested. Such an approach will, probably, solve the throughput problem associated with laser processes and requires more attention.

So far all the laser-processing experiments have been carried out in an apparatus schematically shown in fig. 11.3. The apparatus is simple. Osgood and Gilgen [12] have discussed the limitations of such an apparatus, especially for writing lines with submicrometer widths. When we consider VLSI circuits with 10^5 - 10^7 devices on the chip, laser direct writing, at present is not practical. For smaller and custom circuits, certain types of laser processing have been carried out successfully. For example, photolithographic mask repairs [22], integrated circuit repair [23], circuit optimization [24], interconnecting gate arrays [19,25], and forming ohmic contacts [26] have been demonstrated. However considerable effort will be needed in order to make laser processing practical. In addition, all laser processing does not appear feasible since etching and deposition of ceramics, in-situ cleaning, and large-area annealing will require alternate techniques.

11.2.2 Particle-Beam Processes

Particle-beam processes could be grouped into two categories: a) the electron-beam processes and b) the ion, atom, or molecular-beam processes. The former have been investigated for a considerable period of time and have found application in generating masks for IC fabrication and direct-writing patterns on the wafers. The ion-beam processes are still at the laboratory development stages, except for the use of wide-area beams for ion implantation. Both, however, offer one big advantage over photon beams in their ability to respond to electromagnetic fields. This allows not only for focusing, but also the high-speed computer-controlled maneuverability of the beam. Ionized beams are, therefore, highly suitable for resistless and maskless processing.

11.2.2.1 Electron-Beam Processes

Electron beams have been used for a) pattern generation on a mask, b) direct writing on the wafer, c) testing, d) annealing, and even e) doping. Of these,

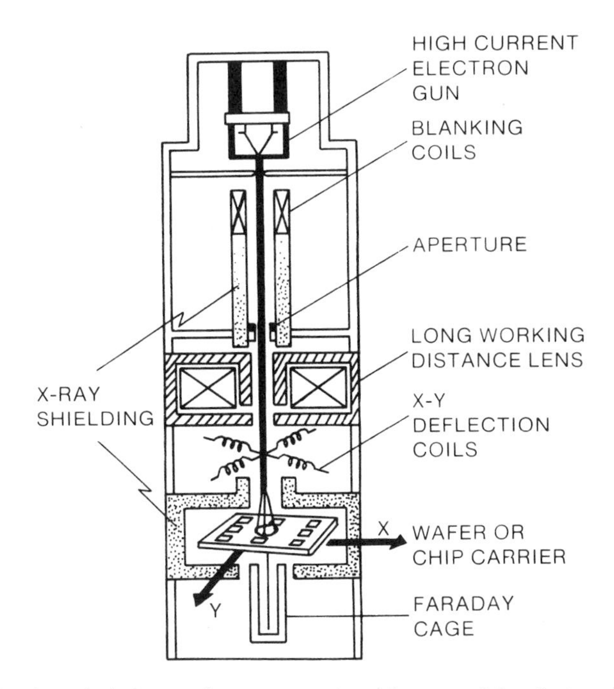

Fig. 11.7. A typical electron beam generating column used for electron-beam annealing experiments. From McMahon, Ahmed [27], with permission from the Electrochemical Society.

generating the pattern to make masks for subsequent pattern transfer using visible, UV, or x-ray exposure systems, has been the most successful application. Direct writing, although very appealing, has not been very popular, due mainly to throughput considerations. A considerable amount of work is being done in this area. Approaches to solving electron-beam problems include a) increasing the brightness, reliability, and the life of the electron source; b) developing a high-speed x-y stage; and c) developing new resists or resist (double- or triple-level) systems. However, the proximity effects associated with electron penetration and scattering are a big concern. Also e-beam lithography is not a resistless and maskless process. Electron-beam lithography has been discussed in detail in Chapters 8 and 9.

The use of electron beams for annealing has also been investigated. Bombarding a substrate with high-energy electrons leads to considerable energy absorption and an increase of the substrate temperature. Very high temperatures can thus be attained. As a result, electron beams can be used to cause annealing. Ion-implantation anneals, silicide formation, and recrystallization of polysilicon films, etc. have all been attempted. In practically all cases the results are similar to those obtained by laser annealings.

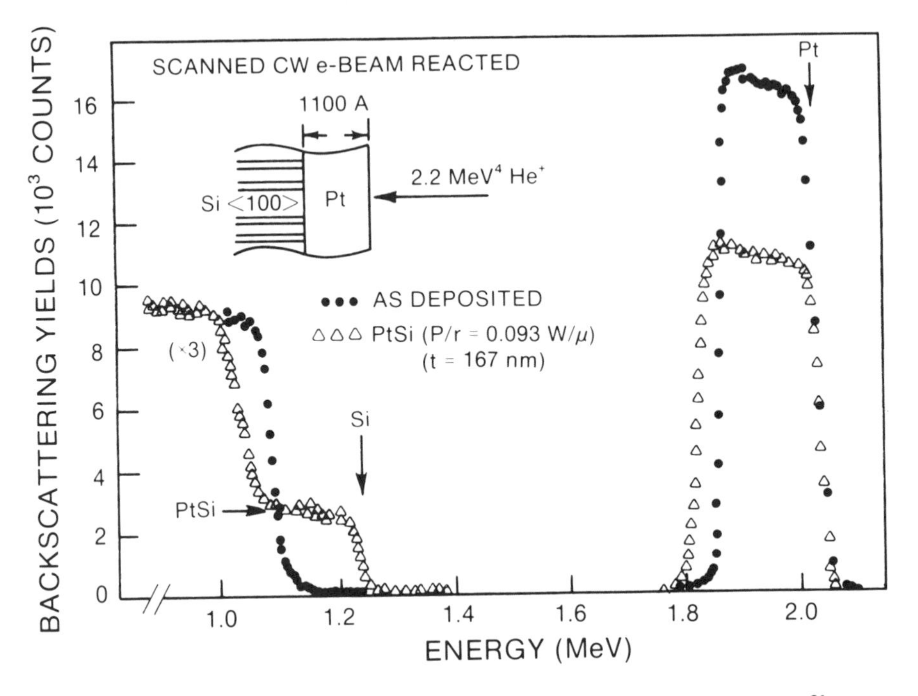

Fig. 11.8. RBS spectra of the as-deposited and e-beam reacted platinum film on silicon. From Shibata et al. [18].

A typical electron-beam annealing apparatus is schematically shown in fig. 11.7 [27]. The apparatus is similar to an electron-beam column of an electron microscope and consists of an electron generator, electromagnetic focusing and deflection lenses, charge and x-ray shields, a sample stage manipulator, and an electron current integrator. In such an apparatus the annealing spot size is small, generally in the range of a few μms. Although such an apparatus makes localized annealing possible, the annealing time for large-area samples is high.

In contrast to using focused small-spotsize e-beams, one can use large-area, defocused (low-resolution) beams. In such cases power per unit area of the spot is small and temperature rise on the sample will be small. However, it has been demonstrated that such annealing systems may be useful for ion-implant activation [28] and possibly for other process anneals.

Figure 11.8 shows formation of PtSi films resulting from focused electron-beam annealing of Pt film on silicon. For the used power (P) to beam radius (r) ratio of 0.093 watt/μm, only PtSi is formed [18]. Figure 11.9 shows the carrier concentration profiles of ion-implanted arsenic in silicon before and after laser, focused e-beam, and thermal annealings [15]. Also shown are the electron mobility values. Laser and e-beam annealings produced similar results, with maximum carrier concentration of about 1.1×10^{21} per cm^3. Standard furnace thermal anneal led to a more uniform carrier

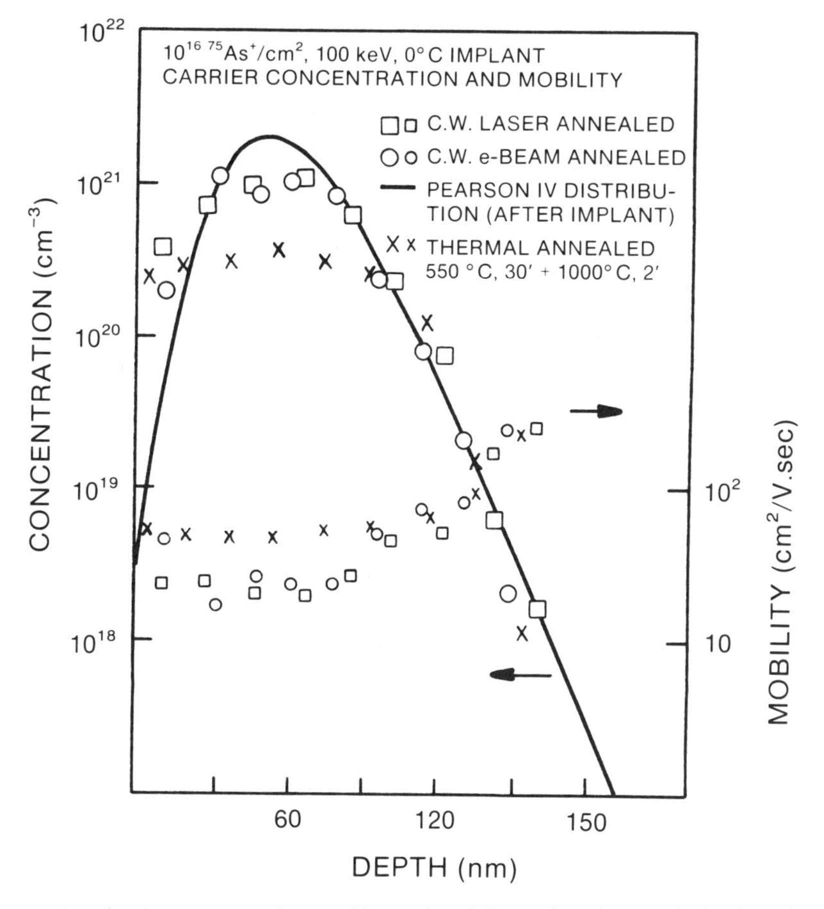

Fig. 11.9. Carrier concentration profiles and mobility values in arsenic implanted silicon before annealing and after e-beam, CW-laser, or furnace thermal annealing. From Lietoila [15].

concentration depth profile in the regions close to surface, but the maximum carrier concentration was only about 3×10^{20} per cm^3. No deep diffusion is seen after these short anneals.

Figure 11.10 shows boron doping profiles of as-implanted and wide-area electron-beam annealed samples [29]. Also shown is the electrically active boron profile. The e-beam annealing is compared with furnace-annealed samples. Electron-beam annealed samples show very small boron diffusion compared to furnaced annealed samples. Although maximum active boron concentration is small, the promise of using broad e-beams has been demonstrated.

Electron-beam testing, at various stages of the processing, is another application that could be considered for in-process testing during in-situ processing.

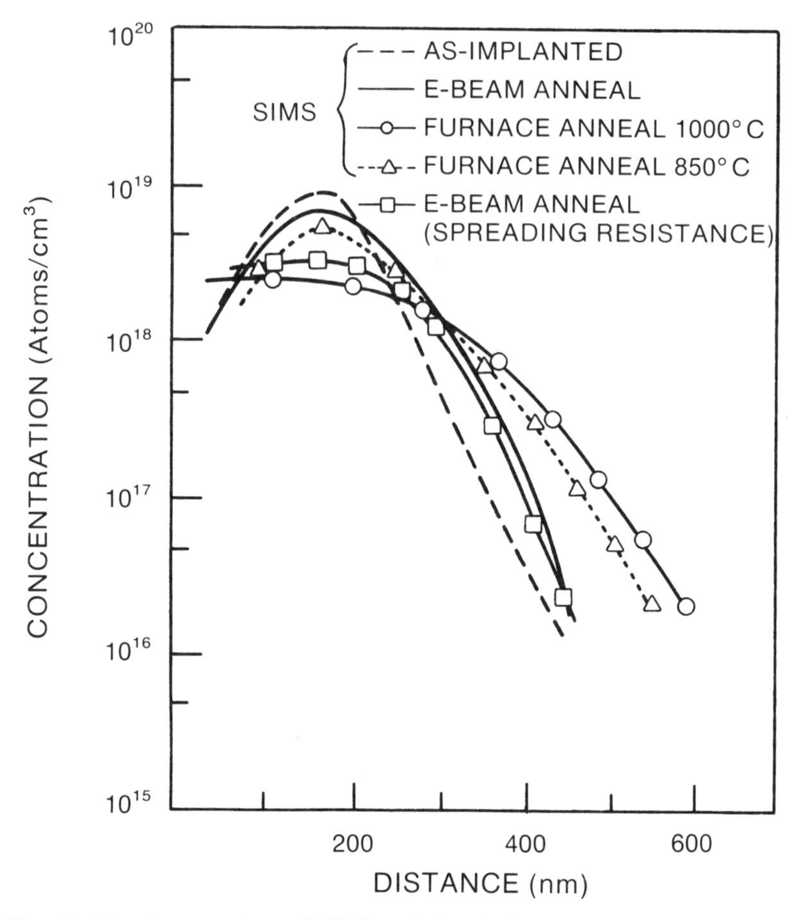

Fig. 11.10. A comparison of SIMS and electrically active doping profiles (latter obtained by spreading resistance measurments) of the boron-implanted and e-beam, and furnace annealed samples. From Yep, Fulks, Powell [29].

11.2.2.2 Ion-Beam Processes

Ion beams have attracted considerable attraction, with potential applications at almost all processing steps. Some of the large-area ion-beam applications, e.g., ion implantation, sputter deposition, plasma-assisted deposition, plasma etching, reactive-ion etching, ion milling, and sputter etching are very common. Several of these have been discussed in earlier chapters. Ion-beam modification of materials has been demonstrated in many cases. Molecular-beam epitaxy is another known application of particle beams. Ion-cluster beam deposition of various materials and of low-temperature epitaxial films has recently shown great promise. Focused ion beams are being extensively explored for practically all aspects of maskless-resistless processing, as well as for conventional lithography.

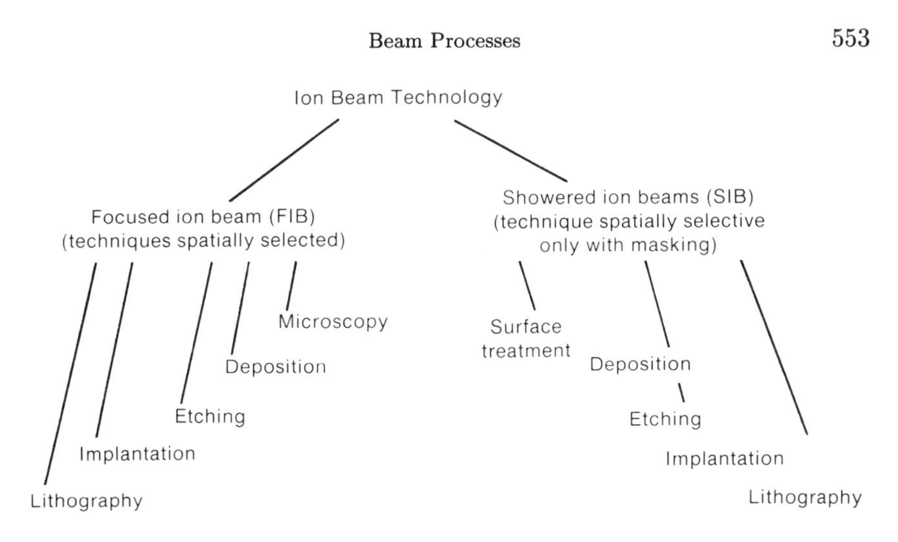

Fig. 11.11. Muray's (Ref. 30) classification of the ion beam technology and applications.

Of critical importance regarding ion beams is their versatility: they can be applied to practically any material. Muray [30] has classified ion-beam technology into two categories: focused and showered ion beams. Figure 11.11 shows this classification and potential applications.

11.2.2.3 Focused Ion Beam (FIB)

Focused ion beams are analogous to electron beams, except for their use of a particle whose mass is about a factor of 1836 or more higher and charge that can be similar to or a multiple of electron charge—positive or negative. Because of their large mass, focused ion beams have a limited penetration depth compared to electron penetration depth. On the other hand, ion beams produce significantly more damage as discussed in Chapter 5. The ion damage is, however, restricted to the ion range in the material. This feature makes FIB very suitable for lithography using thin photoresists.

A focused ion-beam apparatus is similar to an electron-beam system. It consists of a source of ions, a mass separating eletromagnetic filter, accelerating and focusing column, a beam-deflecting mechanism, and a sample table with freedom of translation. Preparing long-lasting high-current sources, especially those required for SIC technology, has been difficult. Table 11.5 gives a list of technologically important liquid-metal ion sources, together with relevent information regarding these sources [31]. In a liquid-metal ion source, the desired material or an alloy is kept molten in a fine capillary. Ions are extracted by applying high fields to the tip of the capillary. All metals do not melt at temperatures that can be employed. To circumvent this problem, a low-melting eutectic of the desired species with one or more other elements is used. For example, $Pd_{70}As_{16}B_{14}$ is used to extract Pd, As,

TABLE 11.5: Partial list of technologically important sources.[a]

Composition	Angular current density (μ A/st)	Current on sample in system (pA)	Current density on sample (A/cm^2)	Lifetime (h)	References
Ga	26	1000	10	200	10, 43, 44
$Au_{65}Si_{27}Be_8$					
$Au_{59}Si_{26}Be_{15}$	20 Be^{++}			100	45, 46, 47, 48
$Pd_{50}Ni_{26}$	...				
$Si_5Be_6B_{13}$		134 Be^{++} 66 Be^+ 150 B^+ 200 Si^{++}			
Pd_2As	5 As^{++}	30 pA As^{++}		150	50
$Pd_{70}As_{16}B_{14}$	1 B^+ 3 As^+ 4 As^{++}			150	50
$Ni_{45}B_{45}Si_{10}$		(25%-35% B^+) 2000(at 100μ A total source)		250	51, 53
$Pd_{64}As_{11}$					
B_9P_{16}	...	(18 atm. % P^+)		25	50
Cu_3P		(10 atm. % P^+)		20	52
Al	20-30			100	54
Au	(1 to 100 μ A total current)			50 50	55 55

[a] From Ref. 31.

or B. The electromagnetic mass separators are used to extract the desired species.

Figure 11.12 describes various applications of FIB technology and required ion fluxes [32]. Applications can be grouped into the following categories: ion implantation for doping, materials modification, deposition, and surface chemistry; ion milling for pattern generation and surface cleaning; lithography for pattern transfer; ion microscopy; and materials analysis. Each of these are discussed in detail by Melngailis [31].

11.2.2.4 Ion Cluster Beam (ICB)

In this technique clusters are created by condensing supersaturated vapor by adiabatic expansion through a small nozzle into a high-vacuum region, where the vapor species are then ionized by electron impact [33]. Ionized clusters can then be accelerated by use of electric potentials. Figure 11.13 shows a typical ion cluster-beam deposition system. It consists of a source, an ionization region, and accelerating electrodes. The source is generally a graphite-crucible that can be heated to temperatures of 2000°C or more to create supersaturated vapor inside. The crucible has a pinhole that is

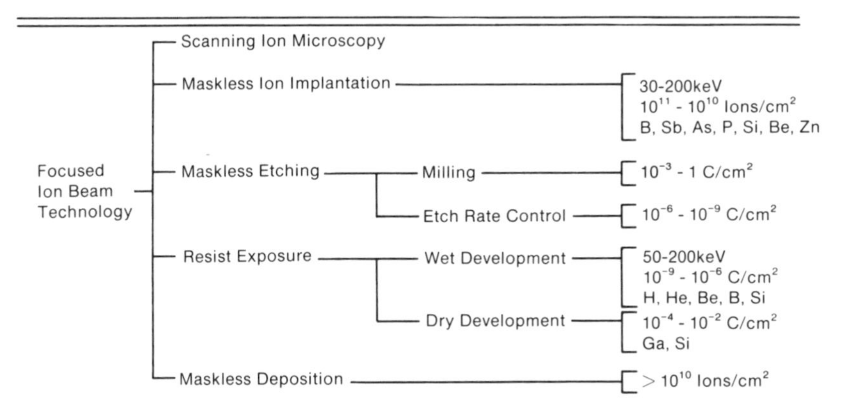

Fig. 11.12. Various applications of focused ion beams. The required fluxes are shown for each application in curies/cm^2 or ions/cm^2. From Kato et al. [32].

normally shuttered and is opened only during deposition. The rest of the system is maintained in a vacuum ranging from 10^{-6} to 10^{-8} Pa.

The unique capabilities of the ICB have been attributed to the properties of the cluster state such as a) the enhancement of the adatom migration, b) low charge to mass ratio, c) low-energy ion-beam transport, and d) cluster substrate impact causing the monentum transfer to the surface atoms. Table 11.6 lists the type of films and materials that have been deposited using this technique. Some films (e.g., Si, Al, Ge, GaAs, etc.) have been deposited epitaxially at significantly lower temperatures.

Figure 11.14 compares ion cluster-beam deposition with other ion-deposition processes [34]. It compares with MBE in the process vacuum conditions. However, particle energies are two to four orders of magnitude higher. Recent demonstrations of the deposition of epitaxial aluminum films on silicon substrate [35] places this deposition system in an extremely important category — with less stringent requirements on vacuum.

In the following applications the ion beams are discussed. Both large area showered beams and FIB are considered, although the discussion of showered beams is limited because they have been covered in chapters on ion implantation, lithography, and pattern transfer.

11.2.2.5 Ion Cleaning

Ion-assisted cleaning of the various surfaces has long been carried out in one form or the other. These has been called a) plasma, b) reactive ion, and c) backsputter cleaning. Focused ion beams can be used in similar fashion to cause localized cleaning. Use of reactive focused ion beams will perhaps minimize or eliminate the ion damage associated with these processes. In any event, a low-temperature (400-700°C) anneal will eliminate the ion damage.

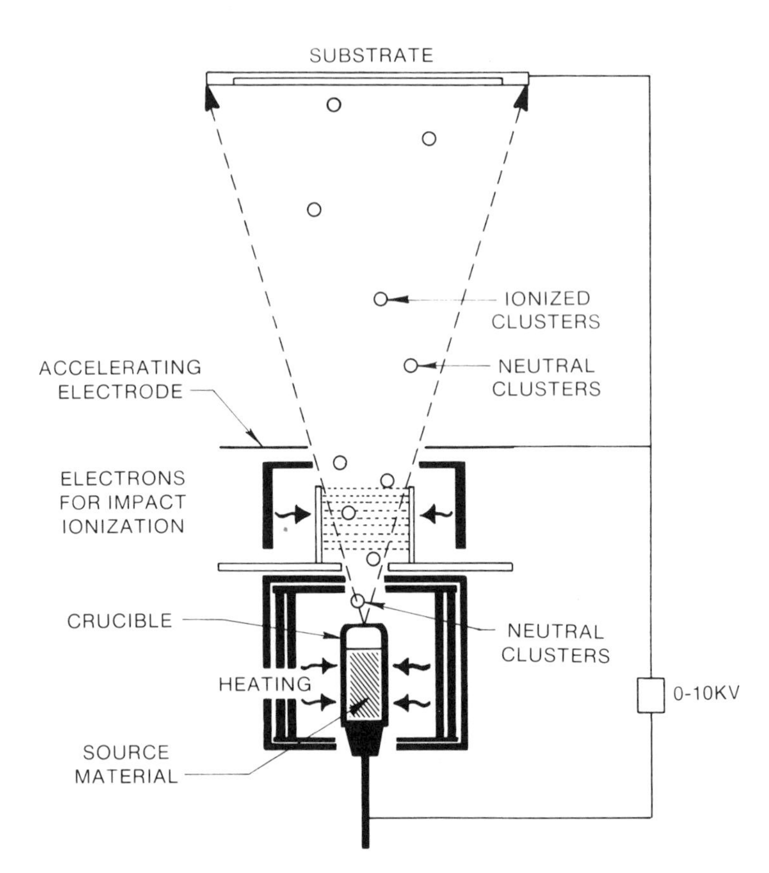

Fig. 11.13. Schematics of a typical ion cluster beam deposition chamber (From Wong [34]). In recent modifications the crucible holding the superheated metal may not be covered with any lid during depositions (See Ref. 35).

Since physical sputtering is involved in such processes, there is no preferential selectivity and practically all surfaces — semiconductor, insulator, and metals — can be cleaned. Even organic materials can be removed under properly controlled backsputtering.

11.2.2.6 Ion-Beam Annealing

High-energy ions, when deposited into a substrate, release energy and raise the temperature of the substrate. The temperature rise of the ion-implanted substrate can be calculated by equating the energy input to the thermal energy redistribution in the substrate [31]. The latter can be obtained by assuming a) a temperature gradient dT/dr from the surface to the bulk of the sample where r is the radius of the hemisphere of a constant temperature in the substrate, and b) that no other heat losses occur. Thermal energy is

TABLE 11.6: Film properties and their features. From Takagi [33].

Films/Substrates	Acel. Volt. (kV)	Subs. Temp. (°C)	Advantages (Applications)
			SEMICONDUCTORS
Si/(111)Si Si/(100)Si Si/(1102)sapphire	6	620	Low temperature epitaxy in a high vacuum (10^{-7}–10^{-6} torr) shallow and sharp p-n junction (semiconductor devices)
Amorphous Si/glass	2	200	Thermally stable (solar cell, thin film transistor)
Ge/(100)Si Ge/(111)Si	0.5–1	300– 500	Low temperature epitaxy in a high vacuum (10^{-7}–10^{-6} torr) shallow and sharp p-n junction (semiconductor devices)
GaAs/Cr-doped GaAs	6	550	Epitaxy in a high vacuum (10^{-7}–10^{-6} torr) (semiconductor devices)
GaP/GaP GaP/Si	4	550 450	Low temperature epitaxy in a high vacuum (low cost LED)
ZnS/NaCl	1	200	Single crystal formation (optical coating)
ZnS:Mn/glass	1	200	(ac-dc electroluminescent cells of low impedance)
CdTe/glass	2	250	Improved monocrystalline domains (infra-red detector)
InSb/sapphire	3	250	Controllable crystal structure (magnetic sensor)
			OXIDES, NITRIDES, AND CARBIDES
Li-doped ZnO/ (1102)sapphire	0.5–1	230	Single crystal formation (optical wave guide)
c-axis preferentially oriented	0	150	Controllable crystal structure Controllable optical transmission (optical wave guide, SAW)
ZnO/glass	0	400	Single crystal formation, transparent film
BeO/(OOO1)sapphire c-axis preferentially oriented	0		(semiconductor device, heat sink,electrical insulator) High electrical resistivity and high thermal conductivity coating.
BeO/glass	0	400	Low temperature growing (semiconductor device, heat sink electrically insulating, and thermal conduction)
PbO/glass	3	RT	Low temperature growth, smooth surface, accurately controllable thickness, good adhesion (superconductor)
SnO_2/glass	0	400	Transparent low resistivity film, strong adhesion (NESA glass)
SiO_2/glass	0–2	200	Extremely low temperature growth (Si devices)
FeO_2/Si,glass	3	250	Controllable optical bandgap (photovoltaic cell)
CaN/ZnO/glass	0	450	Low temperature growth of CaN film on amorphous substrate (low cost LED, phototcathodic electrode)

TABLE 11.6: Film properties and their features (continued)

Films/Substrates	Acel. Volt. (kV)	Subs. Temp. (°C)	Advantages (Applications)
			OXIDES, NITRIDES, AND CARBIDES (continued)
SiC/Si,glass	0–8	600	Controllable crystal state (energy converter, surface protective coating)
			METALS
Au,Cu/glass, Kapton	1–10	RT	Strong adhesion, high packing density, good electrical conduction in a very thin film (high resolution flexible circuits, optical coating)
Pb/glass	5	RT	Controllable crystal structure, strong adhesion, smooth surface, improved stability on thermal cycling (superconductive device)
Ag/Si(n-type)	5	RT	Ohmic contact without alloying
Ag/Si(p-type)		400[a]	Ohmic contact at low temperature (semiconductor metallization)
Ag-Sb/CaP	2	400	Ohmic contact, strong adhesion (semiconductor metallization)
Al/SiO_2,Si	0–5	RT–200	Controllable crystal structure strong adhesion, ohmic contact at low temperature, electro-migration resistant (semiconductor metallization)
c-axis preferentially oriented MnBi/glass	0[b]	300	Spatially uniform magnetic domain, high density optical memory (magneto-optical memory)
CdFe/glass			Thermally stable and uniform, amorphous (magneto-optical memory)
c-axis preferentially oriented PbFe/glass	0–3	200	High Seebeck coefficient, low thermal conductivity (high efficiency thermo-electrical converter)
Preferentially oriented ZnSb/glass	1	140	Thermally stable, amorphous, high Seebeck coefficient (high efficiency thermo-electric converter)
p-$FeSi_2$/glass	0–5	150	Thermally stable, amorphous, p-type or n-type controllable, high Seebeck coefficient (high efficiency thermo-electric converter)
			ORGANIC MATERIALS
Anthracene/glass	0–2	-10	Controllable crystal structure (detectors)
Cu-Phtalocyanine/glass	0–2	200	Controllable crystal structure (photovoltaic devices)
Polyethylene/glass	0–1	0–110	Controllable crystal structure (photovoltaic devices)

[a] Annealing temperature

[b] Ejection velocity only

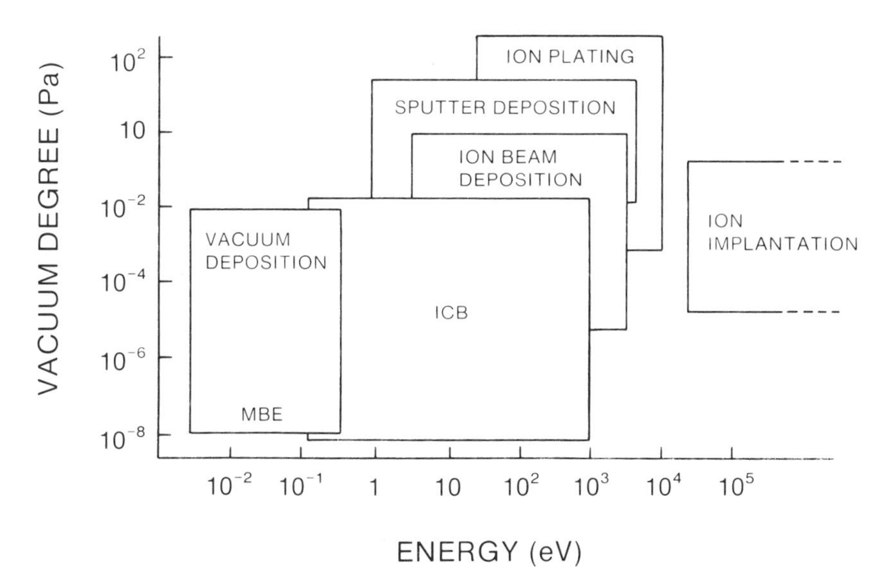

Fig. 11.14. Comparison of typical deposition methods from the viewpoint of the degree of vacuum as a function of the depositing particle energy. From Wong [34].

then given as

$$-2\pi r^2 K \frac{dT}{dr},$$

where K is the thermal conductivity of the substrate. Energy deposited by ion implantation will be simply current I multiplied by the voltage V. If J is the current per unit area and the ion-beam radius is a, then $I = \pi a^2 J$ and deposited power is $\pi a^2 V J$. Then, equating the two quantities one gets

$$dT = \frac{a^2 V J}{2K}\left(\frac{-dr}{r^2}\right). \tag{11.5}$$

By integrating this equation for $T(r)$ at r and $T(\infty)$ at $r = \infty$, one obtains the temperature rise at r, since $T(\infty)$ is the substate temperature far away from the place of ion implantation.

$$\Delta T = T(r) - T(\infty) = \frac{a^2 V J}{2K} \cdot \frac{1}{r}. \tag{11.6}$$

The temperature rise can be significant where ion beam strikes the substrate if K is small.

Problem: *Calculate the temperature rise at the periphery of a 200 nm diameter ion beam in a quartz and a silicon substrate for an ion current density of 10 A per* cm^2. *Ion acceleration voltage is* $90kV$.

K for quartz (SiO_2) = 1.4 x 10^{-2} watt per cm per degree.
K for Si = 1.5 watt per cm per degree.

Solution: $r = a = 10^{-5}$ cm

$$\Delta T \quad \text{(quartz)} = [(10^{-5})^2 90 \times 1000 \times 10/2 \times 0.014][1/10^{-5}] = 321°\text{C}$$
$$\Delta T \quad \text{(Si)} = [(10^{-5})^2 90 \times 1000 \times 10/2 \times 1.5][1/10^{-5}] = 3°\text{C}$$

Thus, there is no significant temperature rise in silicon. However, there is a considerable temperature rise in quartz. In silicon integrated-circuit fabrication, the substrates are oxidized silicon wafers with 10 to 1000 nm of oxide on them. The SiO_2 film thickness is very small compared to the silicon wafer thickness in hundreds of micrometers. Unless the voltages and current densities are significantly higher, no significant temperature rise can be expected. In addition, ion implantation causes considerable damage, as discussed in Chapter 5. Thus, use of ion beams in heating silicon or GaAs ($K = 0.46$ watt per cm per degree) is not practical.

11.2.2.7 Ion-Beam Depositions of Metals, Insulators, and Semiconductors

Several ion beam deposition processes, which fall under shower-ion-beam (SIB) category, have long been used for the deposition of various materials. Sputtering, plasma-assisted CVD, and ion implantation will continue to provide means of forming large-area deposits at low cost. Multiple ion beams can be used to deposit alloys and compounds. FIB has been used to direct write lines and patterns. Table 11.5 listed various ion sources that are available, and the list is growing. Metal or semiconductor patterns, thus, can be created using FIB. Metal-silicon intermetallic lines and patterns could be created by using metal FIB on silicon substrate or silicon FIB on metal substrate. Aluminum lines have been written by localized FIB-induced decomposition of trimethyl aluminum in vapor phase. This latter approach is similar to laser direct writing, except perhaps for the possibility of making very fine lines at lower temperatures.

FIB or SIB has found application in modifying the materials and also in making metastable and new solid-state phases that may have very interesting properties (see Chapter 5). Theoretically, any substrate could be bombarded with any available FIB. One can imagine writing of oxide or nitride isolation patterns in silicon by using focused oxygen or nitrogen beams or creating blanket oxide or nitride layers (to replace oxidation) using oxygen or nitrogen SIBs. Experimental results are available in the latter case for which high-dose oxygen ion implantations were carried out to create oxide layers in silicon. The properties of the oxides or nitrides thus created and the ion-induced damage associated with such large fluxes and their effect on the device and circuit performance will need very careful evaluation.

Ion-beam depositions have an additional advantage, besides those related to beam size, beam maneuverability, and low temperature. This is related to the working of beams, especially FIB, in a very-high-vacuum environment, which eliminates the possibility of contamination. Thus, extremely pure materials can be deposited or formed. Also, as mentioned earlier, multiple beams can be used to deposit compounds and intermetallics [34]. All these features make FIB and SIB very versatile processes.

11.2.2.8 Ion-Beam Doping

Doping of the semiconductors is the simplest and possibly the easiest application of ion beams. Large-area ion implantation of dopants is the standard practice in the industry. Focused ion beams have been used for doping with Ga, Sb, B, and As for silicon, and Be and Si for GaAs.

Direct-write doping of semiconductors to create very narrow and shallow doping profiles will be the most useful application of FIB. Contacting such narrow regions, using direct writing of metal or intermetallic patterns, will be equally important applicantion of FIB. No other technique can match FIB in this respect.

11.2.2.9 Ion-Beam Etching

Large-area plasma or reactive-ion etching of all semiconductor materials has been common. These etching techniques require photolithographically created masks to create the patterns on the substrate. Backsputter etching has been another ion etching technique normally used to clean the surfaces prior to deposition of materials by sputtering. Ion milling, both large-area and focused, has been used to remove material from the desired areas. FIB or FRIB (focused reactive ion beam) etching adds the dimension of direct etching without the need of the mask to create the pattern. FIB etching, in general, is applicable to all materials — semiconductors, metals, insulators, or polymers. FRIB will, in addition, operate with selectivity or the reactive species that are focused on the substrate.

11.2.2.10 Gettering

Gettering of impurities (see Chapter 4) from the active device regions into the regions away from the junction, has been demonstrated using localized ion implantation or ion implantation on the backside of the substrates [36]. Thus, whenever gettering becomes essential this process can be adopted on a localized basis using focused ion beams.

11.2.2.11 Focused Ion-Beam Lithography

Considerable interest has been shown in using the FIB for lithography. The main advantages cited for this application are a) very small or negligible proximity effects — heavier ion like silicon, which will not contaminate SIC, will be among the best, b) shorter range in the resist — thus with the use of very thin upper-level resist in a double- or triple-level structure (see Chapter 9) very good patterns can be created without affecting the substrate, and c) perhaps most significant, FIB technology could be used for mask inspection, mask repair, and repair of other material patterns directly on the chip.

FIB lithography is similar to electron-beam lithography. Since the ion fluxes are, at this time, low in FIB machines, a long exposure to completely develop the resist is necesary. However, because the ion range is extremely low compared to the electron range in a similar resist, only a thin layer of resist needs to be exposed. This reduces the ion flux by a factor of three to five. Also, since ions cause more damage or chemical activity, the resist sensitivity to ions is usually larger by orders of magnitude. For example, the ion dose needed to expose polymethylmethacrylate (PMMA) is nearly a factor of 100 lower than that for electrons. Other resists that have been tried for FIB lithography include chloromethylstyrene (CMS) and polytrimethylsilystyrene-co-chloromethylstyrene. Since the resist thickness that can be employed is limited, the resist must be very good with respect to the pinhole density, other defects, and thickness uniformity on a larger area substrate.

The greatest asset of FIB is, however, in the area of the direct writing of lines or patterns without the use pattern-transfer media such as photoresists. Much research and development has to occur for FIB to find application in semiconductor manufacturing. Needed research efforts include the development of reliable and long-life ion sources of all kinds, process evaluation and reliability, establishing the need for FIB by making unique devices that cannot be made otherwise, investigating the side effects of ion damage, and FIB costs.

11.2.3 Epitaxy of Electronic Materials

A film grows epitaxially if its lattice structure is an extension of the substrate crystal. The process of growing such films is called *epitaxy.* If the film has the same chemical composition as the substrate, for example, silicon on silicon or GaAs on GaAs, the process is termed *homoepitaxy.* On the other hand, if the epitaxial film is a different material, the process is called *heteroepitaxy,* for example, germanium on silicon, GaAs on silicon, or cobalt disilicide on silicon. There are several ways of forming epitaxial films. They include a) deposition from the vapor phase — molecular-beam epitaxy (MBE) and metalorganic chemical-vapor deposition (MOCVD) and b) thermally induced epitaxial growth of amorphous or polycrystalline deposited films.

MOCVD, in recent years, is an important technique for growing high-quality epitaxial films of III-V and II-VI semiconducting materials. This process is similar to the other CVD processes described in Chapter 7 and uses organometallic vapor species such as trimethyl Ga, Al, Sb, In, or Pb and dimethyl tellurium. Arsine, phosphine, stibine, hydrogen selenide, and hydrogen sulphide provide other components. More recently, nontoxic organometallic has become a source of arsenic. Although MOCVD of epitaxial compound semiconductors has become a powerful method, it, like any other CVD technique, will be difficult to incorporate into an in-situ processing scheme or any ultra-high vacuum (UHV) schemes. MBE on the other hand, can easily be incorporated in such schemes.

11.2.3.1 Molecular Beam Epitaxy

MBE is basically a very clean UHV evaporation technique. The essential elements of MBE are schematically shown in fig. 11.15. An MBE machine consists of vapor sources, a substrate, and an electron source and screen for in-situ electron diffraction studies of the growing layers. All these are enclosed in a UHV chamber. The use of 10^{-7} – 10^{-9} Pa vacuums is not only needed to eliminate oxygen, water, and hydrocarbons, but also to produce molecular beams with negligible particle-particle interaction. Note that under such high-vacuum conditions, the mean free path is extremely large and the probability of internal scattering very low (see Chapter 6). Any change in evaporation rate is reflected in the deposition rate. Thus a precise control of deposition rate can be achieved if the evaporation rate can be controlled precisely, as is the practice.

A sophisticated MBE equipment is shown in fig. 11.16. It consists of several UHV chambers that are interconnected with UHV isolation valves, various analytical facilities, sample manipulators and, in this particular case, four different sources. Presently these machines are designed to meet production requirements relating to sample areas, throughput, and reproducibility. For in-situ processing such equipment can be attached to FIB or other UHV or HV stations with appropriate load locks. A load lock is a chamber isolating two chambers with similar or dissimilar vacuums so that sample transfer can be carried out without changing conditions in the individual chambers.

Molecular beam sources are typically heated effusion furnaces containing desired elemental charges. For silicon MBE, an e-gun evaporator is used. For compound semiconductors effusion sources of Ga, As, Al, P, In, etc. are used. For doping during film growth, both evaporation sources, as described above, and ionized sources are available. Ionized sources are basically low-energy

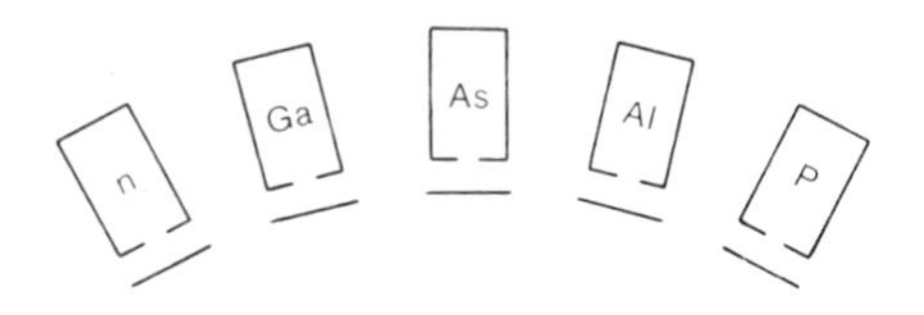

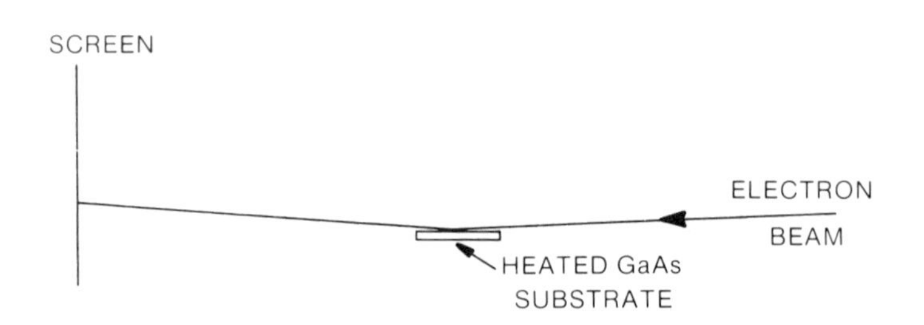

Fig. 11.15. A typical placement of various sources for MBE of III-V semiconductor layers.

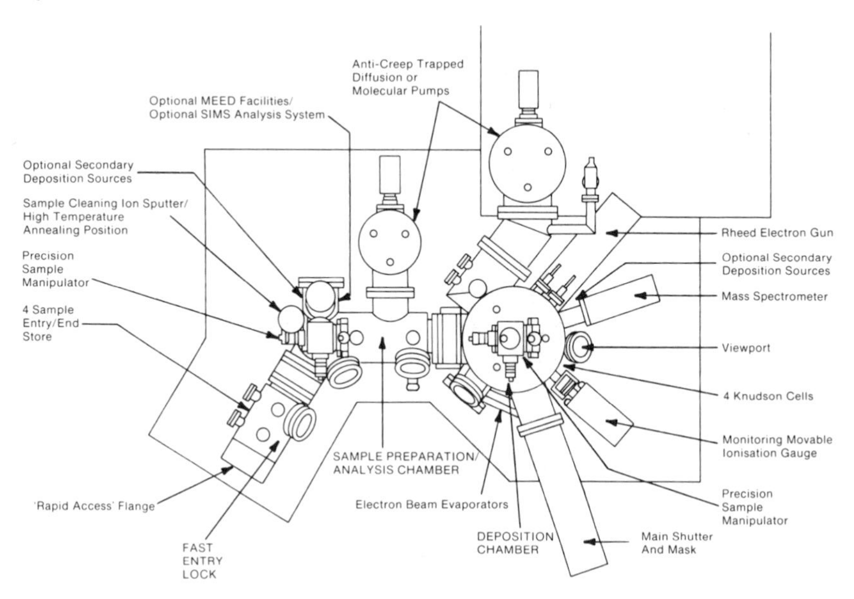

Fig. 11.16. A sophisticated MBE facility.

ion-implantation sources that can be simultaneously used. This provides an accurate control of doping levels and increases the probability of species sticking on to the substrate surface to almost unity. Ionized doping sources have also made it possible to use practically any dopant to any desired concentration in the film. For details see references 37-41.

MBE offers an UHV advantage, leading to in-situ characterization and control of layers during the film growth. This allows the creation of controlled heteroepitaxial layers, strained layer epitaxy, and super lattices. These lead to new forms of engineered materials with extremely interesting properties. Selectively doped heterostructure transistors (SDHT) using AlGaAs and GaAs, $In_{0.52}Al_{0.48}As$ and InP and others [41], Si/$CoSi_2$/Si metal or permeable base transistors [42,43], resonant tunneling bipolar transistors (RTBT) in which the wide band gap of $Al_xGa_{1-x}As$ emitter is graded by chopping the molecular beam from the aluminum effusion cell so that average composition changes from $x = 0.25$ to 0.07 [44], and super lattices using Si/Ge, GaAs/InAs, GaAs/AlAs, and other combinations have been fabricated using MBE systems.

11.2.4 Rapid Thermal Processing

High-temperature processing steps are primarily associated with oxidation, dopant activation and diffusion, P-glass flow, silicide anneal, and gettering. Other comparatively lower temperature processes (400-700°)C are CVD operations and aluminum anneal. Rapid thermal annealing (RTA), using tungsten halogen lamps or similar radiation heating sources, can be used effectively for all those processes, except for thick oxide growth, such as the field oxide.

In a typical rapid thermal annealing setup, a quartz chamber with a gas inlet and outlet is blanketed with banks of tungsten halogen lamps. Wafer is held on a special quartz carrier inside the chamber. By applying power a fast heating of the wafer results. Computerized control of power, temperature, and time is available. A typical temperature-time profile is shown in fig. 11.17. A typical rise time to the desired temperature varies approximately from 1 to 4 seconds. Similarly the cool time is usually a few seconds. Cooling time is generally longer since it depends on several factors including gas flow rate and thermal conductivity of the wafer and surroundings. The time at temperature is determined by the process desired and normally varies from a few seconds to about 2 minutes. Thus RTA takes much longer compared to lasers or electron beams but is much faster compared to furnace annealing. In cases short anneals are necessary — such as to form a shallow p-n junction — RTA is very effective.

Rapid thermal processes (RTP) now include CVD, diffusion, oxidation, implant anneals, damage anneals, metal anneals, silicide formation, and P-glass flow. Equipment has been designed to carry out these processes in all types of gaseous ambients or in vacuum. RTP, however, is a large-area process and equipment that can handle wafers up to six inches in diameter is available. Thus, wherever a whole wafer anneal can be tolerated or is required, RTP is suitable. However, in cases where localized annealing (or

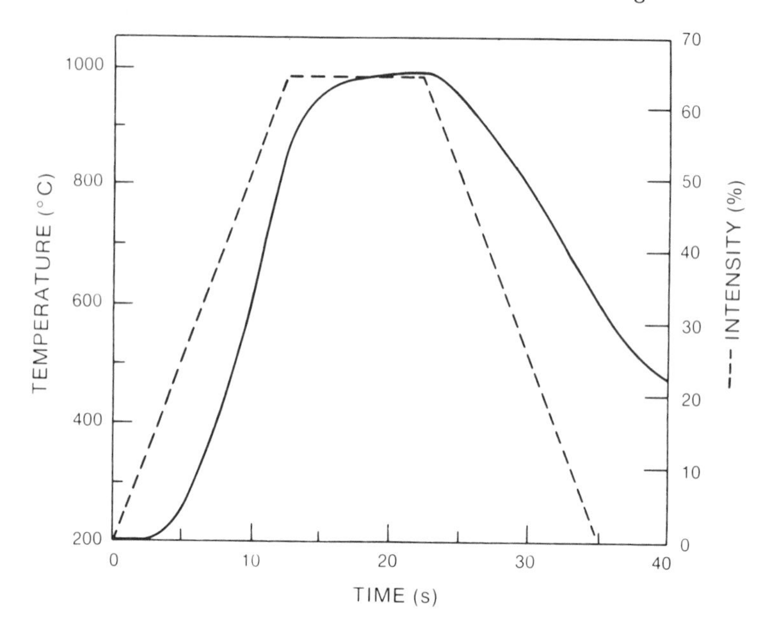

Fig. 11.17. A typical rapid thermal anneal cycle is shown as a temperature-time profile. The lamp intensity profile is also shown for a given rapid thermal annealer. Courtesy of AG Associates, CA.

heating) is necessary, laser-assisted or electron-beam annealing has to be carried out. Localized heating has the benefit of avoiding possible heating effects on other parts of the wafer or integrated circuits.

11.3 Side Effects of Photon and/or Particle-Beam Processes

Defect generation and the effect on the device characteristics are of biggest concern when using beam processes. Yet this aspect of the beam processing has been overlooked to a large extent, although considerable attention was paid to these effects during early days of laser annealing experimentation. Both laser and ion beam processes lead to defect generation by virtue of the thermal, ion, and/or radiation damage associated with exposure of materials to particle beams. *Raising the temperature*, practically instantaneously, and cooling at an equally fast rate leads to thermal damage (such as slip dislocations, quenched-in point defects, bowing, and dislocations caused by uneven thermal expansion of different materials) and also to trapping of impurities. Photon-electron interactions can cause electrical damage as well. In addition, materials of different types could interact leading to undesirable products, e.g., Si and SiO_2 reacting to give SiO, which could vaporize, and metals and silicon reacting to form silicides or undesirable solid solutions. Many of these thermal effects have to be carefully studied, evaluated, and techniques devised to eliminate or anneal them. *Ions or electrons* cause damage by virtue of the charge on the moving species, in addition to the damage caused by the momentum transfer and that associated with whatever heating occurs due to the energy transfer. Such effects lead to dislocation of atoms and electrons from their sites and cause point defect and dislocation generation (the latter especially after annealing). Ions or electrons can impart momentum to surface impurity atoms causing knocked in contamination. Such effects have been characterized to a large extent for large-area implantation and for e-beam processes. However, there is a lack of this understanding in the case of focused ion beams, especially when they are employed to create very small dimensions. Thus, like laser-induced thermal defects, the FIB-introduced defects must be evaluated carefully. *Radiation damage* associated with high-energy radiations, e.g., x-rays, is generally of electrical nature — although some point defects could also be generated.

In all cases, isothermal low-temperature annealing has been known to anneal out most, if not all, of the damage. The effect of such annealings and rapid thermal annealing need more evaluation for different combinations of the processes, i.e., for accumulative effects.

11.4 In-Situ Processing

The key phrase describing in-situ processing is the following: "Evolution of a practical, resistless, maskless, and localized processing sequence". The weakest link in the conventional processing sequence is the creation of patterns using photolithography and patterning techniques. Even for a maskless direct-writing electron beam or ion-beam lithography, a large number of problems remain unsolved, with no answers in sight.

The development of in-situ processing that will utilize beam processes, with or without existing conventional processes discussed in the rest of the book is an attempt to develop fabrication techniques for unique and new devices, circuits, and chips which cannot be made using conventional methods. Table 11.7 suggests the possible elements of in-situ processing that can substitute conventional processing steps and lists concerns associated with such processes. At this writing, research work is carried out to investigate these concerns and to eventually choose optimum in-situ processing sequences.

In-situ processing schemes can be realized in vacuum and near-vacuum environments. Both laser and ion-beam processes could be tried with the priorities suggested above. For direct writing both beam and wafer stage motions will be necessary. Thus, for lasers, accoustically coupled $x - y$ motion generating methods will be essential. The technology for such writing is known and can be easily implemented. A four- to five-chamber intercon-

TABLE 11.7: Comparison of the conventional and in-situ process

Conventional Process	In-Situ Process Elements	Concerns
Surface Cleaning	1. Large area back-sputtering or FIB for localized etching followed by RTA.	Cross-contamination, ion damage. Proper selection of ion energies to do the best job. Annealing of the damage.
	2. RTA – to cause $Si + SiO_2 = SiO(g)$ reaction and leave silicon surface free of native oxide.	Determination of optimum conditions, effect on diffusions and other materials on wafer. Good only for silicon cleaning and not for localized cleaning unless using focused lasers.
	3. Large area RIE or etching contamination	Not good for small plasma localized regions, cross contamination.
Deposition		
Epitaxial Silicon	1. Molecular Beam Epitaxy	Large area techniques
	2. Ion cluster beam	
Polysilicon	1. Laser direct writing	Since molton silicon is deposited $Si + SiO_2 = SiO(g)$ reaction could occur. Is this a reliability problem especially on thin oxide ($\leq$ 100 A) capactiors?
		Could writing speed be mass transport limited?
		Other dopants
		Thermal effects on substrates?
		How to eliminate solidification effects such as cusping?
	2. Large area plasma deposition	Patterning by direct laser etching or ion milling? In-situ doping and grain growth?
	3. FIB direct writing	Ion damage in substrate especially in thin oxide? Ion currents?
Aluminum	1. Direct laser writing	Resistivity, carbon and oxygen contamination, surface morphology.
	2. Large area ICBD	Patterning by laser or ion milling? Stability, step coverage, and crystallinity?
	3. Large area sputtering	Patterning by lasers or ion milling.

TABLE 11.7: Comparison of the conventional and in-situ process (continued)

Conventional Process	In-Situ Process Elements	Concerns
SiO_2 [a]	1. Large area field oxide by ICBD, intermediate dielectric by ICBD, plasma, bias-sputtering, or laser deposition	Oxide quality, step coverage, traps.
	2. FIB direct writing using oxygen (nitrogen) ions followed by RTA	Ion damage, flux, and properties? Ion source?
	3. Direct laser writing of oxides	Not successful yet, but should be tried.
Silicides	1. FIB using metal or silicon ion beams by RTA	Not yet tried. Metal source (Pt) only available at this time.
	2. Self-aligned large area technology using RTA	Will require development of selective dry metal etch.
Doping		
Direct writing very shallow (<1000 Å)	1. Laser assisted doping	Effect of recystallization of molten (doped) silicon layer.
	2. FIB	Ion damage, straggling, and channeling. Beam stability at very low energies.
Junctions with depth 1000 Å or greater.	1. FIB for direct writing	Ion damage, straggling,
	2. Large area ion implantation	Ion damage
Etching		
Selective area etching of Silicon	1. Laser assisted etching etching in presence of reactive component gas.	Redeposition or decomposition of product. Thermal damage.
	2. FIB milling	Ion damage, redeposition of milled species.
Large area etching or Si, SiO_2 or other materials	1. Reactive ion or plasma etching	
Selective area etching of SiO_2	1. FIB milling	Ion damage, redeposition of milled material.
	2. Laser etching	Needs considerable development effort.
Etching of Al and other metals	Selective area etching without a mask - requires considerable development effort.	

TABLE 11.7: Comparison of the conventional and in-situ process (continued)

Conventional Process	In-Situ Process Elements	Concerns
Annealing	1. RTA for all	
	2. Localized annealing by lasers	Selective annealing using proper (absorbing) wave lengths need by explored.
Other metallizations	1. Large area ICB deposition of refractory refractory metals	Step coverage
	2. Localized laser-assisted deposition of other metals	Contamination, higher resistivity

[a] Thick oxides are of importance. Thin gate oxides can be easily done by RTA.

nected equipment will be the best for this purpose. There should be vacuum interlocks that will permit travel from a low vacuum to UHV chamber or vice versa. A vibrationless pumping method is recommended.

One of the chambers will carry out all initial loading, cleaning using RTA following a backsputtering etch, and annealing using RTA. In this chamber a sputtered deposition of large area film could be carried out if necessary.

In the second chamber a FIB system, with easily changeable source head, will be housed, together with analytical capabilities that will provide surface and bulk analyses. In this chamber all suggested FIB operations and necessary materials analysis could be performed. Obviously, this chamber will house a movable stage whose motion can be coupled with the focused ion-beam motion — a technology well known for electron-beam direct-writing equipment.

The third chamber will house a laser direct-writing apparatus. This will be used to try all laser experiments and will require cleaning more often than others. Thus, the deposition chamber should be readily accessible. Two exchangeable loading stages should be considered.

A fourth chamber will be needed for large-area deposition of intermediate dielectrics or metals. We think an ion-cluster beam-deposition system will be highly appropriate, although a plasma deposition or sputter-deposition system could also be used. The latter processes are recommended as alternatives.

The fifth and last chamber will be necessary for MBE applications where both semiconductors and metal silicides can be deposited. Such deposition may not be required for a conventional device but will be invaluable when trying to make certain novel devices such as a metal-base transistor or a three-dimensional device.

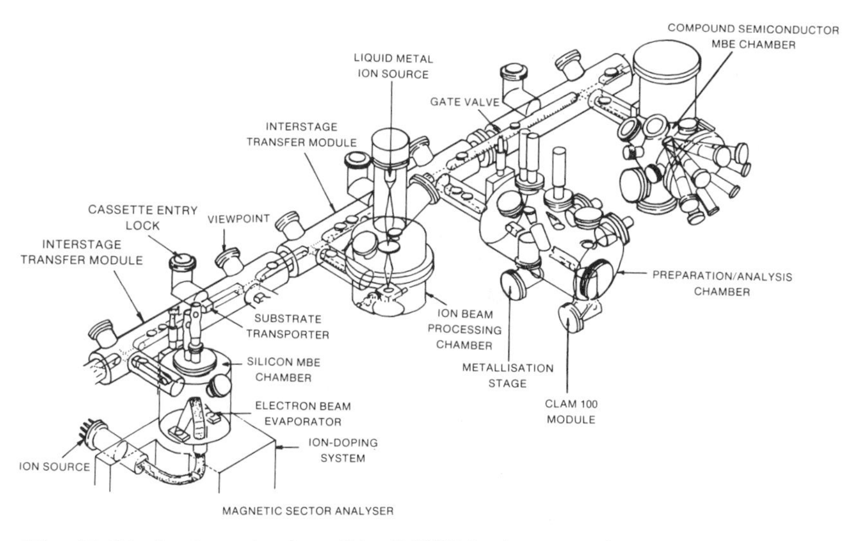

Fig. 11.18. A schematic of possible all-UHV in-situ processing system.

A schematic of an in-situ processing machine, which consists of several of the above-mentioned chambers is shown in fig. 11.18.

Device and circuit applications of in-situ processing fall into a number of categories:

1. Unique device structures;
2. Improved yield technology; and
3. Custom device and circuits.

In the first category are included devices such as a newly demonstrated metal base transistor [45], with a potential of subpicosecond access time, which requires a complete UHV processing. Other examples include hypernarrow base (< 1000 nm wide) lateral pnp or npn bipolar transistor. Extremely tight impurity profiles are necessary, along with narrow metallization schemes and perhaps the use of silicide layers. The realization of such structures require focused ion beams and epitaxial growth. In general, the use of focused ion beams to tailor impurity profiles in all three dimensions should lead to improved device performance. The applications involve the doping of extremely small regions, as mentioned above, but also larger regions easily accessible with in-situ processing, but difficult to implement with conventional processing. An example is the retrograde doping of CMOS wells, which would easily be achieved with a combined FIB/MBE system. Indeed, using FIB, the doping profile or the well could be made retrograde along the vertical boundaries as well. In another area, three-dimensional transistors, with buried epitaxial silicide metallization and heteroepitaxy of silicon on top of such metallization materials, are excellent candidates for in-situ processing.

The second category includes technologies whose yield could be improved substantially through in-situ processing. This covers a broad spectrum of devices. At the high level of integration are the multimegabit "superchips," which will require a defect-free processing environment. At the other end of the spectrum are more immature technologies, such as compound semiconductor devices, which have traditionally suffered from lack of suitable passivation technology. In this case in-situ processing will result in significant yield improvement. For example, maskless GaAs device in-situ fabrication, which included localized FIB doping and MBE multilayer growth, has recently been reported [46] in a system very similar to that of fig. 11.18.

In the category of custom ICs, the ability to specifically tailor the performance of the IC to the needs of the application merits a considerable price premium. In-situ processing can have a tremendous impact here in two ways. First, by being able to process numerous alternatives in parallel, the cost and time for determining the optimum configuration could be greatly reduced. Secondly, by its very nature, in-situ processing involves localized fabrication and thus the ability to customize the process and the device/circuit performance. The question that remains to be answered is whether the overall equipment set necessary for in-situ processing can perform the integrated fabrication functions. Since application-specific ICs (ASICs) represent a growing percentage of the overall IC market, a base of potential uses of in-situ processing exists.

11.5 Summary

In this chapter we examined the processes that may be adopted in the future for device and circuit fabrication. The concept and applicability of the in-situ processing was also discussed. Specifically, individual photon and ion, and molecular beam processes were considered. Lasers and ion beams, coupled with molecular beam epitaxy, show a greater promise for the fabrication of superfast and specialized devices.

References for Chapter 11

1. There are a large number of papers that deal with photon, electron, ion, or atom beam processing and report many new concepts and devices. Most of them have been presented in the symposia arranged by the Materials Research Society, the Electrochemical Society, and the American Vacuum Society. Practically all of these papers are available in the symposia volumes published by these societies.

2. P.H. Singer, Semiconductor International, 8, 68 (1984).
3. Hitachi 1982 Study - Source: G.B. Larrabee's paper presented at the Am. Vac. Soc. Texas Chapter meeting at Dallas, Texas, June 3/4 (1985).
4. A. Tasch, paper presented at the Am. Vac. Soc. Texas Chapter meeting at Dallas, Texas, June 3–4 (1985).
5. Called sulfuric-peroxide or P-clean.
6. W. Kern, D. Puotinen, RCA Rev. 31, 187 (1970).
7. W.J. Bertram, in "VLSI Technology"S. Sze, ed., McGraw-Hill, NY, p. 599 (1983).
8. G.B. Larrabee, Chem. Eng. June 10, 51 (1985).
9. M. Accomazzo, K.L. Rubow, B.Y.H. Liu, Solid State Tech. March, 27, 141 (1984).
10. G.B. Larrabee, Paper presented at the Am. Vac. Soc. Texas Chapter meeting at Dallas, Texas, June 3/4 (1985).
11. D.J. Ehrlich, J.Y. Tsao, in "VLSI Electronics Microstructure Sciences," Vol. 6, N.G. Einspruch, ed., Academic Press, Orlando, p. 129 (1983).
12. R.M. Osgood, H.H. Gilgen, Annual Rev. Mats. Science 15, 459 (1985).
13. A.E. Siegman, "Lasers,"Univ. Science Books, Mill Valley, CA, p. 1004 (1986).
14. J.F. Gibbons, in "Semiconductors and Semimetals, Vol. 17, CW Beam Processing of Silicon and Other Semiconductors,"J.F. Gibbons, ed., Academic Press, Orlando, p. 1 (1984).
15. A. Lietoila, R.B. Gold, J.F. Gibbons, L.A. Christel, in "Semiconductors and Semimetals, Vol. 17, CW Beam Processing of Silicon and Other Semiconductors,"J.F. Gibbons, ed., Academic Press, Orlando, p. 71 (1984).
16. K.F. Lee, T.J. Stulz, J.F. Gibbons, in "Semiconductors and Semimetals, Vol. 17, CW Beam Processing of Silicon and Other Semiconductors,"J.F. Gibbons, ed., Academic Press, Orlando, p. 227 (1984).
17. J.S. Williams, "Laser and Electron Beam Processing of Electronic Materials,"C.L. Anderson, G.K. Celler, G.A. Rozgonyi, eds., The Electrochem. Soc., Pennington, NJ, Vol. 80-1, p. 249 (1980).
18. T. Shibata, A. Wakita, T.W. Sigmon, J.F. Gibbons, in "Semiconductors and Semimetals Vol. 17, CW Beam Processing of Silicon and other Semiconductors,"J.F. Gibbons, ed., Academic Press, Orlando, p. 341 (1984).
19. B. McWilliams, paper presented at the SRC In-Situ Processing Workshop, Chapel Hill, NC, June 17-18 (1982).
20. D.J. Ehrlich, J.Y. Tsao, Appl. Phys. Lett. 41, 297 (1982).
21. R. Srinivasan, IBM-T.J. Watson Research Center, Yorktown Heights, private communication (1985).

22. D.J. Ehrlich, R.M. Osgood, D.J. Silversmith, T.F. Deutsch, IEEE Electron Dev. Lett. 1, 101 (1980).
23. H.H. Gilgen, C.J. Chen, R. Krchnavek, R.M. Osgood in "Laser Diagnostics and Photochemical Processing," D. Bauerle, ed., Springer, Berlin, p. 225 (1985).
24. J.Y. Tsao, D.J. Ehrlich, D.J. Silversmith, R.W. Mountain, IEEE Electron Dev. Lett. 3, 164 (1982).
25. D.J. Ehrlich, J.Y. Tsao, D.J. Silversmith, J.H. Sedlack, R.W. Mountain, W.S. Graber, IEEE Electron Dev. Lett. 5, 164 (1984).
26. D.J. Ehrlich, R.M. Osgood, T.F. Deutsch, Appl. Phys. Lett. 36, 916 (1980).
27. R.A. McMahon, H. Ahmed, in "Laser and Electron Beam Processing of Electronic Materials," C.L. Anderson, G.K. Celler, G.A. Rozgonyi, eds., The Electrochem. Soc., Pennington, NJ, p. 123 (1980).
28. K.N. Ratnakumar, R.F.W. Pease, D.J. Bartelink, N.M. Johnson, J.D. Meindl, Appl. Phys. Lett. 35, 463 (1979).
29. T.O. Yep, R.-T. Fulks, R.A. Powell, in "Laser and Electron Beam Solid Interactions and Materials Processing," J.F. Gibbons, L.D. Hess, T.W. Sigmon, eds., North-Holland, NY, p. 345 (1981).
30. J.J. Muray, Semiconductor International 8, 130 (1984).
31. J. Melngailis, J. Vac. Sci. Technol. *B*5, 469 (1987).
32. T. Kato, H. Morimoto, K. Saitoh, H. Nakata, J. Vac. Sci. Technol. *B* 3, 50 (1985).
33. T. Takagi, Proc. MRS Symp. 27, 501 (1984).
34. J. Wong, Ph.D. Thesis, RPI, Troy, NY (1987).
35. S.-N. Mei, T.M. Lu, J. Vac. Sci. Technol. 6, 9 (1988).
36. S.P. Murarka, M. Kelly, D.S. Yaney, J.H. Levinstein, upublished work.
37. A.Y. Cho, J.R. Arthur, Prog. Solid State Chem. 10, 157 (1975).
38. B.R. Pamplin, ed., "Molecular Beam Epitaxy," Pergamon, Oxford (1980).
39. J.C. Bean, in "Impurity Doping Processes in Silicon," F.F.Y. Wang, ed., North-Holland, NY, p. 175 (1981).
40. A.C. Gossard, in "Preparation and Properties of Thin Films," K.N. Tu, R. Rosenburg, eds., Treatise on Materials Science and Technology, Vol. 24, Academic Press, Orlando, p. 13 (1982).
41. C.W. Tu, R.H. Hendel, R. Dingle, in "Gallium Arsenide Technology," D.K. Ferry, ed., SAMS, Indianapolis, p. 107 (1985).
42. S.M. SZE, "Physics of Semiconductor Devices," Wiley, NY, p. 567 (1969).
43. C.O. Bozler, G.D. Alley, IEEE Elec. Dev. ED-27, 1128 (1980).
44. F. Capasso, S. Sen, A.C. Gossard, A.L. Hutchinson, J.H. English, IEEE Electron Dev. Letters EDL-7, 573 (1986).
45. J.C. Hensel, A.F.J. Levi, R.T. Tung, J.M. Gibson, Appl. Phys. Lett. 47, 151 (1985).
46. E. Miyauchi, T. Morita, A. Takamori, M. Arimoto, Y. Bamba, H. Hashimoto, J. Vac. Sci. Technol. 134, 189 (1986).

Problem Set for Chapter 11

11.1 In Chapter 9, electromigration related failure of metal lines was discussed. Assume that the failure is related to Joule heat generated in the conductor due to the passage of large currents. Formulate an equation similar to eq. (11.5) for the temperature rise in metal lines when (a) there are no heat sinks (like underlying substrate or an overlying capping insulator) and (b) the conductor is properly heat sunk with the substrate and an overlying layer.

11.2 Consider laser direct writing of polysilicon lines on SiO_2. During the deposition process the surface temperatures become so high that the polysilicon deposit melts during the short laser exposure. Comment on the temperature rise in the underlying substrate and the stability of SiO_2 layers on which polysilicon lines are formed.

11.3 A gold interconnection line ($0.5\mu m$ wide and $0.5\mu m$ high) has developed a crack at an angle of 30° to the length of the line. Crack width is $0.5\mu m$. FIB is to be used to repair this line using Au^+ beam. Assuming that (a) the crack is fully developed so that the interconnection line has become discontinuous, (b) crack walls are perpendicular to the surface, and (c) beam diameter is $0.25\mu m$, calculate the time required to completely fill the crack if the beam current is $10\mu a/cm^2$.

11.4 Which of the silicon and GaAs technologies will benefit more from the use of beam processes and in-situ processing? Explain.

Appendix I

A Brief Review of Chemical and Thermodynamic Fundamentals

There are two key realizations in chemistry: The world can be subdivided into basic atomic units of fixed mass and in a chemical reaction, fixed numbers of atoms (reactants) merge with each other to form fixed numbers of products. The reactants and products may form associations called *molecules*. Thus, we can write

$$aA_n + bB_n + \ldots \rightarrow a'A'_n + b'B'_n + \ldots \tag{A.1}$$

where "unprimed" variables are reactants, "primed" variables are products. The subscripts refer to the molecular association. For example

$$2H_2 + O_2 \rightarrow 2H_2O. \tag{A.2}$$

Normally, the quantities of atoms are expressed in moles: one mole corresponds to a number of atoms or molecules equal to Avogadro's number: 6.02×10^{23}. One gram-atomic (or gram-molecular) weight has Avogadro's number of atoms (or molecules) in it. An alternative way of stating reaction (A.2) is:

2 moles of hydrogen react with 1 mole of oxygen to make 2 moles of water.

A gram-atomic (molecular) weight is a number of grams of atoms (molecules) equal to the atomic (molecular) weight of the species considered.

It should be noted that for any reaction, such as the one generally expressed in (A.1), the arrow can point to the left as well as to the right. When hydrogen and oxygen are bled into a common chamber and energy is supplied to the system to surmount the "reaction barrier," the system will react (usually violently) to create water vapor. Eventually, the "nonequilibrium" settles into an "equilibrium" in which some (admittedly small) amount of water will dissociate into constituent reactants. In equilibrium, we can write

$$K_{\text{eq}} = \frac{[H_2O]}{[H_2]^2[O]}, \tag{A.3}$$

where K_{eq} is called an equilibrium constant. The square brackets around a species generally represent the concentration of the species, or number of moles per unit volume present in the reaction chamber. Note, in eq. (A.3) $[H_2]$ is squared because of the 2 pre-factor in eq. (A.2). Usually, for atoms or molecules in the gaseous state, concentrations are substituted for by "partial pressures." The partial pressure is just the fractional contribution of the ith species to the total pressure. For an *ideal* gas

$$N_i = \frac{p_i}{P}. \tag{A.4}$$

where N_i = fraction of atoms (molecules) per unit volume of type i, P = total pressure of the system, and p_i = partial pressure of ith species. Equation (A.3) says that in equilibrium there is a well-defined ratio of products to reactants.

The equilibrium constant is determined experimentally. It is related to thermodynamic properties of the system. To see this, let us briefly review some thermodynamic fundamentals. From the first law of thermodynamics, we know

$$dE = \bar{d}Q - \bar{d}W. \tag{A.5}$$

That is, the change in the internal energy of a system (dE) is the heat flowing into the system *minus* the work done by the system on its surroundings. The bars above the d's in the differentials $\bar{d}Q$, $\bar{d}W$ indicate the differentials are *inexact*. That is, they depend on the specific path taken through pressure, volume, and temperature coordinates. dE, though, is an *exact* differential. It is path independent, a true thermodynamic quantity. Another thermodynamic quantity is the entropy:

$$dS = \frac{\bar{d}Q}{T}. \tag{A.6}$$

Related to (A.5) is a quantity called enthalpy:

$$H \equiv E + PV, \tag{A.7}$$

or, in differential form,

$$dH = dE + PdV + VdP. \tag{A.7a}$$

When the pressure is constant dH is just dQ. dH is sometimes called the *heat-of-formation* of a substance. dH is a measure of the energy necessary to form a substance in a given state. For eq. (A.2) we can define a dH that is related directly to the number of bonds broken (and the energy required to break these bonds), as well as the number of bonds formed (and the net amount of energy released on forming the bonds). It was thought at first that dH was an indicator of the tendency of a reaction to "go"or not to "go."If the system gave up heat to the surroundings (negative dH) the reaction was easily accomplished, a positive dH meant the reaction was hard to accomplish. This is true *only for a very large dH* (>100 kcal/mole). Note: dH, generally referred to as ΔH, is a tabulated thermodynamic quantity. It can be looked up (usually) in handbooks or texts.

The knowledge of dH alone is not sufficient to predict chemical spontaneity. To accomplish this one must also consider the entropy change dS. According to the Boltzmann interpretation, the entropy is related to the total number of energy states available to the system. Boltzmann defined the entropy of a system in terms of the number of states available to the system

$$S = \mathrm{k} ln z,$$

where k is the Boltzmann constant and z is the number of states available for the system to occupy. For example for n dopants on N lattice sites

$$z = \frac{N!}{n!(N-n)!}.$$

Such an expression makes it easy to calculate entropy changes in mixing phenomena such as doping or point-defect introduction.

It is desirable for a system to increase the number of available internal states. An indicator of chemical spontaneity must account for this. Gibbs derived such an indicator relation. He called it *free energy.* It is now called *Gibbs free energy.*

$$G = E + PV + (-TS) = H - TS, \tag{A.8}$$

or

$$dG = dE + PdV + VdP - TdS - SdT. \tag{A.8a}$$

From eq. (A.5)

$$dE = dQ - PdV.$$

From eq. (A.6)

$$dE = -PdV + TdS.$$

Thus

$$dG = VdP - SdT. \tag{A.9}$$

At constant temperature

$$dG = VdP. \tag{A.9a}$$

For an ideal gas

$$V = \frac{RT}{P}. \tag{A.10}$$

Therefore,

$$dG = RTdln(P). \tag{A.11}$$

Thus, for a component, i, of a reaction

$$dG_i = RTln(p_i). \tag{A.11a}$$

Next, let us define the standard state of a substance as the pure liquid, solid, or gas at 1 atmosphere pressure at the temperature under consideration. Let us now integrate (A.11a) from the standard state to some final partial pressure

$$\Delta G_i = G_i - G_i^o = RTln\left(\frac{p_i}{P_o}\right) = RTln(p_i), \tag{A.12}$$

since P_o is 1 atmosphere pressure.

Consider eq. (A.1). If we wish to find the *total* free energy change of reaction ΔG, the free energy changes ($\Delta G_{i's}$) for the reactants on the left-hand side of eq. (A.1) are subtracted from the free energy changes ($\Delta G_{i's}$) for the products on the right of eq. (A.1), i.e.,

$$\Delta G = \sum_{\text{products}} \Delta G_i - \sum_{\text{reactants}} \Delta G_i. \tag{A.13}$$

Using eq. (A.12) for individual $\Delta G_{i's}$, one obtains

$$\Delta G = RTln\left(\frac{p_A^a p_B^b \cdots}{p_{A'}^{a'} p_{B'}^{b'} \cdots}\right) = -RTlnK_{\text{eq}}. \tag{A.14}$$

We've thus related K_{eq} to a *tabulated* thermodynamic quantity ΔG. The deltas on the right-hand side of (A.13) refer to the free-energy changes incurred on taking the substance from the reference or standard state to some final state that differs from the standard state in temperature or pressure. For a reaction to occur K_{eq} must be large, i.e., the concentrations of products are larger than those of reactions and ΔG is negative. There is, thus, a lowering of free energy associated with a reaction leading to products. This is consistent with the generalized thermodynamic concept that the total energy of the system tends to be at a minimum.

One uses the above concept in predicting a thermodynamic feasibility of a reaction of phenomenon by calculating ΔG value from the known ΔG_i values. If ΔG turns out to be negative, the reaction may occur. The more negative is ΔG, the greater is the probability for the reaction to occur. The final word about the occurrence of the reaction, however, rests with the kinetics that determines the rate at which a reaction may occur. Negative ΔG simply indicates thermodynamic probability that the reaction may proceed. Whether detectable amounts of products will form is kinetically controlled. One must, however, be aware of the fact that by changing conditions, such as temperature and pressure, which may make the reaction kinetically favorable, ΔG is also changed. For example, consider reaction (A.2). Thermodynamically this reaction leads to a lowering of free energy, i.e., ΔG is negative. However just by mixing of hydrogen and oxygen does not lead to spontaneous reaction. Raising the temperature of mixture, for example, by igniting a matchstick, is one way of causing this reaction to occur. The reaction is exothermic, leading to considerable release of energy to the system and thus causing the reaction to proceed instantaneously.

At this point we introduce a quantity μ called *chemical potential.* The chemical potential μ_a of the species A present in a mixture of species is defined as

$$\mu_a = \left(\frac{\partial G}{\partial n_{\mathrm{a}}}\right)_{p,T,n_i} \tag{A.15}$$

where the right-hand side represents the change in the total free energy (G) of the system associated with the infinitesimal change in the number of moles (n_{a}) of A when pressure, temperature, and the number of moles of other species are kept unchanged. A result of this definition is that the total free energy G is now given as

$$G = \sum_i n_i \mu_i. \tag{A.16}$$

For a pure material μ is then the total free energy per mole.

Associated with a reaction or a phase change (i.e., melting, solidification, sublimation, etc.) there is heat absorbed or given up. For example, when ice melts, it takes up heat from the surrounding fluid and the fluid temperature lowers. Such a reaction is termed *endothermic* — taking up heat. Other reactions require that the energy is given up to the surroundings, as in the case of reaction (A.2) discussed above. Such processes are called *exothermic.* Energy-flow sign conventions may be somewhat confusing and should also be mentioned. When the system gives up energy (exothermic reactions) to its surroundings the sign of energy or heat dQ is *negative.* For the endothermic reactions the energy from the surroundings is absorbed by the systems and the sign for dQ is positive.

Generally, the rate (r) at which a reaction occurs is related to the free-energy change by an equation that is similar to eq. (A.14):

$$\begin{aligned} r = \text{reaction rate} &= \text{const. } \exp\left(\frac{-\Delta G}{RT}\right) \\ &= \text{const. } \exp\left(\frac{\Delta S}{R}\right) . \exp\left(\frac{-\Delta H}{RT}\right). \end{aligned}$$

ΔH, the change in the enthalpy, is called the *activation energy* of the reaction. The above equation is commonly written as

$$r = r_o \exp\left(\frac{-Q}{RT}\right), \tag{A.17}$$

where r_o now has both the constant and entropy terms in it and Q is equal to ΔH. Eq. (A.17) is commonly known as the Arrhenius equation, reflecting an exponential dependence of reaction rate on Q and $1/T$.

For simplicity, eq. (A.14) was derived for ideal gases. Finally we carry this analysis over to non-ideal gases or liquids using the *fugacity* concept. We demand that the form of eq. (A.11a) be maintained but assume that the partial pressure contributes some altered amount f_i, to the free energy. Thus

$$dG_i = RTln(f_i). \tag{A.11b}$$

We further define the *activity* as the ratio of the fugacity at T to the fugacity in the standard state

$$a_i = \frac{f_i}{f_i^o}. \tag{A.18}$$

Under these conditions

$$\Delta G = -RTln\left(\frac{a_A^{a'} \cdot a_B^{b'} \ldots}{a_A^a a_B^b \ldots}\right) = -RTln(K_{\text{eq}}). \tag{A.19}$$

For the case of aqueous solutions we usually make several assumptions. For an *ideal* solution:

$$a_i = N_i. \tag{A.20}$$

where N_i is the atomic fraction of species i in solution (with respect to the *total* number of atoms). The *solvent* (major constituent) almost always obeys eq. (A.20). Equation (A.20) is called *Raoult's law*. *Dilute* solutions usually follow Henry's law

$$a_i = k_{\text{h}} N_i. \tag{A.21}$$

Here k_{h} is the Henry's law coefficient.

Appendix II

Crystal Structure and Related Concepts

Solids are, generally, grouped in two broad categories: crystalline and non-crystalline. Crystalline solids are those in which atoms or molecules are arranged in some regular repetitious pattern in three-dimensional space. All others, which do not fulfill this criterion, are noncrystalline.

Crystalline solids cannot have infinitely large numbers of atomic or molecule spacial arrangements — a result that is the outcome of the symmetry considerations. This can be understood by visualizing atomic positions as points in space. In one dimension, there is only one way in which identical points can be selected, by placing them at equal distances along a line. Thus there is only one point group arrangement in this case. For a two-dimensional arrangement of points only five unique point groups exist. In three dimensions there are 32 point groups. A point group is a collection of symmetry operations applied about a point that leaves the body invariant. For example, if a rotation of half a turn (i.e, 180°) about a certain axis brings a body into a new position that is indistinguishable from one prior to rotational operation, then the axis is symmetric axis — a twofold rotation axis. Other symmetry elements are reflection across a plane, center of inversion, and a combination of rotation and inversion. For details the student is advised to look into these considerations in a book on crystallography and symmetry operations.

Let us define $\vec{a}$, $\vec{b}$, $\vec{c}$ as three fundamental translational vectors. Any point, representing an atomic or molecular position, in three-dimensional space can then be identified with respect to an origin by a vector $\vec{r}_{123} = n_1 \vec{a} + n_2 \vec{b} + n_3 \vec{c}$. A property of fundamental vectors is such that any other point in space can be located in terms of $\vec{r}_{123}$ and a translation vector defined in terms of $\vec{a}$, $\vec{b}$, $\vec{c}$. These vectors can be chosen such that n_1, n_2, and n_3 are integers. In such a case $\vec{a}$, $\vec{b}$, and $\vec{c}$ are called *primitive vectors.*

Coordinates of a point in a three-dimensional space (or of an atom or a molecule in a structure) are generally described in terms of axes as shown in fig. AII.1. Here the vectors $\vec{a}$, $\vec{b}$, and $\vec{c}$ and the angular convention is defined. In this configuration seven crystal systems, given in Table A-II.1,

TABLE A-II.1: Seven crystal systems

Crystal System	Axial relationships[a]	Example
Cubic	$a = b = c,\ \alpha = \beta = \gamma = 90°$	Si, GaAs
Tetragonal	$a = b \neq c,\ \alpha = \beta = \gamma = 90°$	β-Sn
Orthorhombic	$a \neq b \neq c,\ \alpha = \beta = \gamma = 90°$	Ga
Rhombohedral	$a = b = c,\ \alpha = \beta = \gamma \neq 90°$	As
Hexagonal	$a = b \neq c,\ \alpha = \beta = 90°,\ \gamma = 120°$	Zn
Monoclinic	$a \neq b \neq c,\ \alpha = \gamma = 90° \neq \beta$	β-S
Triclinic	$a \neq b \neq c,\ \alpha \neq \beta \neq \gamma \neq 90°$	K_2CrO_7

[a] a, b, c now represent magnitudes of the vectors $\vec{a}$, $\vec{b}$, $\vec{c}$

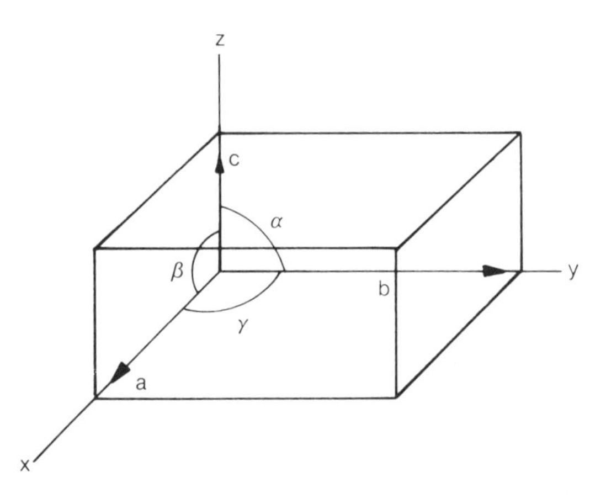

Fig. AII.1. Orthogonal coordinate system of axes.

exist. They are cubic, tetragonal, orthorhombic, rhombohedral, hexagonal, monoclinic, and triclinic. One example of each type is also given. Figure AII.2 shows these systems. Each parallelepiped is called a *unit cell.* The key feature of the unit cell is that it contains full description of the structure as a whole.

A unit cell is always constructed with lattice points on the corners. Such is the case with primitive vectors, and, therefore, these cells are called primitive cells. However, the stacking of atoms leads to atoms or molecules on unit cell faces and at the center of the volume. The presence of atoms on such sites increases the permissible number of space lattices from seven primitive cells to 14 Bravais lattices. On the other hand, the 30 points groups can all

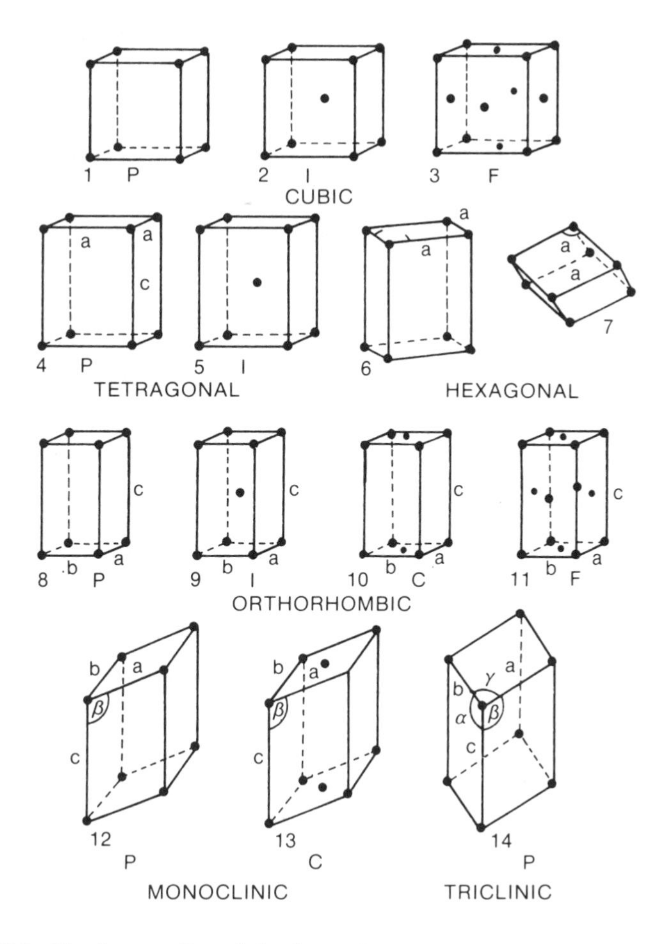

Fig. AII.2. The fourteen Bravais lattices.

be described by these 14 Bravais lattices. These 14 space lattices are shown in fig. AII.2.

Most semiconductors of interest crystallize in a diamond cubic structure. The structure is composed of two face-centered-cubic (FCC) lattices displaced from each other by one quarter of a body diagonal. Thus if one FCC lattice has an atom at $n_1 = 0$, $n_2 = 0$, and $n_3 = 0$, then the second FCC lattice has the equivalent atom at $n_1 = 1/4$, $n_2 = 1/4$, and $n_3 = 1/4$. For elemental semiconductors like Si, Ge, carbon, and grey tin, all positions are occupied by identical atoms. For compound semiconductors like GaAs, Ga atoms are occupying sites of one FCC lattice and As atoms are occupying the other.

Metals generally crystallize in one or the other cubic structure and in a close-packed hexagonal (CPH) structure shown in fig. AII.3. In this case

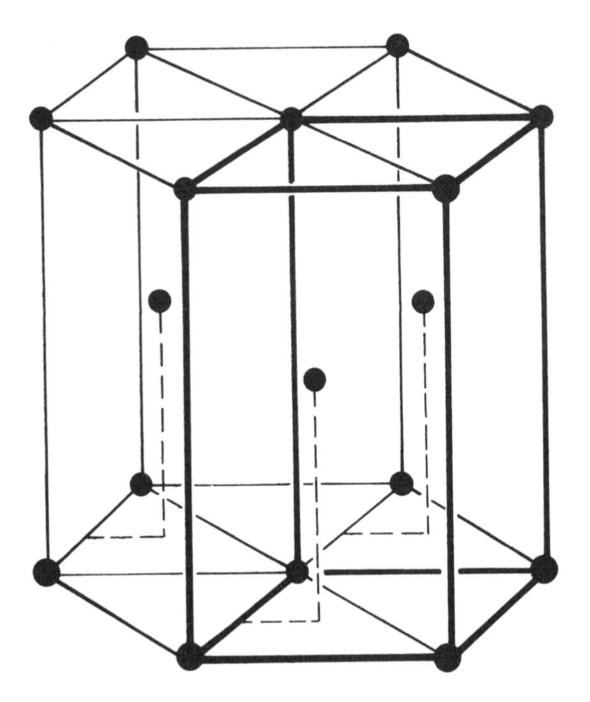

Fig. AII.3. The closed-packed hexagonal structure.

the atomic packing sequence (see Chapter 2) is ABABAB instead of AB-CABCABC for FCC. The cell can be visualized in terms of three axes a_1, a_2, a_3 in the $x - y$ plane and at 120° to each other and the fourth c axis perpendicular to the other three. For perfect CPH structure the $c/a = 1.633$. The primitive cell of CPH is the hexagonal cell shown as dark shaded region.

II.1 Indexing the Crystal Directions and Planes

Lattice directions and planes are indexed with reference to three crystal axes with the origin at one point of the primitive cell. Directions and distances are indexed in terms of the lattice constants a, b, and c (see fig. AII.1). A direction $[n_1, n_2, n_3]$ indicates the direction of a line connecting a point $n_1 a$, $n_2 b$, $n_3 c$ with 0,0,0. The distance between these two points is the magnitude of the vector r_{123} and is given as

$$r_{123} = (n_1^2 + n_2^2 + n_3^2)^{1/2}.$$

A specific process is adopted to index a plane resulting in the so-called Miller indices. These indices are obtained by a) finding the intercepts of a plane with the three crystal axes in units of lattice constants a, b, and c, b) taking the reciprocal of these members and reducing them to the smallest three integers having the same ratio, and c) enclosing them in the parenthesis: (hkl). The h, k, and l representing intercepts on the a, b, and c axes, respectively. Thus a (111) representation indicates a single plane or a set of parallel planes with intercepts on a, b, and c axes of na, nb, and nc. A (123) representation indicates intercepts of a, $1/2b$, and $1/3c$, respectively. On the other hand, (100) indicates planes that intercept at na, ∞, and ∞. If, however, the intercept occurs on the negative side of the origin then a bar is placed on appropriate Miller index or indices to indicate the negative intercept, e.g. $(h\bar{k}l)$ or $(\bar{h}kl)$.

Note that directions are designated by square brackets [ ], whereas planes are indexed by parenthesis ( ). All equivalent planes are represented by curly brackets { } and all equivalent directions are represented by carets $< \;\; >$. Thus all the faces of a cube are represented by (100) and all the cube edges by $< 100 >$.

For hexagonal structure a fourth index i is used to represent the intercept on the third axis (a_3) in the $x-y$ plane. Thus a plane is indexed as $(hkil)$ in a hexagonal structure. It is left to the student to show that $h + k = -i$ for any plane in the hexagonal system. The $(hkil)$ system for hexagonal system is called Miller-Bravais indices.

Appendix III

Useful Acronyms

AC = Alternating current
AMU = Atomic mass unit
AR = Antireflection coating
ASIC = Application specific IC
C-V = Capacitance voltage
CCD = Charge-coupled devices
CD = Critical dimension
CEF = Constant electric field
CEM = Contrast enhancement material
CHE = Channel hot electron
CMOS = Complimentary MOS
CMS = Chloromethylstyrene
CRT = Cathode ray tube
CS = Chip select
CVD = Chemical vapor deposition
CW = Continuous wave
CZ = Czhochralski
DC = Direct current
DF = Depth-of-focus
DIBL = Drain-induced barrier lowering
DLTS = Deep-level transient spectroscopy
DRAM = Dynamic RAM
FB = Flat band
FET = Field effect transistor
FIB = Focused ion beam
FIT = Failure in time (unit as failures in a billion hours)
FPS = Foot pound system
FZ = Float zone

GR = Generation recombination
HCI = High-current implanter
HEI = High-energy implanter
HF = High frequency
HIC = hybrid intergrated circuit
HMDS = Hexamethyldisiloxane
HTSC = High-temperature super conductor
I-V = Current voltage
IC = integrated circuit
ICB = Ion cluster beam
IMPATT = Impact avalanche and transit time
IR = Infrared
JFET = Junction FET
KRPM = Kilo rotation per minute
LAA = Laser-assisted annealing
LAPS = Laser-assisted processing of semiconducting materials
LASER = Light amplification by simulated emission of radiation
LCI = Low-current implanter
LDD = Low-doped drain
LEI = Low-energy implanter
LF = Low frequency
LOCOS = Localized oxidation of silicon
LPCVD = Low-pressure CVD
LPE = Liquid-phase epitaxy
LSI = Large-scale integration
LSS = Lindhard, Scharff, and Schiott
MASER = Microwave amplificiaton by stimulated emission of radiation
MBE = Molecular-beam epitaxy
MCI = Medium-current implanter
MESFET = Metal semiconductor FET
MKS = Metric kilogram system
MOCVD = Metal organic CVD
MODFET= Modulated FET
MOS = Metal oxide semiconductor
MOSFET = Metal oxide semiconductor FET

MSI = Medium-scale integration
MTF = Modulation transfer function
NA = Numerical aperture
NAA = Neutron activation analysis
NMOS = n-Channel MOS
OIDR = Oxidation-induced dopant redistribution
OISF = Oxidation-induced stacking fault
OXIM = Oxide-isolated monolith
PAC = Photoactive senstitizer
PECVD = Plasma-enhanced CVD
PG = Pattern generation
PMMA = Polymethylmethacrylate
PMOS = p-Channel MOS
PR = Photoresist
QCV = Quasiconstant Voltage
R/S = Reset/set
RAM = Random access memory
RBS = Rutherford backscattering
RF = Radio frequency
RIE = Reactive ion etching
ROM = Read-only memory
RTA = Rapid thermal annealing
RTBT = Resonant tunneling bipolar transistor
RTC = Resistor transistor logic
RTO = Rapid thermal oxidation
RTP = Rapid thermal processing
SCR = Silicon controlled rectifier
SDHT = Selectively doped heterostructure transistor
SEM = Scanning electron microscope
SEU = Single event upset
SF = Stacking fault
SHE = Substrate hot electron
SIB = Shower ion beam
SIC = Silicon IC
SILOS = Sealed interface local oxidation of silicon

SIMS = Secondary ion mass spectrometry
SOI = Silicon on insulator
SOS = Silicon-on-sapphire
SPE = Solid-phase epitaxy
SRAM = Static RAM
STEM = Scanning TEM
TC(B) = Thermocompression (bonding)
TEM = Transmission electron microscopy
TF = Thin film
TFT = Thin-film transistor
UHF = Ultra-high frequency
UHV = Ultra-high vacuum
ULSI = Ultra LSI
UV = Ultraviolet
V-W = Volmer-Weber
VHF = Very high frequency
VHSIC = Very-high-speed IC
VLF = Very low frequency
VLSI = Very LSI
VPE = Vapor-phase epitaxy
WE = Write enable
XTEM = Cross-sectional TEM
YAG = Yttrium aluminum garnet
YIG = Yttrium iron garnet
ppb = Parts per billion
ppba = Parts per billion (atomic)
ppm = Parts per million
pmma = Parts per million (atomic)

Appendix IV

Various Constants, Quantities, and Properties

TABLE A-IV.1: Physical constants

Angstron (Å)	$1\text{Å} = 10^{-1}\,\text{nm} = 10^{-4}\,\mu\text{m} = 10^{-8}\,\text{cm}$
Avogadro's Constant (N)	$N = 6.02204 \times 10^{-23}\ \text{mol}^{-1}$
Boltzmann Constant (k)	$k = 1.38066 \times 10^{-23}$ Joules per degree K
	$= 8.6174 \times 10^{-5}\, eV$ per degree K
	$= R/N$
Elementary Charge (q)	q= 1.60218×10^{-19} coulombs
Electron rest mass (m_o)	$m_o = 0.91095 \times 10^{-30}$ kg
Electron Volt (eV)	$eV = 1.60218 \times 10^{-19}$ joules $= 23.053$ kcal/mol
Gas constant (R)	$R = 1.98719$ cal per mole per degree K
Permittivity (vacuum)(ϵ_o)	$\epsilon_o = 8.85418 \times 10^{-14}$ Farad per cm
Plank's constant (h)	$h = 6.62617 \times 10^{-34}$ joule-second
Speed of light in vacuum (c)	$c = 2.99792 \times 10^{10}$ cm per second

TABLE A-IV.2: Various quantities, units, and conversions

Quantity	Generally Used Symbol	International System of Units		Conversion to some commonly used units
		Unit	Symbol	
Length	$1, x$	meter	m	m = 100 cm
Mass	m	kilogram	kg	kg=1000 g
Time	t	second	s	
Frequency	f	hertz	Hz	
Force	F	newton	N	N = 10^{-5}dyn
Compressibility	β	meter^2/newton	m^2N^{-1}	
Pressure	P, p	pascal	Pa	Pa = N/m^2 ≡ 10 dyn/cm^2
Thermal				
Temperature	T	kelvin	K	K = 273.15 + ° C
Energy	E	joule	J	J = 10^7 erg
Free energy	G, F			J = 6.62415 × 10^{18} eV
Enthalpy	H			J = 0.239006 cal·g
Entropy	S	joule per Kelvin	JK^{-1}	
Electric				
Charge	Q, q	Coulomb	C	
Potential	V	Volt	V	
Conductance	G	Siemens	S	
Resistance	R	Ohm	Ω	
Resistivity	ρ	Ohm-cm	Ω-cm	
Current	I	Ampere	A	
Capacitance	C	Farad	F	
Power	P	Watt	W	
Inductance	L	Henry	H	
Magnetic				
Induction		Tesla	T	
Flux	B	Weber	Wb	

TABLE A-IV.3: Other commonly used units and symbols and their conversion into units mentioned above

Angstrom	(Å)	=	10^{-10}m = 10^{-1} nm
Barn	(b)	=	10^{-28} m^2 = 10^{-24} cm^2
Atmosphere	(atm)	=	1.01325×10^5 Pa= 1.01325×10^{-6} dyn/cm^2
Bar	(bar)	=	10^5Pa = 10^6dyn/cm^2
poise	(P)	=	Newton· s/m^2 = 10 dyn·s/cm^2
erg	(erg)	=	10^{-7}J = g · cm^2/s^2

1 torr of Hg as the unit of pressure= 1.32×10^2 Pa.

TABLE A-1V.4: Properties of semiconductors

	Si	GaAs
Breakdown field (V/cm)	$\sim 3 \times 10^5$	$\sim 4 \times 10^5$
Density (g/cm^3)	2.33	5.32
Dielectric constant	11.9	13.1
Electron Affinity (V)	4.05	4.07
Energy gap (eV)	1.12	1.424
Linear thermal expansion Coefficient (°C)$^{-1}$	2.6×10^{-6}	6.86×10^{-6}
Melting point (°C)	1412	1238
Mobility (drift)(cm^2/V-s)		
electrons	1500	8500
holes	475	400
Thermal conductivity (W/cm-°C)	1.5	0.46
Young's Modulus (dyn/cm^2)	1.07×10^{12}	–

Appendix V
Process Modeling

Contributed by Chung-Ting Yao and Jein-Chen Young

As the minimum feature size reduces due to technological advances, many second-order effects arise that affect the performance of an integrated circuit. This is true both at the component and at the device levels. Short- and narrow-channel effects, drain-induced barrier-lowering, subthreshold effects, etc., start to dominate device behavior at the micron minimum feature size. At the materials level, concentration-dependent diffusion coefficients, lateral diffusion, bird's-beak effect, etc. become critical limiting factors influencing device density. Fabrication of very-large-scale devices with acceptable yield becomes practical only if these second-order effects are understood and accounted for. In addition, processes must be designed with sufficient "latitude" that the normal variations in equipment parameters encountered on "real" production lines will not adversely affect device yield.

Since the whole process sequence is very complicated and time consuming, and requires the close interaction of many groups of people, it is desirable to have good projections of the results of each process step before the step is actually performed. Such projections can be made only if sufficiently accurate physical models are employed. Even when closed-form solutions of the relevant equations are available, the number of these equations to be evaluated in simulating a complete process precludes hand calculation. In addition, two-dimensional models frequently have no analytic solutions. Sophisticated numerical techniques must be used to get accurate solutions and/or graphical output. Thus, computer-aided-design (CAD) tools for process development are now an essential part of any process line.

Perhaps the most popular commercially available process simulation programs are SUPREM^{TM} [1] and SUPRA^{TM} [2]. SUPREM^{TM} is a one-dimensional process simulator that can supply information concerning doping concentrations and thermal-oxide and deposited-layer thicknesses along a line that is perpendicular to the wafer surface. Most commonly used processing steps are handled by the program, such as high-temperature diffusion under both inert and oxidizing ambients, ion implantation, epitaxial growth, etching, oxidation, and chemical vapor deposition. SUPRA^{TM} is similar to SUPREM^{TM}, but it can perform two-dimensional simulations of the complete integrated-circuit fabrication process. Both of these programs allow the process engineer to "see" the device cross section as it evolves through the many steps required to make a finished device.

The output from these programs is essential to both the process and device engineer. For the process engineer, these results enable the determination of times, temperatures, and ambients for each process step. In addition, the effect of later process steps on earlier step results can be determined. Process latitude studies can be performed that allow an assessment of how "resilient" a process is with respect to uncontrolled variation in process tool parameters. Junction depths, doping profiles, and threshold information are passed on to the device designer. The device designer uses these data directly in CAD programs for device modeling.

Typical device-level simulators include MINIMOS [3] and PISCES [4]. All of these programs solve the bipolar and/or unipolar current transport equations in two dimensions. They can analyze a variety of physical structures. MINIMOS is limited to the basic MOSFET structure, but PISCES can work on nonplanar as well as planar devices such as bipolar, MOSFET, and trench structures with arbitrary doping profiles. These doping profiles can be entered into the programs using either analytic expressions or data files supplied by SUPREM^{TM} or SUPRA^{TM}. The PISCES programs can handle transient responses, while MINIMOS cannot. While MINIMOS can handle only MOSFET static response, it is easier to use and is very popular. Output from the device-level simulator (device current-voltage characteristics, accurate predictions of thresholds, and subthreshold currents, etc.) can be used in circuit-level simulators like SPICE or ASPECT. These programs predict the performance of integrated-circuit functional blocks, such as inverters, shift registers, amplifiers, etc. As the reader can guess, the CAD tools do not terminate at the circuit level. Practically a continuum of hierarchically arranged design tools exist from the process level to the level of very-large-scale integration.

In this appendix, we will use an n-well, single-level polysilicon, single-level metal CMOS process to illustrate the use of CAD programs in process development. The SUPREM^{TM} III program is used here, but the input format is basically the same as that used in SUPRA^{TM}. The result is used to predict MOSFET performance as simulated by PISCES. The process used is a variant of the process described in Chapter 1. More detail is given on process times and temperatures, as these are essential inputs to the SUPREM^{TM} program. A cross section of the completed structure is shown as fig. A.V.1. First, we present the detailed process flow (table A-V.1).

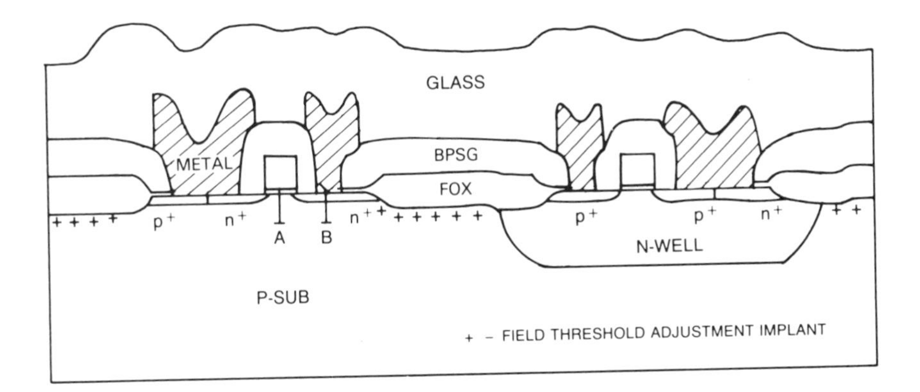

Fig. AV.1. A schematic view of the cross-section of a CMOS device.

Next, we establish the SUPREM^{TM} input data file (table A-V.2). Finally, we show the graphic output from the program (fig. AV.2). Of particular importance to the device engineer are the doping profiles in the transistor source/drain region and in the MOS active channel. These are the data required by PISCES to derive MOSFET current-voltage plots. While the PISCES input file structure is not presented, the MOSFET current-voltage curve provided by this program is shown as fig. AV.3.

Let us focus our attention on table A-V.2 (the SUPREM^{TM} input data file). Note that the files are broken up into blocks separated by spaces. Each line is referred to as a "card."The whole process flow is summarized by the comment cards. Each block summarizes a key process step. The file begins with a title card, which serves as a header for all subsequent output. The next card initializes the program by specifying the starting-material parameters and the grid structure on which the output data is to be presented. For the case at hand, the vertical profile will extend 2 μm below the surface of the silicon, and the process equations will be solved numerically on a mesh whose nodes are 0.005 μm separate in the first 0.02 μm below the surface. The total number of grid units is 300.

TABLE A-V.1: A detailed process sequence for the n-well CMOS process

1. Grow 2500 Å oxide
2. Pattern n-well region (mask #1) and remove oxide within the pattern
3. Implant phosphorous, dose= 4.5×10^{12}, energy= 50keV
4. Strip photoresist
5. High-temperature drive-in, temperature= 1100°C, time= 500min.
6. Strip oxide
7. Grow 450 Å pad oxide
8. Deposit 1000 Å nitride
9. Pattern active region (mask #2), remove nitride
10. Strip photoresist
11. Pattern nMOS channel stop (mask #3)
12. NMOS channel stop implant, BF_2, dose= $3 \cdot 10^{13}$, energy= 45 keV
13. Strip photoresist
14. Grow field oxide to 6000 Å, temperature= 1000°C
15. Strip nitride and pad oxide
16. Grow sacrificial oxide (300 Å)
17. Strip sacrificial oxide
18. Grow gate oxide (300 Å)
19. Threshold adjustment implant (boron, dose= 1.2×10^{12}, energy= 40 keV
20. Gate anneal (950°, 30 min.)

TABLE A-V.1: A detailed process sequence for the n-well CMOS process (continued)

21. Deposit polysilicon (5000 Å)
22. Dope poly (900°, 50 min. $POCl_3$)
23. Pattern gate (mask #4)
24. Strip photoresist
25. Pattern n^+ region (mask #5)
26. As implant, dose= 4×10^{15}, energy= 40 keV
27. Strip photoresist
28. Dry oxygen anneal
29. Pattern p^+ region (mask #6)
30. BF_2 implant, dose= 2×10^{15}, energy= 40 keV
31. Strip photoresist
32. Deposit BPSG
33. Reflow BPSG (900°C, 30 min.)
34. Pattern contact (mask #7), and remove BPSG in the contact region
35. Reflow BPSG (900°C, 15 min.)
36. Strip photoresist
37. Deposit aluminum
38. Pattern and etch metal (mask #8)
39. Strip photoresist
40. Deposit overglass
41. Pattern and etch overglass (mask #9)

TABLE A-V.2: SUPREM data file for simulating the n-well CMOS process

```
Title        CMOS Silicon Active Region
Comment      Active device region initial processing.

Comment      Initialize silicon substrate.
Initialize   <100> Silicon Boron Concentration=3e15
+            Thickness=2.  dx=0.005 Xdx=0.02 Spaces=300

Comment      Grow first oxide (step 1)
Diffusion    Temperature=800 Time=60 T.Rate=5 DryO2
+            Pressure=0.1
Diffusion    Temperature=1100 Time=10 DryO2
Diffusion    Temperature=1100 Time=10.5 WetO2
Diffusion    Temperature=1100 Time=10 DryO2
Diffusion    Temperature=1100 Time=60 T.rate=-5 Nitrogen

Print        Layer

Comment      N-Well (step 2)
$Etch        Oxide
$Implant     Phosphorous Dose=4.5E12 Energy=50 2-Gaussian

Comment      N-Well Drive-in (step 5)
Diffusion    Temperature=800 Time=60 T.Rate=5 DryO2
+            Pressure=0.1
Diffusion    Temperature=1100 Time=300 DryO2
Diffusion    Temperature=1100 Time=200 Nitrogen
Diffusion    Temperature=1100 Time=120 T.Rate=-2.5 Nitrogen

Print        Layer

Comment      Strip Oxide
Etch         Oxide all

Comment      Pad Oxide (step 7)
Diffusion    Temperature=800 Time=40 T.Rate=5 DryO2
+            Pressure=0.1
Diffusion    Temperature=1000 Time=5 DryO2 Pressure=0.1
Diffusion    Temperature=1000 Time=1 DryO2
Diffusion    Temperature=1000 Time=34 DryO2 HCL%=5
Diffusion    Temperature=1000 Time=40 T.Rate=-5 Nitrogen
```

```
Print         Layer

Comment       Nitride Deposition (step 8)
Deposition    Nitride Thickness=0.1  Spaces=20

Comment       Nitride Etch
$Etch         Nitride All

Comment       Field Implant (step 12)
Implant       Boron Dose=3.e13 Energy=11.2

Print         Layer
Plot          Chemical Boron Lp.plot Xmax=0.5

Comment       Field Oxide (step 14)
Diffusion     Temperature=800 Time=40 T.Rate=5 DryO2
+             Pressure=0.1
Diffusion     Temperature=1000 Time=15 DryO2 Pressure=0.25
Diffusion     Temperature=1000 Time=125 WetO2 HCL%=3
Diffusion     Temperature=1000 Time=40 T.Rate=-5 Nitrogen

Print         Layer

Comment       ONO etch (step 15)
Etch          Oxide All $oxide is grown on nitride during oxidation
Etch          Nitride All
Etch          Oxide All

Comment       First Gate Oxide (step 16)
Diffusion     Temperature=800 Time=20 T.Rate 7.5 DryO2
+             Pressure=0.1
Diffusion     Temperature =950 Time=5 DryO2 Pressure=0.1
Diffusion     Temperature =950 Time=5 DryO2
Diffusion     Temperature =950 Time=55 DryO2 HCL%=3
Diffusion     Temperature =950 Time=40 T.Rate=-3.75 Nitrogen

Print         Layer

Comment       Oxide etch (step 17)
Etch          Oxide All

Comment       True gate oxide (step 18)
Diffusion     Temperature=800 Time=20 T.Rate=7.5 DryO2
+             Pressure=0.1
```

```
Diffusion    Temperature=950 Time=5 DryO2 Pressure=0.1
Diffusion    Temperature=950 Time=5 DryO2
Diffusion    Temperature=950 Time=55 DryO2 HCL%=3
Diffusion    Temperature=950 Time=40 T.Rate=-3.75 Nitrogen

Print        Layer

Comment      Blanket Implant (step 19)
Implant      Boron Dose=1.2e12 Energy=40

Print        Layer

Comment      Gate Anneal (step 20)
Diffusion    Temperature=800 Time=20 T.Rate=7.5 Nitrogen
Diffusion    Temperature=950 Time=5 Nitrogen
Diffusion    Temperature=950 Time=20 DryO2
+            Pressure=0.015 HCL%=0.5
Diffusion    Temperature=950 Time=30 T.Rate=-5 Nitrogen

Print        Layer

Comment      Deposit Polysilicon (step 21)
Deposit      Polysilicon Thickness=0.5 Spaces=20 Temperature=620

Comment      POCL3 Dope (step 22)
Diffusion    Temperature=800 Time=20 T.Rate=5 Nitrogen
Diffusion    Temperature=900 Time=20 Dtmin=0.2
+            Phosphor Solidsol
Diffusion    Temperature=900 Time=20 DryO2 Pressure=0.286
Diffusion    Temperature=900 Time=10 Nitrogen
Diffusion    Temperature=900 Time=10 T.Rate=-10 Nitrogen

Print        Layer

Comment      Deglaze (oxide is grown during diffusion)
Etch         Oxide

Comment      S/D implant (step 26)
Implant      Arsenic Dose=4e15 Energy=80 2-Gaussi

Comment      S/D Drive/Reox (step 28)
Diffusion    Temperature=800 Time=20 T.Rate=7.5 Nitrogen
Diffusion    Temperature=950 Time=15 Nitrogen
Diffusion    Temperature=950 Time=10 T.Rate=-5 Nitrogen
```

```
Diffusion     Temperature=900 Time=1 DryO2
Diffusion     Temperature=900 Time=60 DryO2 HCL%=3
Diffusion     Temperature=900 Time=20 T.Rate=-5 Nitrogen

Print         Layer

Comment       S/D Implant (step 30)
$Implant      Boron Dose=2e15 Energy=10

Comment       Deposit BPSG (step 32)
Deposit       Oxide Thickness=0.96 Phosphorous Concentration=1e21
+             Spaces=20

Comment       Densification (step 33)
Diffusion     Temperature=800 Time=20 T.Rate=5 Nitrogen
Diffusion     Temperature=900 Time=30 DryO2
Diffusion     Temperature=900 Time=10 T.Rate=-15 Nitrogen

Print         Layer

Comment       Contact Cut (step 34)

Comment       Second Densification (step 35)
Diffusion     Temperature=800 Time=20 T.Rate=7.5 Nitrogen
Diffusion     Temperature=950 Time=5 DryO2
Diffusion     Temperature=950 Time=10 T.Rate=-15 Nitrogen

Print         Layer

Comment       Deposit Aluminum (step 37)

Comment       Sinter (450)

Comment       Remove BPSG for plot
Etch          Oxide
Print         Layer
Plot          Chemical Net Lp.plot

Save          Structure File=yng
Comment       Threshold voltage calculation
V.Threshold

Stop
```

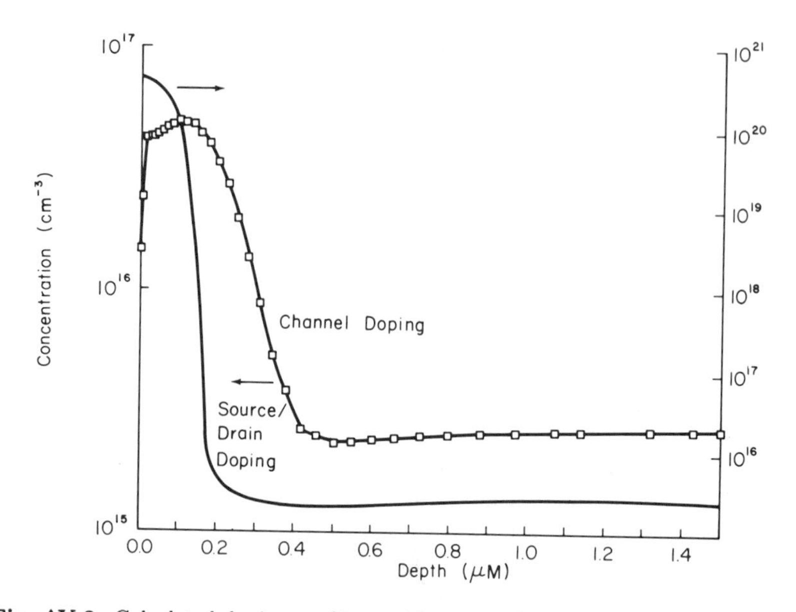

Fig. AV.2. Calculated doping-profile graphic output from the used program.

After titling and initialization the process flow begins. The first block in the process sequence represents the 2500 Å oxidation step. It contains all the necessary information to activate the oxidation model: times, temperatures, and ambients. The "pressure"data refers to the partial pressure of oxygen in the feed stream. The model card allows for a sequencing of ambient conditions. In this case, a dry/wet/dry cycle is specified. Note that the oxidation file is referred to as a "diffusion"on the left-hand side of the data list. This is because the oxidation step is normally a part of every diffusion process and is thus lumped in the diffusion model part of the main program. An example of how to set up an actual diffusion is shown in the polysilicon doping step (step 21). The "Phosphor Solidsol"entry tells the program that the dopant is phosphorous and the surface concentration is pegged at the solid-solubility limit (as is the case during the deposition phase of the diffusion cycle).

The first and last diffusion cards in the bloc contain the entries T.Rate=5 DryO2. Of course, the DryO2 entry refers to the fact that the oxidizer is dry oxygen. The T.Rate specification is a bit less clear. Most oxidation furnaces in use today include a "ramped"entry feature, that is, the temperature at the mouth of the furnace is cooler than the actual "working zone"of the furnace. The wafer is inserted and withdrawn slowly through this ramped-temperature zone. This minimizes wafer damage and breakage. Note that the first diffusion card calls for 800°C, and the second calls for 1100°C. This implies that the temperature ramp the wafer sees extends between these two temperature extremes. The T.Rate=5 entry tells the program that the wafer sees a 5°C temperature increase every minute of the 60-min duration

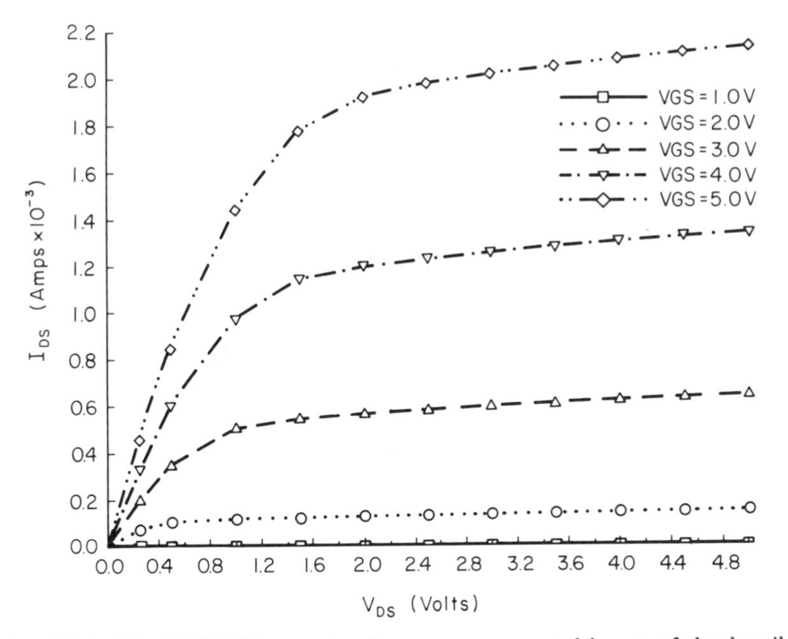

Fig. AV.3. The MOSFET current–voltage curves generated by use of the described program.

of this part of the oxidation. The temperature increases 300°C over this time period, arriving at the final 1100°C. In the last card of the block, the -5 indicates that the temperature is ramped down at the same rate. This is a slow-cool process.

Other model cards include implant, etch, and CVD steps. The implant card requires total implant dose and energy. The implant activation steps are entered with diffusion model cards specifying inert ambients only. The nitride deposition step is represented by a "deposition" card. Merely specifying the thickness tells the program the time and temperature based on a default model or on modifications to the main program. It should be emphasized that, while default parameters and procedures are contained in the "off-the-shelf" program, individual process lines are unique, and some preliminary work should be done to verify values of diffusion coefficients, implant ranges, etc. Other commands include: etch, print, plot, comment, and $. The Etch All command removes a previously grown or deposited layer. The etch command can also be used to determine a final thickness when some etch rate is previously specified. A typical command sequence to do this would be "ER=800 Time=5," for a 5-min etch at a rate equal to 800 Å per minute.

The print and plot commands are used for controlling the hard-copy output from the program. A print/layer statement requests a print-out of the doping concentration in tabular form. The "Chemical Boron Lp.plot

Xmax=0.5"sequence following step 12 requests a semi-logarithmic plot of boron concentration down to a depth of 0.5 μm.

Comment refers to a verbal statement that plays no computational role. The $ before a step disables the step, that is, it turns the card into a comment card. In the program listing provided, the goal is to derive n-channel properties only. The $ entries are distributed to cancel those calculations referring to the p-channel components. The whole process sequence is included for completeness. Removing the $ prefixes would give output on all areas of interest in the device. Note that the "disabling"did not change the apparent temperature history seen by the circuit during the fabrication cycle. This would, of course, alter the projected n-channel doping profiles.

The save command (which can appear anywhere in the program) saves the card stack and the resulting output under the file-name yng. The V.Threshold command calls for an estimation of threshold voltage in the regions for which doping profiles were computed.

A few comments should be made regarding the process shown. First, a cleaning step is required before each high-temperature step. This clean could be either the peroxide clean described in Chapter 2, which may involve a brief buffered HF dip to remove native oxides. In any event, some thinning of oxides present when this step is taken should be anticipated. Thus, an etch card that presents this thinning should be included for more accurate simulation. The sacrificial oxidation step (step 16) is used to prevent the "white ribbon"effect discussed in Chapter 7. The threshold adjustment implant (step 19) is done to both the p- and n-channel active regions. As a result, both thresholds are shifted to more positive values. Polysilicon thicknesses are chosen to be sufficient to stop the self-aligned implants from penetrating into the channel region.

The implant-stopping power of nitride for most relevant implant species is about twice that of oxide. Thus, the oxide/nitride layers used to block the 40-keV active-channel BF_2 implant (step 12) is equivalent to 2450 Å of oxide alone. As a result, the pad oxide can be made thinner in the field region. Thinner pad oxides mean less bird's beak, but very thin pad oxides do not provide sufficient stress relief for the high-stress nitride layer. As a result, substrate damage may occur. Selection of the required layer thicknesses is part of the "art"of processing. In any event, CAD tools aid in final process selection.

The doping profile in the active channel (line A in fig. A.V.1) and in the n-channel source/drain region (line B in fig. A.V.1) are shown in fig. A.V.2. The PISCES output shown in fig. A.V.3 indicates transistor thresholds are 1 volt for the nMOS and 0.7 volts for the pMOS devices. This computation assumes negligible fixed-oxide charge $[Q^f = \mathrm{O})$.

References for Appendix V

1. D.A. Antoniadis, S.E. Hansen, R.W. Dutton, SUPREM II, Tech. Rep. 5019-2, Army Research Office, Contract DAAG 9-77-C-006, June (1978).

2. D. Chin, M. Kup, R.W. Dutton, SUPRA: Stanford University Process Analysis Program, Final Tech. Rep., Army Research Office, Contract DAAG29-80-K-0013 (1980).

3. S. Selberherr, A. Schutz, H.W. Potzl, MINIMOS: IEEE Trans. Electron. Dev. ED-27 (8) 1540-1550 (1980).

4. M.R. Pinto, C.S. Rafferty, R.W. Dutton, PISCES-II: Stanford Electronics Laboratory, Palo Alto, Calif., Technical Report (1985).

Suggested Projects

(For Those with Access to the SUPREM* Program)

1. Set up the SUPREM data file presented in the text. Modify the file by adding or subracting cards. Observe the changes in the output.
2. Construct a similar process sequence for an nMOS device and model it using the techniques described above.
3. Do a "process latitude"study by starting with the data file presented here. Observe the effect of variations in furnace temperature and times on the final predicted doping profiles.

---

* SUPREM, SUPRA, and PISCES are available from the Software Distribution Center, Office of Technology Licensing, 105 Encina Hall Stanford University, Palo Alto California 94305. Commercial packages are also available through Technology Modeling Associates, Inc., Palo Alto, California 94301. MINIMOS is available from A. Schutz, Technische Universitat Wien, Institut für Allgemeine Elektrotechnik und Elektronik, Vienna, Austria.

INDEX

A

B

D

G

H

I

Q

R

S

Z